T0292659

Cyber-Physical Systems

Cyber-Physical Systems
Foundations, Principles and Applications

Edited by

Houbing Song
West Virigina University, USA

Danda B. Rawat
Howard University, USA

Sabina Jeschke
RWTH Aachen University, Germany

Christian Brecher
RWTH Aachen University, Germany

Series Editor Fatos Xhafa
Universitat Politècnica de Catalunya, Spain

AMSTERDAM • BOSTON • HEIDELBERG • LONDON
NEW YORK • OXFORD • PARIS • SAN DIEGO
SAN FRANCISCO • SINGAPORE • SYDNEY • TOKYO
Academic Press is an imprint of Elsevier

Academic Press is an imprint of Elsevier
125 London Wall, London EC2Y 5AS, United Kingdom
525 B Street, Suite 1800, San Diego, CA 92101-4495, United States
50 Hampshire Street, 5th Floor, Cambridge, MA 02139, United States
The Boulevard, Langford Lane, Kidlington, Oxford OX5 1GB, United Kingdom

Library of Congress Cataloging-in-Publication Data
A catalog record for this book is available from the Library of Congress

British Library Cataloguing-in-Publication Data
A catalogue record for this book is available from the British Library

ISBN: 978-0-12-803801-7

For information on all Academic Press publications
visit our website at https://www.elsevier.com/

Working together
to grow libraries in
developing countries

www.elsevier.com • www.bookaid.org

Publisher: Todd Green
Acquisition Editor: Brian Romer
Editorial Project Manager: Amy Invernizzi
Production Project Manager: Punithavathy Govindaradjane
Cover Designer: Victoria Pearson

Typeset by SPi Global, India

Contents

Contributors

T. Abdelzaher
University of Illinois at Urbana-Champaign, Champaign, IL, United States

V. Adamchuk
McGill University, Montreal, QC, Canada

A. Aerts
Eindhoven University of Technology, Eindhoven, The Netherlands

A. Akintayo
Iowa State University, Ames, IA, United States

C. Alcaraz
University of Malaga, Malaga, Spain

E. Almodaresi
Technical University Kaiserslautern, Kaiserslautern, Germany

T.A. Amin
University of Illinois at Urbana-Champaign, Champaign, IL, United States

W. An
Chinese Academy of Sciences, Beijing, China

G. Araniti
University Mediterranea of Reggio, Calabria, Italy

G. Ascheid
RWTH Aachen University, Aachen, Germany

F. Asplund
KTH Royal Institute of Technology, Stockholm, Sweden

P. Bagade
Arizona State University, Tempe, AZ, United States

N. Bajcinca
Technical University Kaiserslautern, Kaiserslautern, Germany

A. Banerjee
Arizona State University, Tempe, AZ, United States

M. Barhoush
RWTH Aachen University, Aachen, Germany

B. Beckmann
GE Global Research, Niskayuna, NY, United States

S. Bensalem
University Grenoble Alpes, Grenoble, France

F. Böhle
Institute for Social Science Research (ISF München), Munich, Germany

E. Bowman
US Army Research Laboratory, Adelphi, MD, United States

C. Brecher
RWTH Aachen University, Aachen, Germany

D. Cancila
CEA, LIST, Gif-sur-Yvette, France

C.G. Cassandras
Boston University, Boston, MA, United States

M. Castillo-Effen
GE Global Research, Niskayuna, NY, United States

L. Cazorla
University of Malaga, Malaga, Spain

Q. Chen
Montana State University, Bozeman, MT, United States

S. Ci
University of Nebraska-Lincoln, Lincoln, NE, United States

T. Citriniti
GE Global Research, Niskayuna, NY, United States

M. Condoluci
King's College London, United Kingdom

C.E. Crawford
Oak Ridge National Laboratory, Oak Ridge, TN, United States

G. Dartmann
University of Applied Sciences Trier, Trier, Germany

J. Deng
University of North Carolina at Greensboro, Greensboro, NC, United States

M. Dohler
King's College London, United Kingdom

C. Ecker
RWTH Aachen University, Aachen, Germany

T.W. Edgar
Pacific Northwest National Laboratory, Richland, WA, United States

M. Erol-Kantarci
Clarkson University, Potsdam, NY, United States

C. Estevez
University of Chile, Santiago, Chile

A. Ferrein
Mobile Autonomous Systems & Cognitive Robotics Science, Aachen University of Applied Sciences, Aachen, Germany

G.A. Fink
Pacific Northwest National Laboratory, Richland, WA, United States

J. George
US Army Research Laboratory, Adelphi, MD, United States

P. Giridhar
University of Illinois at Urbana-Champaign, Champaign, IL, United States

N. Golmie
National Institute of Standards and Technology, Gaithersburg, MD, United States

D. Griffith
National Institute of Standards and Technology, Gaithersburg, MD, United States

F. Guenab
Alstom Transport, Saint-Ouen, France

S.K.S. Gupta
Arizona State University, Tempe, AZ, United States

S.T. Hamman
The Air Force Institute of Technology, WPAFB; Cedarville University, Cedarville, OH, United States

W. Herfs
RWTH Aachen University, Aachen, Germany

M. Hoffmann
RWTH Aachen University, Aachen, Germany

K.M. Hopkinson
The Air Force Institute of Technology, WPAFB, OH, United States

N. Huchler
Institute for Social Science Research (ISF München), Munich, Germany

D.W. Illig
Clarkson University, Potsdam, NY, United States

W.D. Jemison
Clarkson University, Potsdam, NY, United States

S. Jeschke
Institute Cluster IMA/ZLW & IfU, RWTH Aachen University, Aachen, Germany

Z. Jiang
Iowa State University, Ames, IA, United States

L. Kaplan
US Army Research Laboratory, Adelphi, MD, United States

A. Koudri
IRT SystemX, Palaiseau, France

S. Krishnamurthy
Sony Computer Entertainment America, San Diego, CA, United States

G.K. Kurt
Istanbul Technical University, Istanbul, Turkey

G. Lakemeyer
Knowledge-Based Systems Group, RWTH Aachen University, Aachen, Germany

H. Liao
GE Global Research, Niskayuna, NY, United States

C. Lin
Dalian University of Technology; Key Laboratory for Ubiquitous Network and Service Software of Liaoning Province, Dalian, China

G. Liu
Montana State University, Bozeman, MT, United States

Y. Liu
Southeast University, Nanjing, China

J. Lopez
University of Malaga, Malaga, Spain

V. Lücken
RWTH Aachen University, Aachen, Germany

H. Luo
Yahoo, San Jose, CA, United States

D.G. MacDonald
Pacific Northwest National Laboratory, Richland, WA, United States

G.W. Maier
Bielefeld University, Bielefeld, Germany

S. Mallapuram
Towson University, Towson, MD, United States

P.M.N. Martins
Imperial College London, London, United Kingdom

J.A. McCann
Imperial College London, London, United Kingdom

L.A. McCarty
Cedarville University, Cedarville, OH, United States

J. McDermid
University of York, York, United Kingdom

T. Meisen
RWTH Aachen University, Aachen, Germany

P. Moulema
Towson University, Towson, MD, United States

M.R. Mousavi
Halmstad University, Halmstad, Sweden

R. Müller
Technische Universität Dresden, Dresden, Germany

S. Narciss
Technische Universität Dresden, Dresden, Germany

T. Niemueller
Knowledge-Based Systems Group, RWTH Aachen University, Aachen, Germany

M. Obdenbusch
RWTH Aachen University, Aachen, Germany

S.K. Ötting
Bielefeld University, Bielefeld, Germany

R. Passerone
University of Trento, Trento, Italy

H. Pfeifer
fortiss, Munich, Germany

L. Ren
GE Global Research, Niskayuna, NY, United States

M. Reniers
Eindhoven University of Technology, Eindhoven, The Netherlands

S. Reuter
Institute Cluster IMA/ZLW & IfU, RWTH Aachen University, Aachen, Germany

T.R. Rice
Pacific Northwest National Laboratory, Richland, WA, United States

H. Roy
US Army Research Laboratory, Adelphi, MD, United States

L.K. Rumbaugh
Clarkson University, Potsdam, NY, United States

G. Sagerer
Bielefeld University, Bielefeld, Germany

A. Sangiovanni-Vincentelli
University of California, Berkeley, CA, United States

S. Sarkar
Iowa State University, Ames, IA, United States

B. Schätz
Technical University of Munich, Munich, Germany

P. Seetharamu
University of Illinois at Urbana-Champaign, Champaign, IL, United States

Z. Shi
C3 IoT, CA, United States

E. Soubiran
Alstom Transport, Saint-Ouen, France

A. Tewari
ExxonMobil Research & Engineering Company, Annadale, NJ, United States

T. Töniges
Bielefeld University, Bielefeld, Germany

M. Törngren
KTH Royal Institute of Technology, Stockholm, Sweden

L. Urbas
Technische Universität Dresden, Dresden, Germany

D. Wang
University of Notre Dame, Notre Dame, IN, United States

S. Wang
University of Illinois at Urbana-Champaign, Champaign, IL, United States

H. Wang
University of Illinois at Urbana-Champaign, Champaign, IL, United States

L. Wouters
IRT SystemX, Palaiseau, France

B. Wrede
Bielefeld University, Bielefeld, Germany

J. Wu
University of Chile, Santiago, Chile

G. Wu
Dalian University of Technology; Key Laboratory for Ubiquitous Network and Service Software of Liaoning Province, Dalian, China

D. Wu
University of Tennessee at Chattanooga, Chattanooga, TN, United States

F. Xia
Dalian University of Technology; Key Laboratory for Ubiquitous Network and Service Software of Liaoning Province, Dalian, China

Z. Xu
Chinese Academy of Sciences, Beijing, China

B. Xue
Dalian University of Technology; Key Laboratory for Ubiquitous Network and Service Software of Liaoning Province, Dalian, China

Q. Yang
Montana State University, Bozeman, MT, United States

N. Yao
Georgia Institute of Technology, Atlanta, GA, United States

W. Yu
Towson University, Towson, MD, United States

E. Zandi
RWTH Aachen University, Aachen, Germany

F. Zhang
Georgia Institute of Technology, Atlanta, GA, United States

About the Editors

Houbing Song received his Ph.D. degree in electrical engineering from the University of Virginia, Charlottesville, VA, in Aug. 2012, and his M.S. degree in civil engineering from the University of Texas at El Paso, TX, in Dec. 2006.

In Aug. 2012, he joined the Department of Electrical and Computer Engineering, West Virginia University, Montgomery, WV, where he is currently an Assistant Professor and the founding director of the Security and Optimization for Networked Globe Laboratory (SONG Lab, http://www.SONGLab.us), and West Virginia Center of Excellence for Cyber-Physical Systems sponsored by West Virginia Higher Education Policy Commission. He served as an engineering research associate at Texas A&M Transportation Institute in 2007. His research interests lie in the areas of cyber-physical systems, internet of things, cloud computing, big data analytics, connected vehicle, wireless communications, and networking. Dr. Song's research has been supported by the West Virginia Higher Education Policy Commission. Dr. Song was the first recipient of Golden Bear Scholar Award, the highest faculty research award at WVU Tech.

Dr. Song is a senior member of IEEE and a member of ACM. Dr. Song is an associate editor for several international journals, including IEEE Access and KSII Transactions on Internet and Information Systems, and a guest editor of several special issues within leading international journals such as IEEE Transactions on Industrial Informatics. Dr. Song was the general chair of four international workshops, including the first IEEE International Workshop on Security and Privacy for Internet of Things and Cyber-Physical Systems (IOT/CPS-Security), held in London, UK and the first/second/third IEEE ICCC International Workshop on Internet of Things (IOT 2013/2014/2015), held in Xi'an/Shanghai/Shenzhen, China. Dr. Song also served as the technical program committee chair of the fourth IEEE International Workshop on Cloud Computing Systems, Networks, and Applications (CCSNA), held in San Diego, USA. Dr. Song has served on the technical program committee for numerous international conferences, including ICC, GLOBECOM, INFOCOM, WCNC, etc. Dr. Song has published more than 100 academic papers in peer-reviewed international journals and conferences.

Danda B. Rawat is an associate professor in the Department of Electrical Engineering and Computer Science at Howard University, Washington DC, USA. Dr. Rawat's research focuses on wireless communication networks, cybersecurity, cyber physical systems, internet of things, big data analytics, wireless virtualization, software-defined networks, smart grid systems, wireless sensor networks, and vehicular/wireless ad-hoc networks. His research is supported by the U.S. National Science Foundation, University Sponsored Programs, and grants from the Center for Sustainability. Dr. Rawat is the recipient of the NSF Faculty Early Career Development (CAREER) Award. Dr. Rawat has published over 100 research articles and 8 books. He serves as editor/guest editor for over 10 international journals. He serves as a web-chair for the IEEE INFOCOM 2016/2017, served as a Student Travel Grant co-chair of the IEEE INFOCOM 2015, Track Chair for wireless networking and mobility of the IEEE CCNC 2016, Track Chair for Communications Network and Protocols of the IEEE AINA 2015, and so on. He served as a program chair, general chair, and session chair for numerous other international conferences and workshops. He is the recipient of the Outstanding Research Faculty Award (Award for Excellence in Scholarly Activity) 2015 from the Allen E. Paulson College of Engineering and Technology, GSU among others. He is the founder and director of the Cyber-security and Wireless Networking Innovations (CWiNs) Research Lab. Between 2011 and 2016, Dr. Rawat was with Georgia Southern University and Eastern Kentucky University. Dr. Rawat is a senior member of the IEEE and a member of the ACM and the ASEE.

Prof. Dr. rer. nat. Sabina Jeschke is head of the institute cluster IMA/ZLW & IfU at the RWTH Aachen University since 2009. She studied Physics, Computer Science, and Mathematics at the Berlin University of Technology. After research stays at the NASA Ames Research Center/California and the Georgia Institute of Technology/Atlanta, she gained a doctorate on "Mathematics in Virtual Knowledge Environments" in 2004. Following a junior professorship (2005–2007) at the TU Berlin with the construction and direction of its media center, she was the head of the Institute of Information Technology Services (IITS) for electrical engineering at the University of Stuttgart from May 2007 to May 2009, where, during the same period, she was also the director of the Central Information Technology Services (RUS). Her research areas are inter alia distributed artificial intelligence, robotics and automation, traffic and mobility, virtual worlds, and innovation and future research. Sabina Jeschke is vice dean of the Faculty of Mechanical Engineering of the RWTH Aachen University, chairwoman of the board of management of the VDI Aachen and member of the supervisory board of the Körber AG. She is a member and consultant of numerous committees and commissions, alumni of the German National Academic Foundation (Studienstiftung des deutschen Volkes), IEEE Senior and Fellow of the RWTH Aachen University. In Jul. 2014, the Gesellschaft für Informatik (GI) honored her with their award Deutschlands digitale Köpfe (Germany's digital heads). In Sep. 2015 she was awarded the Nikola-Tesla Chain by the International Society of Engineering Pedagogy (IGIP) for her outstanding achievements in the field of engineering pedagogy.

Prof. Dr.-Ing. Christian Brecher has been the Ordinary Professor for Machine Tools at the Laboratory for Machine Tools and Production Engineering (WZL) of the RWTH Aachen as well as the Director of the Department for Production Machines at the Fraunhofer Institute for Production Technology IPT since Jan. 1, 2004. Further, he is CEO of the Cluster of Excellence "Integrative Production Technology for High-Wage Countries," which is funded by the German Research Foundation (DFG). After finishing his academic studies in mechanical engineering, he started his professional career first as a research assistant and later as a team leader in the department for machine investigation and evaluation at the WZL. From 1999 to Apr. 2001, he was responsible for the department of machine tools in his capacity as a Senior Engineer. After a short spell as a consultant in the aviation industry, Professor Brecher was appointed in Aug. 2001 as the Director for Development at the DS Technologie Werkzeugmaschinenbau GmbH, Mönchengladbach, where he bore the responsibility for construction and development until Dec. 2003. Prof. Brecher has received numerous honors and awards including the Springorum Commemorative Coin, the Borchers Medal of the RWTH Aachen, the Scholarship Award of the Association of German Tool Manufacturers (Verein Deutscher Werkzeugmaschinenfabriken VDW) and the Otto Kienzle Memorial Coin of the Scientific Society for Production Technology (Wissenschaftliche Gesellschaft für Produktionstechnik WGP). Currently he is chairman of the scientific group for machines of CIRP, the International Academy for Production Engineering.

Foreword

Cyber-Physical Systems (CPS) have been a critical driving force for economic development during the beginning of the 21st century. An organized effort to define CPS research and development and to establish a sponsored research program commenced 10 years ago. In 2006, the Computer and Network System Division (CNS) of the US National Science Foundation (NSF) reviewed its research programs and decided to initiate a new direction by emphasizing engineering systems that are built from and depend upon the integration of computational and physical components. Consequently, the Computer System Research (CSR) program of CNS was revised and CPS became a thematic research area in CSR program. With the support from the NSF Directorate for Computer & Information Science & Engineering (CISE), the first call-for-proposal of this new research program was announced in Fall 2006 (NSF 07-504).

This first CPS research program received enthusiastic response from all academic, industrial, and governmental sectors. The first workshop about CPS research and development was held in Austin, Texas, in October 2006, with a mission to define the agenda of CPS research and development for the nation. The effort was quickly recognized by the President's Council of Advisors on Science and Technology (PCAST). In PCAST's 2007 report, CPS was given a top priority for substantial research investment. In 2008, the NSF elevated its support for CPS research by launching a fully fledged research program (NSF 08-611), which has now become a core NSF program jointly supported and managed by multiple agencies, including the Department of Homeland Security, the Department of Transportation, the National Aeronautics and Space Administration, the National Institutes of Health, and the Department of Agriculture. Many other countries and organizations have launched similar efforts, triggering a large-scale, globally organized effort on CPS research, education, and development. Consequently, the CPS community has had tremendous growth. Currently, in the United States alone, thousands of researchers and developers are actively working in this emerging field.

This book covers recent advances on Cyber-Physical System research and development, which I believe is the best way to celebrate the 10th anniversary of launching the first CPS program. The papers collected for this book not only report the results in CPS research and development accomplished in the last decade, but they also address open challenging research issues yet to be explored for the success of CPS long-term development.

W. Zhao
Chair professor and rector (president) of the University of Macau
Former director of NSF Computer and Network Systems Division

Preface

Cyber-physical systems (CPSs) are transforming the way people interact with engineered systems, just as the Internet transformed the way people interact with information. CPSs integrate cyber components (namely, sensing, computation, control, and networking) into physical components (namely, physical objects, infrastructure, and human users), connecting them to the Internet and to each other. CPSs are characterized by much higher capability, adaptability, scalability, resiliency, safety, security, and usability. CPS will drive innovation and competition in an ever-growing set of application domains, and enable a smart and connected world to address grand societal challenges.

Tremendous progress has been made in advancing CPS science, technology and engineering over the past decade since the term "CPS" emerged in 2006. An increasing number of scientists and engineers motivated by CPS are building a research community committed to advancing research and education in CPS and to transitioning CPS science and technology into engineering practice. However, there is not a book to present the state-of-the-art and the state of the practice of CPS from the perspective of systems science and engineering. This book serves the purpose of preparing scientists and engineers from various backgrounds for making CPS a reality.

This edited book, *Cyber-Physical Systems*: *Foundations*, *Principles*, *and Applications*, aims to present the scientific foundations and engineering principles needed to realize CPS, and various CPS applications. Towards this goal, this book is organized into three parts: Foundations, Principles, and Applications.

Part 1 is composed of nine chapters. In addition to the opportunities and challenges of CPS (Chapter 1), this part presents various scientific foundations of CPS, including real-time control and adaptation for CPS (Chapters 2 and 3), energy harvesting (Chapter 4), communications and networking (Chapter 5), big data (Chapter 6), computation (Chapter 7), decision-making (Chapter 8), CPS security and privacy (Chapter 9).

Part 2 is composed of 11 chapters. This part presents various engineering principles of CPS, including human-CPS interaction (Chapter 10), signal processing (Chapter 11), system design and verification (Chapters 12 and 19), CPS autonomy (Chapter 13), localization (Chapter 14), green communications and networking (Chapter 15), wireless charging (Chapter 16), game theory (Chapter 17), machine learning (Chapter 18), and smart and connected communities (Chapter 20).

Part 3 is composed of seven chapters. This part presents various CPS applications, spanning agriculture (Chapter 25), energy (Chapters 24 and 27), transportation (Chapters 22 and 23), and manufacturing (Chapters 21 and 26).

This book would not have been possible without the help of many people. First, we would like to thank all the contributors and reviewers of the book from all over the world. Second, we would like to thank our editorial assistants, Ruth Hausmann, Denis Özdemir, Alicia Dröge, all at RWTH Aachen University, who provided indispensable support at all stages of the editorial process of the book. Also we would like to thank our Editorial Project Manager, Amy Invernizzi at Morgan Kaufmann, Imprint of Elsevier, and our Senior Acquisitions Editor, Brian Romer at Elsevier, who helped shepherd us through the book-editing process. Third, we would like to acknowledge the German Research Foundation (DFG) for funding the Cluster of Excellence "Integrative Production Technology for High-Wage Countries" of the RWTH Aachen University within the German Excellence Initiative. Further, we would like to acknowledge the German Federation of Industrial Research Associations (AiF) for funding research projects of the RWTH Aachen University in the context of small- and medium-sized enterprises.

Houbing Song
Danda B. Rawat
Sabina Jeschke
Christian Brecher
May 2016

FOUNDATIONS

CHARACTERIZATION, ANALYSIS, AND RECOMMENDATIONS FOR EXPLOITING THE OPPORTUNITIES OF CYBER-PHYSICAL SYSTEMS

1

M. Törngren*, F. Asplund*, S. Bensalem[†], J. McDermid[‡], R. Passerone[§], H. Pfeifer[¶], A. Sangiovanni-Vincentelli**, B. Schätz[††]

KTH Royal Institute of Technology, Stockholm, Sweden[*] University Grenoble Alpes, Grenoble, France[†] University of York, York, United Kingdom[‡] University of Trento, Trento, Italy[§] fortiss, Munich, Germany[¶] University of California, Berkeley, CA, United States[**] Technical University of Munich, Munich, Germany[††]

1 INTRODUCTION

The findings described in this chapter draw on the CyPhERS project, an EU-funded support action that developed a strategy and agenda for cyber-physical systems (CPS) in Europe. This chapter summarizes the findings, focusing on (1) a characterization of CPS, (2) opportunities and challenges in representative CPS application domains, and (3) recommendations for action resulting from a cross-domain analysis. The interested reader should consult CyPhERS project deliverables for details, starting with deliverables D6.1 and D5.2 (Schätz et al., 2015; Törngren et al., 2014, with further references therein).

As identified during the CyPhERS project, there are different interpretations of what constitutes a CPS, depending on what perspective is taken. The increasing connectivity, and penetration of electronics and software into all facets of our lives, is referred to differently by different research communities, such as CPS, Internet of Things (IoT), ubiquitous computing, fog and swarm; or is labeled under application oriented terms, such as smart cities or Industrie 4.0 (CPS in manufacturing). This led the CyPhERS project to devote a special effort to characterizing CPS to complement existing definitions (Cengarle et al., 2014).

The core members of the CyPhERS project drew on their own expertise, and external expertise. Consequently, the project ran several workshops including consultations with a large number of experts, carried out state-of-the-art surveys, and undertook in-depth analyses, including an analyses of Strengths, Weaknesses, Opportunities and Threats (SWOT), for five domains that were selected for investigation: *manufacturing*, *health*, *smart grid*, *transportation*, and *smart cities*.

CyPhERS had a broad remit and early on decided to go beyond a traditional technological focus. A major reason for the broader scope is due to the perceived disruptive nature of CPS and the potential ways in which CPS technology increasingly affects virtually all aspects of our society.

Cyber-Physical Systems. http://dx.doi.org/10.1016/B978-0-12-803801-7.00001-8

3

Based on the analyses of the five domains, CyPhERS conducted a cross-domain analysis to identify the opportunities, challenges, and strategies found to be common across the domains. This analysis led to the identification of high-level recommendations for action to grasp opportunities and deal with the challenges.

In the remainder of the chapter we first provide the characterization of CPS, followed by an analysis of the five selected domains. We then proceed to describe the synthesized recommendations. Finally, we discuss the results and summarize conclusions. Overall, the chapter contains references for further reading (including the CyPhERS deliverables) and the discussion section provides references to related surveys.

2 CPS CHARACTERIZATION

The term Cyber-Physical Systems (CPS), introduced in the United States in 2006, was prompted by the increase in technical systems in which interactions between interconnected computing systems and the physical world were of primary importance. Early definitions illustrate how CPSs are found both in the small and the large arenas: "Such systems use computations and communication deeply embedded in and interacting with physical processes to add new capabilities to physical systems. These CPS range from minuscule (pace makers) to large-scale (the national power-grid)" (CPS-Summit, 2008).

While such definitions make sense, they are generic; it is becoming increasingly difficult to identify systems that are *not* cyber-physical, given the increasing digitalization with penetration of electronics and software into virtually all facets of our lives. The concept of CPS ranges from massive to minimal systems. The concept is moreover inherently multidisciplinary and multitechnological, and relevant across vastly different domains, with multiple socio-technical implications. The relevance of CPS thus remains difficult to evaluate for the uninitiated with respect to their impact and applicability to particular industrial sectors.

We therefore provide a high-level and minimalistic characterization of CPS using four perspectives that we deem of primary importance, namely, *technical emphasis*, *cross-cutting aspects*, *level of automation,* and *life-cycle integration*. Our intention is to facilitate the description of particular CPS, or classes of CPS of interest, and to provide a checklist to support planning and design of a CPS.

2.1 CPS CHARACTERISTICS

- *Technical emphasis.* CPS represent the integration of *physical* and *embedded systems* with *communication* and *IT* systems. With technical emphasis we refer to the technical part(s) of a CPS that is(are) considered to be of particular importance, e.g., the embedded computing or the IT parts. When designing a CPS, there is a corresponding need to decide where emphasis should be placed, closely related to (1) how physical and embedded system parts are *co-designed* to enable optimizations and synergies, and (2) how communication capabilities are used to off-load systems to enable cost reductions, optimized operations, etc. The most obvious impact of the associated design choices is on the scale and capabilities of a CPS, but there are also indirect business implications. As part of this characteristic we also encompass the considered scale of a CPS to further clarify the focus.

- *Cross-cutting aspects.* With cross-cutting aspects we refer to *system properties* (such as safety and security), *jurisdiction* (i.e., applicable standards and legislation), and *governance* (i.e., where responsibility lies for the safe, efficient, secure operation of the system). These aspects thus refer to the constraints for operation and organizational responsibilities in meeting those constraints. The advances in connectivity make it possible to create new applications that span several traditional application domains. This opens up new business opportunities, but also requires that technical and nontechnical "gaps and barriers" across the domains are dealt with. While connectivity may be desirable, it also necessitates explicit consideration of properties such as security. Adaptability across different environmental contexts and use cases is often driven by business considerations (e.g., reduced maintenance costs and increased availability), and may eventually require dynamic reconfigurability.
- *Level of automation.* Designing a CPS involves careful investigation to ascertain a suitable *level of automation*. With level of automation we refer to what activities are automated and to what degree (Parasuraman et al., 2000). The increasing interest in autonomous vehicles has driven the development of classifications of levels of automation, e.g., the standard for automated driving established by the Society for Automotive Engineers (2014). CPSs are typically designed to act more or less independently of humans, even if they may be triggered by human inputs, *or* interact closely with humans, including shared control. Shared control also has its challenges: it is crucial to clarify who is in control at any point in time to make sure that unintended control does not take place, implying that human-machine interface design is often crucial. The level of automation closely corresponds to notions such as adaptability and, to some extent, corresponds to the "smartness" of a CPS.
- *Life-cycle integration.* Life-cycle integration is driven by quality, cost, and business concerns. With life-cycle integration we refer to a spectrum, from a CPS with no integration in the management of the product, services, and data over the life-cycle, to full integration of development and operations, including capabilities to upgrade and collect data from an operational system. The resulting trade-off concerns the benefits versus the costs of investments to ensure integration between the various IT systems (e.g., the engineering environments) and with the product in operation.

In describing CPS, the characteristics help to clarify what type of system is considered, also with regard to different stakeholders and viewpoints. The characteristics in addition serve to set the ambition in CPS design, e.g., regarding the desired level of automation. The characteristics are applicable for both small and large CPS.

Many terms have been coined to mirror the opportunities enabled by connectivity and computing. These terms largely provide similar messages but from different perspectives. The CyPhERS project contrasted CPS with such other terms. We here provide two examples of this based on the characteristics and refer interested readers to Törngren et al. (2014) for further examples:

- IoT emphasizes sensing of the physical world and uniquely identifiable things with (Internet) connectivity that communicate data with limited or no human interaction. Communication is often considered the key aspect—thus providing a specific *technical emphasis*. CPS differs through a systems perspective, not necessarily requiring Internet connectivity.
- Systems of systems (SoSs) usually address the construction of evolving large-scale systems and the coordination among those systems, specifically focusing on integration and optimization to

satisfy a wide range of objectives. The concept of SoS is independent of the type of system (e.g., organizational or socio-technical). Many SoSs will indeed incorporate CPS, and may also themselves be considered as CPS (as long as one can reconcile the terms system and SoS). The *cross-cutting aspects* of a CPS will largely characterize whether the actual CPS is an SoS or not.

2.2 FURTHER CPS CHARACTERIZATION THROUGH MARKET ANALYSIS

CPSs have the potential to be disruptive—to substantially change the nature of markets. This can be through the creation of new markets or through substantial changes of ecosystems. CyPhERS developed a market analysis method to try to identify the potential for CPS to shape markets—whether disruptive or transformative (McDermid et al., 2014). The method, which complements the previously described characteristics, includes an analysis of opportunities and constraints at each of four "layers": Social, Process, Information, and Technology, see Fig. 1; this model is known as the SPIT model (Sillitto, 2010). CPSs are anchored in the technology and information layers. For example, innovations at the information layer may allow horizontal integration by drawing on the sensing capabilities of a CPS. Such integration may in turn enable new business models at the process level. Innovation and constraints can arise at any level, but the constraints at the social level are important, as what is possible might not be socially acceptable.

An initial analysis of the five domains studied in the CyPhERS project came to the conclusion that disruption is unlikely in the domain of smart grids, but much greater changes are possible in the other domains (McDermid et al., 2014). While these findings need further in-depth analysis to be validated, we found the SPIT model useful for reasoning about the role of CPS in socio-technical systems including assessing potential business impact.

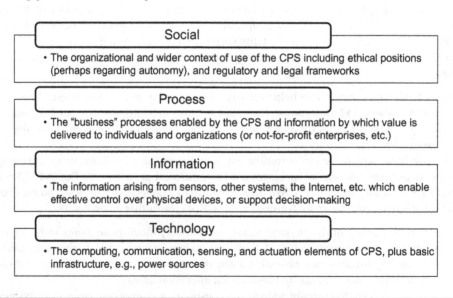

FIG. 1

SPIT model: social, process, information, technology.

3 ANALYSIS OF REPRESENTATIVE CPS DOMAINS

The analysis of opportunities and challenges in the CPS domains was carried out as a comprehensive SWOT analysis (Törngren et al., 2014). This section summarizes opportunities and challenges identified during the analysis, based on the mentioned CPS characteristics.

3.1 CPS IN MANUFACTURING

The industrial domains and processes of manufacturing, representing a major socio-economic force, are strongly characterized by the use of CPS technologies. Manufacturing encompasses CPS with different types of *technical emphases*, from 3D scanners/printers to cloud manufacturing. The increased emphasis on IT integration and openness means that security, as a *cross-cutting property*, is becoming increasingly important. Manufacturing has been a forerunner in *automation* with solutions transferred to other domains, e.g., from industrial robotics to autonomous vehicles. Mass customization is currently driving the development of more flexible and efficient production systems (see, Wang et al., 2015). Advanced industrial companies have already introduced *life-cycle integration*—tracing real operational data back to development and manufacturing. This trend is likely to continue. Manufacturing, as a domain, has also to some extent been integrated with other domains, primarily with transportation for logistics, providing *cross-domain* solutions. Opportunities arise from additive (and distributed) manufacturing, as well as from new business models involving open innovation, paving the way for flexible, customizable distributed manufacturing schemes.

At the societal level, it is essential that sustainability be taken seriously. We note, for example, that about 14% of the total 2652 million tons of waste that were generated in EU-27 countries in 2008 were due to manufacturing (Eurostat, 2011). CPS technology provides solutions that may assist in dealing with sustainability, such as modular architectures to facilitate reuse and recycling.

Complex CPS will feature prominently in future manufacturing systems. The management of such systems, dealing with security and safety risks, and providing efficient interoperability, poses barriers to their successful industrialization. Lack of the new competences required may prevent successful industrial evolution; in particular the provision of additional skill sets encompassing Internet, security, and software are seen as a key enabler.

3.2 CPSs IN HEALTHCARE

CPSs in healthcare have varied applications and *technical emphases*, from medical devices to improve the efficacy of medical treatment and surgery, to remote services based on collected data. The paradigm shift in *level of automation* from what used to be essentially passive devices, controlled by human operators, to IT-enabled devices is significant. Emerging healthcare devices and equipment actively control critical physiological processes and functions. The embedded computing, sensing, modeling, communications, and deep integration with physical elements and processes allow these new CPSs to achieve levels of functionality, adaptability, and effectiveness not possible with simpler passive systems (NITRD, 2009; HMGov, 2013).

A widespread adoption of CPS will be able to provide data of unprecedented size and accuracy regarding the effectiveness of treatment, giving doctors invaluable information for fine-tuning processes and procedures to achieve better *life-cycle integration* for both products and patients. Similarly, a better understanding of the side conditions and real-time information is essential to personalize treatments and achieve better outcomes. In particular, CPSs have the potential to reach the body using minimally or noninvasive techniques, which lower costs and enhance mobility, independence and quality of life. The continuous monitoring of a chronic condition has also the potential of substantially shifting care delivery from inpatient to outpatient services and to the home.

The diversity and interconnection, coupled with the sensitive nature of dealing with life-related conditions, makes the design of CPS in healthcare challenging and leads to severe *cross-cutting* issues. While data collection is essential to improve healthcare services, its security and privacy must be guaranteed, and devices must be immune to attacks, as they may provide access to the body. At the same time, these systems are complex and require new technologies, which have had only limited testing, leading to malfunctions, due primarily to design failures, but also due to materials and components (Admet, 2014). This highlights the challenges faced by the design process, which must be supported by new methodologies and tools (Davare et al., 2013). In addition, healthcare is regulated to guarantee, through certification, the introduction of safe and effective treatments. However, increasingly strict regulations and longer clinical trials could have significant impact on cost and investments. In Europe in particular, regulations are not homogeneous among the member states, leading to uncertainty and risk for the industry.

3.3 CPS AND SMART GRIDS

Infrastructures providing reliable access to energy form a foundation for our industrialized societies. Traditionally, an electric grid is implemented by a small number of high-volume facilities on the production side and large number of generally low-to-medium volume installations on the consumer side, with a varying demand of a factor of four between lows and peaks during the day. The increasingly used renewable energy resources are generally produced by a larger number of facilities with mostly volatile volumes not in synch with the requested consumption, rendering the traditional asymmetric and centralized management scheme of the electric grid increasingly inadequate (see Hashmi, 2011). Here, the new *technical emphasis* of the CPS of a smart grid offers a solution by enabling the decentralized and cooperative coordination of technical and organizational processes, from the control of a photovoltaic installation to the billing and trading of energy. Such decentralized solutions enable microgrids (local grids) that can operate with or without a connection to the main grid.

Similarly, *automation* that monitors and predicts consumption as well as production of renewables, via smart meters on a fine-grained level, facilitates a reliable short-term balancing of demand and supply. The use of intelligent devices and managed installations, including low-volume energy buffers (e.g., batteries), support a shift from a supply- to a demand-side management of the grid. The highly automated control of a CPS allows these processes to be scaled to the required number of participants.

A *life-cycle integration* that allows this control to be updated seamlessly across an entire smart grid promises even further gains. Locally inefficient control can be identified and replaced seamlessly, and additional data or distributed sensing capabilities added in a modular fashion. Furthermore, if data are shared between the manufacturers and users of CPS of smart grids, further opportunities for customization can be identified.

However, this will not happen if the necessary technological (including interoperable, safe, and secure infrastructure) and regulatory prerequisites (including suitable tariffs, market models, and transnational grid operation schemes) cannot be established. The ability to understand the *cross-cutting* implications of producing, trading in and monitoring energy must be solid enough, so that uncertainty and risk does not prevent distributed investments into smart grid infrastructure.

3.4 CPS IN TRANSPORTATION

Transportation encompasses CPS all the way from smart components such as a smart tire to intelligent transportation systems. As transportation systems are growing to meet the future demands of society, they are evolving towards increasingly complex SoSs, as exemplified by the new European Rail Traffic Management System (ERTMS). ERTMS encompasses CPS with a new *technical emphasis* through a dedicated GSM communication system that connects trains, infrastructure systems, and system management.

Transportation systems have a direct relation to *cross-domain* integration. Transportation services require the coordination of processes across sectors like logistics, automotive, and rail. The coordination is influenced both by the corresponding vehicles and infrastructural components (e.g., roads and communications), with differences including speed, capacity, cost, and governance.

With logistics being an integral part of industrial and societal processes, the shift from mobility as the provision of vehicles to mobility as a service is accelerating. Modern mobility solutions will increasingly focus on highly *automated* forms of transport, addressing the need for individual transport without the need for an individually owned vehicle. The vision of automated guided and connected vehicles is necessary not only to cope with the increased demand for mobility, but also to address the additional societal goals of increased safety, efficiency, security, convenience, and the economy (Sussman, 2005). An important factor is the requirement to adapt transport services to an increasingly aging population: quality, reliability, security, accessibility for persons with reduced mobility, and safety are essential to meet this requirement with public transport.

While most CPS domains share similarities in the challenges they face, there are several unique problems linked to transportation CPS. The customization and variability of system architecture weaves a level of complexity rarely observed. The ultra-competitive global landscape mandates ever-evolving requirements for enhanced capabilities, resulting in the need to rapidly adapt systems. The changing landscape leads to an increased importance of *cross-cutting aspects* such as safety and security; and the evolving need for *life-cycle integration* to allow for pattern analysis of accident statistics, continuous updates to vehicles to remove defects or avoid unsuitable driver behavior, etc. In addition to the competition in the pursuit of vehicle consumers, the global competition at the component level is equally fierce, resulting in a diverse and constantly evolving set of component suppliers that must provide products to be integrated into the whole.

3.5 CPS AND SMART CITIES

Smart cities involve the integration of many domains of CPS research and technology. In addition, they touch other domains such as architecture and legal, economics, and social sciences—they truly place emphasis on the *cross-domain* aspects of CPS. A background paper by the UK Department for Innovation and Skills provides a good summary of the challenges and opportunities that cities and business are facing when inserting digital technology into cities (UKBus, 2013).

Opportunities for CPS in smart cities abound due to the present situation. In Europe many of the cities date back to the Roman times. This results in an often chaotic layout of the downtown areas with narrow and winding streets. Traffic is at times unmanageable. In the United States the situation is more critical in other respects, such as the large number of cars in the urban highway systems. Without the determined use of advanced technology with a new *technical emphasis* the situation will be unmanageable. The deployment of a capillary network of sensors and of pervasive communication typical of CPS allows monitoring and controlling security, safety, and efficiency in smart cities as well as the development of new services to make cities more livable. A high degree of *automation* of future city services will provide an unprecedented ability to avoid wasting resources when directing city transports; losses linked to water pollution, fire or the release of hazardous materials; and disturbances due to failing infrastructure.

The most serious challenges are related to decision makers, often unaware of technology. Whenever confronted with a plan to add a CPS-driven infrastructure or service, they may fall prey to unwise designs. Indeed, there have been examples of technology insertions that have ended up being a waste of money and effort (Greenfield, 2013). An increased level of *life-cycle integration*, where data are harvested and certain smart city technology proven to be linked to benefits, will be required to allow decision makers to act (almost) regardless of their level of technological expertise.

4 RECOMMENDATIONS BASED ON A CROSS-DOMAIN ANALYSIS

The SWOT analysis for CPS domains was followed by a cross-domain analysis. This analysis was used to identify patterns (challenges, opportunities, and strategies) common across the domains (Törngren et al., 2014). Due to space limitations, we focus on the recommendations that were common across the domains, summarized in the following:

Strengthen cross-disciplinary research collaboration: The need to strengthen key research in CPS is common across most CPS domains since current approaches to design and verification are already stretching the limits for cost-efficient system development; there is an urgent need to update the engineering methodologies for CPS. There is a need to develop funding schemes that to a greater extent stimulate the creation of truly multidisciplinary consortia, bridging the gaps between traditional disciplines, e.g., embedded systems versus Internet and big-data, and between application domains. Corresponding strategies include support of broader networks of excellence, and to stimulate learning networks among industrial domains.

Foster enabling education and training: Excellence in education and a skilled work force is of paramount importance for exploiting CPS opportunities. The problem is the growing amount of knowledge and skills required for product and service engineering. In order to create engineers capable of building CPS, education must break the disciplinary silos, and provide cross-disciplinary technology and project experiences. Incentives are needed to stimulate academia and industry collaboration in education. To ensure the necessary re-qualification, an academic-industrial alliance should be formed to support engineers in life-long learning.

Stimulate public-private partnerships for CPS technology experimentation to deal with societal challenges and to ensure a dependable information and communication infrastructure: The adoption of key CPS technologies will depend on their maturity, requiring their application in real-world and large-scale installations through maturation initiatives. The level of complexity introduced by CPS further mandates experimental and incremental approaches to system realization. Public-private partnerships are needed to ensure the availability and affordability of dependable and trustworthy information and communication infrastructure.

Promote interoperability of CPS technology through reference platforms and standards: CPSs depend critically on integration. Public incentives are needed to facilitate interoperability *across the engineering life-cycle*, and within and across domains and disciplines. Interoperability goes beyond technology and requires consideration of, for example, concrete business drivers and regulations to make sure that "standards" are developed at the right level. This amounts to providing reference platforms to support the integration of services as well as homogenizing interoperability standards. Whilst this activity must be led by industry, regulators and other public bodies should encourage and support these initiatives.

Prepare for disruption by anticipating new business models and supporting open innovation: New value-added end-user services will become important "products" in the context of CPS and will give rise to new business models and ecosystems (e.g., by selling transport services rather than vehicles). To stimulate such ecosystems, forums should be provided facilitating contacts and collaboration among innovators trying to enter the service ecosystem of a CPS and existing providers of services. Research and innovation should also stimulate the development of, and research into, new business models. The orchestration of basic services will often rely on established and cost-intensive infrastructures where innovation opportunities will depend on easy access to those services. Funding programs must therefore promote open standards, the provision of open-source or open/free license results, and promote interoperability. Opportunities for big-data analytics require well-defined open data access as well. In order to reduce entry barriers for innovative enterprises clear liability regulation frameworks must be provided and corresponding supporting technologies must be put into place that help to identify acceptance and delegation of responsibilities for services provided.

Ensure trustworthiness including safety and security: The pervasiveness of CPS implies that their malfunction or misuse can have dramatic negative effects on society and the economy. Safety and security are exposed as intertwined and truly cross-cutting issues. They require revised standards and regulations and the development of new engineering methodologies to ensure that the implemented systems meet agreed-upon trust levels. Security requires special consideration as previously closed systems become exposed in new ways. Joint public and private investments are needed to assess and improve the security of both public and private information and communication technology to protect these critical infrastructures from cyber-attacks. There is a strong human element here—it is humans who will determine whether or not a CPS-based system is to be trusted.

Ensure that humans are at the center of approaches to CPS: Because societies will rely on CPS it is of paramount importance not only that they are effectively engineered, but also well understood and appropriately used. Overall, ensuring human-centered approaches to CPS requires that related efforts, from training to research and experimentation, need to include and consider a broader set of stakeholders than just engineers and system developers. Essentially, a very broad set of stakeholders, including policy makers and the general public, will need a basic understanding about CPS implications in terms of both opportunities and risks. A further important concern is to address the missing cross-fertilization between engineering sciences and humanities. We cannot afford this gap to continue for societal level CPS systems. This becomes even more important with the increasing level of automation provided by CPS functionalities. Finally, there is an urgent need to pay explicit attention to sustainability and privacy with consideration of related trade-offs, for example, referring to data sharing versus privacy, and openness versus security threats. Economic, social, and environmental sustainability considerations need to be explicitly promoted in CPS initiatives to deal with the embedding of digitization everywhere. The pervasiveness of CPS in social processes demands built-in mechanisms to protect the privacy of its users, but also raises awareness of those users in interacting with CPS. To avoid misuse of sensitive data acquired by CPS, the establishment of regulations clarifying data ownership including granting and revoking access, as well as corresponding technical implementations are necessary.

5 DISCUSSION AND RELATED WORK

The SWOT analysis formed one important background for the recommendations. Several interactions with a wide range of stakeholders and other research initiatives took place to validate the findings. We believe the end results to be valid in that they represent strategic areas for Europe (although not necessarily valid for each European region). We also believe that many of the recommendations would be valid also for regions beyond Europe; however, making such claims for specific regions requires further verification.

Our recommendations were common across all domains albeit with different emphasis in some. The need to achieve better understanding of cross-cutting aspects and domain integration deserves special attention; engaging relevant stakeholders in debate and as part of pilot trials will be very important.

While regulations were not highlighted as an explicit recommendation in this chapter, we would like to emphasize the importance to evolve and harmonize regulations related to CPS, in order not to impose over-constraining barriers.

As CPSs draw upon many different fields of technology, unsurprisingly there is a partial overlap of the identified recommendations with strategic agendas from these domains, most specifically those targeting complex embedded and networked systems, for example ITEA-ARTEMIS (2013), and the ARTEMIS strategic research agenda, ARTEMIS-SRA (2013). Unlike those, CyPhERS took a broader approach including societal, market, and education aspects. There is also a partial overlap concerning the recommendations with other national agendas, most specifically the recommendation from the US CPS-Summit report (CPS-Summit, 2008) and the German agendaCPS (Acatech, 2012). Despite slightly different focus (regions and domains), it is notable that the findings overall point in similar directions.

The interest in CPS is seen from a large number of publications including text books such as Lee and Seshia (2015) and Alur (2015). For the interested reader, the comprehensive survey of CPS technologies and applications by Khaitan and McCalley (2015) provides further useful references. The paper by Fisher et al. (2013) reviews the scientific and engineering challenges of CPS.

As a complement to our high-level characterization of CPS, more detailed frameworks include the ones by Baras and Austin (2013) and the CPS-PWG (2015).

6 CONCLUSIONS

CPSs are characterized by integration, *across technologies*, *industrial domains*, and the *life-cycle*, and by "smartness." CPS can be described using a corresponding set of characteristics: *technical emphasis*, *cross-cutting aspects*, *level of automation*, *and life-cycle integration*.

CPS, intended as the integration of cyber and physical parts, is not a new concept, but is now increasingly manifesting itself in terms of larger scale integrated systems that provide unprecedented opportunities for innovation.

Exploiting the opportunities made possible by CPS requires overcoming a number of challenges including developing scientific and engineering methodologies that cater for the complexity of CPS, providing dedicated education and training to relevant stakeholders, preparing for evolving business models, ensuring trustworthiness, as well as dealing with societal and legislative challenges.

The recommendations we described in this chapter are geared to address these challenges. Electronics is already being embedded "everywhere" in our societies. CPS will pave the way for even more digitalization. CPS further creates important business opportunities for largely automated systems. The implication is that economic, social, and environmental sustainability must be considered now in order to ensure that planning, adoption, and deployments sufficiently consider these aspects, in turn ensuring that humans remain at the center stage of a CPS-based society.

ACKNOWLEDGMENTS

This work was support by the European Commission through the CyPhERS FP7 support action (contract no. 611430) and the CPSE Labs Innovation action (contract no. 644400). We acknowledge contributions from Maria-Victoria Cengarle and Thomas Runkler, who were part of the CyPhERS project, together with inputs and feedback from numerous experts who contributed to the CyPhERS efforts.

REFERENCES

Acatech, 2012. agendaCPS—Integrierte Forschungsagenda Cyber-Physical Systems (Living in a Networked World. Integrated Research Agenda Cyber-Physical Systems (agendaCPS)). In: Geisberger, E., Broy, M. (Eds.), Springer-Verlag, New York. Available from: http://www.acatech.de/de/publikationen/publikationssuche/detail/artikel/living-in-a-networked-world-integrated-research-agenda-cyber-physical-systems-agendacps.html (accessed September 2015).

ADMET, 2014. Medical Device Sector Review. ADMET, Norwood, MA. Available from: http://www.onlinetmd.com/FileUploads/file/ADMET_Medical_Device_Sector_Review.pdf (accessed September 2015).

Alur, R., 2015. Principles of Cyber-Physical Systems. MIT Press, Cambridge, MA.

ARTEMIS ITEA Cooperation Committee, 2013. ITEA ARTEMIS-IA High-Level Vision 2030: Opportunities for Europe, *ARTEMIS Industry Association & ITEA Office Association*.

ARTEMIS-SRA, 2013. Embedded/cyber-physical systems ARTEMIS major challenges: 2014–2020. 2013 Draft Addendum to the ARTEMIS-SRA. Available from: http://www.artemis-ia.eu/publication/download/publication/910/file/ARTEMISIA_SRA_Addendum.pdf (accessed September 2015).

Baras, J., Austin, M., 2013. Development of a framework for CPS open standards and platforms. Technical report 2014-02, Institute for Systems Research, University of Maryland.

Cengarle, M.V., Törngren, M., Bensalem, S., McDermid, J., Sangiovanni-Vincentelli, A., Passerone, R., May 2014. Structuring of CPS Domain: Characteristics, trends, challenges and opportunities associated with CPS. Deliverable D2.2 of the CyPhERS FP7 project. Available from: http://www.cyphers.eu/sites/default/files/D2.2.pdf.

CPS-PWG, 2015. Framework for cyber-physical systems. Draft, Release 0.8, September 2015, Cyber Physical Systems Public Working Group, an open public forum established by the National Institute of Standards and Technology (NIST). Available from: https://pages.nist.gov/cpspwg/ (accessed 20.09.15).

CPS-Summit, 2008. Holistic approaches to cyber-physical integration. CPSWeek report. Available from: iccps2012.cse.wustl.edu/_doc/CPS_Summit_Report.pdf (accessed September 2015).

Davare, A., Densmore, D., Guo, L., Passerone, R., Sangiovanni-Vincentelli, A., Simalatsar, A., Zhu, Q., 2013. metroII: a design environment for cyber-physical systems. ACM Trans. Embed. Comput. Syst. 12 (1s), 1–49.

Eurostat, 2011. Eurostat Pocketbooks – Energy, Transport and Environment Indicators. Technical report, Eurostat, European Union.

Fisher, A., Jacobson, C., Lee, E.A., Murray, R., Sangiovanni-Vincentelli, A., Scholte, E., 2013. Industrial cyber-physical systems—iCyPhy. In: Proceedings of Complex Systems Design & Management (CSD&M), 4 December 2013. Springer, Paris, France, pp. 21–37.

Greenfield, A., 2013. Against the Smart City. DO Projects, New York.

Hashmi, M., 2011. Survey of smart grids concepts worldwide. Technical report, VTT Working Papers 166. Available from: http://www.vtt.fi/inf/pdf/workingpapers/2011/W166.pdf (accessed September 2015).

HMGov, 2013. Strengths and opportunity 2013. Technical report, HM Government. Available from: https://www.gov.uk/government/uploads/system/uploads/attachment_data/file/298819/bis-14-p90-strength-opportunity-2013.pdf (accessed September 2015).

Khaitan, S.K., McCalley, J.D., 2015. Design techniques and applications of cyberphysical systems: a survey. IEEE Syst. J. 9 (2), 350–365.

Lee, E.A., Seshia, S.A., 2015. Introduction to Embedded Systems, A Cyber-Physical Systems Approach, Second ed. ISBN: 978-1-312-42740-2. Publisher: Lee, E.A., Seshia, S.A. Available from: http://LeeSeshia.org.

McDermid, J., Cengarle, M.V., Törngren, M., Runkler, T., 2014. Market and innovation potential of CPS. Deliverable of the CyPhERS FP7 project. Available from: http://www.cyphers.eu/sites/default/files/D3.2.pdf (accessed August 2014).

NITRD, 2009. *High Confidence Software and Systems Coordinating Group*, High-confidence medical devices: cyber-physical systems for 21st century health care. Technical report, Networking and Information Technology Research and Development Program, February 2009.

Parasuraman, R., Sheridan, T.B., Wickens, C.D., 2000. A model for types and levels of human interaction with automation. IEEE Trans. Syst. Man Cybern. Syst. Hum. 30 (3), 286–297.

Schätz, B., Törngren, M., Bensalem, S., Cengarle, M.V., Pfeifer, H., McDermid, J., Passerone, R., Sangiovanni-Vincentelli, A., 2015. Research agenda and recommendations for action. Deliverable of the CyPhERS FP7 project. Available from: http://cyphers.eu/sites/default/files/d6.1+2-report.pdf (accessed March 2015).

Sillitto, H., 2010. Design principles for ultra-large scale (ULS) systems. In: Proceedings of INCOSE International Symposium, pp. 63–82.

Society for Automotive Engineers, 2014. Taxonomy and definitions for terms related to on-road motor vehicle automated driving systems. Surface Vehicle Information Report, J 3016, no. 01.

Sussman, J., 2005. Perspectives on Intelligent Transportation Systems. Springer, New York.

Törngren, M., Bensalem, S., Cengarle, M.V., McDermid, J., Passerone, R., Sangiovanni-Vincentelli, A., 2014. CPS: significance, challenges and opportunities. Deliverable D5.2 of the CyPhERS FP7 project. Available from: http://www.cyphers.eu/sites/default/files/D5.2.pdf (accessed December 2014).

UKBus, 2013. Smart cities: background paper. Department for Business Innovation and Skills of the UK. Available from: https://www.gov.uk/government/uploads/system/uploads/attachment_data/file/246019/bis-13-1209-smart-cities-background-paper-digital.pdf (accessed September 2015).

Wang, L., Törngren, M., Onori, M., 2015. Current status and advancement of cyber-physical systems in manufacturing. J. Manuf. Syst. 37 (Part 2), 517–527.

ADAPTIVE CONTROL IN CYBER-PHYSICAL SYSTEMS: DISTRIBUTED CONSENSUS CONTROL FOR WIRELESS CYBER-PHYSICAL SYSTEMS

G. Dartmann*, E. Almodaresi[‡], M. Barhoush[†], N. Bajcinca[‡], G.K. Kurt[§], V. Lücken[†], E. Zandi[†], G. Ascheid[†]

University of Applied Sciences Trier, Trier, Germany [*] *RWTH Aachen University, Aachen, Germany* [†]
Technical University Kaiserslautern, Kaiserslautern, Germany [‡] *Istanbul Technical University, Istanbul, Turkey* [§]

1 INTRODUCTION

Distributed control and consensus are popular concepts in cyber-physical systems (CPS). Such systems, further, can be seen as systems with multiple agents. An agent is here an independent subsystem (e.g., a robot). All agents work together to achieve a global utility. A typical application of a multiagent system is depicted in Fig. 1. This figure shows four unmanned flying vehicles (quadro-copters) communicating over wireless communication channels. The global goal is to reach a specific formation. This can be achieved, e.g., by an individual position and velocity estimate of each agent. The velocity and the current position are the states of each agent. To keep the formation of the entire group, each agent has to communicate its current state to its neighbors.

The consensus problem depends on the communication topology of the underlying multiagent system. The topology is typically described by a so-called communication graph. In this graph, each vertex corresponds to an agent and each edge corresponds to a communication link. In this chapter, the communication link is considered to be a wireless channel and, therefore, may not always provide reliable communication. In control technology, a reliable communication with low latency is a fundamental requirement of a control system. In this chapter, we will give a tutorial on consensus of multiagent systems. We study the consensus problem with wireless communication links and we investigate the influence of wireless communication parameters, such as signal-to-noise ratio (SNR) and channel fading on convergence of the system. Additionally, we observe, that latency is influenced not only by the mathematical consensus itself, but also by the communication model.

The chapter is organized as follows. In Section 2, we present an introduction to wireless channel models for CPS. In Section 3.1, we present the main findings in this field, which are relevant for

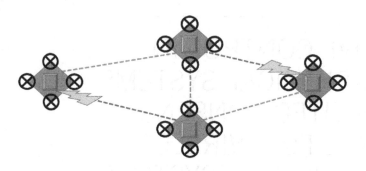

FIG. 1

Example of a multiagent system. Formation control is a typical application where consensus is of central importance (Li and Duan, 2015).

multiagent systems. Multiagent systems can be categorized in static networks without changes of the topology and networks with the switching topology. In Section 3.2, we present the fundamentals of communication protocols for static networks. These fundamentals are required to understand networks with switching topologies, which are presented in Section 3.3. In Sections 4 and 5 we present new ideas for adaptive quantization (AQ) for rate limited CPS.

2 COMMUNICATION CHANNEL OF MULTIAGENT SYSTEMS

The communication between agents constitutes an important part of the distributed control mechanism required for CPS. Conventionally, the communication network that enables agents to interact (and hence enable information exchange) is modeled as a graph (Godsil and Royle, 2001). Thereby, both directed and undirected graphs are frequently utilized. Due to a possible asymmetry in the wireless links considering uplink and downlink communication channels, a directed graph is more realistic to model a wireless CPS. A directed graph $\mathcal{G}$ is defined by its vertices and edges $(\mathcal{V}, \mathcal{E})$. The elements of the nonempty vertex set $\mathcal{V} = \{\boldsymbol{v}_1, \dots, \boldsymbol{v}_n\}$ represent the agents. A total number of n agents is considered. An edge $(\boldsymbol{v}_i, \boldsymbol{v}_j)$, for any $\boldsymbol{v}_i, \boldsymbol{v}_j \in \mathcal{V}$, represents the information exchange between two vertices, $\boldsymbol{v}_i$ and $\boldsymbol{v}_j$, if the corresponding agents can physically communicate. The edge set, $\mathcal{E} \subseteq \mathcal{V} \times \mathcal{V}$, is an ordered set of pairs of nodes. Self-loops, $(\boldsymbol{v}_i, \boldsymbol{v}_j)$, are frequently excluded. If $(\boldsymbol{v}_i, \boldsymbol{v}_j) \in \mathcal{E}$, $\boldsymbol{v}_j$ is a neighbor of $\boldsymbol{v}_i$. The set of neighbors of the agent $\boldsymbol{v}_i$, $\mathcal{N}_i$, and the cardinality of the $\mathcal{N}_i$ is called the degree of $\boldsymbol{v}_i$. A vertex is referred to as an isolated vertex if its neighbor set is empty. A weighted graph is defined as $\mathcal{G} = (\mathcal{V}, \mathcal{E}, A)$. Here, $A = [a_{ij}] \in \mathbb{R}^{n \times n}$ is the weighted adjacency matrix of $\mathcal{G}$. In a weighted graph, a weight is associated with every edge and represents the associated cost of the edge.

A path on $\mathcal{G}$ is defined as a finite sequence of edges that connect a sequence of vertices. When there is an edge from every vertex to every other vertex, the corresponding directed graph is complete. The graph $\mathcal{G}$ defines the topology of the communication network. The frequently used error-free transmission and fixed topology assumptions (Xiao et al., 2005; Ahlswede et al., 2000; Ngai and Yeung, 2004) are valid wired networks. However, these assumptions are overly simplistic for wireless networks.

In wireless networks, the information exchanges among vertices take place over the air. Due to the unguided transmission environment, the transmitted information signals are subject to the impairments of the wireless communication channel, and these impairments are modeled using three main components; the path-loss, the small-scale fading (or multipath fading) and the large-scale fading (or shadowing). The path-loss corresponds to the change in the average received power level related to the distance between the two vertices. The small-scale fading g_{ij} results from the multipath channel between the neighbor vertices and represents rapid fluctuations in the received signal's quality. Finally, as the third factor, the large-scale fading represents the variations in the local mean received power. These three factors significantly affect the transmission quality and cause nonnegligible error rates when transmitting information through an edge. Furthermore, the adjacency matrix becomes time varying due to the dynamic topology due to the impacts of the wireless communication channel. Consequently, with wireless channels the topology of the CPS changes rapidly (Bai et al., 2008, 2011), which can be modeled by a communication graph with $\mathcal{G}_k = (\mathcal{V}, \mathcal{E}, \mathbf{A}_k)$ with a switching signal $k = f(t)$.

3 CONSENSUS CONTROL

In this section, we will present an overview of the mathematical fundamentals of consensus control. Consensus of a multiagent system means the agreement of a common state of all agents and is mainly based on the topology of the communication network. The communication topology is mathematically described with so-called communication graphs. The mathematical theory to quantify the communication abilities of such multiagent systems is based on algebraic graph theory.

3.1 FUNDAMENTALS OF ALGEBRAIC GRAPH THEORY

The fundamentals of algebraic graph theory were already developed in the early 1970s (Fiedler, 1973). Fiedler investigated the algebraic connectivity of graphs. This seminal work was later used in the field of distributed control theory for multiagent systems where communication graphs are used to model the communication among multiple agents. Fiedler's work is only valid for bidirectional communication that corresponds to undirected communication networks. In this case, it can be proved that a connected multiagent system can achieve average consensus. This is not given in the case of directed graphs (Murray, 2007). Therefore average consensus is only possible if the communication graphs are balanced.

A typical example of how to understand algebraic graph theory is depicted in Fig. 2. The figure presents three graphs with four nodes. We will use this example to explain the theory of this chapter. Each

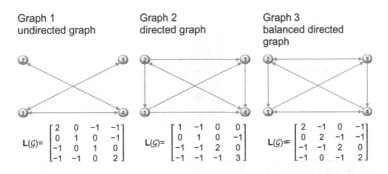

FIG. 2

Example for different types of connected multiagent systems.

node or vertex $\boldsymbol{v}_i$ in the graph has an index $i \in \mathcal{J}$, where $\mathcal{J} = \{1, \ldots, n\}$ denotes the set of all indices that correspond to the set of indices of all agents. Assuming all weights are $a_{ij} \in \{0, 1\}$ have discrete values, the communication graph can be formally defined by the following three definitions (Fiedler, 1973):

Definition 1 Consider a $\mathcal{G} = (\mathcal{V}, \mathcal{E}, \mathbf{A})$ with n nodes. An edge e_{ij} has a nonzero weight a_{ij} if the nodes with index i and j are connected and $i \neq j$, otherwise it has the weight $a_{ij} = 0$ where $i, j \in \mathcal{J}$.

Definition 2 The adjacency matrix $\boldsymbol{A}$, is $[\boldsymbol{A}]_{ij} = a_{ij}$, which denotes the weight of the edge from j to i.

Definition 3 The degree matrix $\boldsymbol{D}$ is a diagonal matrix with $[\boldsymbol{D}]_{ii} = d_{ii} = \sum_{j \neq i} a_{ij}$.

In Graph 1 of Fig. 2, the weight of the edge between Node 1 and Node 3 is given as $a_{13} = a_{31} = 1$ and the weight of the edge between Node 1 and Node 2 is given as $a_{12} = a_{21} = 0$. Based on the definition of the graph, the so-called Laplacian matrix can be associated with this graph (Fiedler, 1973).

Definition 4 The Laplacian matrix $\mathbf{L}(\mathcal{G}) \in \mathbb{R}^{n \times n}$ associated with $\mathcal{G} = (\mathcal{V}, \mathcal{E}, \mathbf{A})$ is $\mathbf{L}(\mathcal{G}) = \mathbf{D} - \mathbf{A}$.

The Laplacian matrix is essential to consensus control. An important parameter of this matrix is the set of eigenvalues. The matrix $\mathbf{L}(\mathcal{G})$ of an undirected graph is symmetric and positive semidefinite, therefore all eigenvalues are also real nonnegative. The second smallest eigenvalue λ_2 of the Laplacian matrix is called the algebraic connectivity of the graph and is often called Fiedler eigenvalue. In case of undirected graphs, the second smallest eigenvalue is larger in case of highly connected graphs compared to sparse graphs (Olfati-Saber and Murray, 2004). We will consider this property later in Section 3.3 when we discuss the convergence rate of consensus with switching topologies. Furthermore, it can be proved that in the case of directed graphs the real part of the eigenvalues is nonnegative, which follows directly from the following theorem.

Theorem 1 (Olfati-Saber and Murray, 2004, Theorem 1) *Let $\eta(\mathcal{G})$ the maximum node degree of a directed communication graph with $\mathcal{G} = (\mathcal{V}, \mathcal{E}, A)$, then all eigenvalues of $L(\mathcal{G})$ are within the following set:*

$$D(\mathcal{G}) = \{z \in \mathbb{C} : |z - \eta(\mathcal{G})| \leq \eta(\mathcal{G})\} \tag{1}$$

The set of possible locations of the complex eigenvalues is depicted in Fig. 3.

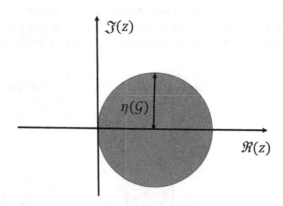

FIG. 3

Complex plane where all eigenvalues of $\mathbf{L}(\mathcal{G})$ are located.

Due to the construction of the Laplacian matrix we have zero row sums, $\sum_j l_{ij} = 0$, hence, $L(\mathcal{G})\mathbf{1} = \mathbf{0}$, which implies that there is at least one eigenvalue $\lambda_1 = 0$.

All other eigenvalues of the symmetric Laplacian matrix satisfy: $0 = \lambda_1 \leq \lambda_2 \leq \cdots \leq \lambda_n$.

The following graph definitions are important to understand the main theorems for consensus in multiagent systems. In directed communication graphs, consensus is mainly associated with balanced and strongly connected graphs.

Definition 5 (Wolfram) A strongly connected digraph is a graph in which it is possible to reach any node starting from any other node by traversing edges in the direction(s) in which they point.

In the example of Fig. 2 the Graphs 1, 2, and 3 are strongly connected.

Definition 6 A graph is balanced if $\sum_j a_{ij} = \sum_j a_{ji}$ for all $i \in \mathcal{J}$.

From our example of Fig. 2 we can see that Graph 1 and Graph 3 are balanced and we can also observe that undirected graphs are always balanced.

Proposition 1 (Oifati-Saber, 2004, Theorem 1) *If $\mathcal{G} = (\mathcal{V}, \mathcal{E}, \mathbf{A})$ is strongly connected, then:*

$$\text{rank}(\mathbf{L}(\mathcal{G})) = n - 1.$$

The following condition associated with a spanning tree. Note, that a strongly connected graph has a spanning tree.

Definition 7 A spanning tree of graph is a tree that contains all nodes of the graph.

Proposition 2 For example Ren and Beard (2008) *Zero is a simple eigenvalue of $\mathbf{L}(\mathcal{G})$ if and only if the communication graph $\mathcal{G} = (\mathcal{V}, \mathcal{E}, \mathbf{A})$ has a directed spanning tree.*

Later in Section 3.2, we will see that the spanning tree inside a communication graph is an important property for consensus in multiagent systems.

For further details of algebraic graph theory we refer to the seminal works of Fiedler (1973) and Wu (2005). Fiedler contributed to this theory for weights $a_{ij} \in \{0, 1\}$. The theory also holds for generalized nonnegative real valued $a_{ij} \in \mathbb{R}^+$.

3.2 CONSENSUS WITH TIME-INVARIANT TOPOLOGIES

In a multiagent system, each node has individual states denoted by the state $x_i(t) \in \mathbb{R}$. In this section, we will consider only scalar states. A generalization to multidimensional state vectors is presented in Li et al. (2010) and Li and Duan (2015).

Roughly speaking the nodes of a multiagent network are said to reach consensus if $x_i = x_j$ for all nodes $i, j \in \mathcal{J}$, (Olfati-Saber and Murray, 2004). For example, the communication of the Graph 2 in Fig. 2 is as follows: in each time instant t, Node 1 forwards its state $x_1(t)$ to Nodes 4 and 3 and Node 4 forwards its states $x_4(t)$ to Node 2. If for $t \to \infty$: $x_1(t) = x_2(t) = x_3(t) = x_4(t)$, we have consensus.

The system node in the scalar case is simply:

$$\dot{x}_i(t) = u_i(t). \tag{2}$$

A protocol per agent achieving consensus in specific cases is given by:

$$u_i(t) = \sum_j a_{ij}(x_j - x_i). \tag{3}$$

The protocol can be presented with the Laplacian in the compact notation:

$$\dot{\mathbf{x}}(t) = -\mathbf{L}(\mathcal{G}) \cdot \mathbf{x}(t) \tag{4}$$

with $\mathbf{x}(t) = [x_1(t), \ldots, x_n(t)]^{\mathrm{T}}$. A formal definition of consensus is presented in Olfati-Saber and Murray (2004):

Definition 8 (Olfati-Saber and Murray, 2004) Let $\mathcal{X} : \mathbb{R}^n \to \mathbb{R}$, we say a protocol solves the χ-consensus problem if and only if there exists an asymptotically stable equilibrium $\mathbf{x}^*$ satisfying $x_i^* = \mathcal{X}(\mathbf{x}(0))$.

Intuitively, consensus is reached if every stationary point satisfies $\dot{\mathbf{x}}(t) = \mathbf{0}$. We have $\dot{\mathbf{x}}(t) = \mathbf{0}$ if, e.g., $\mathbf{x}(t) = \mathbf{1}$. The condition of $\mathbf{L}(\mathcal{G})\mathbf{1} = \mathbf{0}$ of the Laplacian matrix implies that $\sum_{i=1}^n \dot{x}_i(t) = 0$ holds at consensus (Spanos et al., 2005). If consensus is achieved, we have $\mathbf{x}(t) = \alpha \cdot \mathbf{1}$, therefore, we have:

$$\dot{\mathbf{x}}(t) = -\mathbf{L}(\mathcal{G}) \cdot \mathbf{x}(t) = -\mathbf{L}(\mathcal{G}) \cdot \mathbf{1} \cdot \alpha = \mathbf{0}. \tag{5}$$

Theorem 2 (Beard and Stepanyan, 2003) *The multiagent network with protocol(4) achieves consensus if, and only if, the associated communication graph $\mathcal{G} = (\mathcal{V}, \mathcal{E}, \mathbf{A})$ has a spanning tree.*

Hence, consensus is related to the existence of a spanning tree inside the communication graph.

If the communication graph is strongly connected, $\mathbf{L}(\mathcal{G})$ has a simple zero eigenvalue, then due to the property of Theorem 1, $-\mathbf{L}(\mathcal{G})$ has eigenvalues with nonpositive real parts, therefore, the linear system (4) is stable.

A special consensus case is the average consensus, which is defined by $\mathcal{X}(\mathbf{x}) = \frac{1}{n} \sum_{i=1}^n x_i(0)$. Hence, in average consensus the states converge to the average value of the initial states.

The main result concerning average consensus in the literature is presented in the seminal work of Olfati-Saber and Murray (2004). The authors showed that average consensus of a multiagent system is given if the directed graph is strongly connected and if it is balanced (Ren et al., 2007). In this case, $\mathbf{1}$ is the left eigenvector associated with the zero eigenvalue. Olfati-Saber and Murray (2004) proved:

Theorem 3 *A strongly connected multiagent network $\mathcal{G} = (\mathcal{V}, \mathcal{E}, \mathbf{A})$ with protocol (4) achieves average consensus if, and only if, $\mathbf{1}^{\mathrm{T}}\mathbf{L}(\mathcal{G}) = \mathbf{0}$.*

In addition, the following holds also true:

Theorem 4 (Olfati-Saber and Murray, 2004) $\mathcal{G} = (\mathcal{V}, \mathcal{E}, \mathbf{A})$ *is balanced* $\Leftrightarrow \mathbf{1}^{\mathrm{T}}\mathbf{L}(\mathcal{G}) = \mathbf{0} \Leftrightarrow \sum_{i=1}^n \dot{x}_i(t) = 0$.

3.3 CONSENSUS WITH SWITCHING TOPOLOGIES

In practice, static topologies are not always feasible. Especially, in wirelessly linked CPS, effects like, path-loss, shadow-fading, and multipath fading will result in disconnected links within the communication graph.

In specific low fading wireless channels, where no spanning tree exists in the communication graph, consensus is not possible. The consensus protocol of dynamic networks with switching topologies is similar to the consensus protocol for static multiagent networks. The only difference is that the communication graph $\mathcal{G}_k = (\mathcal{V}, \mathcal{E}, \mathbf{A}_k)$ has a time-variant adjacency matrix $\mathbf{A}_k$ with $k = f(t)$ where within a fixed time interval, $f(t)$ switches finitely many times. The function $\mathrm{f} : \mathbb{R}^+ \longrightarrow \{1, \ldots, K\}$ is a switching signal, which indicates the changed topology of the communication graph. Let $t_0, t_1, \ldots$ be a time sequence where $f(t)$ switches. The function $f(t)$ is a piecewise continuous function. The new protocol is, therefore, (Olfati-Saber and Murray, 2004):

$$\dot{\mathbf{x}}(t) = -\mathbf{L}(\mathcal{G}_k) \cdot \mathbf{x}(t). \qquad (6)$$

In case of static, strongly connected and balanced digraphs, average consensus can be achieved. However, for arbitrary digraphs consensus is not always achievable.

Ren and Beard (2005) proved a similar relation between consensus and the spanning tree within the communication graphs as in the static case presented in Theorem 2. They used a union of directed graphs defined and proved that consensus can be achieved asymptotically if the union of the graphs has a spanning tree.

Olfati-Saber et al. proved average consensus for multiagent systems with switching topologies. They used a disagreement value to prove their main result. This mathematical tool is also used in our contribution in Section 4. In the case of an agreement (consensus), we have $\mathbf{x} = \alpha \cdot \mathbf{1}$. Therefore, we can define our disagreement state as follows (Olfati-Saber and Murray, 2004):

$$\mathbf{x} = \alpha \cdot \mathbf{1} + \boldsymbol{\delta}, \qquad (7)$$

where $\boldsymbol{\delta}$ denotes the disagreement vector. Based on the definition in Eq. (7), we can also define the state space representation of the disagreement vector (disagreement dynamics):

$$\dot{\boldsymbol{\delta}} = -\mathbf{L}(\mathcal{G}_k) \cdot \boldsymbol{\delta}. \qquad (8)$$

4 INTERACTION OF CONTROL THEORY AND INFORMATION THEORY

The interaction between control theory (consensus) and information theory is an important research field for the understanding of CPS. Due to a limited data rate on communication links between two nodes, or unreliable communication links, the theory of static consensus as presented in Section 3 must be extended.

In the literature, disturbed states or measurements are often considered in the field of sensor networks (Xiao et al., 2005, 2007) or in consensus control with quantized states (Carli et al., 2008; Frasca et al., 2009):

$$u_i(t) = \sum_j a_{ij} \big(Q(x_j) - Q(x_i) \big). \qquad (9)$$

The function $Q(x_j)$ denotes the quantized state. In Dimarogonas and Johansson (2010) and Guo and Dimarogonas (2013), the authors investigated a quantization protocol where the difference itself is quantized:

$$u_i(t) = \sum_j a_{ij} Q(x_j - x_i). \qquad (10)$$

Dimarogonas and Johansson (2010) proved that the states converge to a ball with a bounded radius in case of a uniform quantization and the states converge to an equilibrium in case of logarithmic quantization. Bauso et al. (2009) considered disturbed measures $\widetilde{Q}(x_j)$ only for the node j. Hence, they consider the following protocol:

$$u_i(t) = \sum_j a_{ij} \Big(\widetilde{Q}(x_j) - x_i \Big). \qquad (11)$$

Note that $\widetilde{Q}$ is in (Bauso et al., 2009) not a quantization function. In their work, the disturbance is $\widetilde{Q}(x_j) = x_j + d_j$, where d_j denotes an unknown but bounded disturbance. Hence, it can be also seen as quantization error. The authors proved that under specific conditions defined in Bauso et al. (2009) an ε-consensus exists.

In Li et al. (2011), the authors made the first connection of consensus and communication theory. They investigated distributed average consensus with limited communication data rates and proved that a faster convergence requires a higher quantization.

In this section, we will start with Shannon's channel capacity and we will directly link the SNR of a channel to the quantization of a link in a communication graph. We will prove that for a single time step the quantization error is bounded by the SNR of the channel. Each link between a node i, and a node j has a specific rate $R_{ij} = \log\left(1 + \gamma_{ij}\right)$, with the instantaneous SNR $\gamma_{ij} = \left|g_{ij}\right|^2 \cdot$ SNR, which is given by the upper bound in the form of the channel capacity on each link. The capacity of the channel is given by Shannon's famous theorem:

Theorem 5 (Cover and Thomas, 2005) *The capacity of a Gaussian channel with an SNR $\gamma_{ij} = g_{ij} \cdot$ SNR is given by:*

$$C_{ij} = \log\left(1 + \gamma_{ij}\right). \tag{12}$$

The available rate is upper bounded by the channel capacity $R_{ij} = C_{ij}$. The maximal quantization error for a given data rate R_{ij} is given by: $e_j = c/2^{R_{ij}+1}$, where $c = \max\left(x_j\right) - \min\left(x_j\right)$. Hence, we can rephrase the quantization error in terms of the current SNR.

Proposition 3 *The worst-case quantization error is given by:*

$$e_j = \frac{c}{2\left(1 + \gamma_{ij}\right)}. \tag{13}$$

Proof The proof follows from Theorem 5 and $e_j = c/2^{N_{Bits}+1}$, hence, $\frac{c}{2e_j} = 2^{N_{Bits}} \Leftrightarrow 1 + \gamma_{ij} = \frac{c}{2e_j}.\square$

In what follows, we consider a strongly connected and balanced topology. Furthermore, we only consider Gaussian noise, then the SNR will be: $\text{SNR} = P/\sigma_n^2$, where P denotes the power of the useful signal and σ_n^2 denotes the variance of the noise. Let us further define the worst-case quantization error for additive white Gaussian noise channels without channel fading $\left(\left|g_{ij}\right|^2 = 1\right)$ as:

$$\varepsilon(\text{SNR}) = \frac{c}{2\left(1 + \text{SNR}\right)}. \tag{14}$$

The following protocol is defined for a communication system where the nodes are transmitting quantized values:

$$u_i(t) = \sum_j a_{ij}\left(Q_j(t) - Q_i(t)\right) = \sum_j a_{ij}\left(x_j(t) + e_j(t) - x_i(t) - e_i(t)\right) \tag{15}$$

where $Q_i(t)$ and $Q_j(t)$ are the quantized values of $x_i(t)$ and $x_j(t)$. Note, that this protocol is a generalization of the following protocol.

If $Q_i(t)$ is a perfect quantization, the error $e_i(t) = 0$. In our communication model we will have a limited capacity form node i to node j and perfect knowledge of the own state $x_i(t)$. In what follows, we use undirected graphs $\mathcal{G}$ to simplify the derivations. Eq. (15) is written as:

$$\mathbf{u}(t) = -\mathbf{L}(\mathbf{x}(t) + \mathbf{e}(t)), \tag{16}$$

where $\mathbf{e} = [e_1, ..., e_n]^{\mathrm{T}}$. It can be simply checked that $\mathbf{1}^{\mathrm{T}}\mathbf{u}(t) = 0$. Then also $\mathbf{1}^{\mathrm{T}}\dot{\mathbf{x}}(t) = 0$ holds. This means $\mathbf{x}(t)$ is a constant over time. Subsequently, $\mathbf{x}(t)$ can be decomposed to two parts $\alpha = \dfrac{1}{n} \mathbf{1}^{\mathrm{T}}\mathbf{x}(0)$, which is the constant mean value of $\mathbf{x}(t)$ over the time, and a deviation vector $\boldsymbol{\delta}(t)$. Hence,

$$\mathbf{x}(t) = \alpha\mathbf{1} + \boldsymbol{\delta}(t). \tag{17}$$

Now, we will prove that $\mathbf{x}(t)$ converges to $\alpha\mathbf{1}$, and $\boldsymbol{\delta}(t)$ is bounded for high values of time. The following proof lines are borrowed from Seyboth et al. (2013). From Eq. (16) it follows:

$$\dot{\boldsymbol{\delta}}(t) = -\mathbf{L}(\mathcal{G})(\boldsymbol{\delta}(t) + \mathbf{e}(t)) \tag{18}$$

Hence:

$$\boldsymbol{\delta}(t) = \mathrm{e}^{-\mathbf{L}(\mathcal{G})t}\boldsymbol{\delta}(0) + \int_0^t \mathrm{e}^{-\mathbf{L}(\mathcal{G})(t-\tau)}\mathbf{L}(\mathcal{G}) \cdot \mathbf{e}(\tau)\mathrm{d}\tau \tag{19}$$

Then, the deviation vector $\boldsymbol{\delta}(t)$ satisfies:

$$\|\boldsymbol{\delta}(t)\| \le \|\mathrm{e}^{-\mathbf{L}(\mathcal{G})t}\boldsymbol{\delta}(0)\| + \int_0^t \|\mathrm{e}^{-\mathbf{L}(t-\tau)}\mathbf{L}(\mathcal{G}) \cdot \mathbf{e}(\tau)\|\mathrm{d}\tau \tag{20}$$

We have $\|\mathrm{e}^{-\mathbf{L}(\mathcal{G})t}\boldsymbol{\delta}(0)\| \le \mathrm{e}^{-\lambda_2 t}\|\boldsymbol{\delta}(0)\|$ (Olfati-Saber and Murray, 2004; Seyboth et al., 2013) and since the Laplacian $\mathbf{L}(\mathcal{G})$ is a Laplacian of an undirected, connected graph $\mathcal{G}$, therefore, $\mathcal{G}$ is also balanced we have $\mathbf{1}^{\mathrm{T}}\mathbf{L}(\mathcal{G})\mathbf{e}(\tau) = 0$. Consequently, with Seyboth et al. (2011, Lemma 1) we have:

$$\|\boldsymbol{\delta}(t)\| \le \mathrm{e}^{-\lambda_2 t}\|\boldsymbol{\delta}(0)\| + \int_0^t \mathrm{e}^{-\lambda_2(t-\tau)}\|\mathbf{L}(\mathcal{G}) \cdot \mathbf{e}(\tau)\|\mathrm{d}\tau \tag{21}$$

where λ_2 is the algebraic connectivity of the graph. Also, $\|\mathbf{e}(\tau)\| \le \Delta(\mathrm{SNR}) \le \sqrt{n} \cdot \varepsilon(\mathrm{SNR})$, where $\Delta(\mathrm{SNR})$ is the maximum quantization error, and $\|\mathbf{L}(\mathcal{G}) \cdot \mathbf{e}(\tau)\| \le \|\mathbf{L}(\mathcal{G})\| \cdot \|\mathbf{e}(\tau)\|$. Then:

$$\|\boldsymbol{\delta}(t)\| \le \mathrm{e}^{-\lambda_2 t}\|\boldsymbol{\delta}(0)\| + \int_0^t \mathrm{e}^{-\lambda_2(t-\tau)}\|\mathbf{L}(\mathcal{G})\|\Delta(\mathrm{SNR})\mathrm{d}\tau \tag{22}$$

Hence:

$$\|\boldsymbol{\delta}(t)\| \le \mathrm{e}^{-\lambda_2 t}\|\boldsymbol{\delta}(0)\| - \frac{\Delta(\mathrm{SNR})\|\mathbf{L}(\mathcal{G})\|}{\lambda_2}\mathrm{e}^{-\lambda_2 t} + \frac{\Delta(\mathrm{SNR})\|\mathbf{L}(\mathcal{G})\|}{\lambda_2}, \tag{23}$$

which means that $\|\boldsymbol{\delta}(t)\|$ converges to $\dfrac{\Delta(\mathrm{SNR})\|\mathbf{L}(\mathcal{G})\|}{\lambda_2}$. This indicates smaller deviation from the consensus value α for smaller values of the quantization error.

We can summarize the previous derivation in the following theorem:

Theorem 6 *Let $\mathcal{G}$ be a static undirected graph, the disagreement vector is bounded as follows:*

$$\|\boldsymbol{\delta}(\infty)\| \le \frac{\Delta(\mathrm{SNR})\|\mathbf{L}(\mathcal{G})\|}{\lambda_2}. \tag{24}$$

If the $\Delta(SNR)$ decreases also the error decreases. The lower the SNR in a time instant, the larger the quantization error. Hence, to control the system we need an adaptation of the quantization error. Concepts for this adaptation are introduced in the next section.

5 CROSS-LAYER DESIGN RESOURCE ALLOCATION FOR DISTRIBUTED CONTROL

In this section, we present transmission protocols to coordinate the transmission of the nodes in case of a wireless channel as a shared medium. Here, we only consider so-called time division multiple access (TDMA) schemes, which means that the channel access is coordinated among different time slots. Due to the limited bandwidth of the shared wireless channel medium, the quantization, which can be used to transmit the state to another user, is limited. Hence, there will be a quantization error as introduced in Section 4.

Communication is structured in layers (Tanenbaum, 2003). Here we use a simplified structure depicted in Fig. 4. The channel resources are also called physical (PHY) layer resource. Here, we consider only time slots that must be shared by different transmitters. The medium access (MAC)—PHY layer coordinates the allocation of resources. Above this layer, there is an interface to the application (APP) layer. The MAC-APP layer adapts the quantization based on the available resources in the MAC-PHY layer. The communication from node i to node j is, therefore, as follows: node i knows the current channel state (proportional to SNR) and, therefore, the available resources. Then, node i can, e.g., choose an adapted quantization to discretize its state, which has to be transmitted to node j. The

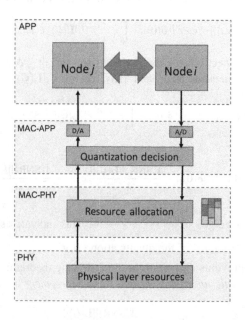

FIG. 4

Communication structure of the consensus system.

MAC-PHY layer assigns resources for the transmission of the quantized state and encodes it for the transmission over the PHY layer by choosing a sufficient channel coding and modulation scheme available for the current SNR.

In the following sections, we will present two simple MAC layer protocols for the adaptation of the quantization or the transmission period. The first scheme is called adaptive quantization (AQ). Here the transmission delay is constant; however, the quantization is adaptive. The second scheme is called adaptive transmission length (ATL). In the ATL scheme, the quantization is constant but the transmission duration is variable.

5.1 ADAPTIVE QUANTIZATION

The idea is very similar to the discussion in Section 4. The transmitted variables are sent via wireless channels with limited capacity. Here, we assume a fading channel with an SNR given by $\gamma_{ij} = g_{ij} \cdot \text{SNR}$ between two nodes i,j. The variable g_{ij} denotes the channel gain. The channel gain is constant during a slot of length T_s. In each time slot we have a different SNR. The transmit scheme is TDMA. Each transmitter has a specific transmit period $T_P = T_s$, which is equal to the slot length. After this transmit period the next transmitter is scheduled. Hence, the shared channel is divided into orthogonal time slots (Fig. 5).

In each slot the transmitter can use the bandwidth B. The upper bound of the achievable rate to transmit a specific state from i to j within this slot period is given by:

$$R_{ij}(k) = B \cdot \log_2\left(1 + \text{SNR} \cdot \left|g_{ij}(k)\right|^2\right) \text{ bits/s}. \tag{25}$$

The transmitter can estimate the SNR of the next slot, e.g., based on the uplink signal. The rate $R_{i,j}$ is variable and depends on the SNR of each link. Therefore the number of bits that can be transmitted during a transmission period T_P is changing. This results in the following protocol:

Protocol 1 *AQ*—The number bits determines the quantization that can be used for the quantization of the state that will be transmitted from node i to node j.

To discretize the system, we use the same discrete time model as presented in Olfati-Saber and Murray (2004) with a step-size $\varepsilon_d > 0$, hence, the update Eq. (2) will be:

$$x_i(k+1) = x_i(k) + \varepsilon_d \cdot u_i(k) \tag{26}$$

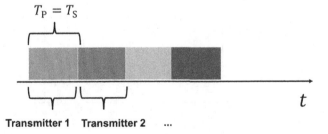

FIG. 5

TDMA transmission.

The quantization $Q(x_j)$ of the state x_j is differential, hence, the node j transmits the quantized difference $Q(x_j(k) - \hat{x}_j(k-1))$. We furthermore assume that the own state $x_i(t)$ is perfectly known $e_i(t) = 0$. The transmitter knows the quantization; therefore the transmitter also knows the signal $\hat{x}_j(k-1)$ updated by the receiving node. The receiving node i then updates its estimation of $\hat{x}_j$ by:

$$\hat{x}_j(k) = \hat{x}_j(k-1) + Q(x_j(k) - \hat{x}_j(k-1)) = x_j(k) + \Delta_j(k). \tag{27}$$

Table 1 presents the first three steps of the proposed protocol with differential quantization.

Based on the estimated SNRs of all links, the transmitter chooses the minimum capacity (and according quantization) over active connections from the transmitting node to all receiving nodes.

The AQ protocol ensures a fixed transmission delay; however, the error due to the quantization in case of low SNR can be very large. In Fig. 6, the quantization function $Q(x)$ for different SNR values is depicted. The larger the SNR, the better the resolution of the quantization function. To simplify the investigations we assume $BT_s = 1$ and $|g_{ij}(k)|^2 = 1$ for all i,j. In case of an SNR of just 5 dB only 2

Table 1 First three steps of the AQ protocol with differential quantization

	Node j			Node i
Time	State	Sent to Node i	Knowledge of Its State at Node i	Updated Value of State j
0	$x_j(0) = \hat{x}_j(0)$	$x_j(0) = \hat{x}_j(0)$	$x_j(0) = \hat{x}_j(0)$	$x_j(0) = \hat{x}_j(0)$
1	$x_j(1)$	$Q(x_j(1) - \hat{x}_j(0))$	$\hat{x}_j(1) = x_j(1) + \Delta_j(1)$	$\hat{x}_j(1) = \hat{x}_j(0) + Q(x_j(1) - \hat{x}_j(0)) = x_j(1) + \Delta_j(1)$
2	$x_j(2)$	$Q(x_j(2) - \hat{x}_j(1))$	$\hat{x}_j(2) = x_j(2) + \Delta_j(2)$	$\hat{x}_j(2) = \hat{x}_j(1) + Q(x_j(2) - \hat{x}_j(1)) = x_j(2) + \Delta_j(2)$
3	$x_j(3)$	$Q(x_j(3) - \hat{x}_j(2))$	$\hat{x}_j(3) = x_j(3) + \Delta_j(3)$	$\hat{x}_j(3) = \hat{x}_j(2) + Q(x_j(3) - \hat{x}_j(2)) = x_j(3) + \Delta_j(3)$
...				

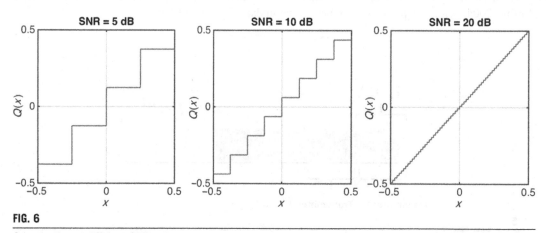

FIG. 6

Quantization at different SNR values.

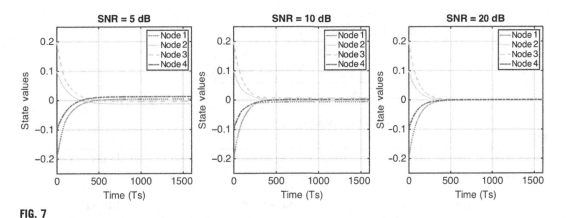

FIG. 7

Simulation results for AQ with a complete graph with four nodes and different SNR values.

bits are available, because $\log_2(1+\text{SNR}) = 2.0574$ bits. The initial states are given by $x(0) = [-0.2, 0.1, 0.2, -0.1]^\text{T}$. The average value is $\alpha = 0$, hence, all states will converge to zero, in case average consensus is achievable.

Fig. 7 presents the simulation results for a network with four nodes. The communication graph is a complete and undirected graph. We can observe that in the case of a low quantization error a final disagreement is the result, while in the case of a high SNR average consensus can be achieved.

5.2 ADAPTIVE TRANSMISSION LENGTH

In this section the SNR is also given by $\gamma_{ij}(k) = g_{ij}(k) \cdot \text{SNR}(k)$ between two nodes i,j. The variable $g_{ij}(k)$ denotes the channel gain. In each time slot k we have a different SNR. The transmit scheme is also TDMA. However, now the quantization is fixed. Hence, the transmitter must use multiple slots in case the achievable rate within a single slot was not sufficient to transmit the state with the given quantization. Therefore, each transmitter has a transmit period that consists of multiple time slots $T_\text{P} = N \cdot T_\text{s}$. After this transmit period the next transmitter is scheduled. The upper bound of the achievable rate is then given by:

$$R_{ij} = \sum_{k=1}^{N} B \cdot \log_2\left(1 + \text{SNR} \cdot \left|g_{ij}(k)\right|^2\right) \text{ bits} \tag{28}$$

The rate R_{ij} is fixed and determined by the used quantization. The protocol is defined as follows:

Protocol 2 ATL—The quantization used for the transmission of the states is fixed. To ensure a fixed quantization for the transmission of the state of a node i to node j, multiple slots must be used. The transmission is complete if all bits for the current quantized states are transmitted. In the case of multiple active links from one transmitting node to multiple receiving nodes, the link with the lowest SNR determines the (largest) number of slots that is necessary to transmit the current state with the given quantization error free to all receivers.

Fig. 8 presents the slot structure of the proposed protocol.

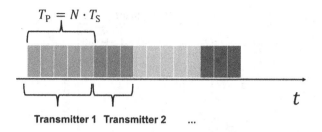

FIG. 8

TDMA scheme used for the transmission. Here each transmitter has a different number of slots for the transmission of its current state.

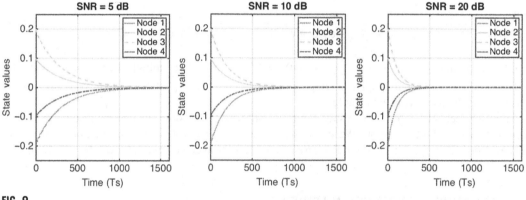

FIG. 9

Simulation results for ATL with a complete graph and different SNR values.

In contrast to the AQ protocol, which has a variable quantization but guarantees a transmission delay due to the fixed transmission period T_P, we now have a variable transmission period $T_P = N \cdot T_s$ and a fixed quantization. The node will continue the transmission until all bits of the state with the given quantization are transmitted. This scheme needs an additional negotiation among the nodes due to the variable transmission periods.

Fig. 9 presents the simulation results for ATL and the same setting as in the previous section. Here, we use a desired quantization of 6 bits. To achieve the desired quantization a different number N of slots is required depending on the given SNR; the higher the SNR, the faster the convergence. Compared to AQ, the final accuracy is the same for all SNR values. In contrast to ATL, AQ has the same convergence rate until the possible accuracy is reached. ATL has for all simulated SNRs the same accuracy, but a different convergence time.

6 CONCLUSIONS AND EMERGING TOPICS

This chapter presented how a limited capacity on the links results in an additive consensus error. This error depends on the current SNR of the communication links. Two communication, protocols AQ and ATL, are proposed for these capacity limited multiagent systems. Comparing the results of AQ and

ATL, we can observe that high SNR can achieve either a high accuracy (AQ) or a fast convergence (ATL) in case of a given, desired quantization.

In addition to the SNR, the network topology is a further performance parameter. Based on the discussion of the previous sections, we can observe that a dense topology is not always advantageous compared to a sparse topology. A dense topology can achieve a fast convergence in a mathematical sense of the consensus system. However, then multiple slots are required increasing the convergence time. Therefore the future work must investigate the trade-off between consensus convergence and sparse allocation of communication resources. The topology must be optimized based on communication channel states. However, the communication channel can be influenced by the system dynamics. Therefore a challenging research topic for future research is the joint control of the system dynamics with respect to the communication abilities resulting from a specific system topology.

REFERENCES

Ahlswede, R., Cai, N., Li, S.-Y., Yeung, R., 2000. Network information flow. IEEE Trans. Inf. Theory 46 (4), 1204–1216.

Bai, B., Chen, W., Cao, Z., Ben Letaief, K., 2008. Achieving high frequency diversity with subcarrier allocation in OFDMA systems. In: IEEE Global Telecommunications Conference (ICC), November.

Bai, B., Chen, W., Ben Letaief, K., Cao, Z., 2011. Diversity-multiplexing tradeoff in OFDMA systems: an h-matching approach. IEEE Trans. Wirel. Commun. 10 (11), 3675–3687.

Bauso, D., Giarré, L., Pesenti, R., 2009. Consensus for networks with unknown but bounded disturbances. SIAM J. Control. Optim. 48 (3), 1756–1770.

Beard, R.W., Stepanyan, V., 2003. Synchronization of information in distributed multiple vehicle coordinated control. In: IEEE Conference on Decision and Control.

Carli, R., Fagnani, F., Frasca, P., Zampieri, S., 2008. The quantization error in the average consensus problem. In: 2008 16th Mediterranean Conference on Control and Automation, 25–27 June. pp. 1592, 1597.

Cover, T.M., Thomas, J.A., 2005. Elements of Information Theory (Wiley Series in Telecommunications and Signal Processing). Wiley-Interscience.

Dimarogonas, D.V., Johansson, K.H., 2010. "Stability analysis" for multi-agent systems using the incidence matrix: quantized communication and formation control. Automatica. 46 (4).

Fiedler, M., 1973. Algebraic connectivity of graphs. Czech. Math. J. 23 (2), 298–305.

Frasca, P., Carli, R., Fagnani, F., Zampieri, S., 2009. Average consensus on networks with quantized communication. Int. J. Robust Nonlinear Control 19 (16), 1787–1816.

Godsil, C., Royle, G., 2001. Algebraic Graph Theory. Springer, New York.

Guo, M., Dimarogonas, D.V., 2013. Consensus with quantized relative state measurements. Automatica 49 (8), 2531–2537.

Li, Z., Duan, Z., 2015. Cooperative Control of Multi-Agent Systems: A Consensus Region Approach (Automation and Control Engineering). CRC Press, Boca Raton, FL.

Li, Z., Duan, Z., Chen, G., Huang, L., 2010. Consensus of multi-agent systems and synchronization of complex networks: a unified viewpoint. IEEE Trans. Circuits Syst. I Regul. Pap. 57 (1), 213–224.

Li, T., Fu, M., Xie, L., Zhang, J.-F., 2011. Distributed consensus with limited communication data rate. IEEE Trans. Autom. Control 56 (2), 279–292.

Murray, R.M., 2007. Recent research in cooperative control of multivehicle systems. J. Dyn. Syst. Meas. Control 129 (5), 571–583.

Ngai, C.K., Yeung, R., 2004. Network coding gain of combination networks. In: Proc. IEEE Inf. Theory Workshop, pp. 283–287.

Olfati-Saber, R., Murray, R.M., 2004. Consensus problems in networks of agents with switching topology and time-delays. IEEE Trans. Autom. Control 49 (9), 1520–1533.

Ren, W., Beard, R.W., 2005. Consensus seeking in multi-agent systems under dynamically changing interaction topologies. IEEE Trans. Autom. Control 50 (5), 655–661.

Ren, W., Beard, R.W., 2008. Distributed Consensus in Multi-vehicle Cooperative Control Theory and Applications (Communications and Control Engineering). Springer, London.

Ren, W., Beard, R.W., Atkins, E.M., 2007. Information consensus in multivehicle cooperative control. IEEE Control Syst. 27 (2), 71–82.

Seyboth, G., Dimarogonas, D.V., Johansson, K.H., 2011. Control of multi-agent systems via event-based communication. In: Proc. 18th IFAC World Congress, Milan, Italy.

Seyboth, G., Dimarogonas, D.V., Johansson, K.H., 2013. Event-based broadcasting for multi-agent average consensus. Automatica 49 (1), 245–252.

Spanos, D.P., Olfati-Saber, R., Murray, R.M., 2005. Dynamic consensus for mobile networks. In: IFAC World Congress.

Tanenbaum, A.S., 2003. Computer Networks, fourth ed. Prentice-Hall, Englewood Cliffs, NJ.

Wolfram (www.wolfram.de). http://mathworld.wolfram.com/StronglyConnectedDigraph.html (accessed 4th January 2016).

Wu, C.W., 2005. Algebraic connectivity of directed graphs. Linear Multilinear Algebra. 53 (3).

Xiao, L., Boyd, S., Lall, S., 2005. A scheme for robust distributed sensor fusion based on average consensus. In: IPSN 2005, Fourth International Symposium on Information Processing in Sensor Networks, 15 April. pp. 63, 70.

Xiao, L., Boyd, S., Kim, S.-J., 2007. Distributed average consensus with least-mean-square deviation. J. Parallel Distrib. Comput. 67 (1), 33–46.

ONLINE CONTROL AND OPTIMIZATION FOR CYBER-PHYSICAL SYSTEMS

3

C.G. Cassandras

Boston University, Boston, MA, United States

1 INTRODUCTION

The term *cyber-physical system* (CPS) (IEEE, 2014) is used to describe dynamic systems that combine components characterized by a physical state (e.g., the location, power level, and temperature of a mobile robot) with components (mostly digital devices empowered by software) characterized by an operational state or mode (e.g., on/off, transmitting/receiving). From a modeling point of view, physical states evolve according to *time-driven* dynamics commonly described through differential (or difference) equations, while operational states have *event-driven* dynamics where "events" may be controllable (e.g., a "turn on" command) or uncontrollable (e.g., a random failure). Imparting intelligence to a CPS implies the presence of multiple additional events that correspond to actions such as "start moving" for a mobile robot or "change sampling rate" for a sensor. These physical and operational states generally interact to give rise to a *hybrid dynamic system*. For example, a sensor with autonomous control capabilities may switch to a data transmitting mode as a result of a particular physical state change (e.g., its residual energy drops below a certain threshold).

As many CPSs have become increasingly networked, wireless, and distributed, traditional time-driven methodologies developed for sampling, estimation, communication, control, and optimization have come to question. The reason is that the use of an underlying clock with time steps dictating state transitions may make it infeasible to guarantee the synchronization of all components of a distributed system to such a clock; nor is it efficient to trigger actions with every time step when such actions may be unnecessary. In fact, in CPSs involving energy-constrained wireless devices, frequent communication among system components can be inefficient, unnecessary, and sometimes infeasible. Thus, rather than imposing a rigid time-driven communication mechanism, it is reasonable to seek instead to define specific events that dictate when a particular node in a network needs to exchange information with one or more other nodes. In other words, we seek to complement synchronous operating mechanisms with asynchronous ones that can dramatically reduce communication overhead without sacrificing adherence to design specifications and desired performance objectives. When, in addition, the environment is stochastic, significant changes in the operation of a system are the result of random event occurrences, so that, once again, understanding the implications of such events and reacting to them is crucial.

Cyber-Physical Systems. http://dx.doi.org/10.1016/B978-0-12-803801-7.00003-1

These developments have led to the emergence of an alternative *event-driven* paradigm whereby a clock should not be assumed to dictate actions simply because a time step is taken; rather, an action should be triggered by an "event" specified as a well-defined condition on the system state or as a consequence of environmental uncertainties that result in random state transitions. Observing that such an event could actually be defined as being a "clock tick," it follows that this framework may in fact incorporate time-driven methods as well. On the other hand, defining the proper "events" requires more sophisticated techniques compared to simply reacting to time steps.

For the class of discrete event system (DES), where the only changes in their state are dictated by event occurrences (e.g., the Internet), a rigorous theory has been in place since the 1980s (see, for example, Cassandras and Lafortune, 2008; Ho and Cassandras, 1983; Ramadge and Wonham, 1989; Baccelli et al., 1992; Moody and Antsaklis, 1998). This paved the way for event-based models of certain dynamic systems and spurred new concepts and techniques for control and optimization. By the early 1990s it became evident that many interesting dynamic systems are in fact "hybrid" in nature, i.e., at least some of their state transitions are caused by (possibly controllable) events (Branicky et al., 1998; Antsaklis et al., 1998; Cassandras et al., 2001; Zhang and Cassandras, 2002; Lemmon et al., 1999; Sussmann, 1999; Alur et al., 1996). This was reinforced by technological advances through which sensing and actuating devices are embedded into systems allowing physical processes to interface with such devices that are inherently event-driven, i.e., what we now call CPSs. A good example is the modern automobile where an event induced by a device that senses slippery road conditions may trigger the operation of an anti-lock breaking system, thus changing the operating dynamics of the actual vehicle. On a larger scale, another CPS example is the emerging smart city, where a network of sensing and actuating devices for monitoring and controlling physical processes is combined with software platforms enforcing requirements for mobility, security, safety and privacy in the processing of massive amounts of information (so-called "big data").

While the importance of event-driven behavior in dynamic systems was recognized as part of the development of DESs and then hybrid systems, more recently there have been significant advances in applying event-driven methods (also referred to as "event-based" and "event-triggered") to classical feedback control systems (see Arzen, 2002; Heemels et al., 2008; Lunze and Lehmann, 2010; Tabuada, 2007; Anta and Tabuada, 2010 and references therein). In distributed systems, event-driven mechanisms have the advantage of significantly reducing communication among networked components without affecting the desired performance objectives (see Wang and Lemmon, 2011; Zhong and Cassandras, 2010; Heemels et al., 2013; Trimpe and DAndrea, 2014; Garcia and Antsaklis, 2014; Liu et al., 2014). In multi-agent systems, on the other hand, the goal is for networked components to cooperatively maximize (or minimize) a given objective; it is shown in Zhong and Cassandras (2010) that an event-driven scheme can still achieve the optimization objective while drastically reducing communication (hence, prolonging the lifetime of a wireless network), even when delays are present (as long as they are bounded). Event-driven approaches are also attractive in receding horizon control, where it is computationally inefficient to re-evaluate a control value over small time increments as opposed to event occurrences defining appropriate planning horizons for the controller, e.g., see Khazaeni and Cassandras (2014).

In the remainder of this chapter, we adopt a view of CPSs as stochastic hybrid systems and consider a general-purpose control and optimization framework where controllers are parameterized and the parameters are adaptively tuned online based on observable data. One way to systematically carry out this process is through gradient information pertaining to given performance measures with respect to these parameters, so as to iteratively adjust their values. When the environment is stochastic, this entails generating gradient estimates with desirable properties such as unbiasedness. This gradient evaluation/estimation approach is based on the infinitesimal perturbation analysis (IPA) theory (Cassandras and Lafortune, 2008; Ho and Cao, 1991) originally developed for DES and now adapted to HS where it results in an "IPA calculus" (Cassandras et al., 2010), which amounts to a set of simple event-driven iterative equations. In this approach, the gradient evaluation/estimation procedure is based on directly observable data and it is entirely event-driven. This makes it computationally efficient, since it reduces a potentially complex process to a finite number of actions. More importantly perhaps, this approach has two key benefits that address the need for scalable methods in large-scale systems and the difficulty of obtaining accurate models especially in stochastic settings. First, being event-driven, it is scalable in the size of the event space and not the state space of the system model. As a rule, the former is much smaller than the latter. Second, it can be shown that the gradient information is often independent of model parameter values, which may be unknown or hard to estimate. In stochastic environments, this implies that complex control and optimization problems can be solved with little or no knowledge of the noise or random processes affecting the underlying system dynamics.

As an application of this control and optimization framework, the second half of the chapter addresses a fundamental problem in many CPSs: that of deploying a mobile multi-agent system which is tasked to cooperatively collect data from various distributed sources and deliver the data to one or more designated bases. The mobile agents should follow an optimal path (in some sense to be defined), which allows visiting each data source frequently enough and within the constraints of a given environment (e.g., that of an urban setting.)

The chapter is organized as follows. A general online control and optimization framework for hybrid systems is presented in Section 2, whose centerpiece is a methodology used for evaluating (or estimating in the stochastic case) a gradient of an objective function with respect to controllable parameters. This event-driven methodology, based on IPA, is described in Section 3. In Section 4, four key properties of IPA are presented. In Section 5, the data harvesting problem is described and an optimal trajectory planning problem is formulated. Its solution is given by using the proposed online optimization framework with the event-based IPA calculus and numerical examples are included.

2 A CONTROL AND OPTIMIZATION FRAMEWORK FOR CPSs AS HYBRID SYSTEMS

As already mentioned, we adopt a general-purpose hybrid system (HS) model for a CPS. The modeling, control, and optimization of these systems is quite challenging (Cassandras and Lygeros, 2007). In particular, the performance of a stochastic hybrid system (SHS) is generally hard to estimate because of the absence of closed-form expressions capturing the dependence of interesting performance metrics on various design or control parameters. Most approaches rely on approximations and/or using computationally taxing methods, often involving dynamic programming techniques. The inherent computational complexity of these approaches, however, makes them unsuitable for online control and optimization. Yet, in some cases, the structure of a dynamic optimization problem solution can be shown to be of parametric form, thus reducing it to a parametric optimization problem. As an example, in a linear quadratic Gaussian (LQG) setting, optimal feedback policies simply depend on gain parameters to be selected subject to certain constraints. Even when this is not provably the case, one can still define parametric families of solutions that can be optimized and yield near-optimal or at least vastly improved solutions relative to ad hoc policies often adopted. For instance, it is common in solutions based on dynamic programming (Powell, 2011) to approximate cost-to-go functions through parameterized function families and then iterate over the parameters involved seeking near-optimal solutions for otherwise intractable problems.

With this motivation in mind, we consider a general-purpose framework as shown in Fig. 1. The starting point is to assume that we can observe state trajectories of a given hybrid system and measure a performance (or cost) metric denoted by $L(\theta)$ where θ is a parameter vector. This vector characterizes a controller (as shown in Fig. 1), but may also include design or model parameters. The premise here is that the system is too complex for a closed-form expression of $L(\theta)$ to be available, but that it is possible to measure it over a given time window. In the case of a stochastic environment, the observable state trajectory is a sample path of a SHS, so that $L(\theta)$ is a sample function and performance is measured through $E[L(\theta)]$ with the expectation defined in the context of a suitable probability space. In addition to $L(\theta)$, we assume that all or part of the system state is observed, with possible noisy measurements. Thus, randomness may enter through either the system process or the measurement process or both.

The next step in Fig. 1 is the evaluation of the gradient $\nabla L(\theta)$. In the stochastic case, $\nabla L(\theta)$ is a random variable that serves as an estimate (obtained over a given time window) of $\nabla E[L(\theta)]$. Note that we require $\nabla L(\theta)$ to be evaluated based on available data observed from a *single* state trajectory (or sample path) of the hybrid system. This is in contrast to standard derivative approximation or estimation methods for $\dfrac{dL(\theta)}{d\theta}$ based on finite differences of the form $\dfrac{L(\theta + \Delta\theta) - L(\theta)}{\Delta\theta}$. Such methods require two state trajectories under θ and $\theta + \Delta\theta$ respectively and are vulnerable to numerical problems when $\Delta\theta$ is selected to be small so as to increase the accuracy of the derivative approximation.

The final step then is to make use of $\nabla L(\theta)$ in a gradient-based adaptation mechanism of the general form $\theta_{n+1} = \theta_n + \eta_n \nabla L(\theta)$, where $n = 1, 2, \ldots$ counts the iterations over which this process evolves and $\{\eta_n\}$ is a step size sequence which is appropriately selected to ensure convergence of the controllable parameter sequence $\{\theta_n\}$ under proper stationarity assumptions. After each iteration, the controller is adjusted, which obviously affects the behavior of the system, and the process repeats. Clearly, in a stochastic setting there is no guarantee of stationarity conditions and this framework is simply one where the controller is perpetually seeking to improve system performance.

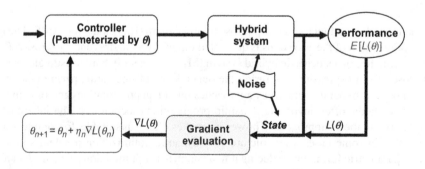

FIG. 1

Online control and optimization framework for CPS as hybrid systems.

The cornerstone of this *online* framework is the evaluation of $\nabla L(\boldsymbol{\theta})$ based *only* on data obtained from the observed state trajectory. The theory of IPA (Cassandras et al., 2002, 2010) provides the foundations for this to be possible. Moreover, in the stochastic case where $\nabla L(\theta)$ becomes an estimate of $\nabla E[L(\theta)]$, it is important that this estimate possesses desirable properties, such as unbiasedness, without which the ultimate goal of achieving optimality cannot be provably attained. As we will see in the next section, it is, indeed, possible to evaluate $\nabla L(\theta)$ for virtually arbitrary SHS through a simple systematic event-driven procedure we refer to as the "IPA calculus". In addition, this gradient is characterized by several attractive properties under mild technical conditions.

In order to formally apply IPA and subsequent control and optimization methods to hybrid systems, we need to establish a general modeling framework. We use a standard definition of a hybrid automaton (Cassandras and Lygeros, 2007). Thus, let $q \in Q$ (a countable set) denote the discrete state (or mode) and $x \in X \subseteq \mathbb{R}^n$ denote the continuous state of the hybrid system. Let $v \in \Upsilon$ (a countable set) denote a discrete control input and $u \in U \subseteq \mathbb{R}^m$ a continuous control input. Similarly, let $\delta \in \Delta$ (a countable set) denote a discrete disturbance input and $d \in D \subseteq \mathbb{R}^p$ a continuous disturbance input. The state evolution is determined by means of:

- a vector field $f : Q \times X \times U \times D \rightarrow X$,
- an invariant (or domain) set $Inv : Q \times \Upsilon \times \Delta \rightarrow 2^X$,
- a guard set $Guard : Q \times Q \times \Upsilon \times \Delta \rightarrow 2^X$, and
- a reset function $r : Q \times Q \times X \times \Upsilon \times \Delta \rightarrow X$.

A trajectory or sample path of such a system consists of a sequence of intervals of continuous evolution followed by a discrete transition. The system remains at a discrete state q as long as the continuous (time-driven) state x does not leave the set $Inv(q, v, \delta)$. If, before reaching $Inv(q, v, \delta)$, x reaches a set $Guard(q, q', v, \delta)$ for some $q' \in Q$, a discrete transition is allowed to take place. If this transition does take place, the state instantaneously resets to (q', x') where x' is determined by the reset map $r(q, q', x, v, \delta)$. Changes in the discrete controls v and disturbances δ are discrete events that either *enable* a transition from q to q' when $x \in Guard(q, q', v, \delta)$ or *force* a transition out of q by making sure $x \notin Inv(q, v, \delta)$. We will also use $\mathcal{E}$ to denote the set of all events that cause discrete state transitions and will classify events in a manner that suits the purposes of perturbation analysis.

In order to illustrate this modeling framework, we provide the example shown in Fig. 2. The underlying system is a machine with three discrete states (modes) {0, 1, 2}. The continuous state vector is

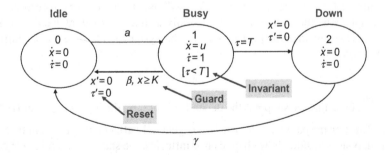

FIG. 2

A hybrid automaton example.

$[x\ \tau]$ where $x(t)$ is a simple physical process whose goal is to drive an initial state $x(0) = 0$ to a final value $x(t) = K$ through the dynamics $\dot{x}(t) = u$ under a given control input $u(t)$, $t \geq 0$. The second state variable is a simple timer initialized at $\tau(0) = 0$ with $\dot{\tau}(t) = 1$. At mode 0, the machine is "idle" and neither the physical process nor the timer are in operation. An exogenous event α causes the transition to mode 1 during which the machine is "busy". There are now three possibilities: (*i*) an exogenous event β forces the machine to return to the idle mode and reset the two state variables as shown: $x' = 0$ and $\tau' = 0$. (*ii*) The physical process reaches (or exceeds) the target state $x(t) = K$ for some $t < T$, in which case the guard condition $x \geq K$ is satisfied and the machine returns to the idle mode with the two state variables being once again reset: $x' = 0$ and $\tau' = 0$. (*iii*) Note that there is an invariant condition $[\tau < T]$ at mode 1; this condition must remain true while the system is in that mode and a transition must occur if the condition is violated. Thus, as soon as the guard condition $\tau = T$ is satisfied, a transition to mode 3 takes place. Mode 3 represents a "down" state, capturing the fact that the physical process has taken longer than expected to satisfy $x \geq K$, therefore a "timeout event" is triggered through $\tau = T$. Finally, the exogenous event γ causes the machine to return to its idle state. This example serves to show the difference between an invariant condition and a guard condition: while the transition from 1 to 0 when $x \geq K$ is optional (i.e., an additional control policy or random mechanism may determine when exactly this transition should occur), the transition due to the violation of $[\tau < T]$ is required to take place.

In what follows, we will provide an overview of the "IPA calculus" and refer the reader to Cassandras et al. (2010) and Kebarighotbi and Cassandras (2012) for more details.

3 INFINITESIMAL PERTURBATION ANALYSIS (IPA): THE IPA CALCULUS

In this section, we describe the general framework for IPA as presented in Wardi et al. (2010) and generalized in Cassandras et al. (2010) and Kebarighotbi and Cassandras (2012). Let $\theta \in \Theta \subset \mathbb{R}^l$ be a global variable, henceforth called the *control parameter*, where Θ is a given compact, convex set. This may include system design parameters, parameters of an input process, or parameters that characterize a policy used in controlling this system. The disturbance input $d \in D$ encompasses various random processes that affect the evolution of the state (q, x) so that, in general, we can deal with a SHS. We will assume that all such processes are defined over a common probability space $(\Omega, \mathcal{F}, P)$. Let us fix a particular value of the parameter $\theta \in \Theta$ and study a resulting sample path of the SHS. Over such a sample path, let $\tau_k(\theta)$, $k = 1, 2, \ldots$, denote the occurrence times of the discrete events in increasing order, and define $\tau_0(\theta) = 0$ for convenience. We will use the notation τ_k instead of $\tau_k(\theta)$ when no confusion arises. The continuous state is also generally a function of θ, as well as of t, and is thus denoted by $x(\theta, t)$. Over an interval $[\tau_k(\theta), \tau_{k+1}(\theta))$, the system is at some mode during which the time-driven state satisfies:

$$\dot{x} = f_k(x, \theta, t) \tag{1}$$

where $\dot{x}$ denotes $\dfrac{\partial x}{\partial t}$. Note that we suppress the dependence of f_k on the inputs $u \in U$ and $d \in D$ and stress instead its dependence on the parameter θ, which may generally affect either u or d or both. The purpose of perturbation analysis is to study how changes in θ influence the state $x(\theta, t)$ and the event times $\tau_k(\theta)$ and, ultimately, how they influence interesting performance metrics, which are generally expressed in terms of these variables. The following assumption guarantees that Eq. (1) has a unique solution w.p.1 for a given initial boundary condition $x(\theta, \tau_k)$ at time $\tau_k(\theta)$:

Assumption 1. W.p.1, there exists a finite set of points $t_j \in [\tau_k(\theta), \tau_{k+1}(\theta)), j = 1, 2, \ldots$, which are independent of θ, such that, the function f_k is continuously differentiable on $\mathbb{R}^n \times \Theta \times ([\tau_k(\theta), \tau_{k+1}(\theta)) \backslash \{t_1, t_2, \ldots\})$. Moreover, there exists a random number $K > 0$ such that $E[K] < \infty$ and the norm of the first derivative of f_k on $\mathbb{R}^n \times \Theta \times ([\tau_k(\theta), \tau_{k+1}(\theta)) \backslash \{t_1, t_2, \ldots\})$ is bounded from above by K.

An event occurring at time $\tau_{k+1}(\theta)$ triggers a change in the mode of the system, which may also result in new dynamics represented by f_{k+1}, although this may not always be the case; for example, two modes may be distinct because the state $x(\theta, t)$ enters a new region where the system's performance is measured differently without altering its time-driven dynamics (i.e., $f_{k+1} = f_k$). The event times $\{\tau_k(\theta)\}$ play an important role in defining the interactions between the time-driven and event-driven dynamics of the system.

We now classify events that define the set $\mathcal{E}$ as follows:

- *Exogenous*: An event is exogenous if it causes a discrete state transition at time τ_k independent of the controllable vector θ and satisfies $\frac{d\tau_k}{d\theta} = 0$. Exogenous events typically correspond to uncontrolled random changes in input processes.

- *Endogenous*: An event occurring at time τ_k is endogenous if there exists a continuously differentiable function $g_k : \mathbb{R}^n \times \Theta \to \mathbb{R}$ such that:

$$\tau_k = \min\{t > \tau_{k-1} : g_k(x(\theta, t), \theta) = 0\} \tag{2}$$

   The function g_k is normally associated with an invariant or a guard condition in a hybrid automaton model.

- *Induced*: An event at time τ_k is induced if it is triggered by the occurrence of another event at time $\tau_m \leq \tau_k$. The triggering event may be exogenous, endogenous, or itself an induced event. The events that trigger induced events are identified by a subset of the event set, $\mathcal{E}_I \subseteq \mathcal{E}$.

Although this event classification is sufficiently general, recent work has shown that in some cases it is convenient to introduce further event distinctions (Wardi et al., 2013). Moreover, it has been shown in Kebarighotbi and Cassandras (2012) that an explicit event classification is in fact unnecessary if one is willing to appropriately extend the definition of the hybrid automaton described earlier. However, for the rest of this chapter we shall only make use of the above classification.

Next, consider a performance function of the control parameter θ:

$$J(\theta; x(\theta, 0), T) = E[L(\theta; x(\theta, 0), T)]$$

where $L(\theta; x(\theta, 0), T)$ is a sample function of interest evaluated in the interval $[0, T]$ with initial conditions $x(\theta, 0)$. For simplicity, we write $J(\theta)$ and $L(\theta)$. Suppose that there are N events, with occurrence times generally dependent on θ, during the time interval $[0, T]$ and define $\tau_0 = 0$ and $\tau_{N+1} = T$. Let $L_k : \mathbb{R}^n \times \Theta \times \mathbb{R}^+ \to \mathbb{R}$ be a function satisfying Assumption 1 and define $L(\theta)$ by:

$$L(\theta) = \sum_{k=0}^{N} \int_{\tau_k}^{\tau_{k+1}} L_k(x, \theta, t) dt \tag{3}$$

where we reiterate that $x = x(\theta, t)$ is a function of θ and t. We also point out that the restriction of the definition of $J(\theta)$ to a finite horizon T, which is independent of θ is made merely for the sake of simplicity.

Returning to Fig. 1 and considering (for the sake of generality) the stochastic setting, the ultimate goal of the iterative process shown is to maximize $E_\omega[L(\theta, \omega)]$, where we use ω to emphasize dependence on a sample path ω of a SHS (clearly, this is reduced to $L(\theta)$ in the deterministic case). Achieving such optimality is possible under standard ergodicity conditions imposed on the underlying stochastic processes, as well as the assumption that a single global optimum exists; otherwise, the gradient-based approach is simply continuously attempting to improve the observed performance $L(\theta, \omega)$. Thus, we are interested in estimating the gradient:

$$\frac{dJ(\theta)}{d\theta} = \frac{dE_\omega[L(\theta, \omega)]}{d\theta}$$

by evaluating $\dfrac{dL(\theta, \omega)}{d\theta}$ based on *directly observed data*. We obtain θ^* (under the conditions mentioned above) by optimizing $J(\theta)$ through an iterative scheme of the form:

$$\theta_{n+1} = \theta_n - \eta_n H_n(\theta_n; x(\theta, 0), T, \omega_n), \quad n = 0, 1, \ldots \tag{4}$$

where $\{\eta_n\}$ is a step size sequence and $H_n(\theta_n; x(\theta, 0), T, \omega_n)$ is the estimate of $\dfrac{dJ(\theta)}{d\theta}$ at $\theta = \theta_n$. In using IPA, $H_n(\theta_n; x(\theta, 0), T, \omega_n)$ is the sample derivative $\dfrac{dL(\theta, \omega)}{d\theta}$, which is an unbiased estimate of $\dfrac{dJ(\theta)}{d\theta}$ if the condition (dropping the symbol ω for simplicity):

$$E\left[\frac{dL(\theta)}{d\theta}\right] = \frac{d}{d\theta}E[L(\theta)] = \frac{dJ(\theta)}{d\theta} \tag{5}$$

is satisfied, which turns out to be the case under mild technical conditions to be discussed later. The conditions under which algorithms of the form in Eq. (4) converge are well-known (e.g., see Kushner and Yin, 1997). Moreover, in addition to being unbiased, it can be shown that such gradient estimates are independent of the probability laws of the stochastic processes involved and require minimal information from the observed sample path.

The process through which IPA evaluates $\dfrac{dL(\theta)}{d\theta}$ is based on analyzing how changes in θ influence the state $x(\theta, t)$ and the event times $\tau_k(\theta)$. In turn, this provides information on how $L(\theta)$ is affected, since it is generally expressed in terms of these variables. Given $\theta = [\theta_1, \ldots, \theta_l]^T$, we use the Jacobian matrix notation:

$$x'(\theta, t) \equiv \frac{\partial x(\theta, t)}{\partial \theta}, \quad \tau_k' \equiv \frac{\partial \tau_k(\theta)}{\partial \theta}, \quad k = 1, \ldots, K$$

for all state and event time derivatives. For simplicity of notation, we will omit θ from the arguments of the functions above unless it is essential to stress this dependence. It is shown in Cassandras et al. (2010) that $x'(t)$ satisfies:

$$\frac{d}{dt}x'(t) = \frac{\partial f_k(t)}{\partial x}x'(t) + \frac{\partial f_k(t)}{\partial \theta} \tag{6}$$

for $t \in [\tau_k(\theta), \tau_{k+1}(\theta))$ with boundary condition:

$$x'(\tau_k^+) = x'(\tau_k^-) + \left[f_{k-1}(\tau_k^-) - f_k(\tau_k^+)\right]\tau_k' \tag{7}$$

for $k = 0, \ldots, K$. We note that whereas $x(t)$ is often continuous in t, $x'(t)$ may be discontinuous in t at the event times τ_k, hence the left and right limits above are generally different. If $x(t)$ is not continuous in t at $t = \tau_k(\theta)$, the value of $x(\tau_k^+)$ is determined by the reset function $r(q, q', x, v, \delta)$ discussed earlier and:

$$x'(\tau_k^+) = \frac{dr(q, q', x, v, \delta)}{d\theta} \tag{8}$$

Furthermore, once the initial condition $x'(\tau_k^+)$ is given, the linearized state trajectory $\{x'(t)\}$ can be computed in the interval $t \in [\tau_k(\theta), \tau_{k+1}(\theta))$ by solving Eq. (6) to obtain:

$$x'(t) = e^{\int_{\tau_k}^t \frac{\partial f_k(u)}{\partial x} du} \left[\int_{\tau_k}^t \frac{\partial f_k(v)}{\partial \theta} e^{-\int_{\tau_k}^t \frac{\partial f_k(u)}{\partial x} du} dv + \xi_k \right] \tag{9}$$

with the constant ξ_k determined from $x'(\tau_k^+)$ in either Eq. (7) or Eq. (8).

In order to complete the evaluation of $x'(\tau_k^+)$ in Eq. (7), we need to also determine τ_k'. Based on the event classification above, $\tau_k' = 0$ if the event at $\tau_k(\theta)$ is exogenous and:

$$\tau_k' = -\left[\frac{\partial g_k}{\partial x} f_k(\tau_k^-) \right]^{-1} \left(\frac{\partial g_k}{\partial \theta} + \frac{\partial g_k}{\partial x} x'(\tau_k^-) \right) \tag{10}$$

if the event at $\tau_k(\theta)$ is endogenous, i.e., $g_k(x(\theta, \tau_k), \theta) = 0$, defined as long as $\frac{\partial g_k}{\partial x} f_k(\tau_k^-) \neq 0$ (details may be found in Cassandras et al. (2010)). Finally, if an induced event occurs at $t = \tau_k$ and is triggered by an event at $\tau_m \leq \tau_k$, the value of τ_k' depends on the derivative τ_m'. The event induced at τ_m will occur at some time $\tau_m + w(\tau_m)$, where $w(\tau_m)$ is a (generally random) variable that is dependent on the continuous and discrete states $x(\tau_m)$ and $q(\tau_m)$, respectively. This implies the need for additional state variables, denoted by $y_m(\theta, t)$, $m = 1, 2, \ldots$, associated with events occurring at times τ_m, $m = 1, 2, \ldots$ The role of each such state variable is to provide a "timer" activated when a triggering event occurs. Triggering events are identified as belonging to a set $\mathcal{E}_I \subseteq \mathcal{E}$ and let e_k denote the event occurring at τ_k. Then, define $F_k = \{m : e_m \in \mathcal{E}_I, m \leq k\}$ as the set of all indices with corresponding triggering events up to τ_k. Omitting the dependence on θ for simplicity, the dynamics of $y_m(t)$ are then given by:

$$\dot{y}_m(t) = \begin{cases} -C(t) & \tau_m \leq t < \tau_m + w(\tau_m), \ m \in F_m \\ 0 & \text{otherwise} \end{cases}$$

$$y_m(\tau_m^+) = \begin{cases} y_0 & y_m(\tau_m^-) = 0, \ m \in F_m \\ 0 & \text{otherwise} \end{cases} \tag{11}$$

where y_0 is an initial value for the timer $y_m(t)$, which decreases at a "clock rate" $C(t) > 0$ until $y_m(\tau_m + w(\tau_m)) = 0$ and the associated induced event takes place. Clearly, these state variables are only used for induced events, so that $y_m(t) = 0$ unless $m \in F_m$. The value of y_0 may depend on θ or on the continuous and discrete states $x(\tau_m)$ and $q(\tau_m)$, while the clock rate $C(t)$ may depend on $x(t)$ and $q(t)$ in general, and possibly θ. However, in most simple cases where we are interested in modeling an induced event to occur at time $\tau_m + w(\tau_m)$, we have $y_0 = w(\tau_m)$ and $C(t) = 1$, i.e., the timer simply counts down for a total of $w(\tau_m)$ time units until the induced event takes place. Henceforth, we will consider $y_m(t)$, $m = 1, 2, \ldots$, as part of the continuous state of the SHS and we set:

$$y'_m(t) \equiv \frac{\partial y_m(t)}{\partial \theta}, \quad m = 1, \ldots, N \tag{12}$$

For the common case where y_0 is independent of θ and $C(t)$ is a constant $c > 0$ in Eq. (11), the following lemma facilitates the computation of τ'_k for an induced event occurring at τ_k. Its proof is given in Cassandras et al. (2010).

Lemma 1. *If in Eq. (11)* y_0 *is independent of* θ *and* $C(t) = c > 0$ *(constant), then* $\tau'_k = \tau'_m$.

With the inclusion of the state variables $y_m(t)$, $m = 1, \ldots, N$, the derivatives $x'(t)$, τ'_k, and $y'_m(t)$ can be evaluated through Eqs. (6)–(11) and this set of equations is what we refer to as the "IPA calculus." In general, this evaluation is recursive over the event (mode switching) index $k = 0, 1, \ldots$ In other words, the IPA estimation process is entirely event driven. For a large class of problems, the SHS of interest does not involve induced events and the state does not experience discontinuities when a mode-switching event occurs. In this case, the IPA calculus reduces to the application of three equations:

1. Eq. (9):

$$x'(t) = e^{\int_{\tau_k}^t \frac{\partial f_k(u)}{\partial x} du} \left[\int_{\tau_k}^t \frac{\partial f_k(v)}{\partial \theta} e^{-\int_{\tau_k}^t \frac{\partial f_k(u)}{\partial x} du} dv + \xi_k \right] \tag{13}$$

which describes how the state derivative $x'(t)$ evolves over $[\tau_k(\theta), \tau_{k+1}(\theta))$,

2. Eq. (7):

$$x'(\tau_k^+) = x'(\tau_k^-) + \left[f_{k-1}(\tau_k^-) - f_k(\tau_k^+) \right] \tau'_k \tag{14}$$

which specifies the initial condition ξ_k in Eq. (9), and

3. Either $\tau'_k = 0$ or Eq. (10):

$$\tau'_k = - \left[\frac{\partial g_k}{\partial x} f_k(\tau_k^-) \right]^{-1} \left(\frac{\partial g_k}{\partial \theta} + \frac{\partial g_k}{\partial x} x'(\tau_k^-) \right) \tag{15}$$

depending on the event type at $\tau_k(\theta)$, which specifies the event time derivative present in Eq. (7).

The last step in the IPA process involves using the IPA calculus in order to evaluate the IPA derivative $dL/d\theta$. This is accomplished by taking derivatives in Eq. (3) with respect to θ:

$$\frac{dL(\theta)}{d\theta} = \sum_{k=0}^N \frac{d}{d\theta} \int_{\tau_k}^{\tau_{k+1}} L_k(x, \theta, t) dt \tag{16}$$

Applying the Leibnitz rule we obtain, for every $k = 0, \ldots, N$:

$$\frac{d}{d\theta} \int_{\tau_k}^{\tau_{k+1}} L_k(x, \theta, t) dt = \int_{\tau_k}^{\tau_{k+1}} \left[\frac{\partial L_k}{\partial x}(x, \theta, t) x'(t) + \frac{\partial L_k}{\partial \theta}(x, \theta, t) \right] dt$$
$$+ L_k(x(\tau_{k+1}), \theta, \tau_{k+1}) \tau'_{k+1} - L_k(x(\tau_k), \theta, \tau_k) \tau'_k \tag{17}$$

where $x'(t)$ and τ'_k are determined through Eqs. (6)–(10). What makes IPA appealing is the simple form the right-hand-side above often assumes. As we will see, under certain commonly encountered conditions, this expression is further simplified by eliminating the integral term.

4 IPA PROPERTIES

In this section, we present four key properties of IPA.

Property 1 (Unbiasedness). This property is important in ensuring that when IPA involves estimates of gradients, these estimates are unbiased under mild conditions. Remember that the IPA derivative $\frac{dL(\theta)}{d\theta}$ is an unbiased estimate of the performance (or cost) derivative $\frac{dJ(\theta)}{d\theta}$ if condition in Eq. (5) holds. In a pure DES, the IPA derivative satisfies this condition for a relatively limited class of systems (see Cassandras and Lafortune, 2008; Ho and Cao, 1991). However, in a SHS, the technical conditions required to guarantee the validity of Eq. (5) are almost always applicable. The following result has been established in Rubinstein (1986) regarding the unbiasedness of IPA.

Theorem 1. *Suppose that the following conditions are in force: (i) for every $\theta \in \Theta$, the derivative $\frac{dL(\theta)}{d\theta}$ exists w.p.1. (ii). W.p.1, the function $L(\theta)$ is Lipschitz continuous on Θ, and the Lipschitz constant has a finite first moment. Then, for a fixed $\theta \in \Theta$, the derivative $\frac{dJ(\theta)}{d\theta}$ exists, and the IPA derivative $\frac{dL(\theta)}{d\theta}$ is unbiased.*

The crucial assumption for Theorem 1 is the continuity of the sample function $L(\theta)$, which in many SHS is guaranteed in a straightforward manner. Differentiability w.p.1 at a given $\theta \in \Theta$ often follows from mild technical assumptions on the probability law underlying the system, such as the exclusion of co-occurrence of multiple events (see Yao and Cassandras, 2011a). Lipschitz continuity of $L(\theta)$ generally follows from upper boundedness of $|\frac{dL(\theta)}{d\theta}|$ by an absolutely integrable random variable: generally a weak assumption. In light of these observations, the proofs of unbiasedness of IPA have become standardized and the assumptions in Theorem 1 can be verified fairly easily from the context of a particular problem.

Property 2 (Robustness to stochastic model uncertainties). The IPA derivatives are "robust" in the sense that they do not depend on specific probabilistic characterizations of any stochastic processes involved in the hybrid automaton model of a SHS. This property holds under certain sufficient conditions that are easy to check. Returning to $\frac{dL(\theta)}{d\theta}$ in Eq. (16), note that, as seen in Eq. (17), it generally depends on information accumulated over all $t \in [\tau_k, \tau_{k+1})$. It is, however, often the case that it depends only on information related to the event times τ_k, τ_{k+1}, resulting in an IPA estimator that is very simple to implement. Using the notation $L_k'(x, t, \theta) \equiv \frac{dL_k(x, t, \theta)}{d\theta}$, we can rewrite $\frac{dL(\theta)}{d\theta}$ in Eq. (16) as:

$$\frac{dL(\theta)}{d\theta} = \sum_k \left[\tau_{k+1}' \cdot L_k\left(\tau_{k+1}^+\right) - \tau_k' \cdot L_k\left(\tau_k^+\right) + \int_{\tau_k}^{\tau_{k+1}} L_k'(x, t, \theta) dt \right] \tag{18}$$

The following theorem provides two sufficient conditions under which $\frac{dL(\theta)}{d\theta}$ involves only the event time derivatives τ_k', τ_{k+1}' and the "local" performance $L_k\left(\tau_{k+1}^+\right), L_k\left(\tau_k^+\right)$, which is obviously easy to observe. The proof of this result is given in Yao and Cassandras (2011b).

Theorem 2. *If condition* (C1) *or* (C2) *below holds, then* $\dfrac{dL(\theta)}{d\theta}$ *depends only on information available at event times* $\{\tau_k\}$, $k = 0, 1, \ldots$

(C1) $L_k(x, t, \theta)$ *is independent of* t *over* $[\tau_k, \tau_{k+1})$ *for all* $k = 0, 1, \ldots$

(C2) $L_k(x, t, \theta)$ *is only a function of x and the following condition holds for all* $t \in [\tau_k, \tau_{k+1})$, $k = 0, 1, \ldots$:

$$\frac{d}{dt}\frac{\partial L_k}{\partial x} = \frac{d}{dt}\frac{\partial f_k}{\partial x} = \frac{d}{dt}\frac{\partial f_k}{\partial \theta} = 0 \tag{19}$$

The implication of Theorem 2 is that Eq. (18), under either (C1) or (C2), reduces to:

$$\frac{dL(\theta)}{d\theta} = \sum_k \left[\tau'_{k+1} \cdot L_k(\tau_{k+1}^+) - \tau'_k \cdot L_k(\tau_k^+) \right]$$

and involves only directly observable performance sample values at event times along with event time derivatives, which are either zero (for exogenous events) or given by Eq. (10). The conditions in Theorem 2 are surprisingly easy to satisfy for many problems of practical interest (a simple example is given in Cassandras, 2015).

Property 3 (State trajectory decomposition). This property is related to the discontinuity in $x'(t)$ at event times, described in Eq. (7). This happens when endogenous events occur, since for exogenous events we have $\tau'_k = 0$. The next theorem identifies a simple condition under which $x'(\tau_k^+)$ is independent of the dynamics f before the event at τ_k. This implies that we can evaluate the sensitivity of the state with respect to θ without any knowledge of the state trajectory in the interval $[\tau_{k-1}, \tau_k)$ prior to this event. Moreover, under an additional condition, we obtain $x'(\tau_k^+) = 0$, implying that the effect of θ is "forgotten" and one can reset the perturbation process. This allows us to decompose an observed state trajectory (or sample path) into "reset cycles", greatly simplifying the IPA process. The proof of the next result is also given in Yao and Cassandras (2011b).

Theorem 3 *Suppose an endogenous event occurs at* $\tau_k(\theta)$ *with a switching function* $g(x, \theta)$. *If* $f_k(\tau_k^+) = 0$, $x'(\tau_k^+)$ *is independent of* f_{k-1}. *If, in addition,* $\dfrac{\partial g}{\partial \theta} = 0$, *then* $x'(\tau_k^+) = 0$.

The condition $f_k(\tau_k^+) = 0$ typically indicates a saturation effect or the state reaching a boundary that cannot be crossed, e.g., when the state is constrained to be nonnegative. This often arises in stochastic flow systems used to model how parts are processed in manufacturing systems or how packets are transmitted and received through a communication network (Yu and Cassandras, 2004; Cassandras, 2006). In such cases, the conditions of both Theorems 1 and 2 are frequently satisfied since (i) common performance metrics such as workload or overflow rates satisfy Eq. (19), and (ii) flow systems involve nonnegative continuous states and are constrained by capacities that give rise to dynamics of the form $\dot{x} = 0$. This class of SHS is also referred to as stochastic flow models (SFMs) and the simplicity of the IPA derivatives in this case has been thoroughly analyzed, e.g., see Cassandras et al. (2002) and Sun et al. (2004).

Property 4 (Scalability). From a computational standpoint, the IPA derivative evaluation process takes place iteratively at each event defining a mode transition at some time instant $\tau_k(\theta)$. At this point in time, we have at our disposal the value of $x'(\tau_{k-1}^+)$ from the previous iteration, which specifies ξ_{k-1} in Eq. (9) applied for all $t \in [\tau_{k-1}(\theta), \tau_k(\theta))$. Therefore, setting $t = \tau_k(\theta)$ in Eq. (9) we also have at our disposal the value of $x'(\tau_k^-)$. Next, depending on whether the event is exogenous or endogenous, the value of τ'_k can be obtained: it is either $\tau'_k = 0$ or given by Eq. (10) since $x'(\tau_k^-)$ is known. Finally, we obtain $x'(\tau_k^+)$ using Eq. (7). At this point, one can wait until the next event occurs at $\tau_{k+1}(\theta)$ and repeat the process that can, therefore, be seen to be entirely event-driven. The implication of this observation is that an IPA estimator scales with the number of events and not the state of the system, which generally increases exponentially with the number of components in a CPS.

5 THE DATA HARVESTING PROBLEM IN CPS

The data harvesting problem arises in many CPS, including Smart Cities where wireless sensor networks are being widely deployed for purposes of monitoring the environment, traffic, infrastructure for transportation and for energy distribution, surveillance, and a variety of other specialized purposes (Roscia et al., 2013). Although many efforts focus on the analysis of the vast amount of data gathered, we must first ensure the existence of robust means to collect all data in a timely fashion when the size of the sensor networks and the level of node interference do not allow for a fully wireless connected system. Sensors can locally gather and buffer data, while mobile elements (e.g., vehicles, aerial drones) retrieve the data from each part of the network. Similarly, mobile elements may themselves be equipped with sensors and visit specific points of interest to collect data that must then be delivered to a given base. These mobile agents should follow an optimal path (in some sense to be defined), which allows visiting each data source frequently enough and within the constraints of a given environment.

The data harvesting problem using mobile agents known as "message ferries" or "data mules" has been considered from several different perspectives; see Akkaya and Younis (2005) and Liu et al. (2011) and references therein. In Chang et al. (2014) algorithms are proposed for patrolling target points with the goal of balanced time intervals between consecutive visits. In Ny et al. (2008) the problem is viewed as a polling system with a mobile server visiting data queues at fixed targets. Trajectories are designed for the mobile server in order to stabilize the system, keeping queue contents (modeled as fluid queues) uniformly bounded. Here, we consider the data harvesting problem as an optimization problem that fits the online framework presented in the previous section. The data collection task is carried out by a team of cooperating mobile agents that periodically visit specific points where data are generated by arbitrary random processes, upload the data, and then occasionally deliver the data accumulated to a base. The ultimate goal is for the data to be collected and delivered with minimum expected delay. Rather than looking at this problem as a scheduling task where visit times for each target are determined assuming agents only move in straight lines between targets, we aim to optimize a two-dimensional trajectory for each agent, which may be periodic and can collect data from a target once the agent is within a given range. We limit ourselves to trajectories with no constraints due to obstacles or other factors (clearly, in some environments this is generally not the case).

5.1 PROBLEM FORMULATION

We consider a data harvesting problem where N mobile agents collect data from M stationary targets in a two-dimensional rectangular mission space $S = [0, L_1] \times [0, L_2] \subset \mathrm{R}^2$. Each agent may visit one or more of the M targets, collect data from them, and deliver them to a base. It then continues visiting targets, possibly the same as before or new ones, and repeats this process. By cooperating in how data are collected and delivered, the objective of the agent team is to minimize a weighted sum of collection and delivery delays over all targets. Let $s_j(t) = [s_j^x(t), s_j^y(t)]$ be the position of agent j at time t with $s_j^x(t) \in [0, L_1]$ and $s_j^y(t) \in [0, L_2]$. We assume that the agent controls its orientation and speed. The position of the agent follows single integrator dynamics with $\dot{s}_j^x(t) = u_j(t) \cos \theta_j(t)$ and $\dot{s}_j^y(t) = u_j(t) \sin \theta_j(t)$ where $0 \le u_j(t) \le 1$

is the scalar speed of the agent and $0 \leq \theta_j(t) < 2\pi$ is the angle relative to the positive direction. An agent is represented as a particle, thus avoiding the need for collision avoidance control.

Consider a set of data sources as points $w_i \in S$, $i = 1,\ldots,M$, with associated ranges r_{ij}, i.e., agent j can collect data from w_i only if the Euclidean distance $D_{ij}(t) = \|w_i - s_j(t)\|$ satisfies $D_{ij}(t) \leq r_{ij}$. Similarly, there is a base at $w_B \in S$ that receives all data collected by the agents. An agent can only deliver data to the base if the Euclidean distance $D_{Bj}(t) = \|w_{Bj} - s_j(t)\|$ satisfies $D_{Bj}(t) \leq r_{Bj}$. We define a function $P_{ij}(t)$ to be the normalized data collection rate from target i when the agent is at $s_j(t)$:

$$P_{ij}(t) = p(w_i, s_j(t)) \tag{20}$$

and we assume that: **(A1)** it is monotonically nonincreasing in the value of $D_{ij}(t) = \|w_i - s_j(t)\|$, and **(A2)** it satisfies $P_{ij}(t) = 0$ if $D_{ij}(t) > r_{ij}$. Thus, $P_{ij}(t)$ can model communication power constraints which depend on the distance between a data source and an agent equipped with a receiver (similar to the model used in Ny et al., 2008) or sensing range constraints if an agent collects data using on-board sensors. For simplicity, we will also assume that: **(A3)** $P_{ij}(t)$ is continuous in $D_{ij}(t)$. Note that $P_{ij}(t)$ may be viewed as a switch activated when $D_{ij}(t) \leq r_{ij}$ to capture the finite range between agent j and target i. Similarly, we define:

$$P_{Bj}(t) = p(w_B, s_j(t)) \tag{21}$$

The data harvesting problem described above can be viewed as a polling system where mobile agents are serving the targets by collecting data and delivering them to the base. The queues at these targets are modeled as flow systems whose dynamics are given next (however, as we will see, the agent trajectory optimization is driven by events observed in the underlying system where queues contain discrete data packets so that this modeling device has minimal effect on our analysis).

There are three sets of queues: (*i*) the data contents $X_i(t) \in \mathbb{R}^+$ at each target $i = 1,\ldots,M$ where we use $\sigma_i(t)$ as the instantaneous inflow rate. In general, we treat $\{\sigma_i(t)\}$ as a random process assumed only to be piecewise continuous. (*ii*) The data contents $Z_{ij}(t) \in \mathbb{R}^+$ onboard agent j collected from target i as long as $P_{ij}(t) > 0$. (*iii*) The queues $Y_i(t) \in \mathbb{R}^+$ containing data at the base, one queue for each target, delivered by some agent j as long as $P_{Bj}(t) > 0$.

Note that $\{X_i(t)\}$, $\{Z_{ij}(t)\}$, and $\{Y_i(t)\}$ are all random processes and the same applies to the agent states $\{s_j(t)\}$, $j = 1,\ldots,N$, since the controls are generally dependent on the random queue states. Thus, we ensure that all random processes are defined on a common probability space. The maximum rate of data collection from target i by agent j is μ_{ij} and the actual rate is $\mu_{ij}P_{ij}(t)$ if j is connected to i. We will assume that: **(A4)** only one agent at a time is connected to a target i even if there are other agents l with $P_{il}(t) > 0$; this is not the only possible model, but we adopt it based on the premise that simultaneous downloading of packets from a common source creates problems of proper data reconstruction at the base. The dynamics of $X_i(t)$, assuming that agent j is connected to it, are:

$$\dot{X}_i(t) = \begin{cases} 0 & \text{if } X_i(t) = 0, \ \sigma_i(t) \leq \mu_{ij}(t)P_{ij}(t) \\ \sigma_i(t) - \mu_{ij}(t)P_{ij}(t) & \text{otherwise} \end{cases} \tag{22}$$

Obviously, $\dot{X}_i(t) = \sigma_i(t)$ if $P_{ij}(t) = 0$, $j = 1,\ldots,N$. In order to express the dynamics of $Z_{ij}(t)$, let:

$$\tilde{\mu}_{ij}(t) = \begin{cases} \min\left(\dfrac{\sigma_i(t)}{P_{ij}(t)}, \mu_{ij}(t)\right) & \text{if } X_i(t) = 0, \ P_{ij}(t) > 0 \\ \mu_{ij}(t) & \text{otherwise} \end{cases}$$

This gives us the dynamics:

$$
\dot{z}_{ij}(t) = \begin{cases} 0 & \text{if } Z_{ij}(t) = 0, \ \tilde{\mu}_{ij}(t)P_{ij}(t) - \beta_{ij}(t)P_{Bj}(t) \le 0 \\ \tilde{\mu}_{ij}(t)P_{ij}(t) - \beta_{ij}(t)P_{Bj}(t) & \text{otherwise} \end{cases}
\tag{23}
$$

where β_{ij} is the maximum rate of data from target i delivered by agent j. For simplicity, we assume that: **(A5)** $\|w_i - w_B\| > r_{ij} + r_{Bj}$ for all $i = 1,\ldots,M$ and $j = 1,\ldots,N$, i.e., the agent cannot collect and deliver data at the same time. Therefore, in Eq. (23) it is always the case that $P_{ij}(t)P_{Bj}(t) = 0$. Finally, the dynamics of $Y_i(t)$ depend on $Z_{ij}(t)$, the content of the on-board queue of each agent j from target i as long as $P_{Bj}(t) > 0$. We define $\beta_i(t) = \sum_{j=1}^{N} \beta_{ij} P_{Bj}(t) \mathbf{1}[Z_{ij}(t) > 0]$ to be the total instantaneous delivery rate for target i data, so that the dynamics of $Y_i(t)$ are:

$$
\dot{Y}_i(t) = \beta_i(t)
\tag{24}
$$

Our objective is to maintain minimal values for all target and on-board agent data queues, while maximizing the contents of the delivered data at the base queues. Thus, we define $J_1(X_1,\ldots,X_M,t)$ to be the weighted sum of expected target queue contents (recalling that $\{\sigma_i(t)\}$ are random processes):

$$
J_1(X_1,\ldots,X_M,t) = \sum_{i=1}^{M} q_i E[X_i(t)]
\tag{25}
$$

where the weight q_i represents the importance factor of target i. Similarly, we define a weighted sum of expected base queues contents:

$$
J_2(Y_1,\ldots,Y_M,t) = \sum_{i=1}^{M} q_i E[Y_i(t)]
\tag{26}
$$

For simplicity, we will assume that $q_i = 1$ for all i. Therefore, our optimization objective may be a convex combination of Eqs. (25) and (26). However, at the same time we need to ensure that the agents are controlled so as to maximize their utilization, i.e., the fraction of time spent performing a useful task by being within range of a target or the base. Equivalently, we aim to minimize the nonproductive idling time of each agent during which it is not visiting any target or the base. Let:

$$
D_{ij}^+ = \max\left(0, D_{ij}(t) - r_{ij}\right), \quad D_{Bj}^+ = \max\left(0, D_{Bj}(t) - r_{Bj}\right)
\tag{27}
$$

so that the idling time for agent j occurs when $D_{ij}^+(t) > 0$ for all i and $D_{Bj}^+(t) > 0$. We define the idling function $I_j(t)$:

$$
I_j(t) = \log\left(1 + D_{Bj}^+(t) \prod_{i=1}^{M} D_{ij}^+(t)\right)
\tag{28}
$$

where $I_j(t) = 0$ if and only if the product term inside the bracket is zero, i.e., agent j is visiting a target or the base; otherwise, $I_j(t) > 0$. $I_j(t)$ is monotonically nondecreasing in the number of targets M. The logarithmic function prevents the value of $I_j(t)$ from dominating those of $J_1(\cdot)$ and $J_2(\cdot)$ when included in a single objective function. We define:

$$
J_3(t) = M_I \sum_{j=1}^{N} E[I_j(t)]
\tag{29}
$$

where M_I is a weight for the idling time effect relative to $J_1(\cdot)$ and $J_2(\cdot)$. Note that $I_j(t)$ is also a random variable. Finally, we define a terminal cost at T capturing the expected value of the amount of data left on board the agents, noting that the effect of this term vanishes as $T \to \infty$ as long as all $E[Z_{ij}(T)]$ remain bounded:

$$J_f(T) = \frac{1}{T} \sum_{i=1}^{M} \sum_{j=1}^{N} E[Z_{ij}(T)] \tag{30}$$

We formulate a stochastic optimization problem with speeds and headings as the control variables denoted by vectors $\mathbf{u}(t) = [u_1(t), \ldots, u_N(t)]$ and $\boldsymbol{\theta}(t) = [\theta_1(t), \ldots, \theta_N(t)]$, respectively (omitting their dependence on the full system state at t). Combining Eqs. (25), (26), (29), and (30) we have:

$$\min_{\mathbf{u}(t), \boldsymbol{\theta}(t)} J(T) = \frac{1}{T} \int_0^T [\alpha J_1(T) + (1-\alpha)J_2(T) + J_3(T)]dt + J_f(T) \tag{31}$$

where $\alpha \in [0, 1]$ is a weight capturing the relative importance of collected data as opposed to delivered data and $0 \le u_j(t) \le 1, 0 \le \theta_j(t) < 2\pi$. To simplify notation, we have also expressed $J_1(X_1, \ldots, X_M, t)$ and $J_2(Y_1, \ldots, Y_M, t)$ as $J_1(t)$ and $J_2(t)$. Since we are considering a finite time optimization problem, instability in the queues is not an issue. However, stability of such a system can, indeed, be an issue in the sense of guaranteeing that $E[X_i(t)] < \infty, E[Z_{ij}(T)] < \infty$ for all i, j under a particular control policy when $t \to \infty$. This problem is considered in Ny et al. (2008) for a simpler deterministic data harvesting model where target queues are required to be bounded.

As formulated, Eq. (31)) is a finite horizon optimal control problem. A standard Hamiltonian analysis when all data arrival processes are assumed to be deterministic is given in Khazaeni and Cassandras (2015a) where it is shown that the optimal speed is $u_j^*(t) = 1$ for all agents $j = 1, \ldots, N$. On the other hand, the optimal heading $\theta_j^*(t)$ can only be obtained by discretizing the problem in time and numerically solving a two point boundary value problem (TPBVP) with a forward integration of the state and a backward integration of the costate. Although intractable as the number of agents and targets grows, one of the insights this analysis provides is that under optimal control the data harvesting process operates as a hybrid system with discrete states (modes) defined by the dynamics of the flow queues in Eqs. (22)–(24), while the agents maintain a fixed speed. The events that trigger mode transitions are defined in Table 1 (the superscript 0 denotes events causing a variable to reach a value of zero from above and the superscript + denotes events causing a variable to become strictly positive from a zero value). Finally, note that all events above are directly observable during the execution of any agent trajectory and they do not depend on our model of flow queues. Dealing with a hybrid dynamic system further complicates the solution of a TPBVP, but it enables us to make use of IPA in the framework of Fig. 1 in order to carry out the parametric trajectory optimization process discussed in the next section.

5.2 AGENT TRAJECTORY PARAMETERIZATION AND OPTIMIZATION

The main idea here is to represent each agent's trajectory through general parametric equations:

$$s_j^x(t) = f(\Theta_j, \rho_j(t)), \quad s_j^y(t) = g(\Theta_j, \rho_j(t)) \tag{32}$$

where the function $\rho_j(t)$ controls the position of the agent on its trajectory at time t and Θ_j is a vector of parameters controlling the shape and location of the agent j trajectory. Let $\Theta = [\Theta_1, \ldots, \Theta_N]$. We now replace the problem in Eq. (31) with a new one:

$$\min_{\Theta \in F_\Theta} J(T) = \frac{1}{T} \int_0^T [\alpha J_1(\Theta, T) + (1-\alpha)J_2(\Theta, T) + J_3(\Theta, T)]dt + J_f(\Theta, T) \tag{33}$$

Table 1 Hybrid System Events

Event Name	Description
1. ξ_i^0	$X_i(t)$ hits 0, for $i = 1,...,M$
2. ξ_i^+	$X_i(t)$ leaves 0, for $i = 1, ..., M$.
3. ζ_{ij}^0	$Z_{ij}(t)$ hits 0, for $i = 1, ..., M, j = 1, ..., N$
4. δ_{ij}^+	$D_{ij}^+(t)$ leaves 0, for $i = 1, ..., M, j = 1, ..., N$
5. δ_{ij}^0	$D_{ij}^+(t)$ hits 0, for $i = 1, ..., M, j = 1, ..., N$
6. Δ_j^+	$D_{Bj}^+(t)$ leaves 0, for $j = 1, ..., N$
7. Δ_j^0	$D_{Bj}^+(t)$ hits 0, for $j = 1, ..., N$

which, recalling that we allow for arbitrary stochastic data arrival processes $\{\sigma_i(t)\}$, is a parametric stochastic optimization problem with F_Θ appropriately defined depending on Eq. (32). The cost function in Eq. (33) is $J(\Theta, T; X(\Theta, 0)) = E[L(\Theta, T; X(\Theta, 0))]$, where $L(\Theta, T; X(\Theta, 0))$ is a sample function defined over $[0, T]$ and $X(\Theta, 0)$ is the initial value of the state vector. For convenience, in the sequel we will use L_1, L_2, L_3, L_f to denote sample functions of J_1, J_2, J_3 and J_f, respectively. Note that in Eq. (33) we suppress the dependence of the four objective function components on the controls $u(t)$ and $\theta(t)$ and instead stress their dependence on the parameter vector Θ. In what follows, we will limit ourselves to the family of elliptical trajectories using an approach similar to the one used in the multi-agent persistent monitoring problem in Lin and Cassandras (2015) (a more general Fourier series trajectory representation may be found in Khazaeni and Cassandras, 2015b,a). The hybrid dynamics of the data harvesting system allow us to apply IPA to obtain online the gradient of the sample function $L(\Theta, T; X(\Theta, 0))$ with respect to Θ. As already described in the previous section, the value of the IPA approach is twofold: (i) the sample gradient $\nabla L(\Theta, T)$ can be obtained online based on an observable sample path data only, and (ii) $\nabla L(\Theta, T)$ is an unbiased estimate of $\nabla J(\Theta, T)$ under mild technical conditions. Therefore we can use $\nabla L(\Theta, T)$ in a standard gradient-based stochastic optimization algorithm:

$$\Theta^{l+1} = \Theta^l - \nu_l \nabla \mathcal{L}(\Theta^l, T), \quad l = 0, 1, ... \tag{34}$$

to converge (at least locally) to an optimal parameter vector Θ^* with a proper selection of a step-size sequence $\{\nu_l\}$.

Based on the events defined in Table 1, we can specify event time derivative and state derivative dynamics for each mode of the hybrid system by directly applying the IPA calculus Eqs. (13)–(15). We will use the notation introduced in Eqs. (13)–(15) so that $\tau_k' = \frac{d\tau_k}{d\Theta}, X'(t) = \frac{d\mathcal{X}}{d\Theta}$ are the Jacobian matrices of partial derivatives with respect to all components of the controllable parameter vector Θ. We will also use $f_k(t) = \frac{d\mathcal{X}}{dt}$ to denote the state dynamics in effect over an interevent time interval $[\tau_k, \tau_{k+1})$.

Table 1 contains all possible endogenous event types for our hybrid system. To these, we add exogenous events $\kappa_i, i = 1,...,M$, to allow for possible discontinuities (jumps) in the random processes $\{\sigma_i(t)\}$, which affect the sign of $\sigma_i(t) - \mu_{ij}P_{ij}(t)$ in Eq. (22). We will use the notation $e(\tau_k)$ to denote the event type occurring at $t = \tau_k$ with $e(\tau_k) \in E$, the event set consisting of all endogenous and exogenous events. Finally, we make the following assumption that is needed in guaranteeing the unbiasedness of the IPA gradient estimates: **(A6)** two events occur at the same time w.p.0 unless one is directly caused by the other.

5.2.1 Objective function gradient

The sample function gradient $\nabla \mathcal{L}(\Theta, T)$ needed in Eq. (34) is obtained from Eq. (33) assuming a total of K events over $[0\ T]$ with $\tau_{K+1} = T$ and $\tau_0 = 0$:

$$\nabla \mathcal{L}(\Theta, T; \mathbf{X}(\Theta, 0)) = \frac{1}{T} \int_0^T [\alpha \nabla \mathcal{L}_1(\Theta, T) + (1-\alpha) \nabla \mathcal{L}_2(\Theta, T) + \nabla \mathcal{L}_3(\Theta, T)] dt + \nabla \mathcal{L}_f(\Theta, T) \qquad (35)$$

Note that $\nabla \mathcal{L}(\Theta, T)$ does not have any direct dependence on any τ'_k; this dependence is indirect through the state derivatives involved in the four individual gradient terms. Referring to Eq. (25), the first term involves $\nabla \mathcal{L}_1(\Theta, t)$, which is as a sum of $X'_i(t)$ derivatives. Similarly, $\nabla \mathcal{L}_2(\Theta, t)$ is a sum of $Y'_i(t)$ derivatives and $\nabla \mathcal{L}_f(\Theta, T)$ requires only $Z'_i(T)$. The third term, $\nabla \mathcal{L}_3(\Theta, t)$, requires derivatives of $I_j(t)$ in Eq. (28) which depend on the derivatives of the max function in Eq. (27) and the agent state derivatives $s'_j(t)$ with respect to Θ. Possible discontinuities in these derivatives occur when any of the last four events in Table 1 takes place. The evaluation of Eq. (35) requires the state derivatives $X'_i(t)$, $Z'_{ij}(t)$, $Y'_i(t)$, and $s'_j(t)$. The latter, $s'_j(t)$, are easily obtained for any specific choice of f and g in Eq. (32) and may be found in Khazaeni and Cassandras (2015a). The former require a rather laborious use of the IPA equations provided earlier for the state and event time derivatives that, however, reduce to a simple set of state derivative dynamics. The detailed analysis may be found in Khazaeni and Cassandras (2015a) and we limit ourselves here to the final results.

The first key result provides the state derivatives $X'_i(\tau_k^+), Y'_i(\tau_k^+), Z'_{ij}(\tau_k^+)$ with respect to the controllable parameter Θ after an event occurrence at $t = \tau_k$:

$$X'_i(\tau_k^+) = \begin{cases} 0 & \text{if } e(\tau_k) = \xi_i^0 \\ X'_i(\tau_k^-) - \mu_{il}(t)P_{il}(\tau_k)\tau'_k & \text{if } e(\tau_k) = \delta_{ij}^+ \\ X'_i(\tau_k^-) & \text{otherwise} \end{cases}$$

where $l \neq j$ with $P_{il}(\tau_k) > 0$ if such l exists and $\tau'_k = \dfrac{\partial D_{ij}(s_j)}{\partial s_j} \dfrac{\partial s_j}{\partial \Theta} \left(\dfrac{\partial D_{ij}(s_j)}{\partial s_j} \dot{s}_j(\tau_k) \right)^{-1}$.

$$Y'_i(\tau_k^+) = \begin{cases} Y'_i(\tau_k^-) + Z'_{ij}(\tau_k^-) & \text{if } e(\tau_k) = \zeta_{ij}^0 \\ Y'_i(\tau_k^-) & \text{otherwise} \end{cases}$$

$$Z'_{ij}(\tau_k^+) = \begin{cases} 0 & \text{if } e(\tau_k) = \zeta_{ij}^0 \\ Z'_{ij}(\tau_k^-) + X'_i(\tau_k^-) & \text{if } e(\tau_k) = \xi_i^0 \\ Z'_{ij}(\tau_k^-) & \text{otherwise} \end{cases}$$

where $e(\tau_k) = \xi_i^0$ occurs when j is connected to target i:

This result shows that only three of the events in E can actually cause discontinuous changes to the state derivatives. Further, note that $X'_i(t)$ is reset to zero after a ξ_i^0 event. Moreover, when such an event occurs, note that $Z'_{ij}(t)$ is coupled to $X'_i(t)$. Similarly for $Z'_{ij}(t)$ and $Y'_i(t)$ when event ζ_{ij}^0 occurs, showing that perturbations in Θ can only propagate to an adjacent queue when that queue is emptied.

The second key result provides the state derivatives $X_i'(\tau_{k+1}^-)$, $Y_i'(\tau_{k+1}^-)$ with respect to the controllable parameter Θ after an event occurrence at $t = \tau_k$:

$$X_i'(\tau_{k+1}^-) = \begin{cases} 0 & \text{if } e(\tau_k) = \xi_i^0 \\ X_i'(\tau_k^+) - \int_{\tau_k}^{\tau_{k+1}} \mu_{ij} P_{ij}'(u)du & \text{otherwise} \end{cases}$$

$$Y_i'(\tau_{k+1}^-) = Y_i'(\tau_k^+) + \int_{\tau_k}^{\tau_{k+1}} \beta_i'(u)du$$

where j is such that $P_{ij}(t) > 0$, $t \in [\tau_k, \tau_{k+1}]$.

The final result gives the state derivatives $Z_{ij}'(\tau_{k+1}^+)$ with respect to the controllable parameter Θ after an event occurrence at $t = \tau_k$:

(1) If j is connected to target i,

$$Z_{ij}'(\tau_{k+1}^-) = \begin{cases} Z_{ij}'(\tau_k^+) & \text{if } e(\tau_k) = \xi_i^0, \ \zeta_{ij}^0 \text{ or } \delta_{ij}^+ \\ Z_{ij}'(\tau_k^+) + \int_{\tau_k}^{\tau_{k+1}} \mu_{ij} P_{ij}'(u)du & \text{otherwise} \end{cases}$$

(2) If j is connected to B with $Z_{ij}(\tau_k) > 0$, $Z_{ij}'(\tau_{k+1}^-) = Z_{ij}'(\tau_k^+) - \int_{\tau_k}^{\tau_{k+1}} \beta_{ij} P_{Bj}'(u)du$
(3) Otherwise, $Z_{ij}'(\tau_{k+1}^-) = Z_{ij}'(\tau_k^+)$.

An important corollary of these results is the following: the state derivatives $X_i'(t)$, $Z_{ij}'(t)$, $Y_i'(t)$ with respect to the controllable parameter Θ are independent of the random data arrival processes $\{\sigma_i(t)\}$, $i = 1, ..., M$. As this statement asserts, one can apply IPA regardless of the characteristics of the random processes $\{\sigma_i(t)\}$. This robustness property does not mean that these processes do not affect the values of the $X_i'(t)$, $Z_{ij}'(t)$, $Y_i'(t)$; this happens through the values of the event times τ_k, $k = 1, 2, ...$, which are observable and enter the computation of these derivatives as seen above. Also, as already pointed out in the previous section, the IPA estimation process is event-driven: $X_i'(\tau_k^+)$, $Y_i'(\tau_k^+)$, $Z_{ij}'(\tau_k^+)$ are evaluated at event times and then used as initial conditions for the evaluations of $X_i'(\tau_{k+1}^-)$, $Y_i'(\tau_{k+1}^-)$, $Z_{ij}'(\tau_{k+1}^-)$ along with the integrals appearing in the expressions above, which can also be evaluated at $t = \tau_{k+1}$. Consequently, this approach is scalable in the number of events in the system as the number of agents and targets increases. In addition, despite the elaborate derivations in Khazaeni and Cassandras (2015a), the actual implementation is simple. Finally, returning to Eq. (35), note that the integrals involving $\nabla \mathcal{L}_1(\Theta, t)$, $\nabla \mathcal{L}_2(\Theta, t)$ are directly obtained from $X_i'(t)$, $Y_i'(t)$, the integral involving $\nabla \mathcal{L}_3(\Theta, t)$ is obtained from straightforward differentiation of Eq. (28), and the final term is obtained from $Z_{ij}'(T)$.

5.2.2 Objective function optimization

The trajectory optimization process is carried out using Eq. (34) with an appropriate step size sequence. For the elliptical trajectories we are considering here, there are five parameters involved: the center coordinates, minor and major axes and orientation. Agent j's position $s_j(t) = [s_j^x(t), s_j^y(t)]$ follows the general parametric equation of the ellipse:

$$s_j^x(t) = A_j + a_j \cos \rho_j(t) \cos \phi_j - b_j \sin \rho_j(t) \sin \phi_j$$

$$s_j^y(t) = B_j + a_j \cos \rho_j(t) \sin \phi_j + b_j \sin \rho_j(t) \cos \phi_j$$

and in this case the controllable parameter vector is $\Theta_j = [A_j, B_j, a_j, b_j, \phi_j]$ where A_j, B_j are the coordinates of the center, a_j and b_j are the major and minor axis respectively while $\phi_j \in [0, \pi)$ is the ellipse orientation defined as the angle between the x axis and the major axis of the ellipse. Finally, $\rho_j(t)$ is the eccentric anomaly of the ellipse whose dynamics are given in Khazaeni and Cassandras (2015a).

To capture multiple visits to the base we allow an agent trajectory to consist of a sequence of admissible ellipses. For each agent, we define E_j as the number of ellipses in its trajectory. The parameter vector Θ_j^κ with $\kappa = 1, \ldots, E_j$, defines the κth ellipse in agent j's trajectory and T_j^κ is the time that agent j completes ellipse κ. Since we cannot optimize over all possible E_j for all agents, an iterative process needs to be performed in order to find the optimal number of segments in each agent's trajectory. We can now formulate the following parametric optimization problem with $\Theta_j = [\Theta_j^1, \ldots, \Theta_j^{E_j}]$ and $\Theta = [\Theta_1, \ldots, \Theta_N]$:

$$\min_{\Theta \in F_\Theta} J(T) = \frac{1}{T} \int_0^T (\alpha J_1(\Theta, T) + (1-\alpha) J_2(\Theta, T) + J_3(\Theta, T)) dt$$
$$+ M_C \sum_{i=1}^N C_j(\Theta_j) + J_f(\Theta, T)$$

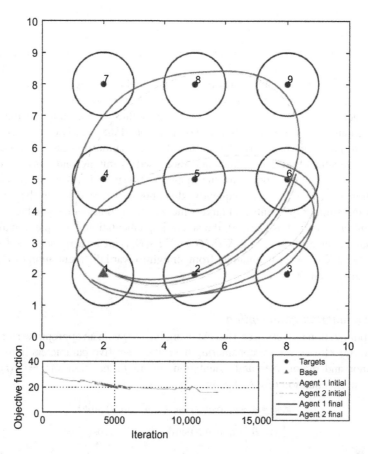

FIG. 3

TPBVP trajectories ($T = 30$): $J = 15.82$.

where M_C is a large multiplier and $C_j(\Theta_j) = \left(1 - f_j^1 \cos^2\phi_j - f_j^2 \sin^2\phi_j - f_j^3 \sin 2\phi_j\right)^2$ is a constraint term to ensure trajectories pass through the base, with f_j^1, f_j^2 and f_j^3 being constants. The evaluation of ∇C_j does not depend on any event (details are given in Khazaeni and Cassandras, 2015a).

We conclude with a numerical example where we consider eight targets, two agents and a base as shown in Fig. 3. We assume deterministic arrival processes with $\sigma_i = 0.5$ for all i. For Eqs. (20) and (21) we have used $p(w,v) = \max\left(0, 1 - \dfrac{D(w,v)}{r}\right)$ where r is the corresponding value of r_{ij} or r_{Bj}. We set $\mu_{ij} = 50$ and $\beta_{ij} = 500$ for all i and j. Other parameters used are $\alpha = 0.5$, $r_{ij} = r_{Bj} = 1$, $M_I = 1$ and $T = 100$ except for the TPBVP case where $T = 30$. In Fig. 3 results of the TPBVP are shown, which depend heavily on the initial trajectory, and this is the best result among several initializations. These results are after 10,000 iterations of the TPBVP solver. In Fig. 4 the results are shown for the (locally) optimal trajectory with two ellipses in each agent's trajectory ($E_j = 2$). In this case, note that the elliptical trajectories provide better performance ($-50.9 < 15.82$), a consequence of the fact that the TPBVP solver is only able to yield a locally optimal solution.

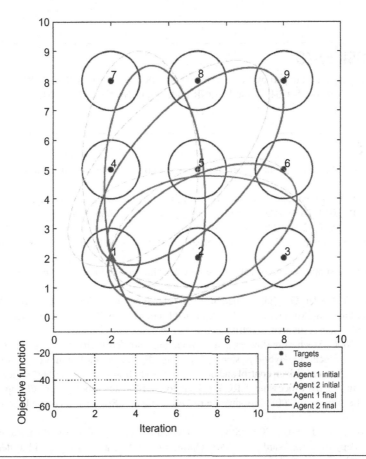

FIG. 4

Elliptical trajectories ($T = 100$): $J = -50.9$.

6 CONCLUSIONS

CPSs may be modeled as stochastic hybrid dynamic systems for which we have provided a framework enabling the online solution of control and optimization problems where controllers are parameterized and the parameters are adaptively tuned based on observable real-time data. This is possible through the use of IPA in order to estimate gradients of performance metrics with respect to various parameters. The event-driven nature of IPA-based gradient estimation implies its scalability in the size of an event set, as opposed to the system state space. We have applied this framework to the data harvesting problem, a fundamental component of many CPS where a number of mobile agents cooperate in order to collect the data generated by various sources distributed in a given region and then deliver them to a base. The goal is to minimize expected delays at the data sources or while the agents are in transit. We have formulated this as an agent trajectory optimization problem and showed how to solve it online through the use of IPA. We have shown robustness of the solution with respect to stochastic data generation processes by considering stochastic data arrivals at targets. The results may be extended to more complex trajectory families and future research aims at incorporating obstacles in the planning of optimal trajectories.

ACKNOWLEDGMENTS

The authors' work is supported in part by NSF under grants CNS-1239021, ECCS-1509084, and IIP-1430145, by AFOSR under grant FA9550-15-1-0471, and by ONR under grant N00014-09-1-1051.

REFERENCES

Akkaya, K., Younis, M., 2005. A survey on routing protocols for wireless sensor networks. Ad Hoc Netw. 3, 325–349.

Alur, A., Henzinger, T.A., Sontag, E.D. (Eds.), 1996. Hybrid Systems. Springer-Verlag, New York/Berlin.

Anta, A., Tabuada, P., 2010. To sample or not to sample: self-triggered control for nonlinear systems. IEEE Trans. Automat. Control 55 (9), 2030–2042.

Antsaklis, P., Kohn, W., Lemmon, M., Nerode, A., Sastry, S. (Eds.), 1998. Hybrid Systems. Springer-Verlag, New York/Berlin.

Arzen, K.E., 2002. A simple event based PID controller. In: Proceedings of 14th IFAC World Congress, pp. 423–428.

Baccelli, F., Cohen, G., Olsder, G.J.(Eds.), Quadrat, J.P., 1992. Synchronization and Linearity. Wiley, New York.

Branicky, M.S., Borkar, V.S., Mitter, S.K., 1998. A unified framework for hybrid control: model and optimal control theory. IEEE Trans. Automat. Control 43 (1), 31–45.

Cassandras, C.G., 2006. Stochastic flow systems: modeling and sensitivity analysis. In: Cassandras, C.G., Lygeros, J. (Eds.), Stochastic Hybrid Systems. Taylor and Francis, Boca Raton, FL, pp. 139–167.

Cassandras, C.G., 2015. Event-driven control and optimization in hybrid systems. In: Miskowicz, M. (Ed.), Event-Based Control and Signal Processing. CRC Press/Taylor and Francis, Boca Raton, FL, pp. 21–36.

Cassandras, C.G., Lafortune, S., 2008. Introduction to Discrete Event Systems, second ed. Springer, New York/Berlin.

Cassandras, C.G., Lygeros, J. (Eds.), 2007. Stochastic Hybrid Systems. CRC Press, Boca Raton, FL.

Cassandras, C.G., Pepyne, D.L., Wardi, Y., 2001. Optimal control of a class of hybrid systems. IEEE Trans. Automat. Control 46 (3), 398–415.

Cassandras, C.G., Wardi, Y., Melamed, B., Sun, G., Panayiotou, C.G., 2002. Perturbation analysis for on-line control and optimization of stochastic fluid models. IEEE Trans. Automat. Control 47 (8), 1234–1248.

Cassandras, C.G., Wardi, Y., Panayiotou, C.G., Yao, C., 2010. Perturbation analysis and optimization of stochastic hybrid systems. Eur. J. Control 16 (6), 642–664.

Chang, C., Yu, G., Wang, T., Lin, C., 2014. Path construction and visit scheduling for targets using data mules. IEEE Trans. Syst. Man Cybern.: Syst. 44 (10), 1289–1300. [Online]. Available: http://wireless.cs.tku.edu.tw/cychang/SMC_TCTP.pdf.

Garcia, E., Antsaklis, P.J., 2014. Event-triggered output feedback stabilization of networked systems with external disturbance. In: Proceedings of 53rd IEEE Conference On Decision and Control, pp. 3572–3577.

Heemels, W.P., Sandee, J.H., Bosch, P.P., 2008. Analysis of event-driven controllers for linear systems. Int. J. Control 81 (4), 571.

Heemels, W.P.M.H., Donkers, M.C.F., Teel, A.R., 2013. Periodic event-triggered control for linear systems. IEEE Trans. Automat. Control 58 (4), 847–861.

Ho, Y.C., Cao, X., 1991. Perturbation Analysis of Discrete Event Dynamic Systems. Kluwer Academic Publishers, Boston.

Ho, Y.C., Cassandras, C.G., 1983. A new approach to the analysis of discrete event dynamic systems. Automatica 19, 149–167.

IEEE, 2014. Special issue on goals and challenges in cyber-physical systems research. IEEE Trans. Automat. Control 59 (12).

Kebarighotbi, A., Cassandras, C.G., 2012. A general framework for modeling and online optimization of stochastic hybrid systems. In: Proceedings of 4th IFAC Conference on Analysis and Design of Hybrid Systems.

Khazaeni, Y., Cassandras, C.G., 2014. A new event-driven cooperative receding horizon controller for multi-agent systems in uncertain environments. In: Proceedings of 53rd IEEE Conference On Decision and Control, pp. 2770–2775.

Khazaeni, Y., Cassandras, C.G., 2015a. An optimal control approach for the data harvesting problem. Available at arXiv:1503.06133.

Khazaeni, Y., Cassandras, C.G., 2015b. An optimal control approach for the data harvesting problem. In: 54th IEEE Conference on Decision and Control, December 2015, pp. 5136–5141.

Kushner, H.J., Yin, G.G., 1997. Stochastic Approximation Algorithms and Applications. Springer-Verlag, New York/Berlin.

Lemmon, M., He, K.X., Markovsky, I., 1999. Supervisory hybrid systems. IEEE Control Syst. Mag. 19 (4), 42–55.

Lin, X., Cassandras, C.G., 2015. An optimal control approach to the multi-agent persistent monitoring problem in two-dimensional spaces. IEEE Trans. Automat. Control 60 (6), 1659–1664.

Liu, M., Yang, Y., Qin, Z., 2011. A survey of routing protocols and simulations in delay-tolerant networks. Lect. Notes Comput. Sci. 6843, 243–253.

Liu, T., Cao, M., Hill, D.J., 2014. Distributed event-triggered control for output synchronization of dynamical networks with non-identical nodes. In: Proceedings of 53rd IEEE Conference on Decision and Control, pp. 3554–3559.

Lunze, J., Lehmann, D., 2010. A state-feedback approach to event-based control. Automatica 46 (1), 211–215.

Moody, J.O., Antsaklis, P.J., 1998. Supervisory Control of Discrete Event Systems Using Petri Nets. Kluwer Academic Publishers, Boston.

Ny, J.L., Dahleh, M.a., Feron, E., Frazzoli, E., 2008. Continuous path planning for a data harvesting mobile server. In: Proceedings of the IEEE Conference on Decision and Control, pp. 1489–1494.

Powell, W.B., 2011. Approximate Dynamic Programming, second ed. John Wiley and Sons, New York.

Ramadge, P.J., Wonham, W.M., 1989. The control of discrete event systems. Proc. IEEE 77 (1), 81–98.

Roscia, M., Longo, M., Lazaroiu, G., 2013. Smart City by multi-agent systems. In: International Conference on Renewable Energy Research and Applications, IEEE, October 2013, pp. 20–23. [Online]. Available: http://ieeexplore.ieee.org/xpls/abs_all.jsp?arnumber=6749783.

Rubinstein, R., 1986. Monte Carlo Optimization, Simulation and Sensitivity of Queueing Networks. John Wiley and Sons, New York.

Sun, G., Cassandras, C.G., Panayiotou, C.G., 2004. Perturbation analysis of multiclass stochastic fluid models. J. Discret. Event Dyn. Syst.: Theory Appl. 14 (3), 267–307.

Sussmann, H.J., 1999. A maximum principle for hybrid optimal control problems. In: Proceedings of 38th IEEE Conf. On Decision and Control, December 1999, pp. 425–430.

Tabuada, P., 2007. Event-triggered real-time scheduling of stabilizing control tasks. IEEE Trans. Automat. Control 52 (9), 1680–1685.

Trimpe, S., DAndrea, R., 2014. Event-based state estimation with variance-based triggering. IEEE Trans. Automat. Control 49 (12), 3266–3281.

Wang, X., Lemmon, M., 2011. Event-triggering in distributed networked control systems. IEEE Trans. Automat. Control 56 (3), 586–601.

Wardi, Y., Adams, R., Melamed, B., 2010. A unified approach to infinitesimal perturbation analysis in stochastic flow models: the single-stage case. IEEE Trans. Automat. Control 55 (1), 89–103.

Wardi, Y., Giua, A., Seatzu, C., 2013. IPA for continuous stochastic marked graphs. Automatica 49 (5), 1204–1215.

Yao, C., Cassandras, C.G., 2011a. Perturbation analysis and optimization of multiclass multiobjective stochastic flow models. J. Discret. Event Dyn. Syst. 21 (2), 219–256.

Yao, C., Cassandras, C.G., 2011b. Perturbation analysis of stochastic hybrid systems and applications to resource contention games. Front. Electr. Electron. Eng. China 6 (3), 453–467.

Yu, H., Cassandras, C.G., 2004. Perturbation analysis for production control and optimization of manufacturing systems. Automatica 40, 945–956.

Zhang, P., Cassandras, C.G., 2002. An improved forward algorithm for optimal control of a class of hybrid systems. IEEE Trans. Automat. Control 47 (10), 1735–1739.

Zhong, M., Cassandras, C.G., 2010. Asynchronous distributed optimization with event-driven communication. IEEE Trans. Automat. Control 55 (12), 2735–2750.

ENERGY-HARVESTING LOW-POWER DEVICES IN CYBER-PHYSICAL SYSTEMS

4

M. Erol-Kantarci, D.W. Illig, L.K. Rumbaugh, W.D. Jemison
Clarkson University, Potsdam, NY, United States

1 INTRODUCTION

The advances in cyber-physical systems (CPSs) and Internet of Things (IoT) are leading to a world with hyper-connected objects, machines and humans. Most of the time, connectivity is wireless since wireless connectivity comes with a number of benefits, including the opportunity to provide anytime anywhere access. On the other hand, the wireless medium is shared among devices, and it is well-known that multiple access to the shared medium may result in collision, unless access is properly coordinated. This problem escalates for densely located devices where frequent packet collisions are inevitable even with range adjustments. As a result of this handicap, packet delivery success rate reduces, latency increases and battery lifetime of sensors decreases. Considering that CPSs are expected to house a large mass of sensors, actors and computing units, their reliability, latency and lifetime become important concerns.

In most CPSs, sensors and actors are usually not plugged in an electrical outlet and operate on batteries, since they are required to be close to the monitored area or target. Their capability to work on batteries provides ease of deployment and access even in hard-to-reach locations. On the other hand, sensors' lifetime is limited by the capacity of the battery. Replacing sensor batteries every now and then causes undesired maintenance cost, while sensors located in hazardous or hard-to-reach areas make battery replacement difficult if not impossible; in addition, disposal of used batteries has considerable negative impacts on the environment. Energy conservation has always been useful to extend the battery life of sensors, however most of the energy conservation techniques have been able to support a sensor network for no more than several years (Anastasi et al., 2009).

As an alternative to battery replacement, rechargeable batteries can be used in combination with energy conservation techniques. Rechargeable batteries can be topped up from ambient energy resources such as indoor or outdoor light, thermal gradients, mechanical pressure or motion, and even wind or aquatic motion (e.g., ocean waves, water currents), depending on the sensor deployment area. Extending the battery life of sensors through energy harvesting is possible through various means, but the intermittency of the above-mentioned energy resources raises reliability and availability issues.

Cyber-Physical Systems. http://dx.doi.org/10.1016/B978-0-12-803801-7.00004-3

55

Recently, energy harvesting from ambient electromagnetic waves emitted from dedicated transmitters, WiFi access points and LTE base stations has been explored. This concept is usually referred to as radio frequency (RF) energy harvesting or RF energy transfer. RF energy transfer relies on wireless power transmission where electromagnetic (EM) signals from a transmitter can be received by one or more sensors and the sensors can store the harvested energy in their capacitors. The range of wireless power transmission is limited due to attenuation and multi-path effects. RF-powered low-power sensors are still in their infancy and further research is needed to improve rectenna efficiency, power transmission schedules and the relaying of energy from one device to another.

In this chapter, we aim to motivate the use of RF energy harvesting sensors and computing devices in CPSs. We start by providing a basic understanding of the RF energy harvesting technique. Then, we discuss opportunities for RF energy harvesting in sensor networks and wireless heterogenous networks (HetNets), which is a common architecture for fourth generation and beyond wireless cellular networks. Today, most demonstrated commercial RF energy harvesting circuits consider single hop energy transfer, while a few academic studies have focused on relayed energy harvesting. We will first present some of the state of the art work in direct (single-hop) RF energy transfer, and then present preliminary results on relayed (multi-hop) energy transfer. Proof-of-concept simulation results indicate that energy harvesting nodes may be able to receive more RF energy using the relayed approach.

The rest of the chapter is organized as follows. In Section 2, we begin with a basic introduction to CPSs and RF energy harvesting. In Section 3, we present two schemes; one scheme on direct energy transfer in sensor networks and another one in direct energy transfer in HetNets. Section 4 presents preliminary results on relayed energy transfer. We summarize the chapter and discuss the open issues in Section 5.

2 CYBER-PHYSICAL SYSTEMS AND ENERGY-HARVESTING LOW-POWER SENSORS

2.1 BACKGROUND ON CYBER-PHYSICAL SYSTEMS

CPSs bridge the gap between cyber and physical space, and redefine the way humans interact with and control the physical world (Rajkumar et al., 2010; Wolf, 2009; Lu et al., 2008). A CPS houses thousands to millions of embedded computers that are networked and tightly integrated with feedback loops. Those embedded computers sense, communicate, compute and control physical processes in real-time (Lee and Seshia, 2011). Typical examples of CPSs are medical systems, transportation systems, energy systems, aerospace systems, process control and automation, and smart spaces (Guturu and Bhargava, 2011; Ilic et al., 2008).

CPSs employ a wide variety of communication technologies due to their large scale and presence in heterogeneous environments. At the high-end, embedded computers are usually hard-wired and use the traditional communication technologies such as controller area network (CAN) and FlexRay, meanwhile at the low-end, flexibility and ubiquity mandates wireless communications. In this sense, two-way interaction between humans and processes, parallel operation of multiple processes in real-world and feedback loops from multiple sources require strong communication and networking algorithms that can work in ultra large-scale and densely deployed networks. In CPSs, wireless sensors, actors and devices are densely deployed, such that packet collisions and overhearing problems are magnified by several orders of magnitude in comparison to conventional wireless networks. As a result, energy consumed for transmission and reception of packets increases dramatically. Wirelessly connected embedded computing devices in a CPS (for example, untethered medical robotic devices, mobile furniture in a smart office, or smart cargo in a supply chain) operate on batteries and have tight energy budgets. In order to extend the lifetime of these devices, as well as to provide the basis for future battery-free CPSs, energy-efficiency and energy replenishment are crucial.

2.2 BACKGROUND ON ENERGY-HARVESTING LOW-POWER SENSORS

Energy efficiency and energy scavenging for low-power sensor devices have been broad research areas in the two decades. Anastasi et al. (2009) provides a rigorous review of energy conservation techniques for wireless sensor networks (WSNs). While energy conservation can certainly extend the battery life, energy replenishment provides opportunities for recharging a battery, hence promises longer operation when combined with energy-efficient protocols. Ambient energy replenishment sources include the sun, vibration, body heat, foot strike, mechanical pressure on switches, temperature gradients in the soil, river currents, ocean waves, electromagnetic induction, and so on (Witricity Corp, 2016; Sudevalayam and Kulkarni, 2011). On the other hand, energy scavenging from electromagnetic waves, or in other words RF signals, requires little investment since wireless technologies are pervasive. WiFi access points, small cell base stations, and, indeed, any RF transmitter can be a source of wireless power. Wireless power transfer using RF has been implemented and commercialized for on-body implants and WSNs (Powercast Corporation, 2016). There is strong evidence that RF-powered computing will become widespread in the near future (Gollakota et al., 2014).

In RF energy transfer, a transmitter Tr_s emitting EM waves can charge the battery of a receiver, Rv_s that is located within its charging range, R_c, assuming a circular disk transmission model. When operating in free space, the received power is inversely proportional to d_{rt}^2 for omni-directional antennas, and it is assumed to be 0 when $d_{rt} > R_c$, where d_{rt} is the distance between Tr_s and Rv_s. Here, the received power is given by:

$$P_r = P_t G_t G_r \left(\frac{\lambda}{4\pi d_{rt}}\right)^2 \tag{1}$$

where λ is the wavelength, G_r is the linear receiver gain and $P_t G_t$ is the transmitted RF signal power times the linear transmitter gain. For instance, for $\lambda = 0.328$ m at 915 MHz and $G_r = 3.98$, a transmitted power of 3 W will be received as 0.325 mW at a distance d_{rt} of 5 m. Upon reception of the RF signal, the receiver converts the received power to a DC voltage and stores energy in a capacitor. Hence, a sensor equipped with a receiver, converter and a capacitor can harvest energy and store it in its battery for future use. Note that, in practice, the amount of received power will vary depending on the propagation properties of the environment.

3 DIRECT RF ENERGY HARVESTING

Direct RF energy harvesting has been studied in several previous works for recharging sensors, low-power devices and RFID tags. Those studies perform both simulations and hardware tests. Testbed studies on sensors and low-power devices have widely utilized Powercast Corp.'s (Powercast Corporation, 2016) wireless power transmitter and harvester circuits. Meanwhile, the University of Washington's Wireless Identification and Sensing Platform (WISP) uses RF energy harvesting for powering RFID tags (He et al., 2011).

In Mohammady et al. (2010), the authors proposed a charging-aware routing protocol and an optimization framework in order to determine optimal charging and transmission cycles in a sensor network. (The authors assumed that charging and communications take place in the industrial, scientific and medical (ISM) band, i.e., the 2.4 GHz band. Although this is not the case for Powercast equipment, which operates at 915 MHz, simulation studies have enforced in-band energy transfer to test worst-case scenarios.) The designed routing protocol coordinates charging with the data communications such that simultaneous charging and data transmission do not disrupt communications. The authors have included a "charging duration" field in the routing packets of the ad hoc on-demand distance vector (AODV) protocol. The modified AODV selects the path with minimum charging duration. Then, the transmission and charging schedules of the nodes that are on the forwarding path are determined using an optimization framework. The work in Mohammady et al. (2010) assumes stationary chargers, while in Shi et al. (2011) the authors have studied charging via mobile chargers. The authors assume that the mobile charging vehicle visits and supplies energy to each sensor in the field. The batteries are charged such that the minimum available energy at each sensor is higher than a given threshold within one cycle of charging. When the objective is maximizing the ratio of the docking time over cycle time, it is shown that the optimal traveling path for the vehicle is the shortest Hamiltonian cycle. Even though having the charger move close to the sensor is expected to supply energy effectively, as the number of devices grow, such as in applications of IoT, this approach becomes impractical.

In Erol-Kantarci and Mouftah (2012c), the authors make use of omni-directional RF energy transfer and assume a mobile charger stopping at a landmark charges the batteries of sensors within its transmission range. The authors have investigated ways of selecting best landmarks and then optimal ways of traversing those landmarks. A priority-aware RF energy harvesting mechanism has been proposed in Erol-Kantarci and Mouftah (2012a) for the smart grid environment and in Erol-Kantarci and Mouftah (2012b) the authors have considered recharging sensors in a utility-maximizing way. In CPSs, mission-awareness is significant since sensors are expected to be assigned to various missions. Maximizing the utility of sensors while maximizing the harvested energy is highly desired. In the next section, we will introduce mission-aware placement of wireless power transmitters (MAPIT), which optimizes the placement of RF-based chargers in order to achieve utility-energy maximization in detail.

The above-mentioned works consider RF energy harvesting for sensors. RF-energy transfer from RF identification (RFID) readers to RFID tags has been studied in He et al. (2011). Assuming stationary RFID readers that can both read and charge mobile RFID tags, the problem is formulated as minimizing the number of readers in the network. When mobile tags are considered, RFID tags can receive power from different readers as they move. In this case, the problem turns into selecting optimal reader locations that provide adequate charging for the tags along their path.

CPSs will accommodate objects with RFID tags, sensors, embedded computers and so on. Most of these devices will need power on-the-go. Therefore, RF energy harvesting as well as other ambient energy harvesting techniques will be highly desirable for CPSs. It is also possible to harvest energy from base stations of cellular networks. In the following sections, RF-energy harvesting in HetNets is introduced.

3.1 MISSION-AWARE DIRECT RF-ENERGY HARVESTING IN WIRELESS SENSOR NETWORKS

MAPIT optimizes the placement of RF chargers in a wireless rechargeable sensor network (WRSN) such that the number of charged sensors is maximized, and those sensors participate in profit maximizing missions. MAPIT was initially proposed in Erol-Kantarci and Mouftah (2012b). The locations where wireless power transmitters park and transmit wireless power to the sensor nodes are called "landmarks". A harvester device can receive power if $d_{rt} < R_c$, i.e., the sensor is in transmission range. It has been shown in previous studies that having a mobile transmitter visit landmark locations, rather than densely deploying fixed transmitters, decreases the deployment cost and provides flexibility (Erol-Kantarci and Mouftah, 2012c).

Mission assignment in MAPIT works as follows. MAPIT assumes that sensors share the missions that are expected to be performed by the sensor network. A sensor that is close to a mission ideally participates in that particular mission, while other missions are handled by the other sensor nodes. For instance, a group of sensor nodes may be engaged in target tracking since a target may be in the event sensing range of those sensors, while others may measure ambient temperature, etc. A sensor participating in the target tracking mission may be able to provide more accurate measurements based on its proximity to the event than the other nodes. This means selecting a sensor close to the monitored phenomena increases the profit achieved by the target tracking mission. This notion can be generalized to most sensing missions.

MAPIT is formulated as an ILP model with the objective function given in Eq. (2). The objective function aims to maximize the number of sensors that are replenished from the selected landmarks, and to maximize the profit of the whole network by the participation of the sensors that are charged from those landmarks:

$$\text{maximize} \quad \sum_j \sum_i \sum_x \sum_y \theta_{xy}^{ij} \frac{\sigma_{ij}}{s_j} \tag{2}$$

Here θ_{xy}^{ij} is a binary variable that is equal to 1 if sensor i is charged from a landmark at (x, y) and sensor i is participating in mission j and $\theta_{xy}^{ij} = z_{xy}^i r_{ij}$. σ_{ij} is the utility of sensor i to mission j, s_j is sensing demand of mission j, and τ_S denotes the energy limit of power transmitter. The binary variables are as follows: z_{xy}^i is 1 if sensor i is receiving power from a landmark at (x, y), r_{ij} is 1 if sensor i is participating to mission j. θ_{xy}^{ij} needs to be reformulated in a linear way in order to be used in the ILP model. We use a simple linearization technique that works as follows: when a and b are binary variables and c is a product of the two variables, i.e., $c = a \cdot b$, $c \leq a$, $c \leq b$, and $c - b - a \geq -1$ holds.

In a WRSN, most of the energy is consumed during packet transmission, hence the frequency of forwarding events to the sink and relaying the packets of the neighbors determines the amount of energy required for topping up the battery of a sensor node in each charging cycle. Demand intensity quantifies the packet transmission energy requirements on each sensor node. This demand intensity varies depending on whether the sensor is located on a busy path toward the sink or not. In the ILP formulation, we denote the demand intensity as E_i. On the other hand, the energy supply of the charger is limited by the capacity of its battery. We assume that at each landmark, the power transmitter is able provide τ_s units of energy. This energy should be greater than or equal to the total demand intensity of the sensors receiving power from the transmitter when it is at that landmark. Denoting the landmark position as (x, y), this constraint is formulated by Eq. (3):

$$\sum_i z_{xy}^i E_i \leq \tau_S \quad \forall x, y \tag{3}$$

We assume that a sensor node i is allowed to receive power from one and only one landmark location, which is assured by Eq. (4):

$$\sum_x \sum_y z_{xy}^i = 1 \quad \forall i \tag{4}$$

We consider that landmarks are selected such that when the transmitter parks at the landmark, it can transmit power to at least one sensor node. Eqs. (5) and (6) ensure that a transmitter located at (x, y) is able to transmit power to at least one sensor where δ_{xy} is a binary variable that is 1 if there is a landmark located at (x, y) and δ_{xy}^i is 1 if the distance between the landmark at (x, y) and sensor i is less than the transmitter's power transmission range R_c:

$$z_{xy}^i \leq \delta_{xy} \quad \forall i, x, y \tag{5}$$

$$\delta_{xy} \leq \delta_{xy}^i \quad \forall i, x, y \tag{6}$$

The maximum number of landmarks is limited to N_l as ensured by Eq. (7):

$$\sum_x \sum_y \delta_{xy} \leq N_l \tag{7}$$

MAPIT allows a sensor i to participate in one and only one mission at a time. This constraint is formulated by Eq. (8):

$$\sum_j r_{ij} \leq 1 \quad \forall i \tag{8}$$

Furthermore, the utility provided by the sensor nodes should be able to satisfy the requirement of the mission as given by Eq. (9). We assume that missions are accomplished only when the participating sensors have adequate resources for the mission. In the literature, partial satisfaction of missions has also been considered (La Porta et al., 2011):

$$\sum_i r_{ij} \sigma_{ij} \geq s_j \quad \forall j \tag{9}$$

3.2 DIRECT RF ENERGY HARVESTING IN HETEROGENOUS NETWORKS

Recently, Wireless Power Harvesting for Cell Phones (2009) and Huang and Lau (2014) have considered RF energy harvesting for cellular networks where mobile user devices harvest energy from base stations. This is also very relevant to IoT devices as they may receive power from small cell base stations. It also can be useful in CPSs where sensors and actors moving in a cell coverage area can top their batteries up. Nokia researchers have showed that scavenging energy from base stations, TV broadcasts, and WiFi can provide several milliwatts of power, which may be used toward stand-by (Wireless Power Harvesting for Cell Phones, 2009). To operate a user equipment (UE) device or to place a call, more power would need to be harvested.

In Huang and Lau (2014), the authors have proposed to utilize power beacons (PBs), which are omnidirectional (isotropic) power emitting antennas in a cellular network's coverage area. The deployment of PBs under an outage constraint has been explored by the authors. A different perspective has been suggested in Zhou et al. (2013), for simultaneously transferring information and energy. Simultaneous transfer is challenging due to the harmonics coming from the information transfer; in addition, the intermittence of information transfer impacts energy harvesting.

In this section, we introduce direct RF energy harvesting with the help of dedicated antennas. We call these dedicated energy-transmitting towers (ETTs). In a HetNet, a mix of macro and small cell base stations (BSs) are available and power is scavenged from the already existing small cell base stations in addition to the ETTs. The proposed ILP model selects active base stations and ETTs in an optimal way. These ideas have been initially presented in Erol-Kantarci and Mouftah (2014a,b).

Fig. 1 shows how dedicated antennas and small cell base stations contribute to energy harvesting by UEs in a HetNet. Efficiency of wireless power transfer increases as the number of users in the cell increases, due to the broadcast nature of wireless power transfer when omni-directional antennas are used. We call a power dissipating pico cell base station (PBS) an active PBS.

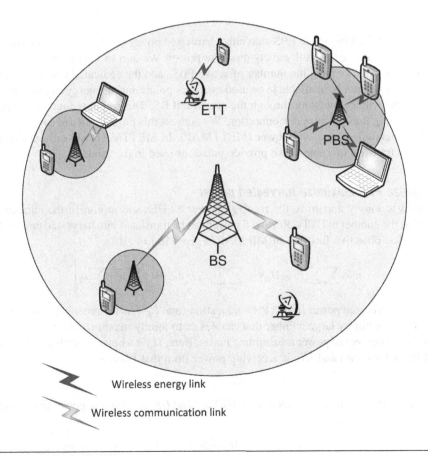

FIG. 1

Wireless energy transfer in HetNets.

Table 1 Notations

i	UE identifier
j	Location identifier
Π_{ij}	Binary variable that is 1 if there is an active PBS at j and UE_i is receiving power from that PBS
R_j	Binary value denoting the location of a PBS
Δ_{ij}	Binary variable that is 1 if there is an ETT at j and UE_i is receiving power from that ETT
γ_{ij}	Binary variable that is 1 if UE_i is receiving power from the PBS at j
θ_{ij}	Binary variable that is 1 if UE_i is receiving power from the ETT at j
μ_j	Binary variable that is 1 if there is an ETT at j
a_j	Binary variable that is 1 if there is an active PBS at j
d_{i,PBS_j}	Distance of UE_i to PBS_j
d_{i,ETT_j}	Distance of UE_i to ETT_j
R_{PBS_j}	Power transmission range of PBS_j
R_{ETT_j}	Power transmission range of an ETT_j

In one scheme called minimize PBS-maximize harvested power (MIPMAP), we consider that active PBSs and dedicated ETTs collectively dissipate power. We aim to maximize the total received power by the UEs while keeping the number of active PBSs and the dedicated towers to a minimum. Since active PBSs will not be available to be used as access points during energy transfer, the UEs will need to access the wireless network through the macro cell BS. This might reduce the capacity or, in some cases, this might even cause disconnection. We address this problem in a second scheme called minimize ETT-maximize harvested power (METTMAP). In METTMAP, we only utilize ETTs for wireless energy transfer. In Table 1, we provide notations used in the models.

3.2.1 Minimize PBS-maximize harvested power

MIPMAP aims to jointly maximize the received power by UEs and minimize the number of active PBSs as well as the number of ETTs. Received power can be translated into harvested power after some circuitry loss. The objective function of MIPMAP is given in Eq. (10):

$$\max \sum_i \sum_j P_{PBS_j} \Pi_{ij} R_j + \sum_i \sum_j P_{ETT_j} \Delta_{ij} - M \left[\sum_j \mu_j + \sum_j a_j \right] \qquad (10)$$

where P_{PBS_j} is the received power from a PBS at location j and P_{ETT_j} is the received power from ETT at location j. M is an arbitrary large number that enables us to jointly maximize the received power and minimize the total number of power transmitting nodes. Here, Π_{ij} is a binary variable that is 1 if there is an active PBS at location j and UE_i is receiving power from that PBS, ie:

$$\Pi_{ij} = \gamma_{ij} a_j \qquad (11)$$

and Δ_{ij} is a binary variable that is 1 if there is an ETT at j and UE_i is receiving power from that ETT that is given by:

$$\Delta_{ij} = \theta_{ij} \mu_j \qquad (12)$$

MIPMAP is solved subject to the following constraint set:

$$\gamma_{ij}d_{i,PBS_j} \leq R_{PBS_j}R_j, \quad \forall i,j \tag{13}$$

$$\theta_{ij}d_{i,ETT_j} \leq R_{ETT_j}, \quad \forall i,j \tag{14}$$

The above equations guarantee that a UE can receive power from a PBS or an ETT when it is within the energy transmission range, which is denoted by R_{PBS_j} for a PBS and R_{ETT_j} for an ETT. For simplicity we assume the maximum transmission power of PBSs are identical, thus their range is identical. We assume that the ETTs' maximum transmission power is lower than that of the PBSs, and that the range is identical for all ETTs. R_j denotes whether there is a pre-deployed PBS at j or not.

$$\sum_i \gamma_{ij} \geq a_j, \quad \forall j \tag{15}$$

$$\sum_i \theta_{ij} \geq \mu_j, \quad \forall j \tag{16}$$

Eqs. (15) and (16) ensure that if a PBS is active there is at least one UE within its range and an ETT is deployed if there is at least one UE that can receive power within its range, respectively:

$$\gamma_{ij} \leq a_j, \quad \forall i,j \tag{17}$$

$$\theta_{ij} \leq \mu_j, \quad \forall i,j \tag{18}$$

Eq. (17) ensures that a UE can receive power from a PBS if it is active and Eq. (18) ensures that a UE can receive from an ETT if there is an ETT at location j:

$$\mu_j + R_j \leq 1, \quad \forall j \tag{19}$$

An ETT and a PBS cannot be practically placed at the same location which is denoted by Eq. (19):

$$\sum_j a_j \leq MAX_{PBS}, \quad \forall i \tag{20}$$

The total number of active PBSs are less than the available PBSs in the network, MAX_{PBS}, as given in Eq. (20):

$$\sum_j \theta_{ij} + \sum_j \gamma_{ij} = 1, \quad \forall i \tag{21}$$

A UE is allowed to receive from either a PBS or an ETT. This condition is essential to limit the number of power transmitting towers and given by Eq. (21):

$$\gamma_{ij} > \theta_{ij}, \quad \forall i,j \tag{22}$$

Eq. (22) guarantees that if a UE is the range of both a PBS and a ETT, it is associated with a PBS to avoid the extra cost of tower deployment.

The constraints (23) through (28) are linearization constraints:

$$\Pi_{ij} \leq \gamma_{ij} \quad \forall i,j \tag{23}$$

$$\Pi_{ij} \leq a_j \quad \forall i,j \tag{24}$$

$$\Pi_{ij} - a_j - \gamma_{ij} \geq -1, \quad \forall i,j \tag{25}$$

$$\Delta_{ij} \le \theta_{ij} \quad \forall i, j \tag{26}$$

$$\Delta_{ij} \le \mu_j \quad \forall i, j \tag{27}$$

$$\Delta_{ij} - \mu_j - \theta_{ij} \ge -1, \quad \forall i, j \tag{28}$$

MIPMAP optimizes energy harvesting UEs while minimizing the number of active PBSs and ETTs in a network.

3.2.2 Minimize ETT-maximize harvested power

METTMAP avoids using PBSs for power transmission since this may cause unavailability of the pico cell due to the time-sharing nature of in-band energy transfer. To address this, METTMAP assumes only dedicated ETTs are used for RF energy harvesting. The objective of METTMAP is given in Eq. (29) and it aims to maximize the power received from ETTs while keeping the number of ETTs at a minimum. Minimizing number of ETTs also enforces the selection of optimal locations, similar to how MAPIT selects best landmark locations:

$$\max \sum_i \sum_j P_{ETT_j} \Delta_{ij} - M \sum_j \mu_j \tag{29}$$

Eq. (29) is solved subject to the constraints of Eqs. (26)–(28), which are inherited from MIPMAP. Additional constraints are:

$$\theta_{ij} d_{i,ETT_j} \le R_{ETT_j}, \quad \forall i, j \tag{30}$$

$$\sum_i \theta_{ij} \ge \mu_j, \quad \forall j \tag{31}$$

$$\theta_{ij} \le \mu_j, \quad \forall i, j \tag{32}$$

$$\sum_j \theta_{ij} = 1, \quad \forall i \tag{33}$$

Eqs. (30) and (31) enforce the power transmission limits of ETTs and make sure ETTs cover at least one UE. Eq. (32) ensures that a UE can receive power from an ETT if there is an ETT at location j, and each UE receives power from one and only one ETT as given in Eq. (33).

3.3 NUMERIC RESULTS FOR DIRECT RF ENERGY HARVESTING

We evaluate the performance of MAPIT by considering a square field of 50 m × 50 m where $N_s = 30$ sensor nodes are randomly deployed. We assume that M missions can occur simultaneously in the sensor network and we vary M from 5 to 15. The landmark limit N_l varies between 10 and 20 (Erol-Kantarci and Mouftah, 2012b). All missions are assumed to require an identical amount of sensing resources, which is $s_j = 10$. We assume that the maximum battery capacity of a sensor node is 10 kJ (Shi et al., 2011). We evaluate the performance of MAPIT under various demand intensities. The battery capacity of the power transmitter is assumed to be 20 kJ. We set the wireless energy transfer range as $R_c = 4$ m. We use CPLEX optimization software to determine the optimal landmark locations by MAPIT, and we average the results obtained for all investigated values of the demand intensity.

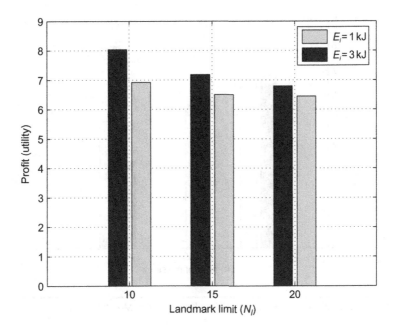

FIG. 2

Profit achieved under varying landmark limits (Erol-Kantarci and Mouftah, 2012b).

In Fig. 2, we present the profit achieved by performing the missions. The number of missions is set to $M = 5$ in this set of results. The landmark limit N_l varies from 10 to 20, and we observe the profit for two different values of energy replenishment demand of the sensors. Note that profit is defined in units of utility. The gray bar denotes the case where each sensor have 1 kJ of energy demand and the blue bar denotes 3 kJ of energy demand. When $E_i = 1$ kJ, for a landmark limit of ten, the profit is almost 7 units of utility. As the landmark limit increases, the profit reduces slightly while under fixed N_l, increasing demand intensity increases the number of sensors that receive power from a certain landmark location. When N_l is increased, the sensor nodes receive power from different landmarks. The profit is defined as a function of z^i_{xy}, which is the number of nodes receiving power from the landmark at (x, y). Hence the profit reduces as the number of sensors charged from the same landmark reduces. When $E_i = 3$ kJ, profit is higher than the case for $E_i = 1$ kJ.

In Fig. 3, we present the profit under a varying number of missions that are set between 5 and 15 and the landmark limit is $N_l = 15$. We consider two different demand profiles. For $E_i = 1$ kJ, the profit achieved from the utility of sensor nodes reduces as the number of missions increases. When there are five missions the profit is below 7 units of utility and it reduces to less than 4 units of utility for $M = 15$. As the number of missions increases, the distance between the respective missions and the sensors is reduced. Since the profit is related to the distance through the definition of utility, profit reduces as the number of missions increases. For $M = 15$ and $E_i = 3$ kJ profit drops even lower than $E_i = 1$ kJ, which can be explained as follows. Increasing the number of missions reduces the number of sensors that receive power from the same landmark location and also participate in the same missions.

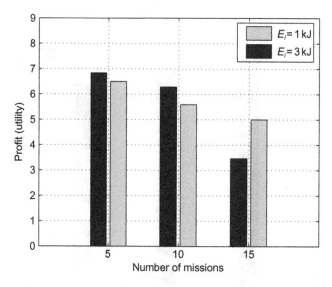

FIG. 3

Profit achieved under varying number of missions (Erol-Kantarci and Mouftah, 2012b).

To evaluate the performance of RF energy transfer in HetNets, as per the proposed techniques in Section 3.2, we consider a single cell area with 1 km radius. We initially assume PBSs are randomly deployed across the cell, and that they operate in 2.6 GHz band. The power transmission range of a PBS is assumed as 100 m while the range of an ETT is set to 50 m to avoid excessive interference. We assume the UEs are equipped with antennas with antenna gain equal to 6 dBi (which is attainable through patch antennas, Powercast Corporation, 2016). We vary the number of PBSs from 10 to 50 and the number of UEs are selected to be either 100 or 200. (In urban scenarios, the number of UEs can go as high as 200 according to the statistical data from Paul et al., 2011 and Mindspeed White Paper, 2016). UEs are randomly deployed across the cell. Our performance metrics are as follows: number of active PBSs, number of ETTs, and power received from ETTs and PBSs. We solve MIPMAP and METTMAP using CPLEX and give the averaged results over 10 runs, where each run has a different random device geometry.

In the next set of results we compare the performance of MIPMAP and METTMAP for a varying number of UEs. The number of UEs vary from 50 to 250. In Fig. 4, we present the number of deployed ETTs. METTMAP deploys more ETTs since it cannot take advantage of PBSs. In Fig. 5, we compare the received power for MIPMAP and METTMAP when MIPMAP has 10 deployed PBSs. We show that MIPMAP outperforms METTMAP in terms of delivered power.

4 RELAYED RF ENERGY HARVESTING

Relayed RF energy harvesting is a relatively new approach in which both a source node and intermediate relay node(s) transfer energy to an energy harvesting node. Proof-of-concept experimental results indicate that the energy harvesting node will be able to receive more RF energy using this relayed approach (Kaushik et al., 2013). This section demonstrates that increasing the time that an RF synthesizer

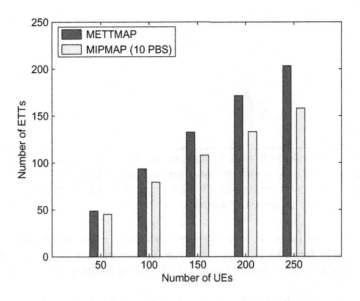

FIG. 4

Energy-transmitting towers' efficiency under varying number of user equipment (number of PBSs set to 10) (Erol-Kantarci and Mouftah, 2014a).

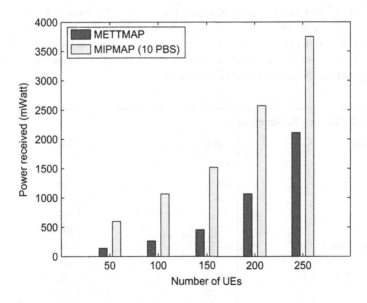

FIG. 5

Energy-transmitting towers' efficiency under varying number of user equipment (number of PBSs set to 10) (Erol-Kantarci and Mouftah, 2014a).

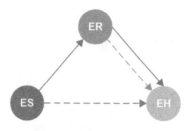

FIG. 6

Scenario under test: energy source (ES) node transmits data at 20 packets/second directly to energy harvester (EH) node (dashed blue; dark gray in print version) as well as forwarding the data through the energy relay (ER) node (solid blue; dark gray in print version). The ER has its own data stream to transmit to the EH (dashed orange; light gray in print version) at a variable data rate.

spent transmitting allowed for an increase in energy harvesting, as would be expected. A new set of simulations has been performed which expands upon the scenario modeled in Kaushik et al. (2013) by modeling relayed RF energy harvesting in a CPS with reasonably long distances between the sensor nodes.

The scenario under test is shown in Fig. 6. For proof-of-concept purposes, the wireless rechargeable network consists of three nodes separated by 5 m. The energy source (ES) node transmits data directly to the energy harvester (EH) node at a fixed packet rate. In addition to the direct path between the ES and EH, a second path may be formed through the energy relay (ER) node. A secondary path works well in many cases for protection and survivability. The ER forwards the packets of ES to the EH, and it also has its own data stream to transmit. The EH does not transmit any data back to either the ER or ES nodes.

We use QualNet to simulate network performance under realistic conditions (QualNet, 2016). The main parameters used in the simulations are summarized in Table 2. The transmitter parameters are modeled following (Kaushik et al., 2013). All three nodes are modeled as Berkeley MICA motes, using IEEE 802.11b for the physical and medium access control (MAC) layer. Sensor nodes perform periodic monitoring and generate constant bit rate (CBR) traffic.

Table 2 Selected QualNet Parameters

Physical Parameter	Value	Unit
Node separation	5.0	m
Antenna frequency	2.4	GHz
ES transmit power	+13	dBm
ES antenna gain	0	dBi
ER transmit power	+3.0	dBm
ER and EH antenna gain	6.1	dBi
Receiver sensitivity	−91.0	dBm
ES data rate	20	pkt/s
ER data rate	10–30	pkt/s
Packet size	512	B

4.1 NUMERIC RESULTS FOR RELAYED RF ENERGY HARVESTING

Simulations were performed with three different packet rates and two different packet sizes. For a packet size of 512 bytes, packet rates of 10, 20, and 30 packets per second were explored for the ER. An additional simulation was performed for a packet size of 4096 bytes using a rate of 30 packets per second for the ER in order to explore performance under high network utilization. In both cases, the ES packet rate was fixed at 20 packets per second. All conditions were run for 10 independent trials simulating 1 hour of network activity. As network utilization increases, the amount of RF energy available for the EH node to harvest should also increase.

Table 3 summarizes the results from the simulations investigating energy harvesting as a function of packet rate with fixed packet size of 512 bytes, where all data are presented as mean $\pm$ standard deviation. Note that the network is significantly underutilized in these conditions, and as a result virtually all packets destined for the EH node were successfully received. The high packet delivery ratios indicate that the extra packets being generated by the ER are not adversely affecting network performance under these conditions.

RF power profiles for power received at the EH for each packet rate are shown in Fig. 7 through Fig. 8 using a minute-by-minute time resolution. Each marker is the mean value of 10 independent trials, with error bars representing the standard deviation. Fig. 7 shows results for the case $R = 10$ pkts/s, with the average power harvested per minute at approximately 8.1 μW. A larger value is present in the first minute, as the EH is able to harvest RF power from the burst of traffic associated with the network discovery and initialization operations; this also implies that the channel utilization during initialization actually exceeds the utilization during regular activity for this configuration. A similar trend is seen in Fig. 8, for the $R = 30$ pkts/s where the average power received per minute has risen to approximately 9.2 μW. These profiles confirm that more RF power can be harvested at higher packet rates.

Table 3 also summarizes the energy harvesting results. These show that the energy harvested increases as the packet rate increases, as would be expected since the amount of packets transmitted, i.e., RF energy transmitted, increases as a function of packet rate. Energy consumed by the EH in these scenarios ranges from about 65 to 70 J, while energy harvested ranges from 21 to 30 J. While there is still a net energy deficit at the EH node, these results show that the relayed RF energy harvesting should extend the EH node's lifetime by 32–44% depending on packet rate.

Table 3 Packet Rate Simulation Results (512 bytes)

Parameter	10 pkts/s	20 pkts/s	30 pkts/s
Network utilization	12%	27%	39%
Packet delivery ratio (%)	100.00 ± 0.00	99.9999 ± 0.0006	99.9999 ± 0.0003
Total RF power received (mW)	29.08 ± 0.20	31.12 ± 0.19	32.93 ± 0.20
Time receiving (s)	726.90 ± 0.46	825.69 ± 0.42	926.61 ± 0.46
RF energy harvested (J)	21.13 ± 0.16	25.69 ± 0.17	30.51 ± 0.20
EH energy consumed (J)	65.85 ± 0.006	67.58 ± 0.006	69.37 ± 0.007
Net energy (J)	-44.72 ± 0.156	-41.89 ± 0.162	-38.84 ± 0.19
Lifetime extension	1.32	1.38	1.44

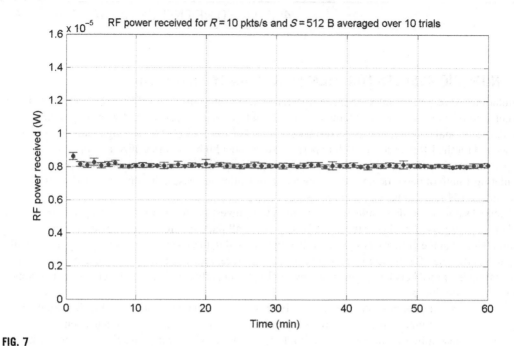

FIG. 7

Radio frequency power harvested each minute for 10 pkt/s, 512 B scenario.

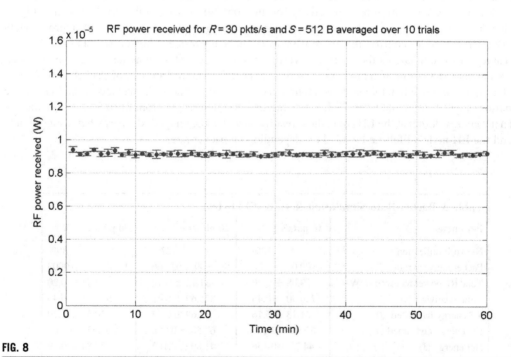

FIG. 8

Radio frequency power harvested each minute for 30 pkt/s, 512 B scenario.

Table 4 Packet Size Simulation Results (30 pkts/s)

Parameter	512 B	4 kB
Network utilization	39%	97%
Packet delivery ratio (%)	99.9999 ± 0.0003	93.12 ± 21.75
Total RF power received (mW)	32.93 ± 0.20	48.90 ± 12.91
Time receiving (s)	926.61 ± 0.46	2671.92 ± 120.28
RF energy harvested (J)	30.51 ± 0.20	132.07 ± 44.32
EH energy consumed (J)	69.37 ± 0.007	95.97 ± 2.722
Net energy (J)	−38.84 ± 0.19	+36.09 ± 41.59
Lifetime extension	1.44	2.36

Table 4 summarizes the investigation of increasing packet size while using a constant packet rate, where all data are presented as mean ± standard deviation. Here packet rate is kept fixed at 30 pkts/s, with packet size varied from 512 bytes to 4 kilobytes. The network utilization is approaching full capacity for the 4 kB packet size case, which results in more dropped packets as shown by the packet delivery ratio decreasing from 99.9999% for 512 B packets to 93% for 4 kB packets. The RF power profile for the 4 kB scenario is shown in Fig. 9, showing that as a result of the increased packet size the

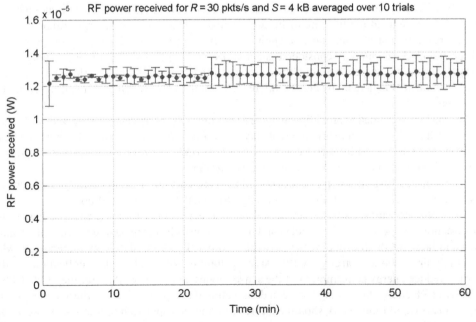

FIG. 9

Radio frequency power harvested each minute for 30 pkt/s, 512 B scenario.

averaged power harvested per minute increased to about 12 μW. Channel utilization during regular transmission now exceeds that of initialization, such that the power received in the first minute is now slightly below the average power received during the rest of the hour.

As the channel utilization has increased, the potential for RF energy harvesting has also increased as shown in Table 4. The 97% utilization in the 4 kB packet size case indicates that there is almost always traffic being transmitted through the network, so that the EH node has a greater opportunity to harvest energy. The EH node spends about 75% of the simulated hour in the receiving state for the 4 kB packet size, compared to about 25% for the 512 B packet size. Although these packet sizes are not applicable for IEEE 802.15.4, sensor networks using low-power WiFi or other standards can make use of larger packets. In the 512 B packet size scenario the EH consumed about 69 J and harvested about 30 J, leading to a potential lifetime extension of 44%. For the 4 kB packet size, due to the increased time spent in the receiving state versus the idle state, the EH now consumes about 96 J. However, the EH is potentially able to harvest 132 J, such that there is theoretically an energy surplus at the EH node. This surplus could more than double the lifetime of the EH node, showing that the benefits of relayed RF energy harvesting greatly increase as channel utilization increases, at the cost of reliable packet delivery.

5 SUMMARY AND OPEN ISSUES

RF energy harvesting is an invaluable source of energy replenishment for sensors and actors that are used in CPSs. Traditionally, energy conservation or ambient energy harvesting techniques have been widely utilized to extend the lifetime of the sensor nodes. Duty-cycling, energy-efficient medium access, energy-efficient routing and scavenging energy from sun, wind, mechanical motion, etc., have been among the commonly considered techniques for this purpose. Nevertheless, none of those answer the long-term requirements for CPSs, which can have many-year monitoring and control missions, and which are challenged by energy-intensive communications demands resulting from dense deployments.

In this chapter, we aimed at providing an understanding of RF energy harvesting, and we presented recent research results on the topics that might play a significant role in extending the lifetime of CPSs. We presented schemes used for direct energy transfer in wireless sensor networks and HetNets, and demonstrated preliminary results on relayed energy transfer.

RF energy harvesting for low-power devices of CPSs is in its infancy. There are many open issues in this area. In wireless energy transfer, losses due to signal attenuation and power conversion are inevitable. Effort toward minimizing these losses is an active field of research. Another open research direction is to design medium access and routing protocols that make the best use of RF energy transfer and increase the lifetime of the network without degrading data communications. Meanwhile, combining mobile charging and mission awareness with mobile data collection solutions can further reduce energy consumption of the sensor nodes and improve the lifetime of the CPS. Finally, relayed RF energy harvesting has been demonstrated to extend network lifetime in a proof-of-concept scenario, and future work should explore applying this approach to realistic network topologies and applications.

REFERENCES

Anastasi, G., Conti, M., Di Francesco, M., Passarella, A., 2009. Energy conservation in wireless sensor networks: a survey. Ad Hoc Netw. 7 (3), 537–568.

Erol-Kantarci, M., Mouftah, H.T., 2012a. DRIFT: differentiated RF power transmission for wireless sensor network deployment in the smart grid. In: Proceedings of IEEE GLOBECOM Workshop on Smart Grid Communications: Design for Performance, Anaheim, CA, December 3–7, 2012, pp. 1491–1495.

Erol-Kantarci, M., Mouftah, H.T., 2012b. Mission-aware placement of RF-based power transmitters in wireless sensor networks. In: Proceedings of IEEE Symposium on Computers and Communications (ISCC), Cappadocia, Turkey, July 1–4, 2012, pp. 12–17.

Erol-Kantarci, M., Mouftah, H.T., 2012c. SuReSense: sustainable wireless rechargeable sensor networks for the smart grid. IEEE Wirel. Commun. Mag. 19 (3), 30–36.

Erol-Kantarci, M., Mouftah, H.T., 2014a. Radio-frequency-based wireless energy transfer in LTE-A heterogeneous networks. In: Proceedings of IEEE Symposium on Computers and Communications (ISCC), June 2014.

Erol-Kantarci, M., Mouftah, H.T., 2014b. Challenges of wireless power transfer for prolonging user equipment (UE) lifetime in wireless networks. In: Proceedings of IEEE International Symposium on Personal, Indoor and Mobile Radio Communications (PIMRC)—Workshop on Current Challenges for Wireless Power Transfer, Washington, DC, USA.

Gollakota, S., Reynolds, M.S., Smith, J.R., Wetherall, D.J., 2014. The emergence of RF-powered computing. IEEE Comput. Mag. 47 (1), 32–39.

Guturu, P., Bhargava, B., 2011. Cyber-physical systems: a confluence of cutting edge technology streams. In: International Conference on Advances in Computing and Communication, April 2011.

He, S., Chen, J., Jiang, F., Yau, D.K.Y., Xing, G., Sun, Y., 2011. Energy provisioning in wireless rechargeable sensor networks. In: Proceedings of IEEE INFOCOM, Shanghai, China, April 2011, pp. 2006–2014.

Huang, K., Lau, V.K.N., 2014. Enabling wireless power transfer in cellular networks: architecture, modeling and deployment. IEEE Trans. Wirel. Commun. 13 (2), 902–912.

Ilic, M.D., Xie, L., Khan, U.A., Moura, J.M.F., 2008. Modeling future cyber-physical energy systems. In: IEEE Power and Energy Society General Meeting, July 2008, pp. 1–9.

Kaushik, K., Mishra, D., De, S., Basagni, S., Heinzelman, W., Chowdhury, K., Jana, S., 2013. Experimental demonstration of multi-hop RF energy transfer. In: Proceedings of IEEE 24th International Symposium on Personal Indoor and Mobile Radio Communications (PIMRC), London, UK, pp. 538–542.

La Porta, T., Petrioli, C., Spenza, D., 2011. Sensor-mission assignment in wireless sensor networks with energy harvesting. In: Proceedings of 8th Annual IEEE Conference on Sensor, Mesh and Ad Hoc Communications and Networks (SECON), June 27–30, 2011, pp. 413–421.

Lee, E.A., Seshia, S.A., 2011. Introduction to Embedded Systems—A Cyber-Physical Systems Approach. LeeSeshia.org.

Lu, S., Gopalakrishnan, S., Liu, X., Wang, Q., 2008. Cyber-physical systems: a new frontier. In: IEEE International Conference on Sensor Networks, Ubiquitous and Trustworthy Computing, June 2008, pp. 1–9.

Mindspeed White Paper, 2016. Estimation of Potential Deployment of LTE Small Cell Base Stations in 2015. [Online]. www.mindspeed.com/assets/001/36058.pdf.

Mohammady, R.D., Chowdhury, K., Di Felice, M., 2010. Routing and link layer protocol design for sensor networks with wireless energy transfer. In: IEEE GLOBECOM, Miami, December 2010.

Paul, U., Subramanian, A.P., Buddhikot, M.M., Das, S.R., 2011. Understanding traffic dynamics in cellular data networks. In: IEEE INFOCOM, April 2011, pp. 882–890.

Powercast Corporation, 2016. [Online]. http://www.powercastco.com/.

QualNet, 2016. Scalable Network Technologies. http://web.scalable-networks.com/content/qualnet.

Rajkumar, R., Lee, I., Sha, L., Stankovic, J., 2010. Cyber-physical systems: the next computing revolution. In: Proceedings of the ACM/IEEE Design Automation Conference (DAC), June 2010, pp. 731–736.

Shi, L., Xie, L., Hou, Y.T., Sherali, H.D., 2011. On renewable sensor networks with wireless energy transfer. In: Proceedings of the IEEE INFOCOM, Shanghai, China, April 10–15, 2011, pp. 1350–1358.

Sudevalayam, S., Kulkarni, P., 2011. Energy harvesting sensor nodes: survey and implications. IEEE Commun. Surv. Tutorials 13 (3), 443–461.

Wireless Power Harvesting for Cell Phones, 2009. [Online]. http://www.technologyreview.com/news/413744/wireless-power-harvesting-for-cell-phones/. (accessed July 2016).

Witricity Corp, 2016. http://witricity.com/ (accessed July 2016).

Wolf, W., 2009. Cyber-physical systems. Computer 42 (3), 88–89.

Zhou, X., Zhang, R., Ho, C.K., 2013. Wireless information and power transfer: architecture design and rate-energy tradeoff. IEEE Trans. Commun. 61 (11), 4757–4767.

MACHINE-TYPE COMMUNICATIONS OVER 5G SYSTEMS: CHALLENGES AND RESEARCH TRENDS FOR SUPPORTING INDUSTRIAL CPS APPLICATIONS

5

M. Condoluci*, M. Dohler*, G. Araniti[§]

King's College London, United Kingdom[*] University Mediterranea of Reggio, Calabria, Italy[§]

1 INTRODUCTION

One of the key goals for the next-to-come fifth generation (5G) wireless systems is the natively support of a novel and an unprecedented communication paradigm, referred to as *machine-to-machine (M2M)*. The M2M communications refer to scenarios where machines (sensors, actuators, etc.) are connected to each other (or to remote servers) and communicate without (or with a minimal) human interaction. The amount of M2M traffic is constantly growing and, as analyzed by Goncalves and Dobbelaere (2010), has a promising economic and strategic value in the 5G mobile market ecosystem as M2M communications open unprecedented business models to telco operators.

M2M is expected to play a crucial role in the effective deployment and development of *cyberphysical systems (CPS)*. In detail, CPS technology will transform the way people interact with engineered systems and will drive innovation and competition on a large scale in different domestic (smart homes, remote surveillance, etc.) and industrial (e.g., transport and logistics, smart power grids, agriculture, automation, healthcare, and manufacturing) fields. A detailed analysis of the role of M2M ecosystem is given by Palattella et al. (2016). The advances of CPS aim to enable *capability, adaptability, scalability, resiliency, safety, security,* and *usability* that will far exceed the simple embedded systems of today. In this scenario, M2M communications represent a viable solution to meet the expected benefits of CPS also in critical industrial scenarios with strict requirements in terms of reliability and security.

The effective provisioning of M2M applications in industrial environments is still challenging and this limits the effectiveness of developing M2M-based CPS solutions. Nowadays, industrial M2M applications typically adopt low-power wireless (e.g., WirelessHART and ISA100.11a) solutions, which are designed to offer long lifetimes. Unfortunately, as discussed for instance by Olyaei et al. (2013), the main drawback is related to the fact that, due to the short-range of these links, previously mentioned wireless solutions require the adoption of mesh networks which, even with the best design available today, seriously jeopardize end-to-end reliability and delay. This, consequently, limits the reach of the expected benefits of CPS.

To circumvent the previously mentioned issues, cellular communications have recently been taken into consideration by standardization bodies, research communities, and industries handling M2M communications. In particular, as mentioned by Astely et al. (2013), the third Generation Partnership Project (3GPP) is aiming to standardize viable solutions to effectively support *machine-type communications (MTC)* over long-term evolution (LTE), LTE-advanced (LTE-A) and beyond cellular systems. The interest of 3GPP is to design a viable MTC solution guaranteeing *low-complexity*, *low-cost*, *low-energy consumption*, and *short delays*. In this scenario, several challenges still need to be considered and solved. Firstly, due to the high (and unpredictable) number of MTC devices expected to simultaneously access the cellular network, congestion and overloading of radio access and core networks are the prime issues to be solved in order to guarantee low-latency and low-energy MTC. Another important issue is related to the observation that many machines are geographically located in a very confined and coverage-limited area, in particular when considering industrial CPS environments. As a consequence, the radio access network (RAN) of LTE/LTE-A and beyond networks should be able to efficiently manage several hot-spots, many of which might be located in challenging positions (e.g., indoors or at the cell-edge). Another research focus is on the random access (RA) in the uplink of the machines, due to the fact that many MTC/CPS applications are mainly based on asymmetric uplink transmissions by sensors/machines to remote servers and actuators. A proper management of uplink MTC should aim to minimize the energy consumption, delay, and complexity.

The aim of this chapter is to offer an exhaustive overview on the standardization activities and research efforts in the field of MTC to support industrial CPS applications over 5G systems. Firstly, the chapter will focus on the application scenarios for MTC in industrial CPS and, according to such scenarios, the characteristics and requirements of MTC traffic will be outlined. Then, the scope of the chapter will be to critically discuss the challenges related to the congestion and overloading of radio access and core networks that MTC pose on 5G systems. Finally, the chapter will take into account the most promising proposals in literature in terms of network design and data transmission procedures (with particular attention to alarm messages) and will discuss the benefits introduced by such solutions in supporting industrial CPS applications.

2 APPLICATION SCENARIOS FOR 5G MTC IN INDUSTRIAL CPS APPLICATIONS

The current mobile market scenario demonstrates the accurateness of the predictions that in recent years saw an exponential increase in the traffic generated by sensors and similar machines. This situation is analyzed for instance by Zheng et al. (2014), who discuss the role of MTC in the future 5G ecosystem. In particular, MTC is expected to play a key role in industrial CPS applications, such as that, for instance, explored by Palattella et al. (2016). Indeed, industries have the need to automate (in the most efficient and reliable way) their *real-time monitoring* and *control processes*. Nevertheless, the

MTC traffic is characterized by unique and unprecedented traffic patterns that require substantial enhancements to current cellular systems (such as LTE and LTE-A) to achieve an effective management of MTC traffic. The characteristics of MTC traffic and its challenges in industrial CPS applications will be the focus of this section.

2.1 CHARACTERISTICS AND REQUIREMENTS OF MTC TRAFFIC

The set of MTC applications is currently evolving, and this makes it hard to give a clear overview of application scenarios for 5G MTC/CPS traffic. Nevertheless, several examples can be provided to gain an exhaustive overview. Examples are: *automotive industry*, which utilizes sensors to monitor the status of critical car components; *smart grid industry*, which monitors critical points in the power transportation and distribution networks; *smart city market*, which provides innovative services to citizens by using real-time sensory data from the streets; *logistic*, where sensors provide information about the position and additional information on some items of interest.

The applications considered above have a set of unique characteristics that are totally different from the human-type communications (HTC) typically handled by cellular networks. Indeed, while HTC is typically a bursty-based traffic where users mainly require high data rates to transmit large packets, MTC devices usually send only a few bytes. Furthermore, to save battery, MTC terminals are typically in *idle mode* (i.e., the radio interface is turned-off) and switch to *active mode* for short intervals to transmit data as quickly as possible. The features of MTC traffic are discussed by Laya et al. (2013), while Zheng et al. (2014) considers the issues of MTC traffic over LTE/LTE-A networks in detail. These aspects are summarized in Table 1.

3 NETWORK ARCHITECTURE FOR ULTRA-DENSE MTC

The number of MTC devices that simultaneously access the cellular network is expected to become high (and unpredictable), as foreseen for instance by Ericsson (2011). In these future scenarios, as discussed by Goncalves and Dobbelaere (2010), *congestion and overloading of radio access and core networks* are the prime issues to be solved in order to guarantee low-latency and low-energy MTC and to minimize the impact of MTC on these network segments. Another issue to be taken into account, and one analyzed by Astely et al. (2013), is that many machines are expected to be geographically located in a *very confined and coverage-limited area* (e.g., sensors/actuators in a hospital or a refinery).

The issues outlined earlier present several challenges that need to be considered in the proper design of MTC-oriented cellular systems. In particular, the RAN should efficiently manage several clustered devices mainly located in challenging positions (e.g., indoors or at the cell-edge). Furthermore, network drives are exacerbated by considerations such as the MTC terminals need to be low-cost (i.e., low-complexity and with limited computational capabilities) and they should have low-energy consumption.

3.1 LIMITATION OF TRADITIONAL MACRO-CELL DEPLOYMENTS

In the traditional deployment of LTE/LTE-A cellular systems, as presented by 3GPP (2014), the evolved NodeBs (eNBs)—the LTE's base station (BS)—offer control/data planes connectivity to

Table 1 A Summary of the Features and Challenges of MTC Traffic (Laya et al., 2013)

	Features	Challenges
Traffic direction	• Mainly uplink data to report sensed information • Symmetric uplink/downlink transmissions are needed for some applications to allow dynamic interaction between sensors and actuators	• The uplink direction becomes overloaded • Also in case of asymmetric uplink/downlink configurations, adequate resources for both directions are needed
Message size	• Generally very small (e.g., few bits of the reading of a meter, or even just 1 bit to inform of the existence/absence of a given event)	• High overhead (i.e., the amount of control bits needed to transmit a small data is close to the amount of data bits)
Connection and access delays	• Mainly applications based on duty-cycling, i.e., devices sleep and just wake up to transmit data on a trigger/periodic basis • The connection delays should be very short to guarantee quick access to the network and low-energy consumption in active mode	• The access delay is drastically influenced by the number of accessing devices
Transmission periodicity	• Very wide range of alternatives	• Semipersistent scheduling should take into account the different transmission periodicity/priority of MTC applications
Mobility	• For a large ratio of applications, mobility is not a major concern (i.e., fixed MTC devices)	• In case of mobile sensors (e.g., sensors of cars or trains), ad-hoc solutions are needed to optimize the paging and handover mechanisms
Information priority	• Some applications involve the transmission of critical information (such as critical and emergency alarms) thus requiring very high-priority	• Priority should be considered in access procedures to guarantee differentiation in access delays
Load of devices	• Higher than in human-based communications • Hundreds or thousands of devices per connection point is expected	• The network has to manage a huge amount of devices which may wake up simultaneously or in a very short interval times
Lifetime and energy consumption	• As longer as possible • Some devices may require to operate for years or decades without maintenance	• Access and data transmission procedures should be optimized to reduce the energy consumption by cutting the delays and the overhead

LTE-equipped devices. The eNBs are directly attached to two entities of the core network (a.k.a. System Architecture Evolution, SAE): the Mobility Management Entity (MME) and the Serving Gateway (S-GW). The former is a control-plane unit that handles authentication, authorization, handovers, and idle/connected switching operations. The S-GW is a user plane entity that manages routing/forwarding of data packets to/from the eNBs.

The exploitation of macro-cell eNBs (MeNBs) may involve severe limitations when considering MTC traffic, as analyzed by Condoluci et al. (2015). In the RAN, this is mainly due to the challenging locations of MTC devices within the coverage area of the MeNB: this deteriorates the performance in terms of latency and the number of devices properly managed by the MeNBs. In the core network, the traditional macro-cell deployments suffer from several inefficiencies in terms of MME/S-GW over-load when the number of MeNBs (as well as the number of connections per MeNB) becomes large. This aspect is drastically challenging as the overload of MME/S-GW may involve a performance degradation in a large portion of the network (i.e., the overload caused by a single MeNB may affect the performance of a large set of MeNBs due to the MME/S-GW's slowdown caused by the overload).

3.2 SMALL CELL-BASED SCALABLE NETWORK ARCHITECTURE

The limitations of traditional macro-cell deployment in MTC environments dictate a novel network design able to efficiently support the unique patterns of MTC traffic. With this aim, Condoluci et al. (2015) proposed a small cell-based 3GPP architecture, as depicted in Fig. 1. The key idea is to exploit the benefits of local-area access, as for instance discussed by Andrews et al. (2012) and Sun et al. (2012) in scenarios relevant to human traffic. In the MTC system, the idea proposed by Condoluci et al. (2015) is that LTE-capable MTC devices communicate directly with home-eNodeBs (HeNBs), i.e., low-cost femtocells for local-area access with low-power transmission (<100 mW) capability. Such HeNBs are managed by the HeNB-Gateway (HeNB-GW).

In the architecture detailed in Fig. 1, the HeNB-GW has the key role of being a concentrator of several HeNBs for both control and user planes. This has obvious advantages in terms of *scalability, capacity* and *load reduction* in the core network. Indeed, the exploitation of HeNBs, instead of MeNBs, with the joint use of HeNB-GW achieves the following goals:

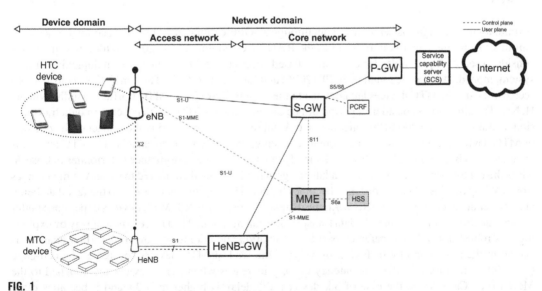

FIG. 1

Enhanced network architecture proposed by Condoluci et al. (2015) for ultra-dense MTC access to the 3GPP LTE/LTE-A core via HeNBs/HeNB-GWs.

- *Separation of MTC and human-type traffic.* HeNBs may be installed by individual/company/industry customers whose main aim is to interconnect their own MTC devices in specific areas. MeNBs are instead exploited by the network provider for HTC traffic, which are consequently not affected by MTC traffic load.
- *Reduced intra- and inter-cell interference.* As addressed by Astely et al. (2013) for human-type traffic, traffic separation between macro and femto-cell can be obtained through *frequency-separated deployment* (for instance, 2 GHz for MeNBs and 3 GHz for HeNBs); this allows the avoidance of inter-cell interference. Intra-femto interference cancellation in scenarios with multiple deployed femtocells can be properly managed by the proposals in literature such as the one of Sun et al. (2012).
- *Closed access.* Access to HeNB(s) is only given to those machines that belong to a given closed subscriber group. According to this solution, *security* is guaranteed as a customer can admit only its own trusted devices through its own HeNB(s).
- *Coverage extension.* HeNBs represent a viable solution to extend the coverage to MTC devices located in challenging positions (e.g., rural areas, indoor-deployments, smart meters in the basements of the buildings) without requiring network re-planning.
- *One stream control transmission protocol (SCTP) connection in the control-plane between the HeNB-GW and MME.* This is due to the fact that, from a core network point of view, the HeNB-GW acts as a concentrator of HeNBs. As a consequence, if the number of HeNBs grows, *the overload at the MME is avoided* as only the HeNB-GW sends SCTP heartbeat messages to the MME.
- *Reduction of GPRS tunneling protocol (GTP) and UDP/IP connections between HeNB-GW and S-GW.* In this way, the *S-GW scalability* is granted: the number of HeNBs may increase without involving an increase in the number of UDP/IP paths and GTP echo messages managed by the S-GW.

A simulation campaign has been carried out through a 3GPP-calibrated system level simulator, with the aim of showing the benefits introduced to the RAN by Condoluci et al. (2015) solution. Simulations consider a set of 54 preambles for contention-based access in a 5 MHz time division duplexing (TDD) system, more details can be found in 3GPP (2006) and Laya et al. (2013). The simulations consider two scenarios: *Case A*, MTC devices are attached to the MeNB; *Case B*, the MTC traffic is handled via the HeNB. The simulation considers a period of 60s during which MTC terminals perform data transmission, composed of one 200 bytes long message. With the aim of considering the impact of the position of MTC devices, results show the performance when varying the mean distance of 1 k MTC terminals from the MeNB (left-side plot in Fig. 2). It clearly emerges that the performance deteriorates in Case A, while the exploitation of HeNB allows a latency gain to be obtained compared to Case A, which ranges from 14% up to 30% when the distance becomes higher. The improvement is due to the fact that, being at a shorter distance from the HeNB w.r.t. the distance from the MeNB, MTC devices experience better channel qualities and consequently data transmission can be handled in a more efficient way by exploiting less robust transmission parameters (i.e., modulation and coding schemes). A further analysis is shown in the right-side plot of Fig. 2, by varying the load of MTC devices located at the cell-edge (i.e., 800 m from the MeNB). The latency quickly increases when MTC devices are attached to the MeNB (i.e., Case A): in the case of 3 k devices, the delay is higher than 2 s and it becomes close to 14 s for the heavy case of 30 k machines. On the contrary, the exploitation of HeNB (i.e., Case B) guarantees a delay of about 60 ms until the case of 20 k machines.

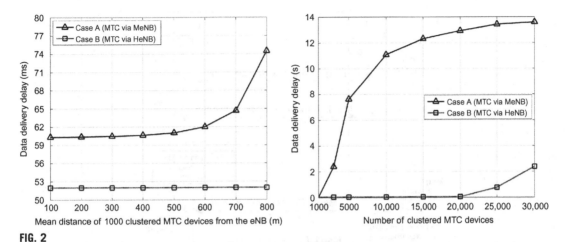

FIG. 2

Latency of MTC traffic by varying the distribution *(left)* and the load *(right)* of MTC devices.

4 THE RA FOR MTC

The RA procedure, as for instance discussed in depth by Laya et al. (2013), is considered the most challenging issue related to MTC traffic. Indeed, the 3GPP standardization body is focusing on the design of improvements for the access mechanisms of cellular systems, especially when the number of subscribers increases to tens of thousands per cell. The 3GPP RA procedure with its limitations and the related studies in literature are summarized in this section.

4.1 THE 3GPP RA

The access procedure defined by 3GPP (2013) is performed by MTC devices in the following situations:

- Upon initial access to the network
- For the reception/transmission of new data in case the device is not synchronized
- Upon transmission of new data in case of no scheduling request resources are configured on the uplink control channel
- During handover (i.e., change of associated BS) to avoid a session drop
- For connection reestablishment after a radio link failure.

The *contention-based 3GPP-RACH* procedure (depicted in Fig. 3) starts with the transmission of a preamble (Msg1) on the Physical Random Access Channel (PRACH). The PRACH is a periodic sequence of reserved uplink time-frequency resources (a.k.a. RA slots) whose periodicity is broadcasted by the BS in the PRACH Configuration Index. The preamble is randomly chosen among a predefined set of orthogonal pseudo-random preambles. A collision occurs if two or more MTC devices transmit the same preamble in the same RA slot. If Msg1 is successfully decoded, the BS sends the Random Access Response (RAR, a.k.a. Msg2); the RAR contains information about the detected preamble, uplink timing alignment, and the grant for the transmission of the Connection Request (Msg3) on the Physical

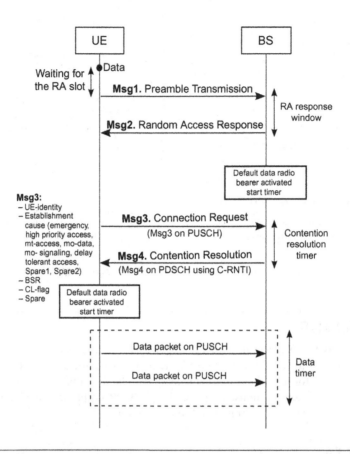

FIG. 3

The legacy 3GPP-RACH procedure.

Uplink Shared Channel (PUSCH). An undetected collision of preambles could also involve an Msg3 collision. The Msg3 also lists the device identifier and the buffer state report useful to the BS for scheduling the following data transmission. Finally, a Contention Resolution message (Msg4) terminates the RA procedure and confirms the grant for the subsequent data transmission on PUSCH.

4.1.1 Limitations of 3GPP RA

The contention-based 3GPP-RACH is an *ALOHA-based radio access*, where devices start the access procedure in the first available opportunity; this could involve performance degradations such as the high probability of collision in the case of a huge load of simultaneous access requests. To better understand this aspect, 3GPP (2011) performed some studies on the capacity limitations of the LTE RACH. By considering that RA slots are available every 5 ms and 54 preambles are used for contention-based access, the system offers 200 access opportunities per second, i.e., a capacity of about 11 k preambles per second. By the way, it is worth noting that this number represents the absolute *maximum capacity*, i.e., the capacity in case of absence of collisions; in real scenarios, the effective capacity is severely reduced. Furthermore, it is also worth mentioning that preamble collisions require that colliding devices perform a novel RACH procedure and this involves additional access delays (which may be unacceptable for particular industrial CPS applications) and, consequently, battery consumption (which limits the lifetime of MTC devices).

The works in literature which aim at overcoming the limitations of 3GPP-RACH are surveyed in the following of this section. Detailed discussion about the start of the art on RA procedure is given by Laya et al. (2013).

4.2 STATE OF THE ART ON RA PROCEDURE

The recent advances that aim to boost the performance of 3GPP-RACH in terms of access delay, energy consumption and overhead can be summarized as follows:

- *Optimized medium access control (MAC)*. Cheng and Wang (2010) suggested transmitting the data embedded into the RA process by attaching data to either the preambles (i.e., Msg1) or the Msg3 of the RA process. This option may significantly reduce the amount of control information exchanged between the MTC devices and the BS.
- *Access class barring (ACB)*. It is defined by 3GPP (2012) with the aim of introducing prioritization in the RA. In the case of network overload, the BS transmits a set of ACB parameters (in particular, a probability factor and the barring timer relevant to the predefined ACB classes). Accessing devices will draw a random number; if this number is lower than the probability factor, the device is able to attempt access, otherwise the access is barred and the device performs a random backoff time (according to the related barring timer value) before scheduling the preamble transmission. The ACB approach may guarantee short access delays to high-priority devices at the expense of a higher delay for other devices. Exhaustive studies on ACB in literature are summarized by Laya et al. (2013).
- *Virtual resource allocation*. Cheng et al. (2011), with the aim of avoiding the negative impact that massive MTC traffic may involve on HTC, proposed splitting the available RA resources into two subsets, one reserved for HTC and the other one for MTC devices. This can be achieved by splitting the set of available preambles or by allocating different RA slots to HTC and MTC devices. In contrast, Lee et al. (2011) considered assigning a restricted portion of RA resources only to MTC devices, while HTC terminals continue to exploit the whole set of available resources. The main issue of this approach is that the introduced benefits are limited and the access delay may increase due to the fact that available RA resources are severely reduced for MTC devices.
- *Dynamic allocation of RACH resources*. 3GPP (2010) considered allocating additional RA slots to MTC devices in the case of congestion. Although this solution is effective in coping with scenarios with huge access load, it is worth noting that the increase in the available RA slots reduces the amount of resources available for data transmission and this may involve performance degradation for both HTC and MTC traffic.
- *Backoff adjustment*. Yang et al. (2012) considered exploiting different backoff timers to introduce access delay differentiation among MTC devices. The main drawback of this scheme is the average access delay will be severely degraded without substantially improving the access probability, and this becomes more evident in cases of huge MTC load.

- *Slotted access.* 3GPP (2011) evaluated the performance of RA procedure when dedicated RA slots are defined for each MTC device; the idea is that MTC devices calculate their corresponding RA slot (based, for instance, on their identity and additional parameters broadcasted by the BS). The side effect of this approach is that, in order to allocate a dedicated RA slot per device in the case of huge load, larger RA cycles are needed thus increasing the access delays.
- *Code-expanded.* Thomsen et al. (2013) proposed a mechanism aiming at increasing the number of available access codewords available for the RA procedure. The code-expanded RA consists of transmitting a codeword composed by several preambles instead of a single preamble, as in the current 3GPP RACH. This allows the number of contention resources to be drastically increases and, therefore, the number of collisions in scenarios of huge MTC traffic to be reduced. The only noticeable negative aspect of this proposal is that, as multiple preambles need to be transmitted, the energy consumption increases.

5 TRANSMISSION OF INDUSTRIAL ALARM MESSAGES

Special classes of MTC applications, with particular reference to control industries and CPS eco-system, deal with the transmission of *alarm messages*, i.e., high-priority trigger-based data transmitted by MTC devices (usually fixed) in the case of alerts (e.g., overheating, pressure overflow). Alarm messages, as discussed for instance by Zheng et al. (2014) and Palattella et al. (2016), have unique traffic characteristics (please, refer also to Section 2.1 of this chapter), which dictate strict requirements: *reliability*, *security/authenticity*, and *low-latency*. In particular, as for instance considered by Stenumgaard et al. (2013), the latter parameter is actually considered the most challenging as alarm messages ask for very strict data latency requirements ranging from a few up to tens of milliseconds.

5.1 ENHANCED RA FOR EMERGENCY ALARM MESSAGES

As mentioned throughout this chapter, enabling an effective transmission of MTC alarm messages in industrial CPS applications would enable the design of effective MTC business models. This aspect becomes more challenging when taking into account special classes of alarm, namely *emergency alarm messages (EAMs)*, which, as analyzed by Laya et al. (2013), are related to trigger-based emergency alerts requiring an immediate action (for instance, to solve industry chain instabilities). Obviously, the strict delay requirement of EAMs drastically exacerbates the limitations of legacy 3GPP RACH. Condoluci et al. (2016) proposed a novel RA mechanism, namely the *EAM-RACH*, where the EAM is conveyed to the BS thanks to a quick and secure sequence of access preambles, generated according to the device's cyphering key.

The EAM-RACH procedure is a *two-message handshake* between the device and the BS and is based on the following features:

- The EAM is associated to an *EAM preamble sequence*, s, generated through a hash function H according to the device's key K_{CLT}. The sequence s is composed of L different preambles chosen from those (already) reserved for noncontention-based access.
- The EAM-RACH exploits a cloud-based RAN, which, according to some target values measured by sensors, switches the *PRACH periodicity to 1 ms* (i.e., RACH configuration index #14) when the

probability of having an EAM transmission increases. This allows a quicker transmission of the preamble sequence. Techniques such as the dynamic TDD, introduced by Astley et al. (2013), can be used to dynamically change the PRACH configuration.

- To guarantee security, the BS stores the granted preamble sequence for each MTC device allowed to transmit EAMs in its cell. The mapping between the UE-identity and the related preamble sequence is stored at the BS in a *hash table HT*.

The time diagram of EAM-RACH is depicted in Fig. 4. When an MTC device triggers the transmission of an EAM, it consecutively transmits in the next L RA slots the preambles of the access sequence s. Once the reception of a sequence composed of L different preambles belonging to the set of noncontention preambles, the BS performs the authentication check by comparing the received sequence with the HT entries. In the case of success, it means that the access sequence was transmitted by an authorized MTC device, and the BS transmits the Msg2 to the temporary address relevant to the last preamble in the sequence. Once an EAM reception (or several unauthorized attempts), the security keys ought to be updated for security reasons.

A simulation campaign has been carried out to assess the expected benefits of the EAM-RACH procedure proposed by Condoluci et al. (2016). The same simulation settings as in Section 3.2 are assumed. Arrival rates of the nonEAM devices are uniformly distributed over 20s. The number of preambles for noncontention-based access is set to 10 while the preamble sequence length L to 6; we consider that only one MTC device transmits EAMs in the whole simulation period. The performance comparison (shown in Fig. 5) with respect to the legacy 3GPP-RACH is in terms of delay between the EAM generation and its reception at the BS. The novel EAM-RACH scheme has the most interesting performance: the transmission of the EAM preamble sequence takes approximately 10 ms (in both Cases A and B) if the channel condition between the transmitter and the BS is good. In the case of reception failure, an additional delay (up to 26 ms in the worst case) is introduced and this leads to an average EAM delay of 15 ms in Case A, caused by occasional retransmission in the macro-cell environments. The use of small cells (Case B) leads to small coupling losses; this results in a low average delay of about 10 ms.

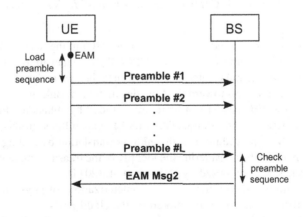

FIG. 4

The EAM-RACH procedure proposed by Condoluci et al. (2016).

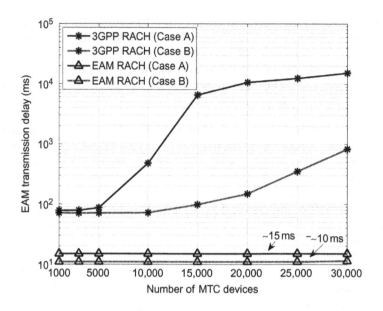

FIG. 5

Delay analysis of EAM-RACH w.r.t. 3GPP RACH -RACH.

6 FUTURE TRENDS

The design of an MTC-oriented network architecture is still to be accomplished, as several challenges are still open to further optimize the management of data sent/received by MTC devices. Indeed, the RA procedure involves a very large overhead/latency before the effective data transmission; this aspect is further exacerbated by considering that MTC terminals usually send information composed of few bits/bytes.

To enhance the performance of MTC traffic, the use of a *cloud-RAN architecture*, may yield several advantages:

- The transmission of full robust header compression (ROHC) in the data packets may be avoided. Usually, ROHC is used to compress the large TCP/UDP/IP header to avoid the transmission of large packet headers over the radio interface. The ROHC header is decompressed in the core network, to route the data packets toward the final destination. Thanks to the exploitation of cloud-RAN architecture, the ROHC header can be totally avoided by enhancing the functionalities of the cloud-RAN with network address translation (NAT) capabilities. In this scenario, the cloud-RAN may perform a NAT procedure, i.e., the address translation by adding the address of the remote server to be reached, by considering the identity of the transmitting device and the data type. A similar approach has been proposed by Dawson et al. (2014).
- Dynamic network configuration by tuning the parameterizations of physical and MAC layers according to the traffic measurement monitored by the cloud-RAN.
- Exploitation of caching capabilities to store and process information for signaling minimization and reliability maximization.

Further solutions are currently under investigation, as discussed by Dahlman et al. (2014), which proposed a *flexible network architecture* allowing core network functions to be run at (or close to) the BSs. Thus, system functionalities can be flexibly allocated to different execution nodes through software defined network (SDN) and network functions virtualization (NFV) approaches.

7 CONCLUSIONS

This chapter offered an exhaustive overview on the standardization activities in the field of MTC to support CPSs over 5G systems. It also outlined the characteristics and requirements, with regard to the related challenges, of MTC traffic, with particular attention paid to the issues to be solved when considering industrial ecosystems with strict reliability, delay and security requirements.

We discussed the drivers in terms of network design and access/data transmission procedures. In particular, we analyzed the challenges relevant to the congestion and overloading of radio access and core networks and we presented the recent advances in literature that aim to effectively and efficiently solve such issues. Furthermore, we provided an overview on the legacy data transmission procedures standardized by 3GPP and we analyzed the pros and cons of related works in literature. Finally, we focused on a communication scenario of main importance in the CPS deployment, i.e., the transmission of trigger-based alarm messages that require strict (from tens down to few milliseconds) delays. We analyzed in depth the issues posed by this particular and challenging scenario, and the most promising recent results present in the literature.

The crucial role of MTC as an enabler of industrial CPS applications was also analyzed. The study of the state of the art testified that: (i) recent advances are taking into account the requirements that CPSs demand from MTC by thus demonstrating the interest of research and standardization bodies in enabling CPS-oriented MTC over 5G systems; (ii) effective solutions are already compatible with the current LTE/LTE-A wireless systems; (iii) the further benefits in terms of network design/paradigms, such as C-RAN, SDN, and NFV expected in 5G systems may definitively open the market to CPS-oriented MTC traffic with high-performance in terms of capability, adaptability, scalability, resiliency, safety, security, and usability.

REFERENCES

3GPP, 2006. Physical layer aspect for evolved universal terrestrial radio access (UTRA). Technical Report 25.814.

3GPP, 2010. MTC simulation results with specific solutions. TSG RAN WG2 #71 R2-104662.

3GPP, 2011. Study on RAN improvements for machine-type communications. Technical Report 37.868.

3GPP, 2012. Evolved Universal Terrestrial Radio Access (E-UTRA); Radio Resource Control (RRC). Technical Specification 36.331.

3GPP, 2013. Study on enhancements to machine-type communications (MTC) and other mobile data applications; radio access network (RAN) aspects. Technical Report 37.869.

3GPP, 2014. Evolved Universal Terrestrial Radio Access (E-UTRA) and Evolved Universal Terrestrial Radio Access Network (E-UTRAN). Technical Specification 36.300.

Andrews, J.G., Claussen, H., Dohler, M., Rangan, S., Reed, M.C., 2012. Femtocells: past, present, and future. IEEE J. Sel. Areas Commun. 30 (3), 497–508.

Astely, D., Dahlman, E., Fodor, G., Parkvall, S., Sachs, J., 2013. LTE release 12 and beyond. IEEE Commun. Mag. 51 (7), 154–160.

Chen, Y., Wang, W., 2010. Machine-to-machine communication in LTE-A. In: IEEE 72nd Vehicular Technology Conference, Fall (VTC 2010-Fall).

Cheng, J.P., Lee, C.H., Lin, T.M., 2011. Prioritized random access with dynamic access barring for RAN overload in 3GPP LTE-A networks. In: IEEE GLOBECOM Workshops (GC Wkshps).

Condoluci, M., Dohler, M., Araniti, G., Molinaro, A., Sachs, J., 2016. Enhanced radio access and data transmission procedures facilitating industry-compliant machine-type communications over LTE-Based 5G networks. IEEE Wirel. Commun. 23 (1), 56–63.

Condoluci, M., Dohler, M., Araniti, G., Molinaro, A., Zheng, K., 2015. Toward 5G DenseNets: architectural advances for effective machine-type communications over femtocells. IEEE Commun. Mag. 53 (1), 134–141.

Dahlman, E., Mildh, G., Parkvall, S., Peisa, J., Sachs, J., Selén, Y., Sköld, J., 2014. 5G wireless access— requirements and realization. IEEE Commun. Mag. 52 (12), 42–47.

Dawson, A., Marina, M., Garcia, F., 2014. On the benefits of RAN virtualisation in C-RAN based mobile networks. In: Third European Workshop on Software Defined Networks (EWSDN), pp. 103–108.

Ericsson, 2011. More than 50 billion connected devices. White paper.

Goncalves, V., Dobbelaere, P., 2010. Business scenarios for machine-to-machine mobile applications. In: International Conference on Mobile Business and Ninth Global Mobility Roundtable (ICMB-GMR), pp. 394–401.

Laya, A., Alonso, L., Alonso-Zarate, J., 2013. Is the random access channel of LTE and LTE-A suitable for M2M communications? A survey of alternatives. IEEE Commun. Surv. Tutorials 16 (1), 4–16.

Lee, K.D., Kim, S., Yi, B., 2011. Throughput comparison of random access methods for M2M service over LTE networks. In: IEEE GLOBECOM Workshops (GC Wkshps), pp. 373–377.

Olyaei, B.B., Pirskanen, J., Raeesi, O., Hazmi, A., Valkama, M., 2013. Performance comparison between slotted IEEE 802.15.4 and IEEE 802.11ah in IoT based applications. In: IEEE 9th WiMob, pp. 332–337.

Palattella, M., Dohler, M., Grieco, A., Rizzo, G., Torsner, J., Engel, T., Ladid, L., 2016. Internet of things in the 5G era: enablers, architecture and business models. IEEE J. Sel. Areas Commun. 34 (3), 510–527.

Stenumgaard, P., Chilo, J., Ferrer-Coll, P., Angskog, P., 2013. Challenges and conditions for wireless machine-to-machine communications in industrial environments. IEEE Commun. Mag. 51 (6), 187–192.

Sun, Y., Jover, R.P., Wang, X., 2012. Uplink interference mitigation for OFDMA femtocell networks. IEEE Trans. Wirel. Commun. 11 (2), 614–625.

Thomsen, H., Pratas, N.K., Stefanović, C., Popovski, P., 2013. Code-expanded radio access protocol for M2M communications. Trans. Emerg. Telecommun. Technol. 24 (4), 355–365.

Yang, X., Fapojuwo, A., Egbogah, E., 2012. Performance analysis and parameter optimization of random access backoff algorithm in LTE. In: IEEE Vehicular Technology Conference (VTC Fall).

Zheng, K., Ou, S., Alonso-Zarate, J., Dohler, M., Liu, F., Zhu, H., 2014. Challenges of massive access in highly dense LTE-advanced networks with machine-to-machine communications. IEEE Wirel. Commun. 21 (3), 12–18.

ABOUT THE AUTHORS

Massimo Condoluci (massimo.condoluci@kcl.ac.uk) received his M.Sc. degree in telecommunications engineering and Ph.D. degree in information technology in 2011 and 2016, respectively, from the University Mediterranea of Reggio Calabria, Italy. He is currently a research associate at the Centre for Telecommunications Research (CTR), King's College London, UK. His current research interests include virtualization and softwarization of mobile core, functionality split, multicasting, and machine-type communications, in 5G cellular networks.

Mischa Dohler (mischa.dohler@kcl.ac.uk) is a full Professor in Wireless Communications at King's College London, Head of the Centre for Telecommunications Research, co-founder and member of the Board of Directors of the smart city pioneer Worldsensing, Fellow and Distinguished Lecturer of the IEEE, and Editor-in-Chief of the Transactions on Emerging Telecommunications Technologies. He is a frequent keynote, panel and tutorial speaker. He has pioneered several research fields and contributed to numerous wireless broadband and IoT/M2M standards, holds a dozen patents, has organized and chaired numerous conferences, has more than 200 publications, and has authored several books. He has a citation h-index of 37. He acts as policy, technology and entrepreneurship adviser, examples being Richard Branson's Carbon War Room, the House of Lords UK, the EPSRC ICT Strategy Advisory Team, the European Commission, the ISO Smart City Working Group, and various start-ups. He is also an entrepreneur, angel investor, passionate pianist and is fluent in six languages. He has talked at TEDx. He has contributed to national and international TV and radio; and these have featured on BBC News and in the Wall Street Journal.

Giuseppe Araniti (araniti@unirc.it) is an Assistant Professor of Telecommunications at the University Mediterranea of Reggio Calabria, Italy. From the same University he received the Laurea (2000) and the Ph.D. degree (2004) in Electronic Engineering. His main areas of research include personal communications systems, enhanced wireless and satellite systems, traffic and radio resource management in 4G mobile radio systems, multicast and broadcast services, and digital video broadcasting-handheld. He is a member of IEEE.

DATA RELIABILITY CHALLENGE OF CYBER-PHYSICAL SYSTEMS

D. Wang

University of Notre Dame, Notre Dame, IN, United States

1 THE AGE OF SOCIAL SENSING IN CYBER PHYSICAL SYSTEMS

1.1 OVERVIEW

A growing number of cyber-physical systems (CPSs) application domains, such as transportation, energy, sustainability, health, and disaster response, involve *humans* in nontrivial ways. An emerging application paradigm along with this trend is the use of *humans as sensors*, which is also commonly known as *social sensing*, a widely used practice in many CPS applications. For example, drivers may contribute data through their smartphones to report the state of traffic congestion at various locales. Survivors may contribute data to online social media to document the damage in the aftermath of a natural disaster. In these applications, large numbers of individuals serve as inexpensive, ubiquitous, and versatile sensors to report the states of the physical world at scale. However, a critical challenge in the context of using humans as sensors is that data sources may be unreliable. In fact, the reliability of individual observers in social sensing applications is generally unknown a priori. A common thread in CPS research focuses on the *reliability* of CPSs. Current research has mostly focused on two aspects of CPS reliability; namely, the correctness of *temporal behavior* and correctness of *software function*. In order for social sensing to become a viable component in CPS feedback loops, it is crucial to understand the correctness of collected observations from unreliable individuals as well. We call this latter challenge the *data reliability challenge*. This chapter will review recent progress and the state-of-the-art techniques developed in various communities to address the emerging data reliability challenge in CPS.

1.2 BACKGROUND AND MOTIVATION

The proliferation of a wide variety of sensors in the possession of average people (e.g., smartphones), the ubiquity of mobile Internet access (e.g., 4G and WiFi) and the advent of online social media (e.g., Twitter and Flickr) allow humans to create and disseminate a deluge of information about the physical world. This opens up unprecedented opportunities and challenges in *social sensing*, a key emerging field at the intersection of CPS and big data, where the goal is to distil accurate and credible information from large amounts of unfiltered, unstructured, and unvetted data generated by social sources (e.g., humans or devices in their possession). Little is analytically known about data validity in this new sensing paradigm, where sources are noisy, unreliable, erroneous, and largely unknown. This motivates a closer look into recent advances in social sensing with an emphasis on the key problem faced by

Cyber-Physical Systems. http://dx.doi.org/10.1016/B978-0-12-803801-7.00006-7

application designers; namely, how to distil reliable information from the big data collected from largely unknown and possibly unreliable sources. Novel solutions that leverage techniques from estimation theory, machine learning, information fusion, and data mining recently offer significant progress on this problem and are reviewed in this chapter.

In situations, where the reliability of sources is known, it is easy to compute the probability of correctness of different observations. For example, one can use, say, Bayesian analysis to fuse data from sources of different (known) degrees of reliability. The distinguishing challenge in social sensing applications is that the reliability of sources is often unknown. For example, much of the chatter on Twitter might come from users who are unknown to the data collection system. Hence, it is hard to assess the reliability of their observations. The same is true of situations where individuals download a smartphone app that allows them to contribute to a social sensing data collection campaign. If anyone is allowed to participate, the pool of sources is unvetted and the reliability of individual observers is generally unknown to the data collector. It is in this context that the problem of distilling reliable information becomes challenging. The challenge arises from the fact that one can neither identify the sources, nor immediately verify their claims. What can be rigorously said, in this case, about the correctness of collected data? More specifically, how can one jointly ascertain data correctness and source reliability in big data social sensing applications? The problem is of importance in many domains and, as such, touches upon several areas of active research.

In sensor networks, an important challenge has always been to derive accurate representations of physical state and physical context from possibly unreliable, nonspecific, or weak proxies. Often one trades off quantity and quality. While individual sensors may be less reliable, collectively (using the ingenious analysis techniques published in various sensor network venues) they may yield reliable conclusions. Much of the research in that area focused on physical sensors. This includes dedicated devices embedded in their environment, as well as human-centric sensing devices such as cell-phones and wearables. Recent research proposed challenges with the use of humans as sensors. Clearly, humans differ from traditional physical devices in many respects. Importantly in relation to the reliability analysis, they lack a design specification and a reliability standard, making it hard to define a generic noise model for sources. Each human is an individual with different model parameters that predict how good that individual person's observations are. Hence, many techniques that estimate probability of error for sensors do not apply, since they assume the same error model for all sensors.

Humans also exhibit other interesting artifacts not common to physical sensors, such as gossiping. It is usually hard to tell where a particular observation originated. Even if we are able to unambiguously authenticate the source, who we received some observation from, it is hard to tell if the source made that observation themselves, or obtained it from another. Hence, the original provenance of observations may remain uncertain. Techniques reviewed in this chapter offer analytic means to determine source reliability and mitigate uncertainty in data provenance.

In CPS research, an important emphasis has always been on ensuring validity and on proving that systems meet specifications (Rajkumar et al., 2010). The topics of reliability, predictability, and performance guarantees receive much attention. While past research on CPS addressed correctness of software systems (even in presence of unverified code), in today's data-driven world, a key emerging challenge becomes to ascertain correctness of data (even in the presence of unverified sources). The challenge is promoted by the need to account for the humans in the loop. Humans are the drivers in transportation systems, the occupants in energy management systems, the survivors and first responders in disaster response systems, and the patients in medical systems. It makes sense to utilize their input when trying to assess system state. For example, one can get a more accurate account of the current vehicular traffic situation and a more accurate prediction of its future evolution, if driver input is taken into account in some global, real-time, and automated fashion. This is assuming that the inputs are reliable, which is not always the case. The data reliability problem, if solved, would enable the development of dependable applications in domains of transportation, energy, disaster response, and military intelligence, among others, where correctness is guaranteed despite reliance on the collective observations of untrained, average, and largely unreliable sources.

Reputation system is another area of research, where source reliability is the issue. The assumption is that, when sources are observed over time, their reliability is eventually uncovered. Social sensing applications, however, often deal with scenarios, where a new event requires data collection from sources who have not previously participated in other data collection campaigns, or perhaps not been "tested" in the unique circumstances of the current event. For example, a hurricane strikes New Jersey. This is a rare event. We do not know how accurate the individuals who fled the event have been in describing the damage left behind. No reputation is accumulated for them in such a scenario. Yet, it would be desirable to leverage their collective observations to deploy help in a more efficient and timely manner. How do we determine which observations to believe?

The techniques reviewed in this chapter are also of relevance to intelligence applications, where one is interested in making sense out of large amounts of unreliable data. These techniques can thus serve applications in social networks, big data, and human-in-the-loop systems, and leverage the proliferation of computing artifacts that interact with or monitor the physical world. The goal of this chapter is to review the needed theoretical foundations that exploit advances in social sensing analytics to support emerging data-driven applications.

2 REVIEW OF THE STATE-OF-THE-ART

Social sensing has received significant attention due to the great increase in the number of mobile sensors owned by individuals (e.g., smart phones with GPS, camera, etc.) and the proliferation of Internet connectivity to upload and share sensed data (e.g., WiFi and 4G networks). A broad overview of social sensing applications is presented in Aggarwal and Abdelzaher (2013). Some early applications include CenWits (Huang et al., 2005), a participatory sensor network to search and rescue hikers in emergency situations; CarTel (Hull et al., 2006), a vehicular sensor network for traffic monitoring and mitigation; and BikeNet (Eisenman et al., 2009), a bikers sensor network for sharing cycling related data and mapping the cyclist experience. More recent work has focused on addressing the challenges of preserving privacy and building general models in sparse and multidimensional social sensing space (Ahmadi et al., 2011; Wang et al., 2011c). Social sensing is often organized as "sensing campaigns" where participants are recruited to contribute their personal measurements as part of a large-scale effort to collect data about a population or a geographical area (Mun et al., 2009). In addition, social sensing can also be triggered spontaneously without prior coordination (e.g., via Twitter and Youtube). Recent research attempts to understand the fundamental factors that affect the behavior of these emerging social sensing applications, such as analysis of characteristics of social networks, information propagation, and tipping points (Hui et al., 2010).

Human-in-the-loop cyber-physical systems (HiLCPSs) incorporate a challenging and promising class of CPS applications that augment and facilitate human interaction with the physical world

(Schirner et al., 2013). Some examples of these applications include energy management, health care, automobile systems, and disaster response. Many interesting research challenges have been studied in HiLCPSs applications. For example, Wolpaw et al. developed a noninvasive brain computer interface to efficiently measure electric potential on the scalp for the inference of human's intent (Wolpaw and Wolpaw, 2012). Lu et al. designed a smart thermostat system by leavening hidden Markov model to measure the occupancy and sleep patterns of the residents in a home for energy savings (Lu et al., 2010). In this chapter, we focused on the set of works that address the *data reliability* problems in CPS where humans play the role of sensors or sensor carriers and where the reliability of data sources and the collected data is in general unknown a priori.

To assess the credibility of facts reported in information networks, a relevant body of work in the machine learning and data mining communities performs trust analysis. Hubs and Authorities (Kleinberg, 1999) developed the basic fact-finder framework that computes the credibility of claims and the reliability of sources in an iterative fashion. Pasternack et al. extend the fact-finder framework by incorporating prior knowledge into the analysis and proposes several extended algorithms: *Average. Log, Investment, Pooled Investment* (Pasternack and Roth, 2010). Yin et al. introduce *TruthFinder* as an unsupervised fact-finder for trust analysis on a providers-facts network (Yin et al., 2008). Other fact-finders enhance the basic framework by incorporating analysis on properties or dependencies within assertions or sources. Galland et al. (2010) take the notion of hardness of facts into consideration by proposing their algorithms: *Cosine, 2-Estimates, 3-Estimates*. The source dependency detection problem has been discussed and several solutions have been proposed (Dong et al., 2010). Additionally, trust analysis has been done both on a homogeneous network (Balakrishnan and Kambhampati, 2011) and a heterogeneous network (Sun et al., 2009). Fact-finding in the case of social sensing is more challenging due to the highly dynamic nature of social sensing applications (Aggarwal and Abdelzaher, 2013). Moreover, the outputs of fact-finders are generally rankings of credibility values of sources and facts, which cannot be used to directly *quantify* the participant reliability or measurement correctness for social sensing (Wang et al., 2011a).

In the information fusion community, belief theory provides the mechanism to combine evidence from multiple possibly conflicting sources (Shafer, 1976; Yager et al., 1994; Smarandache and Dezert, 2006). The concept of discounting beliefs based upon source reliability before fusion goes back to Shafer (1976). Recently, subjective logic has emerged as a means to reason over conflicting evidence (Jøsang et al., 2006). Subjective opinions are formed from evidence observed from individual sources. When incorporating multiple opinions, the subjective opinions need to be discounted similar to Dempster-Shafer theory before consensus fusion. In essence, this form of discount fusion can be interpreted as a weighted sum of evidence where the weights are proportional to the source reliabilities. The consensus fusion operation in subjective logic assumes the evidence used to form the subjective opinions of the sources is independent. Current research is investigating the proper fusion rule when the sources incorporate correlated evidence.

Since people are an indispensable element in social sensing, some popular attacks originated from human (or source) interactions are interesting to investigate. Collusion attack is carried out by a group of colluded attackers who collectively perform some malicious (sometimes illegal) actions based on their agreement to defraud honest sources or obtain an objective forbidden by the system. This attack could be mitigated by monitoring the interactions or relationships among colluded attackers or identifying the abnormal behavior within the group (Lian et al., 2007). Sybil attack is another related attack carried out by a single attacker who intentionally creates a large number of pseudonymous entities and uses them to gain a disproportionately large influence on the system. This attack could be mitigated by certifying trust of identity assignment, increasing the cost of creating identities, limiting the resources the attacker can use to create new identities, etc. (Yu et al., 2006). Problems become more interesting when sources are not just duplicates but actually linked through some orthogonal information network (e.g., social network).

The data reliability problem we studied in this chapter also bears resemblance to reputation systems. The basic idea of reputation systems is to let entities rate each other (e.g., after a transaction) or review some objects of common interests (e.g., products or dealers), and use the aggregated ratings or reviews to derive trust or reputation scores, which can help other entities in deciding whether or not to trust a given entity or purchase a certain object (Jøsang et al., 2007). Different types of reputation systems are being used successfully in commercial online applications. For example, eBay is a type of reputation system based on homogeneous peer-to-peer systems, which allows peers to rate each other after each pair of them conduct a transaction (Houser and Wooders, 2006). Amazon on-line review system represents another type of reputation systems, where different sources offer reviews on products (or brands, companies) they have experienced. Customers are influenced by those reviews (or reputation scores) when making purchase decisions. It turns out that the data reliability problem in social sensing fits better into these types of reputation systems and has the potential to provide more refined and confident results for the reputation computation. Additionally, reputation systems are in general vulnerable to several attacks: self-promoting, slandering, denial of service, etc. (Hoffman et al., 2009). Many of the attacks actually originate from collusion and Sybil attacks as mentioned earlier.

3 AN ANALYTICAL FRAMEWORK TO ADDRESS DATA RELIABILITY IN CPS

This section reviews a comprehensive analytical framework that takes the first step to optimally (in the sense of maximum likelihood estimation (MLE)) solve the data reliability problem in CPSs (social sensing in particular) and rigorously analyze the accuracy of the estimation results (Wang et al., 2012a,b, 2013a,b,c, 2014a,b,c,d; Zhao et al., 2012, 2014; Qi et al., 2013; Huang and Wang, 2015, 2016; Huang et al., 2015; Tanvir Al Amin et al., 2015). It offers a principled approach and a solid theoretical foundation for future research work on the topic of addressing the data reliability problem in CPS.

In the remainder of this section, we will focus on the social sensing applications of CPS. To that end, we adopt some CPS community friendly terminologies to denote several concepts we will use in the problem statement. In particular, we denote sources as participants, assertions as measured variables, claims as observations, and source truthfulness as participant reliability. Similarly, as we mentioned before, we consider a social sensing application model where a group of M participants, $S_1, S_2, ..., S_M$ make individual observations about a set of N measured variables $C_1, C_2, ..., C_N$ in their environment. For example, a group of tourists might join a geo-tagging campaign to report litter locations in the park. Hence, each measured variable denotes the existence or lack thereof of litter at a given location. We assume that locations are discretized, and therefore finite. We consider only binary variables and assume, without loss of generality, that their "normal" state is negative (e.g., no litter on the ground). Hence, participants report only when a positive value is encountered.

Let us first define some notations we will use in this section, let $P(C_j = 1)$ and $P(C_j = 0)$ denote the probability that the actual variable C_j is indeed true and false, respectively. $S_i C_j = 1$ when source S_i

reports claim C_j to be true and $S_iC_j = 0$ otherwise. Different participants may make different numbers of observations. Let the probability that participant S_i makes an observation be s_i. Further, let the probability that participant S_i is right be t_i and wrong be $1 - t_i$. Note that, this probability depends on the participant's reliability, which is not known *a priori*. Formally, t_i is defined as:

$$t_i = P\left(C_j = 1 | S_iC_j = 1\right) \qquad (1)$$

Let us also define a_i as the (unknown) probability that participant S_i reports a variable to be true when it is indeed true, and b_i as the (unknown) probability that participant S_i reports a variable to be true when it is in reality false. Formally, a_i and b_i are defined as follows:

$$a_i = P\left(S_iC_j = 1 | C_j = 1\right)$$
$$b_i = P\left(S_iC_j = 1 | C_j = 0\right) \qquad (2)$$

The input to the algorithm is the matrix SC, which is referred to as the *observation matrix* in the context of social sensing. We also denote the overall prior probability that a randomly chosen measured variable is true as d. Note that, this value can be known from past statistics. It does not indicate, however, whether any particular claim about a specific measured variable is true or not.

Let us denote $d = P\left(C_j = 1\right)$ and $s_i = P\left(S_iC_j = 1\right)$. Plugging these, together with t_i into the definition of a_i and b_i, we get the relationship between t_i, a_i, and b_i through Bayes' theorem:

$$a_i = \frac{t_i \times s_i}{d}$$
$$b_i = \frac{(1 - t_i) \times s_i}{1 - d} \qquad (3)$$

The goal of the algorithm is to compute (i) the best estimate h_j of the value of each variable C_j and (ii) the best estimate t_i of the reliability of each participant S_i. Let us denote the sets of the estimates by vectors H and E, respectively. The goal is to find the optimal H^* and E^* vectors in the sense of being most consistent with the observation matrix SC. Formally, this is given by:

$$<H^*, E^*> \ = \ \underset{<H,E>}{\text{argmax}} P(SC|H,E) \qquad (4)$$

It turns out we are able to find the optimal solution (in the sense of MLE) of the above equation by intelligently casting it as an expectation maximization (EM) problem.

In particular, we first introduce a latent variable Z for each measured variable to indicate whether it is true or not. Specifically, we have a corresponding variable $z_j = 1$ for the jth measured variable such that: $z_j = 1$ when the measured variable C_j is true and $z_j = 0$ otherwise. We further denote the observation matrix SC as the observed data X, and take $\theta = (a_1, a_2, ..., a_M; b_1, b_2, ..., b_M, d)$ as the parameter of the model that we want to estimate. The goal is to get the MLE of θ for the model containing observed data X and latent variables Z. We run the EM algorithm by iterating between two main steps of EM (namely, the E-step and the M-step) until the estimation converges (i.e., the likelihood function reaches the maximum). The MLE directly leads to an accurate quantification of measurement correctness as well as participant reliability. The above MLE framework has been evaluated through both simulation/emulation and real world social sensing case studies and was shown to outperform the-state-of-art baselines (Wang et al., 2012b, 2014d).

In the above discussions, we reviewed a maximum likelihood estimator based on EM to estimate the reliability of participants and determine the correctness of facts concluded from the data. However, an important problem that remains unanswered from the EM scheme is: what is the confidence of the resulting participant reliability estimation? Only by answering this question, can we completely characterize estimation performance, and hence participant reliability in social sensing applications. In the remainder of this section, we review an analytically founded bound that quantifies the accuracy of such MLE in social sensing (Wang et al., 2012a).

In particular, the goal of the confidence bound derivation is to demonstrate, in an analytically founded manner, how to compute the confidence interval of each participant's reliability. Formally, this is given by:

$$\left(\hat{t}_i^{\text{MLE}} - c_p^{\text{lower}}, \quad \hat{t}_i^{\text{MLE}} + c_p^{\text{upper}}\right) c\% \quad i = 1, 2, 3, \ldots, M \tag{5}$$

where $\hat{t}_i^{\text{MLE}}$ is the MLE on the reliability of participant S_i, $c\%$ is the confidence level of the estimation interval, c_p^{lower} and c_p^{upper} represent the lower and upper bound on the estimation deviation from the MLE $\hat{t}_i^{\text{MLE}}$ respectively. We target to find c_p^{lower} and c_p^{upper} for a given $c\%$ and an observation matrix SC.

The maximum likelihood estimator has a number of attractive asymptotic properties. One of them is called *asymptotic normality*, which basically states the MLE estimator is asymptotically distributed with Gaussian behavior as the data sample size goes up, in particular (Casella and Berger, 2002):

$$\sqrt{n}\left(\hat{\theta}_{\text{MLE}} - \theta_0\right) \xrightarrow{d} N\left(0, I^{-1}\left(\hat{\theta}_{\text{MLE}}\right)\right) \tag{6}$$

where n is the sample size, θ_0 and $\hat{\theta}_{\text{MLE}}$ are the true value and the MLE of the parameter θ respectively. $I^{-1}\left(\hat{\theta}_{\text{MLE}}\right)$ is the inverse of the Fisher information matrix at the point of MLE, which is also defined as the Cramer-Rao lower bound (CRLB) of the MLE (Hoel, 1954). Hence, the asymptotic normality property means that in a regular case of estimation and in the distribution limiting sense, the maximum likelihood estimator $\hat{\theta}_{\text{MLE}}$ is unbiased and its covariance reaches the CRLB (i.e., an efficient estimator).

It turns out we are able to compute an approximation of the above CRLB for the MLE of EM scheme (Wang et al., 2011b). It is an approximation approach because we assume that the truthfulness of each hidden variable defined in the likelihood function of EM scheme is correctly estimated. By plugging the approximate CRLB into Eq. (6), we obtain the covariance matrix $Cov\left(\hat{\theta}_{\text{MLE}}\right)$ of the asymptotic normal distribution for the MLE of EM scheme, which can be given by:

$$\left(Cov\left(\hat{\theta}_{\text{MLE}}\right)\right)_{i,j} = \begin{cases} 0 & i \neq j \\ \dfrac{\hat{a}_i^{\text{MLE}} \times \left(1 - \hat{a}_i^{\text{MLE}}\right)}{N \times d} & i = j \in [1, M] \\ \dfrac{\hat{b}_i^{\text{MLE}} \times \left(1 - \hat{b}_i^{\text{MLE}}\right)}{N \times (1 - d)} & i = j \in (M, 2M] \end{cases} \tag{7}$$

where $\hat{a}_i^{\text{MLE}}$ and $\hat{b}_i^{\text{MLE}}$ are the MLE on the parameter a_i and b_i defined earlier.

Recall the relation between participant reliability and estimation parameter a_i is $a_i = \dfrac{t_i \times s_i}{d}$. For a given topology, s_i and d are known constants, $\left(\hat{t}_i^{\text{MLE}} - t_i^0\right)$ also follows a norm distribution with 0 mean and variance given by:

$$Var\left(\hat{t}_i^{\text{MLE}}\right) = \left(\frac{d}{s_i}\right)^2 Var\left(\hat{a}_i^{\text{MLE}}\right) \tag{8}$$

Hence, we are able to derive the confidence interval that can be used to quantify the estimation accuracy of the MLE from the EM scheme. The confidence interval of the reliability estimation of participant S_i (i.e., $\hat{t}_i^{\mathrm{MLE}}$) at confidence level p is given by:

$$\left(\hat{t}_i^{\mathrm{MLE}} - c_p \sqrt{Var\left(\hat{t}_i^{\mathrm{MLE}}\right)}, \ \hat{t}_i^{\mathrm{MLE}} + c_p \sqrt{Var\left(\hat{t}_i^{\mathrm{MLE}}\right)} \right) \tag{9}$$

where c_p is the standard score (z-score) of the confidence level p. For example, for the 95 % confidence level, $c_p = 1.96$. Note that the derived confidence interval of the participant reliability can be computed by simply using the converged MLE of the EM scheme. The performance of the techniques we reviewed in this chapter have been evaluated using real-world social sensing applications and shown to outperform many state-of-the-art baselines from network sensing, data mining and machine learning communities (Wang et al., 2012b; Zhao et al., 2012, 2014; Qi et al., 2013; Tanvir Al Amin et al., 2015; Huang and Wang, 2016).

4 SUMMARY AND IMPACT

The data reliability research challenge reviewed in this chapter, if addressed, will enable the development of CPS applications in domains of transportation, energy, disaster response, and healthcare, among others, where correctness is guaranteed despite reliance on the collective observations of untrained, average, and largely unreliable sources. It will also contribute to the building of new CPSs that effectively combine inputs from humans, machines, and the physical world, while offering rigorous reliability guarantees, similar to those obtained from physical (hard) data fusion and signal analysis. The growing need to understand and optimize complex CPSs with humans in the loop have already motivated several industry and government initiatives such as IBM's Smarter Planet, Berkeley's TerraSwarm, Amsterdam Smart City, Microsoft Sensor Map, and China's Internet of Things. Those and other initiatives together with many other CPS applications (e.g., smart cities) will help to eventually materialize a vision of a smarter planet where CPSs, in collaboration with humans in the loop, make life easier, more efficient, and safer than ever before.

REFERENCES

Aggarwal, C.C., Abdelzaher, T., 2013. Social sensing. Managing and Mining Sensor Data. Springer, New York, NY.

Ahmadi, H., Abdelzaher, T., Han, J., Pham, N., Ganti, R.K., 2011. The sparse regression cube: a reliable modeling technique for open cyber-physical systems. In: Proceedings of the 2011 IEEE/ACM Second International Conference on Cyber-Physical Systems. IEEE Computer Society, Chicago, IL, USA, pp. 87–96.

Balakrishnan, R., Kambhampati, S., 2011. SourceRank: relevance and trust assessment for deep web sources based on inter-source agreement. In: Proceedings of the 20th International Conference on World Wide Web. ACM, New York, NY, pp. 227–236.

Casella, G., Berger, R.L., 2002. Statistical Inference. Duxbury, Pacific Grove, CA.

Dong, X.L., Berti-Equille, L., Hu, Y., Srivastava, D., 2010. Global detection of complex copying relationships between sources. Proc. VLDB Endowment 3, 1358–1369.

Eisenman, S.B., Miluzzo, E., Lane, N.D., Peterson, R.A., Ahn, G.-S., Campbell, A.T., 2009. BikeNet: a mobile sensing system for cyclist experience mapping. ACM Trans. Sens. Netw. 6, 6.

Galland, A., Abiteboul, S., Marian, A., Senellart, P., 2010. Corroborating information from disagreeing views. In: Proceedings of the Third ACM International Conference on Web Search and Data Mining. ACM, New York, NY, pp. 131–140.

Hoel, P.G., et al., 1954. Introduction to Mathematical Statistics, second ed. John Wiley & Sons, Inc., New York. Chapman & Hall, Ltd., London.

Hoffman, K., Zage, D., Nita-Rotaru, C., 2009. A survey of attack and defense techniques for reputation systems. ACM Comput. Surv. 42, 1.

Houser, D., Wooders, J., 2006. Reputation in auctions: theory, and evidence from eBay. J. Econ. Manag. Strateg. 15, 353–369.

Huang, C., Wang, D., 2015. Spatial-temporal aware truth finding in big data social sensing applications. In: The 9th IEEE International Conference on Big Data Science and Engineering (BigDataSE 15). IEEE, Helsinki, Finland.

Huang, C., Wang, D., 2016. Topic-aware social sensing with arbitrary source dependency graphs. In: Proceedings of the 15th International Symposium on Information Processing in Sensor Networks (IPSN 16). IEEE, Vienna, Austria.

Huang, J.-H., Amjad, S., Mishra, S., 2005. Cenwits: a sensor-based loosely coupled search and rescue system using witnesses. In: Proceedings of the 3rd International Conference on Embedded Networked Sensor Systems. ACM, New York, NY, pp. 180–191.

Huang, C., Wang, D., Chawla, N., 2015. Towards time-sensitive truth discovery in social sensing applications. In: The 12th IEEE International Conference on Mobile Ad-hoc and Sensor Systems (IEEE MASS 2015). IEEE, Dallas, TX, USA.

Hui, C., Goldberg, M., Magdon-Ismail, M., Wallace, W.A., 2010. Simulating the diffusion of information: an agent-based modeling approach. Int. J. Agent Technol. Syst. 2, 31–46.

Hull, B., Bychkovsky, V., Zhang, Y., Chen, K., Goraczko, M., Miu, A., Shih, E., Balakrishnan, H., Madden, S., 2006. CarTel: a distributed mobile sensor computing system. In: Proceedings of the 4th International Conference on Embedded Networked Sensor Systems. ACM, New York, NY, pp. 125–138.

Jøsang, A., Marsh, S., Pope, S., 2006. Exploring different types of trust propagation. Trust Management. Springer, Berlin.

Jøsang, A., Ismail, R., Boyd, C., 2007. A survey of trust and reputation systems for online service provision. Decis. Support. Syst. 43, 618–644.

Kleinberg, J.M., 1999. Authoritative sources in a hyperlinked environment. J. ACM 46, 604–632.

Lian, Q., Zhang, Z., Yang, M., Zhao, B.Y., Dai, Y., Li, X., 2007. An empirical study of collusion behavior in the Maze P2P file-sharing system. In: 27th International Conference on Distributed Computing Systems, 2007. ICDCS'07. IEEE, Toronto, ON, p. 56.

Lu, J., Sookoor, T., Srinivasan, V., Gao, G., Holben, B., Stankovic, J., Field, E., Whitehouse, K., 2010. The smart thermostat: using occupancy sensors to save energy in homes. In: Proceedings of the 8th ACM Conference on Embedded Networked Sensor Systems. ACM, New York, NY, pp. 211–224.

Mun, M., Reddy, S., Shilton, K., Yau, N., Burke, J., Estrin, D., Hansen, M., Howard, E., West, R., Boda, P. Peir, 2009. The personal environmental impact report, as a platform for participatory sensing systems research. In: Proceedings of the 7th International Conference on Mobile Systems, Applications, and Services. ACM, New York, NY, pp. 55–68.

Pasternack, J., Roth, D., 2010. Knowing what to believe (when you already know something). In: Proceedings of the 23rd International Conference on Computational Linguistics. Association for Computational Linguistics, Stroudsburg, PA, pp. 877–885.

Qi, G.-J., Aggarwal, C.C., Han, J., Huang, T., 2013. Mining collective intelligence in diverse groups. In: Proceedings of the 22nd International Conference on World Wide Web, International World Wide Web Conferences Steering Committee, pp. 1041–1052.

Rajkumar, R.R., Lee, I., Sha, L., Stankovic, J., 2010. Cyber-physical systems: the next computing revolution. In: Proceedings of the 47th Design Automation Conference. ACM, New York, NY, pp. 731–736.

Schirner, G., Erdogmus, D., Chowdhury, K., Padir, T., 2013. The future of human-in-the-loop cyber-physical systems. Computer 46, 36–45.

Shafer, G., 1976. A Mathematical Theory of Evidence. Princeton University Press, Princeton, NJ.

Smarandache, F., Dezert, J., 2006. Advances and Applications of DSmT for Information Fusion (Collected Works), vol. 2 Infinite Study, Porto, Portugal.

Sun, Y., Yu, Y., Han, J., 2009. Ranking-based clustering of heterogeneous information networks with star network schema. In: Proceedings of the 15th ACM SIGKDD International Conference on Knowledge Discovery and Data Mining. ACM, New York, NY, pp. 797–806.

Tanvir Al Amin, M., Li, S., Rahman, M.R., Seetharamu, P.T., Wang, S., Abdelzaher, T., Gupta, I., Srivatsa, M., Ganti, R., Ahmed, R., 2015. Social trove: a self-summarizing storage service for social sensing. In: 2015 IEEE International Conference on Autonomic Computing (ICAC). IEEE, pp. 41–50.

Wang, D., Abdelzaher, T., Ahmadi, H., Pasternack, J., Roth, D., Gupta, M., Han, J., Fatemieh, O., Le, H., Aggarwal, C.C., 2011a. On Bayesian interpretation of fact-finding in information networks. In: Proceedings of the 14th International Conference on Information Fusion (FUSION). IEEE, pp. 1–8.

Wang, D., Ahmadi, H., Abdelzaher, T., Chenji, H., Stoleru, R., Aggarwal, C.C., 2011b. Optimizing quality-of-information in cost-sensitive sensor data fusion. In: International Conference on Distributed Computing in Sensor Systems and Workshops (DCOSS). IEEE, pp. 1–8.

Wang, D., Abdelzaher, T., Kaplan, L., Aggarwal, C.C., 2011c. On quantifying the accuracy of maximum likelihood estimation of participant reliability in social sensing. Urbana 51, 61801.

Wang, D., Kaplan, L., Abdelzaher, T., Aggarwal, C.C., 2012a. On scalability and robustness limitations of real and asymptotic confidence bounds in social sensing. In: 9th Annual IEEE Communications Society Conference on Sensor, Mesh and Ad Hoc Communications and Networks (SECON). IEEE, pp. 506–514.

Wang, D., Kaplan, L., Le, H., Abdelzaher, T., 2012b. On truth discovery in social sensing: a maximum likelihood estimation approach. In: Proceedings of the 11th International Conference on Information Processing in Sensor Networks. ACM, New York, NY, pp. 233–244.

Wang, D., Abdelzaher, T., Kaplan, L., Aggarwal, C.C., 2013a. Recursive fact-finding: a streaming approach to truth estimation in crowd sourcing applications. In: 2013 IEEE 33rd International Conference on Distributed Computing Systems (ICDCS). IEEE, pp. 530–539.

Wang, D., Abdelzaher, T., Kaplan, L., Ganti, R., Hu, S., Liu, H., 2013b. Exploitation of physical constraints for reliable social sensing. In: IEEE 34th Real-Time Systems Symposium (RTSS). IEEE, pp. 212–223.

Wang, D., Kaplan, L., Abdelzaher, T., Aggarwal, C.C., 2013c. On credibility estimation tradeoffs in assured social sensing. IEEE J. Sel. Areas Commun. 31, 1026–1037.

Wang, D., Abdelzaher, T., Kaplan, L., 2014a. Surrogate mobile sensing. IEEE Commun. Mag. 52, 36–41.

Wang, D., Amin, M., Abedlzaher, T., Roth, D., Voss, C., Kaplan, L., Tratz, S., Laoudi, J., Briesch, D., 2014b. Provenance-assisted classification in social networks. IEEE J. Sel. Top. Sig. Process. 8, 624–637.

Wang, D., Amin, M.T., Li, S., Abdelzaher, T., Kaplan, L., Gu, S., Pan, C., Liu, H., Aggarwal, C.C., Ganti, R., Wang, X., Mohapatra, P., Szymanski, B., Le, H., 2014c. Using humans as sensors: an estimation-theoretic perspective. In: Proceedings of the 13th International Symposium on Information Processing in Sensor Networks. IEEE Press, Berlin.

Wang, D., Kaplan, L., Abdelzaher, T.F., 2014d. Maximum likelihood analysis of conflicting observations in social sensing. ACM Trans. Sens. Netw. 10, 30.

Wolpaw, J., Wolpaw, E.W., 2012. Brain-Computer Interfaces: Principles and Practice. Oxford University Press, New York, NY.

Yager, R., Fedrizzi, M., Kacprzyk, J., 1994. Advances in the Dempster-Shafer Theory of Evidence. John Wiley & Sons, New York, NY.

Yin, X., Han, J., Yu, P.S., 2008. Truth discovery with multiple conflicting information providers on the web. IEEE Trans. Knowl. Data Eng. 20, 796–808.

Yu, H., Kaminsky, M., Gibbons, P.B., Flaxman, A., 2006. Sybilguard: defending against sybil attacks via social networks. ACM SIGCOMM Comp. Commun. Rev. 36, 267–278.

Zhao, B., Rubinstein, B.I., Gemmell, J., Han, J., 2012. A Bayesian approach to discovering truth from conflicting sources for data integration. Proc. VLDB Endowment 5, 550–561.

Zhao, Z., Cheng, J., Ng, W., 2014. Truth discovery in data streams: a single-pass probabilistic approach. In: Proceedings of the 23rd ACM International Conference on Conference on Information and Knowledge Management. ACM, New York, NY, pp. 1589–1598.

NETWORK-WIDE PROGRAMMING CHALLENGES IN CYBER-PHYSICAL SYSTEMS

7

P.M.N. Martins, J.A. McCann

Imperial College London, London, United Kingdom

1 OUTLOOK AND CHALLENGES

The ubiquitous computing revolution has brought with it profound and widespread changes to our daily lives. It has affected how we interact with others, the role of memory when we can constantly access digital storage, how products are marketed and sold effectively through new channels, how we perceive and elect our leaders, amongst many other aspects. These changes all emerge from a combination of embedded technologies and the availability of mobile connected devices, and the services being offered to improve our individual lives. This emergent revolution poses significant questions regarding the way in which we design the spaces that we inhabit. Such questions pertain both to how we can optimize existing services so they function at their best given our new expectations, as well as how to integrate these new models of individuality, interaction, commerce, etc., into the spaces and services that the old models used to occupy.

The design of urban environments along these lines relate to smart city initiatives and are already starting to happen worldwide, in pilot programs for instance in Santander (Smart Santander, 2015), Singapore (Massachusetts Institute of Technology, 2015), and London (ICRI Cities, 2015). On the commercial side, we have also seen the appearance of companies such as StreetLine Networks (Streetline, Inc., 2015) and Worldsensing (World Sensing, 2015), providing practical solutions with current technologies. A key common factor in all these enterprises is the need to understand the city better, in terms of how individuals use the city and how the city is affected by them. For example, we can think of sensing the movement of people through the city to build population density models and thus know what areas attract attention and in what circumstances (Smith-Clarke et al., 2014). Another example is where we sense the impact on the environment that the city services are having, for example, by monitoring hot spots of air pollution and using this information to guide traffic policies and road design and investment. One example of an existing project along these lines is the London Air Quality Network, collecting air quality information in and around Greater London (KCL, 2015). Sensing is thus key to inform the design of future systems and to enable automation of services within the city infrastructure. Unsurprisingly then, the first stage in all the smart city initiatives has been to instrument the city with sensing devices.

Cyber-Physical Systems. http://dx.doi.org/10.1016/B978-0-12-803801-7.00007-9

The vision is thus to have widespread sensing of city conditions, and using this to guide infrastructure, policies and services. However, one must be careful in the way in which one instruments the city and not lose track of long-term consequences. For instance, there are complexities and costs associated with the infrastructures (physical and communications) that will be deployed. However, this is nothing compared with the potential costs of maintaining such infrastructures. Further, if sensing truly becomes ubiquitous, bringing with it an explosion of devices communicating data, then existing network infrastructures will surely become saturated. This has been foreseen by the research community, and hard limits have been established some time ago (Gupta and Kumar, 2000).

The idea of programming a network as a whole has been around for a while in the WSN area, and is termed macroprogramming. This idea is used for instance for distributed data query and processing, for instance in tinyDB (Madden et al., 2005) and Regiment (Newton et al., 2007). While the concepts are similar, macroprogramming in WSN focuses on the aspect of computation that pertains to data collection sensor networks, i.e., querying sensed data for a network as a whole and performing operations on them. Moreover, traditionally in WSN research the focus is on heavily constrained devices. As a result, most of the existing solutions have restrictions on the types of computations that can be performed to make the problem more tractable. While we also tackle the issue of data query when specifying the sources to be used for an application, in this article we propose a mechanism for extensions to a general purpose language, whereby the exact location in the network where to run code is determined based on an external specification of requirements and data queries. These applications handle the whole spectrum of computation, including sensing, processing, and actuation duties. Moreover, we can run any program that can be run on the existing operating system. In research on mobile computation and sensing offloading (Kumar et al., 2013; Rachuri et al., 2013) we see a similar move toward executing equivalent types of tasks in both fixed and mobile nodes. This requires the same features and abstractions as in our work in conditionally deploying computation throughout the (cloud to edge) network of heterogenous nodes, and we were inspired by this body of work.

Throughout this chapter we will present a brief overview of the challenges in programming cyber-physical systems (CPSs) and then present a programming model that tackles these challenges, through a practical case study.

2 PROGRAMMING CYBER-PHYSICAL SYSTEMS

CPSs are typified by the integration of computation, networking and physical sensing and actuation. The challenges in the development of applications for large scale CPSs therefore come from achieving efficient utilization and orchestration of networked resources to achieve the aim of the application. The abstractions that we will create will therefore be focused on those aspects. Firstly, we represent sensing and actuation as the production and consumption of streams of data. Sensing is therefore represented by a data query, and is analogous to performing queries in real-time on a distributed database. The result set of a data query is then a producer of a stream of data. These data are streamed from the nodes possessing the required sensors.

Actuation is dually represented as the consumers of data streams. The endpoints for actuation are also represented as queries, this time for actuation capabilities. The return set of these queries will then comprise a stream sink for actuation commands. Sending commands to this sink will then activate the necessary actuators to perform this command.

The third component of CPS programming is then computation. We can think of computation as the core functionality of our system, integrating the data processing and sensing/actuation capabilities and closing the control feedback loop. The computational capabilities of resources in a large-scale CPS can vary wildly. Moreover, the computational requirements of different parts of the application also vary. In order to efficiently utilize the network resources these requirements must be made explicit.

This level of abstraction for specifying behavior at the network-level requires a different paradigm of programming from traditional software development. The application behavior is defined not in terms of node behavior but in terms of network behavior. The orchestration of all the sensing, actuation, and processing elements in the network quickly becomes cognitively overwhelming. In order to program a CPS we, indeed, need a more declarative approach abstracting away from the individual components into the manipulation of semantic network-wide concepts.

2.1 ISSUES OF SCALE

These issues are compounded when we consider the scalability of CPS to a city scale. In this setting, we can imagine standard nodes delivering high-resolution sensor data for multiple parameters such that collecting all the available data in a cloud computing paradigm would be infeasible and cause congestion in the network. In recent years, there has been a research movement to do in-network processing in sensor networks (Kolcun and McCann, 2014), to optimize their functioning. Indeed, in the previous example, if we perform the valve actuation calculations inside the network we are able to not only reduce the delay in actuating the valve, but also decrease the amount of information that is communicated to the cloud, freeing the communication medium. We believe this is essential to allow the city to be further instrumented and automated without interfering with existing services. Complementary to the in-networking approach is network off-loading. Here we minimize the communications range of the devices and instead of building ridged routing trees to multi-hop sensed data around or have each device send data directly to its sinks, we empower mobile devices to collect the data and relay it. This might harness the physical movement of the device to carry out the communication (in data mules, for example) or off-load the traditional network by using "free" communications such as Bluetooth or WiFi-Direct to multi-hop data between mobile devices or piggyback sensed data over other regular communications (Yang et al., 2013).

So we can now think of a continuum where data are processed and relayed (also a form of processing) from source devices via edge devices to the cloud. The question is where do we carry out processing, with what nodes and how? In Fig. 1 we can see an example network comprised of several sensors and actuators, as well as some cloud services. In this network we are running three different services, affecting different subsets of the nodes. The first application is the leak detection application from before, and uses data from sensors S_1 and S_4 to affect the state of the valve actuator A_2. The second application is an image processing application and uses data from sensor S_2 to guide the operation of actuator A_1. The third application, visualizing temperature sensor data on a map, merely requires the output from sensor S_3. On the left hand side we see a diagram of data transfer when all the data are being transferred to the cloud for analysis and processing, before being used to affect the actuators. This has several negative consequences for the performance of the network. First of all, there is a considerable amount of data that are being relayed via nodes that are not involved in a particular application, leading to communication congestion in those nodes (for instance, S_2 is relaying data from two different applications, and is not involved in one of them). Secondly, we can imagine that the delay between reading sensor data from S_1 and actuating A_2 will be quite high as the data need to traverse the whole

The computing continuum:

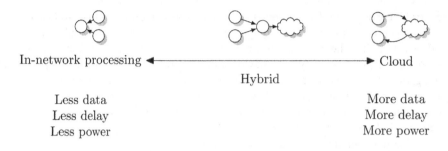

In-network processing ◄──────────────────────────► Cloud

Hybrid

Less data   More data
Less delay   More delay
Less power   More power

Application example for the computing continuum design principle:

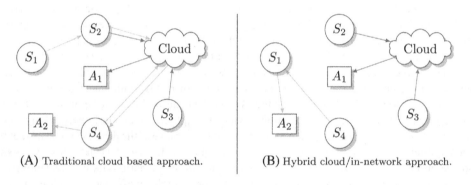

(A) Traditional cloud based approach.   **(B)** Hybrid cloud/in-network approach.

Legend:

Applications:
Nodes: ───── - Leak detection application
S_n - Sensor   ───── - Image processing application
A_n - Actuator   ───── - Temperature data mapping application

FIG. 1

The computing continuum as a design principle for Internet of Things applications.

network and be processed in the cloud. Since leak detection is time-sensitive, as we explained before, this could be disastrous. On the right hand side we see a more balanced distribution, where we have eliminated some of the congestion by doing the computations for the leak detection application within the network. We have eliminated delay for this application, and reduced congestion overall. We have not, however, changed the operation of the image processing application, as it requires more processing power than would be available in edge nodes. Thus, where to place a particular module of computation depends on the requirements for it, e.g., in terms of timeliness, redundancy, and accuracy.

We believe such hybrid approaches will be essential to allow smart city CPS applications to both scale and be maintainable and we should start considering these problems when designing development environments for them. In this way, we should think of the whole network, from edge to cloud, as a programmable device and design technologies to facilitate this view when designing and deploying applications. These technologies will have to take into account requirements that are essential for smart city CPS applications. For instance, if we are using shared infrastructure to build smart city applications, there are fundamental problems in terms of security and privacy that need to be resolved. Moreover, if nodes within the network are going to perform data processing for various stakeholders, we need to be able to ensure isolation between these various applications. Techniques need to be in place to efficiently and fairly utilize the shared networked resources in order to be able to deliver necessary data in a timely manner. A registry needs to be provided in order to be able to discover resources and define these applications. Finally, in terms of city applications, we need to be able to support applications that could have significant real world consequences, in terms of property damage, such as the pipe leak example, or even lives, if we think about automating the healthcare systems. Thus, we need to be able to provide strong guarantees in terms of the application's behavior when needed.

3 NETWORK-WIDE PROGRAMMING

As mentioned previously, the behavior of a CPS is better understood at a global system level. In order to reflect this from a programming language abstraction standpoint we rely on network-wide programming, or macroprogramming as it is called in the sensor network field. This paradigm of programming allows the system developer to write one piece of code for the network, specifying the application at a global semantic level. The task of the compiler is then to not only produce a machine translation of the code, but also decide how to split this code into several images to run on many devices. These images should set up the necessary communication channels, buffering and orchestration between processing devices.

In this way, the role of the compiler is not to produce an executable but to produce a set of deployable software images, along with their deployment requirements. A simple example is: if the application in question was "when the temperature in this room reaches 30°C, turn on the light above the door," the application would be split into a sensing component that can be run on the node connected to the temperature sensor, a computation component to be run on any node capable of executing that operation in real time, and an actuation component to be run on the node attached to the light. These components need not run on different nodes; for example, one can imagine that the computation component would be co-located with the sensing or actuation component in this case.

Some existing macroprogramming frameworks in the wireless sensor network domain include tinyDB (Madden et al., 2005), Kairos (Gummadi et al., 2005), and Pleiades (Kothari et al., 2007). In both, the application is specified at a system level and then resolved to individual node programs. Their abstractions range from device specific to more semantic ones, and from domain specific to general purpose. However, they do not handle actuation or computation requirements, being more focused on the problem of expressing data queries and disseminating them.

More recently, Fog Computing (Bonomi et al., 2014), WuKong (Lin et al., 2013), and Scaffold (Martins and McCann, 2015a) attempt to provide a solution that is more suitable to city CPS, by integrating sensing, actuation, and processing into one unified framework. However, whereas Fog does the application distribution at runtime, Scaffold and WuKong attempt to determine the target nodes necessary for the application and only program those. The static approach provides more intelligibility to the network, at the potential cost of some dynamicity, that cannot be predicted at deploy time.

4 VIRTUALIZATION ON THE EDGE AND NETWORK ISSUES

This paradigm of computation brings with it several challenges pertaining to application development and deployment. In the previous setup, we motivate nodes running multiple applications that might belong to different stakeholders. This is not a new problem in the field of cloud computing, where computing power is shared between different users, and the users have no control over where their application is running. These applications need to be run in isolation as they might belong to different stakeholders. The current solution for this problem in the cloud computing sector is virtualization. Indeed, through the usage of virtual machines in our case, applications can be deployed on any physical node, and transferred without issue. Each individual node can run several applications, while every particular application views the physical hardware as its own and is not affected by the operation of other applications. However, the virtualization concept is not trivial to extend to the edge of the network, where devices are low powered, have low processing capabilities and have stringent battery constraints. One solution we have explored for this problem is to use lighter forms of virtualization, such as operating system-level virtualization. Operating system-level virtualization is a technique whereby all the application instances, called containers, share the kernel of the operating system. Thus, the kernel is responsible for providing isolation between these user-space instances, for instance in terms of file system, access to devices and delimiting CPU quotas. This approach to virtualization has the advantage of not requiring several full instances of the operating system to be running, thus providing sub-second start-up times and being less CPU and memory intensive on the host. The disadvantage is that all the applications need to be running on the same operating system. This approach to virtualization was added to the linux kernel by the linux containers project (LXC, 2015) and support for easier development, deployment, and instance management is offered by the rkt platform (CoreOS, 2015). It should be noted, however, that this approach can extend to any operating system that supports isolation of processes in the same manner.

Another challenge is that the decision regarding where to process data needs to account for the heterogeneous processing capabilities available to the continuum of devices/middle boxes/systems, and the fact that some nodes may be battery powered, and we would like to maximize their lifetime. It is tempting to then adapt solutions reminiscent of desktop computing, whereby a runtime is used on all nodes to abstract away the underlying hardware. This is the approach taken in platforms such as Fog Computing (Bonomi et al., 2014). However, we believe that given that the nodes are always online and deployment can be initiated remotely and at any time, it makes sense to instead adapt compiler-based approaches where optimization is done when the application is compiled/initially deployed. This is aligned with the *constrained dynamicity* design principle. That is, we believe that in order to cope with the large heterogeneity present in our target networks we should aim to have controlled variation, rather than arbitrary homogeneity. Enforcing homogeneity on heterogeneous devices can lead to problems in understanding the behavior of applications, as we can never know for sure which modules will be running and on which platforms they will end upon. This makes the behavior of the system less intelligible. Nevertheless, runtime approaches also have their advantages. It is much easier to provide tools and development support once the hardware differences have been abstracted away. Moreover, runtime environments are more adaptable as the runtime itself can handle migration and determine what nodes should be running the application at any given time depending on changes in available resources, etc. However, the heterogeneity of devices and environments already poses significant difficulties to application design and development, and abstracting this away can be harmful, by either precluding platform-specific optimizations, or hiding these in a way that makes the behavior of the system less intelligible. Intelligibility and manageability are extremely important to make sure that application development is feasible in this complex environment. Otherwise, complex interactions involving the placement of computation and communications can be hard to foresee. For example, migration of components in a highly distributed application is nontrivial, as it has to be handled by all the components that communicate with the migrated component. Otherwise, in a time-critical application like the pipe leak example, it can happen that the system takes a long time to propagate the updated addresses for the application. This can lead to delays in the sensing-actuation loop, which are not due to the system malfunctioning, but just the difficulty of accurately predicting delays in a system that is highly dynamic by default. This link between dynamicity and lack of intelligibility motivates our choice of deployments remaining static by default and only dynamic when necessary. Reasoning in static systems is easier, and thus it is also easier to guarantee their correctness.

5 REGISTRIES AND RESOURCE INTEGRATION

The cognitive overwhelm produced by the development of city scale CPS is not just about orchestrating the communication between nodes. It is also about managing a large network of systems owned by different stakeholders, and visualizing all the data available and correlations, as well as network topologies and node capabilities. Registries take a fundamental role in alleviating this problem, by presenting metadata about all the available functionality in the system.

Thus, a CPS needs to provide functionality for a new node to register its sensing, actuation, and processing capabilities. These capabilities will then become available to a system developer at a semantic level, no longer being tied to the physical nodes that are able to fulfill them. Moreover, after an application has been developed, it should also be possible to present virtual resources, be they composites of sensor data or actuation commands. This allows the potential reuse of calculations and intermediate results in programs.

This is all achieved by allowing a node to register its capabilities in a registry. After the application compiler has produced the set of images to be deployed, the information on this registry can be used to determine a deployment plan. Registries are also a part of all smart city initiatives, with the NGSI standard (NGSI, 2015) being a part of the FIWARE suite of applications for standardized smart cities (FIWARE, 2015). Moreover, on the open data front, CKAN allows device owners to publish their open data and application designers to subscribe to those sources.

6 SCAFFOLD: A FRAMEWORK FOR NETWORK-WIDE PROGRAMMING IN CYBER-PHYSICAL SYSTEMS

In this section we will examine Scaffold (Martins and McCann, 2015b) in more detail, and provide a case study that showcases the application of the network-wide programming principles to a simple example application.

6.1 SYSTEM MODEL

If we take a system-level approach to the development of applications the issues of data provision and actuation are largely orthogonal to the traditional control flow issues in programming languages. This suggests that we can define these primitives in a way not tied to a particular programming language or paradigm but as a framework for augmenting existing programming languages to provide these system-level features. Scaffold does this, through the Scale compiler.

If we consider the target programming language expressions ranged by e, Scale introduces constructs d, for querying data, $d[e]$ for scoped queries of data, when we require caching of a particular data point, c for actuation commands, and $with\ r\ e$ for scoping computation requirements. The behavior of the Scale compiler can then be described informally as follows:

- For every d create a program that broadcasts d on a node that can provide the sensor readings for d
- For every c create a program that listens for the command to perform c on a node that is able to perform it
- For every $with\ r\ e$ create a program that performs the operation e on a node that is capable of satisfying r. In order to then split the program, we need to broadcast the necessary scoped variables before the $with$ block, and after, thus snapshotting the state. This is currently handled at a language-specific level.

Thus, for every system-level program, Scale is going to produce a set of images and deployment requirements, to be resolved and optimized with online information at deploy time. This feature will be provided by the Scanner registry, which allows all nodes to register their capabilities and performs online optimization on the images produced by the Scale compiler (Fig. 2). It will also augment the images with runtimes to provide the required dynamicity, according to the constrained dynamicity design principle.

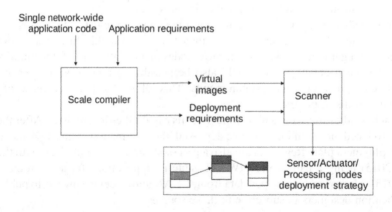

FIG. 2

Scaffold architecture.

7 A CASE STUDY

In this section we will look at a practical case study for the usage of Scaffold in defining and deploying a city CPS application. We will develop and deploy a light switch application, whereby we make the nodes available on a locally run register and then develop the application based on it.

7.1 BACKEND

For the backend, we need to host scanner and start the communication broker. We can do so by typing on the host:

```
$ scn init
```

This will host the necessary broker and the scanner register.

7.2 SENSORS

For the sensors, we will use the Intel Edison platform, and program the sensors individually. First of all, we need to define a driver for Scaffold to use. In this case, we will have installed an application for sending the commands. The scaffold driver can be selected in the configuration command for the node registration. We also advertise our capabilities, in this case that one of the Edison nodes will control a switch and the other will control a light:

```
$ scn register --node <sensor ip> --sense switch --driver /bin/switch
$ scn register --node <actuator ip> --actuate turnOnLight --driver /bin/turnOnLight
$ scn register --node <actuator ip> --actuate turnOffLight --driver /bin/turnOffLight
```

The two executables should provide the reading for the sensor as a numeric value on stdout. This is all the necessary infrastructure setup for the sensors.

7.3 APPLICATION

As an application developer, we can now visualize all the resources available in the network:

```
$ scn display
```

After identifying the application we want to build, we can then define the network-wide code using Scale as such:

```
switch[if ?switch == 0 then turnOffLight else turnOnLight]
```

In the code snippet, ?switch corresponds to a reading of the switch's value and turnOnLight and turnOffLight correspond to command to activate and deactivate the light, respectively. When compiled, this program will produce three target platform artifacts: one for the sensing part, one for the computation part and one for the actuation part.

We can then deploy the application. The default requirements in scaffold are for in-network processing:

```
$ scn deploy <app>
```

If everything went well, the application should be deployed correctly. and we can then observe the messages exchanged by the application by listening on the broker.

8 CONCLUSION

The issues of scale, both from the point of view of the sheer amount of data being produced and the cognitive overload of reasoning about all the data available necessitate new programming approaches.

The programming model we described throughout the chapter abstracts away the requirements of the application and automatically fully exploits all the computing modalities available in the computing continuum. This is done in a manner orthogonal to the programming language. Thus, the framework we presented can be used to augment various programming languages to support scalable system-wide programming of CPSs. The focus on constrained dynamicity and static by default allows us to fulfill the validation requirements that one would have when defining city applications. This research area is still in its infancy as it has just recently been made necessary by the shortcomings in using the established programming models in practical CPSs. However, we foresee that issues raised in this chapter will become much more relevant as more of these systems are implemented and improved.

REFERENCES

Bonomi, F., Milito, R., Natarajan, P., Zhu, J., 2014. Fog computing: a platform for Internet of Things and analytics. In: Bessis, N., Dobre, C. (Eds.), Big Data and Internet of Things: A Roadmap for Smart Environments, Studies in Computational Intelligence, vol. 546. Springer International Publishing, New York/Berlin, pp. 169–186 http://dx.doi.org/10.1007/978-3-319-05029-4_7.

CoreOS, 2015. Rkt, a fast, composable, and secure App Container runtime for Linux. http://rkt.io.

FIWARE, 2015. FIWARE. http://fiware.org.

Gummadi, R., Gnawali, O., Govindan, R., 2005. Macro-programming wireless sensor networks using kairos. In: Distributed Computing in Sensor Systems. Springer, New York/Berlin, pp. 126–140.

Gupta, P., Kumar, P.R., 2000. The capacity of wireless networks. IEEE Trans. Inf. Theory 46 (2), 388–404. ISSN 0018-9448. http://dx.doi.org/10.1109/18.825799.

Cities, ICRI, 2015. ICRI Cities. http://cities.io.

KCL, 2015. London Air Quality Network. http://www.londonair.org.uk/.

Kolcun, R., McCann, J.A., 2014. Dragon: data discovery and collection architecture for distributed IoT. In: Internet of Things 2014—The 4th International Conference on the Internet of Things (IoT 2014), Cambridge, USA, October 2014.

Kothari, N., Gummadi, R., Millstein, T., Govindan, R., 2007. Reliable and efficient programming abstractions for wireless sensor networks. In: ACM SIGPLAN Notices, vol. 42. ACM, New York, NY, USA, pp. 200–210.

Kumar, K., Liu, J., Lu, Y.H., Bhargava, B., 2013. A survey of computation offloading for mobile systems. Mob. Netw. Appl. 18 (1), 129–140.

Lin, K.J., Reijers, N., Wang, Y.C., Shih, C.S., Hsu, J.Y., 2013. Building smart M2M applications using the WuKong profile framework. In: Green Computing and Communications (GreenCom), 2013 IEEE and Internet of Things (iThings/CPSCom), IEEE International Conference on and IEEE Cyber, Physical and Social Computing, August 2013, pp. 1175–1180.

LXC, 2015. Linux Containers. http://linuxcontainers.org.

Madden, S.R., Franklin, M.J., Hellerstein, J.M., Hong, W., 2005. Tinydb: an acquisitional query processing system for sensor networks. ACM Trans. Database Syst. 30 (1), 122–173. ISSN 0362-5915. http://dx.doi.org/10.1145/1061318.1061322.

Martins, P.M.N.,McCann, J.A., 2015a. The programmable city. Procedia Comput. Sci. 52, 334–341, ISSN 1877-0509. http://dx.doi.org/10.1016/j.procs.2015.05.104. The 6th International Conference on Ambient Systems, Networks and Technologies (ANT-2015), the 5th International Conference on Sustainable Energy Information Technology (SEIT-2015). http://www.sciencedirect.com/science/article/pii/S1877050915009047.

Martins, P.M.N., McCann, J.A., 2015b. Scaffold. http://scaffold.tech.

Massachusetts Institute of Technology, 2015. Senseable City Lab. http://senseable.mit.edu.

Newton, R., Morrisett, G., Welsh, M., 2007. The regiment macroprogramming system. In: Proceedings of the 6th International Conference on Information Processing in Sensor Networks, IPSN '07. ACM, New York, NY, USA, pp. 489–498. ISBN 978-1-59593-638-7. http://dx.doi.org/10.1145/ 1236360.1236422.

NGSI, 2015. OMA Next Generation Services Interface V1.0. http://technical.openmobilealliance.org/Technical/technical-information/release-program/current-releases/ngsi-v1-0.

Rachuri, K.K., Efstratiou, C., Leontiadis, I., Mascolo, C., Rentfrow, P.J., 2013. Metis: exploring mobile phone sensing offloading for efficiently supporting social sensing applications. In: 2013 IEEE International Conference on Pervasive Computing and Communications (PerCom). IEEE, pp. 85–93.

Santander, Smart, 2015. Smart Santander. http://smartsantander.eu.

Smith-Clarke, C., Mashhadi, A., Capra, L., 2014. Poverty on the cheap: estimating poverty maps using aggregated mobile communication networks. In: Proceedings of the SIGCHI Conference on Human Factors in Computing Systems, CHI '14. ACM, New York, NY, USA, pp. 511–520, http://dx.doi.org/10.1145/2556288. 2557358.

Streetline, Inc, 2015. Streetline, Inc. http://streetline.com.

World Sensing, 2015. World Sensing. http://worldsensing.com.

Yang, S., Adeel, U., McCann, J.A., 2013. Selfish mules: social profit maximization in sparse SENSORNETS using rationally-selfish human relays. IEEE J. Sel. Areas Commun. 31 (6), 1124–1134. ISSN 0733-8716. http://dx.doi.org/10.1109/JSAC.2013.130614.

CYBER-PHYSICAL SYSTEMS AND HUMAN ACTION

8

A RE-DEFINITION OF DISTRIBUTED AGENCY BETWEEN HUMANS AND TECHNOLOGY, USING THE EXAMPLE OF EXPLICIT AND IMPLICIT KNOWLEDGE

F. Böhle, N. Huchler

Institute for Social Science Research (ISF München), Munich, Germany

In cyber-physical systems (CPSs), an essential challenge is seen in the link between (virtual) objects of information processing and real (physical) objects processing. For instance, Edward Lee tackles this new challenge, called networking embedded CPS, by concentrating upon the timing of interacting components (Lee, 2008). Lee states that the "physical world," as opposed to the cyber world of software, is "not entirely predictable"; for CPS there is no "controlled environment" (Lee, 2008, p. 4). Nevertheless, according to Lee an effort should be made to produce reliability and predictability in CPS as well, although this effort necessarily has to be accompanied by a strategy of robustness (Lee, 2008). In our view, the challenge of CPS described by Lee is even more complicated, since CPS is usually embedded both in the physical and the social environment. Just as the physical world cannot be changed and rebooted until it fits in with the world of software, it is also impossible to exchange and adjust damaged or nonfitting parts of the social world for adaptation to the cyber world. Autonomous vehicles, smart buildings and factories, intelligent networks and service systems all communicate with human developers, coders, controllers, maintenance staff, and, last but not least, different kinds of operators and users. But social interaction of this sort is in essential aspects different, both from the logic of software communication and from the logic of interaction with physical objects.

This chapter focuses on the challenge of the social embedding of CPS. We shall distinguish different forms of relations between CPS and the social environment (Section 1), pointing out that both the communication in social relationships and the communication between humans and technology can assume different forms (Section 2). These different forms of communication are based both on explicit and on implicit (tacit) knowledge and rules (Section 3). This leads to an extension and diversification of forms of communication but also produces new uncertainty (Section 4). Against this backdrop, both extensions and limits of communication between humans and technology will be discussed (Section 5). And finally, we suggest an extended concept of distributed agency that implies a readjustment of the

Cyber-Physical Systems. http://dx.doi.org/10.1016/B978-0-12-803801-7.00008-0

division of labor between humans and technology, along the lines of a differentiation between explicit and implicit knowledge. A "reloaded" concept of distributed agency will, in our view, provide a good orientation guide for future development.

1 RELATIONSHIPS BETWEEN CPS AND HUMAN ACTORS

CPSs are installed and used in industrial production and administration, in services and also in the private sphere. In practice, they are linked not only to the physical world but also to social environments, for instance when used in the traffic system or in health services. What is more, not only do they provide certain services for the social environment, but their very functionality is usually directly linked to and dependent on human action and actors. In this respect, CPSs are not principally different from other technological tools and innovations. However, in the face of selfregulating technologies, multiagent systems and robots, the relationship between man and technology as such undergoes a process of change. In sociological discussions, the traditional contrast of human actors and passive technological objects is dismissed and the actor's role is also attributed to technology (Rammert and Schulz-Schaeffer, 2002; cf. also Latour, 1987).

It is no longer the handling, managing, and controlling of technology by human beings or the increasing replacement of human workforces by technology that is the focus of these contributions, but rather a "distributed" or "hybrid agency" between humans and technology (Rammert, 2003, 2009). This may be exemplified by a quotation concerning the control of an airplane: "In the cockpit, the pilots and a variety of assistance systems like TCAS work together in order to find a solution for a given situation" (Weyer, 2008). Within many fields of application, the "cooperation" of humans and technology is intensifying and the boundaries between both actors are becoming blurred, so it is often said that "man-machine interaction" is in the process of turning into "man-machine cooperation."

This change of the connection between humans and technology, however, allows for a number of different relationships of cooperation between human and technological actors:

- The CPS forms a stand-alone system or a framework for human living and action, following the pattern of a functional division of labor. (An example is an autonomous self-regulating production system or an "intelligent house.")
- The CPS flexibly responds to human needs and human behavior, following either the pattern of an assistance system or the pattern of customer-oriented supply and provisioning. (Examples are augmented-reality tools or the smart refrigerator.)
- The CPS cooperates directly with human actors following the pattern of cooperation in teams at work or customer-orientated interaction in personal services. (Examples are robots in working groups or robots used in healthcare: "carebots.")

In all of these relationships, there is a communication between the technological system and technological actors as well as the social system and human actors. This is most obvious in the direct form of cooperation but is also present in CPS of the self-regulated autonomous and assistance types. Again, this communication may take place in different forms. It is this aspect of communication, and the particular role of implicit and explicit knowledge within communication, which is discussed in the following sections.

2 COMMUNICATION WITHIN THE COOPERATION BETWEEN HUMAN AND TECHNOLOGICAL ACTORS

2.1 VERBAL AND SYMBOLIC COMMUNICATION

Both in common and scientific language, communication is generally understood as verbal information and understanding. Words and also signs used in communication are symbols that refer to defined things or ideas. These symbols (words, etc.) are bound to have the same meaning for all participants of the communication. Thus, a basic requirement for communication is a "common language." Misunderstandings may be caused by acoustic disturbances in oral communication or else by the failure to recognize verbal statements as words and sentences with a meaning, as in unwonted dialects or foreign languages. But communication problems will especially emerge if words and signs are linked to different meanings by the participants respectively, or if it is not clearly and distinctly understandable or understood which meaning is intended in a given case. In order to avoid such communication problems, recommendations are to limit and narrow down the informational contents of verbal symbols, using a language as simple and distinct as possible. Acoustic or written messages should be easily perceptible, even under aggravated circumstances (noise, poor sight, etc.), and their meaning should be immediately understandable and unambiguous (Silberstein and Dietrich, 2003; Sexton and Helmreich, 2003).

Verbal-symbolic communication is also possible between humans and technology. The use of keyboards and screens for information-based technologies is based upon this form of communication. In this context it is negligible which kind of medium is used, whether the communication takes place in oral or written form and whether it is standardized or relatively open. In all cases, humans give commands or relay requests to the technological system by means of words or signs. Conversely, humans receive information from or about the technological system by means of verbal symbols, and possibly also instructions and requests. Moreover, human action and behavior may be described in a verbal-symbolic way, thus serving as information for the technological system in order to adapt its functions.

2.2 ACTION-RELATED COMMUNICATION

Communication between human beings, however, does not only assume a verbal or symbolic form (which is the form that is usually addressed by the term communication) but also takes place "through action." In action-related communication, information and messages are not represented by means of symbols but are relayed directly through forms of practical action, with specific meanings being attributed to these forms of action. The relevance of nonverbal communication (gestures, facial expressions, body posture) is a well-known phenomenon (Knapp and Hall, 2006). It often accompanies verbal communication (in the form of meta-communication) but also has an important stand-alone function in situations where verbal communication is impossible. For instance, the opponents in a boxing match won't tell each other by words or signs how they are going to punch and which strategy they will pursue. If you don't want to be hit you have to observe and perceive the movements of your opponent and recognize the tactics, the punch, etc., expressed by them. In a team sport like soccer the task of the players is not only to adapt their own actions to the actions of the other team, but also to successfully communicate within the team about its collective actions, largely nonverbally (Wiemann and Harrison, 1983).

Action-related communication is also present between human actors and technology. A traditional example is machine control by means of a mechanical handwheel or else the control of a car by means of a steering wheel. Modern procedures and tools like teach-in procedures in robot programming or else touch-screens rely to some extent on the same principle, sometimes using new methods of eye-tracking and recognizing gestures. However, in these cases communication predominantly occurs via direct physical impact, which is rather a border case of action-related communication.

As shown earlier, action-related communication is essentially based upon perceiving an others' behavior and recognizing their intentions relative to the effects of this behavior. The development and improvement of (self-) regulation of robots upon the basis of visual and haptic sensors is a major field of application. In this context, interaction with the social as well as physical environment also constitutes an area of experience and learning for the development of one's own behavior. This can be seen in the improvement of so-called intelligent systems, programmed top-down as well as ex ante, by means of bottom-up learning processes, as in behavior-based robotics which are associated with the principles of subsumption architecture (Brooks, 1999, p. 11). This approach transfers insights originating from the theory of embodied cognition, notably the idea that the body plays an essential role in the ability to perceive and learn. Thus, instead of programming being orientated to the model of the human "mind," experience orientated to the human body and the human senses comes to the fore. In this context, key technologies for action-related communication are sensorics and distributed artificial intelligence. Recent developments in the design of human-technology communication refer to the automatic detection of movement patterns, gestures, facial expressions, etc.

2.3 COMMUNICATION THROUGH ANTICIPATION

Communication between human beings may also occur independently of verbal-symbolic and action-related information and messages. This is the case if the actors know how "others" behave in certain situations. This may be called a borderline case of communication since there is no actual process of communicating but, all the same, comparable results are produced and processes are initiated. A well-known example is that of standardized behavior where actions can be anticipated in a certain situation and on certain occasions before they are actually performed. This presents more difficulties if the situation is open and allows for alternative forms of action. But even in this case it seems possible to anticipate human action through the reconstruction of motives and principles of decision-making.

A theoretical foundation for this kind of communication is offered by theories of human action that primarily rely on rational decision as an explanation (Coleman, 1990). The model of *homo economicus* as it is used in economics is based on this type of theory. It implies four fundamental assumptions: (1) before an action is executed, a decision has been made about the goals of action, the ways to achieve them, the means to be used and, possibly, the potential consequences of the action. (2) Decisions are better when more information is available and more uncertainties can be eliminated. (3) Perception and processing of information as well as decision making are better when they are intellectually and rationally guided. (4) Human action follows the principle of utility maximization. According to this model of action, if it is known in which situation "others" are, an intelligent technical system is capable of anticipating which decisions will be made and which action will ensue.

A different but very similar principle is the recognition of patterns based on statistical regularities or the matching of probabilities in order to derive assumptions of future action (such as in big data algorithms or IBM's artificially intelligent computer system "Watson" with its massively parallel probabilistic evidence-based architecture, DeepQA).

On the basis of such models of action, technological systems are capable of making their own decisions and anticipating how people will decide and act in open situations (Fink and Weyer, 2014).

To sum up, we can say that the different forms of communication between human actors in social relationships, as outlined earlier, are also possible between human actors and technological systems. Subsequently a further differentiation of social action will be pointed out that is indicative of special difficulties and challenges as well as limits of communication between human and technological actors. These problems have to be considered especially in socially embedded CPS in order to tap their full potentials.

3 EXPLICIT AND IMPLICIT KNOWLEDGE

A wide-spread assumption is that knowledge and rules can generally be represented and communicated in the form of language or symbols, regardless of the concrete given conditions to which they refer. Classical examples are scientific compendia, encyclopedias, manuals, and codified bodies of rules. When describing the three forms of communication in the previous section, it was more or less taken for granted that explicit knowledge is used. Hence, it was assumed that the meanings of words and signs, of actions, and of rational decisions can be represented in explicit form and can, if necessary, be taken down in written form.

However, in scientific discussion another type of knowledge is also addressed: implicit knowledge or tacit knowing (Polanyi, 1985). During the last decades, the notion of implicit knowledge has especially been used in the context of knowledge management in enterprises (Nonaka and Takeuchi, 1995). Implicit knowledge is linked to practical action; it cannot be represented or communicated in an isolated, abstracted form. Thus, it is neither explicitly available nor explicitly recognizable as knowledge. Current pertinent discussions distinguish a "soft" and a "hard" understanding of implicit knowledge (Adloff et al., 2015). The "soft" understanding assumes that implicit knowledge indeed exists in practice and plays an important role but can principally be made explicit. According to this position, the meanings of behavior in action-related communication predominantly fall in the realm of implicit knowledge; however, these meanings are amenable to description and can be made explicit in this way, if need be. By contrast, the "hard" understanding assumes that implicit knowledge exists that is not amenable to explication, so a transformation from implicit to explicit knowledge will be impossible or at least only possible to a very limited extent.

Subsequently, we shall refer to this "hard" understanding of implicit knowledge and point out which role it plays in the forms of communication outlined earlier. This will contribute to an understanding of human communication, revealing aspects hitherto hardly noticed with respect to the interaction between human and technological actors, aspects that present a new challenge for the design and development of CPS.

3.1 IMPLICIT KNOWLEDGE IN VERBAL OR SYMBOLIC COMMUNICATION

In scholarly discussions about language and symbols, very early on a differentiation was made between denotative and connotative meanings of words and sentences (Bloomfield, 1933). A denotative meaning refers to an explicit and decidedly conceptual definition as it is taken down in dictionaries and encyclopedias. Connotative meanings, by contrast, refer to valuations, emotional qualities and also associations. Different styles or forms of language may lend a specific weight to denotative and connotative meanings respectively. For instance, poetic language consciously integrates connotative meanings of words and phrases whereas technical language primarily targets and uses denotative meanings. However, empirical studies have found that in practice even technical or technological issues are communicated by way of "story telling" (Orr, 1998; Fahrenwald, 2011). The meaning of such "stories told" predominantly emerges from shared experiences of the participating actors (Pfeiffer and Treske, 2004).

Moreover, verbal communication is often complemented by a kind of "meta-communication" in the form of gestures, looks, or sounds such as clearing one's throat. This meta-communication may support and amplify the verbal message but it may also contain independent or even contradictory messages. In written communication, this may refer to the characters and typefaces used, underlining and highlighting, or graphic addenda. Meta-communication also uses symbols but they are mostly "presentative symbols" rather than "discursive symbols," as in words or signs (Langer, 1965/1984). With presentative symbols, there is an immediate analogous relation between the symbol itself and the issue to which it refers. Examples are pictograms that emphasize specific features of the reference object and represent them in an abstract fashion. They are often in use in signposts at railway stations or airports. On the other hand, pictograms also testify to the effort to define the meanings of presentative symbols as unambiguously as possible. However, presentative symbols, in contrast to discursive symbols, are generally characterized by a tendency to ambiguity and a certain openness to interpretation and associations. An outstanding example for this quality of presentative symbols is music. In "meta-communication," this corresponds to properties such as pitch, but also speech rhythm and melody.

The general assumption is that in communication between human actors and technology, ambiguous presentative symbols are disturbing rather than useful. However, empirical studies show that experts at work with technological systems tend to take their bearing not only from technical displays but also from noises. They perceive a sound "like music," as warm, harmonious or strange. They even talk about the "melody" of a technological system (Bauer et al., 2006; Carus and Schulze, 1995; Böhle and Milkau, 1988). This perception of noises, as with the perception and application of presentative symbols in social interaction, is based upon an implicit experience-based knowledge acquired in practice, a knowledge that cannot be completely explicated. Hence, this kind of implicit knowledge in verbal-symbolic communication is hardly amenable and addressable for technological systems.

3.2 IMPLICIT KNOWLEDGE IN ACTION-ORIENTED COMMUNICATION

Human behavior and action that seems repetitive and appears simplistic at first sight, often proves to be very variable and complicated if one looks closely at it. Even a comparatively simple movement like climbing stairs requires a situational and flexible adaptation to different spatial dimensions and also to environmental influences. Such situational and flexible adaptations of essentially equal or similar processes are both frequent and necessary. For instance, in a boxing or soccer match this ability is indispensable. To be sure, ex post it is possible to establish by systematic observation, perhaps supported by video recordings, which practices and tactics the opponent prefers and maybe also what he or she intends by using them. But in the actual fight or match the player has to act immediately and flexibly according to the given situation. Moreover, it is anything but certain that the opponent will act always in the same manner since she or he is also able to adapt his or her tactics to the situation. This ability to adapt one's action situationally and flexibly to variable conditions without lengthy planning and meditation has been called a special kind of "mastership" (Könnerschaft, Neuweg, 2015) essentially based on implicit knowledge. This ability is also required in the context of work.

Thus, even when performing and re-performing actions of principally the same character, the actors have to be capable of orienting their actions situationally and flexibly to variable concrete conditions and also of relating their respective actions to each other. Scientific discussions about this issue use the terms "practical intelligence" and "practical sense," properties that enable the actors to situationally and flexibly recognize their counterparts' behavior and react to it (Alkemeyer, 2009). Another term to explain this ability is "embodied communication." This means that observation and identification of rules and patterns for behavior are replaced by an empathic "sensing" and subjective re-enacting of an others' behavior (Schmitz, 1985; Böhle and Fross, 2009). This ability allows the actors to sense and recognize "in their own body" how others behave. Empirical studies have found evidence that this kind of flexible action-related communication is especially required for mutual coordination in team or group work (Pfeiffer, 2010; Porschen-Hueck, 2010).

This kind of communication also plays an important role when dealing with technological systems. In this way, experienced actors are able to assess the state of technological systems by means of perceiving "good" or "bad" smells, "sound" and "harmonious" or "unsound" and "discordant" noises, etc. These expressions are registered in the same way as information or memos, e.g., relayed by machine displays.

These insights also offer an enhanced perspective upon the reconstruction and anticipation of action in open situations where different alternatives of action emerge.

3.3 IMPLICIT KNOWLEDGE IN COMMUNICATION THROUGH ANTICIPATION

The assumption that in open situations—characterized by different alternatives for action—human beings will decide rationally on their targets and means, corresponds to the notion of autonomous intellect-driven action, a notion that has emerged mainly in Western societies. However, in practice the preconditions for an explicit rational decision are often missing or given only to a limited extent. Thus, the anticipation of an others' behavior by referring to rational decisions is not always a promising procedure. As early as in the 1950s, the concept of "bounded rationality" (Simon, 1957, 1982) has been deemed more adequate to capture reality in economics. This concept starts from the premise that for a given actor in a concrete situation of decision, the information necessary to eliminate uncertainty is not completely available. Nevertheless, it still adheres to the idea that a rational-intellectual decision is generally necessary and possible. Recent research and discussion, by contrast, increasingly focuses on the fact that there are actually decisions that are not primarily made on a rational-intellectual basis, and that these decisions can actually be successful and adequate. Intuition, emotions and "gut feelings" are considered important bases of such decisions (Gigerenzer and Reinhard, 2007; Nippa, 2001). This is not only true for decisions in the private sphere but also for decisions in economic and technological contexts. This insight amounts to a considerable modification of the model of rational decision. The assumption that especially task-oriented decisions primarily rely on explicit knowledge is challenged. The same objection applies to a primary orientation to statistical probabilities and typical patterns of action. Models of action based on such presuppositions are ultimately equivalent to a theoretical removal of the human ability to actually make situational decisions.

Further studies and approaches also question the strict separation between decision and planning on one hand and practical implementation of action on the other. They draw attention to the phenomenon of stepwise and incremental decision-making (Quinn, 1980). The main reason for this phenomenon is that possibilities and obstacles of action often cannot be predicted in advance but only become recognizable in the practical process of acting itself. Moreover, research about real decision processes shows that human beings are actually capable of attaining goals and solving problems without previous decision-making and planning. This is especially true in situations with different alternatives for action where the informational and temporal resources necessary for rational decision are not available or else the information is inconsistent or even contradictory. There are a number of labels for this kind of action in recent approaches, as "situated action" (Suchman, 1987/2007), "intuitive-improvisatory action" (Volpert, 2003) or a certain "mastership" (Könnerschaft, Neuweg, 2015).

In studies within the context of the sociology of work, the concept of "experience-based subjectifying action" (Böhle et al., 2004; Böhle, 2009) has been developed to grasp this phenomenon. It mainly refers to coping with requirements in the world of work where a high level of uncertainty exists. In experience-based subjectifying action, the practical process of acting is not the result of previous decisions and plans but serves in itself to find and establish goals and possibilities for action. A characteristic feature of this kind of action is an exploratory, interactive-dialogical proceeding, not only in the relationships to other people but also in the dealings with material conditions and things. This proceeding is coupled with a "feeling" and "sensing" kind of perception which permits an appreciation of complex and diffuse properties of given conditions, e.g., noises or an "atmosphere," and to make use of them as "information" and a guide for action. Another component of experience-based subjectifying action comprises particular mental processes such as associative, metaphorical or pictorial thinking. Finally, a special relationship to the environment is characteristic. It is not based on "objectifying" from a distance, but on closeness, connectedness and affinity. Material things or conditions are perceived as (or like) subjects that have a life of their own and are neither totally calculable nor unilaterally manipulable.

Within experience-based subjectifying action, a special experiential knowledge, incorporated into practical action and neither extractable from it nor explicable, is both acquired and applied (Böhle, 2015).

A number of empirical surveys have found that experience-based subjectifying action has an important part to play in dealing with technological systems. In these cases, it especially serves to cope with unpredictable events and imponderabilities as well as irregularities in scientific-technological solutions (Böhle and Rose, 1992; Pfeiffer, 2007; Carus and Schulze, 1995).

4 EXTENSIONS AND LIMITS OF HUMAN-TECHNOLOGY COMMUNICATION

As pointed out in Section 2, the forms of verbal-symbolic communication, action-related communication, and communication through anticipation are all principally possible also in human-technology communication. However, looking at the development of information and communication technology, a bias toward verbal-symbolic communication is observable. Only relatively recently have efforts to enable action-related communication taken place. Well-known instances are the computer mouse, touchscreens, facial recognition systems, and the like (see Burns and Hajdukiewicz, 2004); a similar idea is the (technological) monitoring of user behavior with the intention of orienting technological systems to this behavior. Meanwhile, there are also attempts to include and enable communication through anticipation.

These extensions of human-technology communication play a significant role in CPS. They permit the use of the specific potentials of different forms of communication. For instance, action-related communication proves to be adequate both for routine activities and time-sensitive issues (Rasmussen et al., 1994).

However, the distinction between explicit and implicit knowledge throws light upon another kind of differentiation that is, as yet, hardly taken into consideration in the design of human-technology communication, but is very relevant especially with reference to CPS.

It is often taken for granted that action-related communication generally relies on implicit knowledge, since this kind of communication is not necessarily tied to explicit verbal-symbolic descriptions. Thus, action-related communication is frequently called "intuitive." But, as shown earlier, action-related communication can also be based upon explicit knowledge, as in the case of sensomotorical abilities that have been learned on the basis of explicit rules and now, because of routinization, are no longer in need of conscious regulation (Rasmussen et al., 1987). Consequently, action-related communication does not necessarily rely on implicit knowledge and is not always "intuitive."

Against this backdrop, it becomes evident that the design of human-technology communication is essentially oriented to explicit and explicable knowledge and is usually restricted to this kind of knowledge. This is not only true if communication is limited to simple standardized procedures. Even in cases where a varied and complex communication between humans and technology is possible, as in assistance systems, it is based upon explicit knowledge and formalized procedures. For instance, the attempts to permit an intuitive dealing with technological systems at the hands of their human users, or to enable technological systems to detect human gestures, facial expressions, and emotions, are basically efforts to extend the range of explicit knowledge by discovering explicit or explicable knowledge and formalized procedures within those phenomena. Other examples of a similar kind include the teaching of Artificial Intelligence systems by integrating interpretative data and symbolic interpretations and actions on the basis of probabilities (e.g., Watson) or integration of bio-feedback and other data referring to emotional states and human behavior by means of sensors in order to anticipate human actions. However, a closer look shows that all these designs of human-technology communication are usually restricted to the single aspects of the phenomena they refer to, in the sense of the "tip of the iceberg."

Within the scope of this text, a principal and conclusive answer to the question of whether it is possible to really register and simulate implicit knowledge and the correspondent form of communication by means of technology, cannot be given. But it can be stated as a result of our argumentation that there are principally different challenges to this task and that it is much more difficult to obtain success here than in communication on the basis of explicit knowledge.

Against this backdrop, it appears necessary for the development of CPS to reflect and bring to mind not only the potentials but also the limits of technization. Instead of a one-dimensional focus upon the technological simulation of human thinking and action which implies an assimilation of human actors and technology, a new perspective of the difference between humans and technology is indicated. This should not be approached in terms of deficit analysis but, on the contrary, with the intention of bringing to bear the specific potentials both of technology and of humans respectively, and to make use of them in their mutual complementarity and cooperation.

5 REDEFINITION OF HUMAN-TECHNOLOGY COOPERATION: DISTRIBUTED AGENCY THROUGH DISTRIBUTED KNOWLEDGE

As technological systems are increasingly interlinked and directly integrated into both natural and social systems, it becomes more and more difficult to exclude imponderabilities and uncertainties. But the capacity to act, to attain goals and solve problems under conditions of imponderabilities and uncertainties is essentially dependent on implicit knowledge (see Section 3 and Böhle and Busch, 2012; Böhle, 2013). In the foreseeable future, however, technological systems will only to a limited degree (at best) be capable of registering and making use of implicit knowledge. This leads to a statement that at first view may appear paradoxical: the more the potentials of CPS are developed and used, the more a systematical re-investigation of the necessary cooperation between humans and technology is indicated. Growing interlinkage, complexity and autonomy of technological systems does not reduce but rather increases their dependency on human action as their complement.

A number of concepts have been developed to determine the relationship of humans and technology. The simple confrontation of humans on one hand and technology on the other hand has been replaced by more sophisticated and differentiated descriptions. Thus, so-called MABA-MABA models ("men are better at" vs. "machines are better at") (cf. Fitts, 1951) were criticized because of their quantifying perspective and their reduction to isolated functions, and the necessity was emphasized to consider the mutual interactions of humans and technology (Dekker and Woods, 2002). The cooperative relationship between humans and technology was specifically focused upon using the concepts of socio-technical systems (Grote, 2009) and distributed respectively hybrid agency (Rammert, 2003, 2009).

The distinction between explicit and implicit knowledge allows a determination of precisely how such a notion of distributed agency should be conceived for the future. A concept of distributed agency should refer both to the specific potentials of technology and to the specific potentials of human labor capacity. The potentials of technology are mainly located in the field of explicit knowledge, objectification, and formalization. The potentials of human labor capacity, by contrast, refer to the human ability to develop and make use of both explicit and implicit knowledge. Whilst implicit knowledge is widely used in practice, it usually escapes systematic observation and is applied only tacitly. But considering the rapid progress in the development of technological potentials, it is crucial for the future of human-technology cooperation to recognize and promote not only explicit knowledge but first and foremost the forms of human action that rely on implicit knowledge.

This insight yields consequences for the design of human-technology communication: the primary focus is not (any longer) the technological simulation and replacements of human action, as in the visions of artificial intelligence, but rather precisely the difference between humans and technology. The mode of mutual adaptation and adjustment has to be replaced by the mode of structural coupling (in terms of system theory) of different forms of knowing and acting. Human beings cooperating with technological systems are bound to deal with the functional logic of technology. But conversely, technological systems cooperating with humans are bound to accept and cope with the fact that human beings act and communicate in a fashion that they cannot comprehend (or, at best, only to a limited degree).

The perspective outlined here may be illustrated using the example of (partially) automated flight control. In situations with a high degree of imponderabilities and uncertainty, as during the landing, the autopilot autonomously (!) switches off, passing the control over to the human pilot with the message "take over." This means that technology simultaneously changes its role from an autonomous actor to an assistant. Another example is a voice portal in a call center that reacts to an unexpected course of events in a conversation, unexpected customer behavior and similar disturbances by autonomously passing over to a human employee who can help "with pleasure."

Seen in this perspective, acceptance and success of cooperation between CPS and humans are much better warranted by perceptible usefulness, usability, and conditioned controllability than by the similarity between the respective actions and actors, which is always restricted and can never be really authentic. Moreover, human beings are much better capable of reconstructing and understanding the processes of an autonomous technological system (as an industrial robot) if these processes follow the rational logic of explicit knowledge, without wearing the mask of a virtual humanity. And what is more, this perspective is apt to set boundaries to a one-sided adaptation of human nature and society to technological requirements, to preventively break apparent path dependences and to permit innovation and diversity.

REFERENCES

Adloff, F., Gerund, K., Kaldewey, D. (Eds.), 2015. Revealing Tacit Knowledge. Embodiment and Explication. transcript, Bielefeld.

Alkemeyer, T., 2009. Handeln unter Unsicherheit – vom Sport aus beobachtet. In: Böhle, F., Weihrich, M. (Eds.), Handeln unter Unsicherheit. VS Verlag für Sozialwissenschaften, Wiesbaden, pp. 183–202.

Bauer, H.G., Böhle, F., Munz, C., Pfeiffer, S., Woicke, P., 2006. Hightech-Gespür. Erfahrungsgeleitetes Arbeiten und Lernen in hoch technisierten Arbeitsbereichen. Bertelsmann, Bielefeld.

Bloomfield, L., 1933. Language. Hold, Reinhard & Winston, New York.

Böhle, F., 2009. Erfahrungswissen – Erfahren durch objektivierendes und subjektivierendes Handeln. In: Bolder, A., Dobischat, R. (Eds.), Eigen-Sinn und Widerstand. Kritische Beiträge zum Kompetenzentwicklungsdiskurs. VS Verlag für Sozialwissenschaften, Wiesbaden.

Böhle, F., 2013. Handlungsfähigkeit mit Ungewissheit – Neue Herausforderungen und Ansätze für den Umgang mit Ungewissheit. Eine Betrachtung aus sozioökonomischer Sicht. In: Jeschke, S., Jakobs, E.-M., Dröge, A. (Eds.), Exploring Uncertainty. Springer-Gabler, Wiesbaden, pp. 281–293.

Böhle, F., 2015. Erfahrungswissen jenseits von Erfahrungsschatz und Routine. In: Dietzen, A., Powell, J., Bahl, A., Lassnigg, L. (Eds.), Soziale Inwertsetzung von Wissen, Erfahrung und Kompetenz in der Berufsbildung. first ed. Beltz Juventa (Bildungssoziologische Beiträge), Weinheim, pp. 34–63.

Böhle, F., Busch, S. (Eds.), 2012. Management von Ungewissheit. Neue Ansätze jenseits von Kontrolle und Ohnmacht. transcript, Bielefeld.

Böhle, F., Fross, D., 2009. Erfahrungsgeleitete und leibliche Kommunikation und Kooperation in der Arbeitswelt. In: Alkemeyer, T., Brümmer, K., Kodalle, R., Pille, T. (Eds.), Ordnung in Bewegung. Choreographien des Sozialen. Körper in Sport, Tanz, Arbeit und Bildung. transcript, Bielefeld, pp. 107–126.

Böhle, F., Milkau, B., 1988. Vom Handrad zum Bildschirm – Eine Untersuchung zur sinnlichen Erfahrung im Arbeitsprozeß. Campus, Frankfurt et al..

Böhle, F., Rose, H., 1992. Technik und Erfahrung. Arbeit in hochautomatisierten Systemen. Campus, Frankfurt a. M., New York.

Böhle, F., Pfeiffer, S., Sevsay-Tegethoff, N. (Eds.), 2004. Die Bewältigung des Unplanbaren. VS Verlag für Sozialwissenschaften, Wiesbaden.

Brooks, R., 1999. Cambrian Intelligence: The Early History of the New AI. The MIT Press, Cambridge, MA.

Burns, C.M., Hajdukiewicz, J., 2004. Ecological Interface Design. CRC Press, Boca Raton.

Carus, U., Schulze, H., 1995. Technikbedarf aus der Perspektive erfahrungsgeleiteter Arbeit in der industriellen Produktion. In: Rose, H. (Ed.), Nutzerorientierung im Innovationsmanagement. Campus, Frankfurt a. M, New York, pp. 123–149.

Coleman, J.S., 1990. Foundation of Social Theory. Belknap Press of Harvard University Press, Cambridge, MA.

Dekker, S., Woods, D.D., 2002. MABA-MABA or abracadabra? Progress on human-automation co-ordination. Cogn Technol. Work 4 (4), 240–244.

Fahrenwald, C., 2011. Erzählen im Kontext neuer Lernkulturen. Eine bildungstheoretische Analyse im Spannungsfeld von Wissen, Lernen und Subjekt. VS Verlag für Sozialwissenschaften, Wiesbaden.

Fink, R.D., Weyer, J., 2014. Interaction of human actors and non human agents. A sociological simulation model of hybrid systems. Sci. Technol. Innov. Stud. 10 (1), 47–64.

Fitts, P.M. (Ed.), 1951. Human Engineering for an Effective Air Navigation and Traffic Control System. National Research Council, Washington, DC.

Gigerenzer, G., Reinhard, S., 2007. Bauchentscheidungen. Die Intelligenz des Unterbewussten und die. Macht der Intuition, Bertelsmann, München.

Grote, G., 2009. Die Grenzen der Kontrollierbarkeit komplexer Systeme. In: Weyer, J., Schulz-Schaeffer, I. (Eds.), Management komplexer Systeme – Konzepte für die Bewältigung von Intransparenz, Unsicherheit und Chaos. Oldenbourg-Verlag, München, pp. 149–168.

Knapp, M., Hall, J.A., 2006. Nonverbal Communication in Human Interaction. Zur nonverbalen Kommunikation gehören Körpersprache, Bildsprache, Symbolik, Berührung, Musik und verschiedenste Formen, sich ohne Worte auszudrücken. Thomson Wadsworth, Belmont.

Langer, S., 1965/1984. Philosophie auf neuem Weg. Das Symbol im Denken, im Ritus und in der Kunst. Fischer, Frankfurt a.M.

Latour, B., 1987. Science in Action: How to Follow Scientists and Engineers Through Society. Open University Press, Milton Keynes.

Lee, E.A., 2008. Cyber physical systems: design challenges. University of California, Berkeley Technical Report No. UCB/EECS-2008-8. Retrieved 06.07.08.

Neuweg, G.H., 2015. Das Schweigen der Könner. Gesammelte Schriften zu implizitem Wissen. Waxmann, Münster, New York.

Nippa, M., 2001. Intuition und Emotion in der Entscheidungsforschung. State of the Art und aktuelle Forschungsrichtungen. In: Schreyögg, G., Sydow, J. (Eds.), Emotion ist Management. Managementforschung 11. Gabler, Wiesbaden, pp. 213–247.

Nonaka, I., Takeuchi, H. (Eds.), 1995. The Knowledge Creating Company. How Japanese Companies Create the Dynamics of Innovation. Oxford University Press, New York.

Orr, J.E., 1998. Images of work. Sci. Technol. Hum. Values 23 (4), 439–455.

Pfeiffer, S., 2007. Montage und Erfahrung. Warum Ganzheitliche Produktionssysteme menschliches Arbeitsvermögen brauchen. Hampp, München, Mering.

Pfeiffer, S., 2010. Leib und Stoff als Quelle sozialer Ordnung. In: Böhle, F., Weihrich, M. (Eds.), Die Körperlichkeit sozialen Handelns. Soziale Ordnung jenseits von Normen und Institutionen. transcript, Bielefeld, pp. 129–161.

Pfeiffer, S., Treske, E., 2004. Erfahrung lernen. Gestaltungsperspektiven (nicht nur) für (Tele-)Service. In: - Böhle, F., Pfeiffer, S., Sevsay-Tegethoff, N. (Eds.), Die Bewältigung des Unplanbaren. Fachübergreifendes erfahrungsgeleitetes Lernen und Arbeiten. VS Verlag für Sozialwissenschaften, Wiesbaden, pp. 245–266.

Polanyi, M., 1985. Implizites Wissen. Suhrkamp, Frankfurt.

Porschen-Hueck, S., 2010. Andere Form – anderer Rahmen. Körper- und gegenstandsvermittelte Abstimmung in Arbeitsorganisationen. In: Böhle, F., Weihrich, M. (Eds.), Die Körperlichkeit sozialen Handelns. Soziale Ordnung jenseits von Normen und Institutionen. transcript, Bielefeld, pp. 207–227.

Quinn, J.B., 1980. Strategies for Change. Logical Incrementalism. Irwin, Homewood, IL.

Rammert, W., 2003. Technik in Aktion: Verteiltes Handeln in soziotechnischen Konstellationen. In: Christaller, T., Wehner, J. (Eds.), Autonome Maschinen. Westdeutscher Verlag, Wiesbaden, pp. 289–315.

Rammert, W., 2009. Hybride Handlungsträgerschaft: Ein soziotechnisches Modell verteilten Handelns. In: Herzog, O., Schildhauer, T. (Eds.), Intelligente Objekte. Technische Gestaltung, wirtschaftliche Verwertung, gesellschaftliche Wirkung. Springer, Berlin, pp. 23–33.

Rammert, W., Schulz-Schaeffer, I., 2002. Können Maschienen handeln? Soziologische Beiträge zum Verhältnis von Mensch und Technik. Campus, Frankfurt a. M., New York.

Rasmussen, J., Leplat, J., Duncan, K. (Eds.), 1987. New Technology and Human Error. Wiley, Chichester u.a..

Rasmussen, J., Pejtersen, A.M., Goodstein, L.P., 1994. Cognitive Systems Engineering. Wiley, New York.

Schmitz, H., 1985. Phänomenologie der Leichtigkeit. In: Petzold, H. (Ed.), Leiblichkeit. Philosophische, gesellschaftliche und therapeutische Perspektiven. Junfermann Verlag, Paderborn, pp. 71–106.

Sexton, B.J., Helmreich, R.L., 2003. Using language in the cockpit: relationships with workload and performance. In: Dietrich, R. (Ed.), Communication in High Risk Environments. Helmut Buske Verlag, Hamburg, pp. 57–73.

Silberstein, D., Dietrich, R., 2003. Cockpit communication under high cognitive workload. In: Dietrich, R. (Ed.), Communication in High Risk Environments. Helmut Buske Verlag, Hamburg, pp. 9–56.

Simon, H.A., 1957. Models of Man. Wiley, New York.

Simon, H.A., 1982. Models of Bounded Rationality. MIT Press, Cambridge.

Suchman, L.A., 1987/2007. Plans and Situated Actions. The Problem of Human-machine Communication. Cambridge University Press, Cambridge et al..

Volpert, W., 2003. Wie wir handeln – was wir können. Ein Disput als Einführung in die Handlungspsychologie. Artefact, Sottrum.

Weyer, J., 2008. Mixed Governance. Das Zusammenspiel von menschlichen Entscheidungen und autonomer Technik im Luftverkehr der Zukunft. In: Matuschek, I. (Ed.), Luft-Schichten. Arbeit Organisation und Technik im Luftverkehr. Edition Sigma, Berlin, pp. 205–226.

Wiemann, J.M., Harrison, R.P. (Eds.), 1983. Nonverbal Interaction. Sage, Beverly Hills.

SECURITY AND PRIVACY IN CYBER-PHYSICAL SYSTEMS

G.A. Fink[†], T.W. Edgar[†], T.R. Rice[†], D.G. MacDonald[†], C.E. Crawford*

Oak Ridge National Laboratory, Oak Ridge, TN, United States Pacific Northwest National Laboratory, Richland, WA, United States[†]*

1 INTRODUCTION

Cyber-physical systems (CPSs) are broadly used across technology and industrial domains to enable process optimization and previously unachievable functionality. However, CPSs have been key targets in some of the most highly publicized security breaches over the last decade. Neither cyber nor physical security concepts alone can protect CPS because the cross-over effects can introduce unexpected vulnerabilities: physical attacks may damage or compromise the information system on the device, and cyber attacks can cause physical malfunctions. Because of the many critical applications where CPSs are employed, either kind of attack can result in dire real-world consequences. As a result, security and privacy must be key concerns for CPS design, development, and operation.

In this chapter, we discuss CPS from a security perspective. We explain classical information and physical security fundamentals in the context of CPS deployed across application domains. We give examples where the interplay of functionality and diverse communication can introduce unexpected vulnerabilities and produce larger impacts. We discuss how CPS security and privacy are inherently different from pure cyber or physical systems and what may be done to secure these systems, considering their emergent cyber-physical properties. Finally, we discuss security and privacy implications of merging infrastructural and personal CPS. While helping general users cope with the risks inherent in existing products is important, our goal is to help designers of emerging CPS to build more secure, privacy-enhanced products in the future by incorporating lessons learned from the present.

2 DEFINING SECURITY AND PRIVACY

Before we can discuss security and privacy of CPS, it is crucial to understand the definitions and intricacies of the terms.

☆Corresponds to PNNL report number: PNNL-SA-114171; Contract No. DE AC05-76RL01830.

Security comes from Latin roots that mean "without care" and is a quality of systems that enables people to be free of concern. Security is a set of measures to ensure that a system will be able to accomplish its goal as intended, while mitigating unintended negative consequences. When features are added to a system, security is applied to ensure that the additions do not compromise intended functionality.

The National Institute of Standards and Technology (NIST) defines *privacy* as, "Assurance that the confidentiality of, and access to, certain information about an entity is protected" (Barker et al., 2013, p. 94). "Entity," in this case, can be a corporation or facility as well as an individual person. "Certain information" may refer to any sensitive information such as personally identifiable information.

Security and privacy include the concepts of appropriate use and protection of information. Privacy is often thought of as freedom from observation, disturbance, or unwanted public attention and the ability of an individual or group to limit its selfexpression. Privacy is often seen as an aspect of security because a secure system should protect the privacy of its users. At the same time, security may be considered contrary to privacy. For instance, politicians and industry leaders endure reduced privacy to protect the public trust they hold.

2.1 INFORMATION SECURITY AND PRIVACY

NIST defines *information security* as:

> A condition that results from the establishment and maintenance of protective measures that enable an enterprise to perform its mission or critical functions despite risks posed by threats to its use of information systems. Protective measures may involve a combination of deterrence, avoidance, prevention, detection, recovery, and correction that should form part of the enterprise's risk management approach.
>
> **Kissel (2013, p. 94)**

Information security is generally characterized by core principles, which Pfleeger and Pfleeger (2007) and Cherdantseva and Hilton (2013) defined as:

- *Confidentiality*—computer-related assets accessible only by authorized parties.
- *Integrity*—assets that can be modified only by authorized parties or only in authorized ways.
- *Availability*—assets accessible to authorized parties at appropriate times.
- *Authentication*—verifies the identity, often as a prerequisite to access (Committee on National Security Systems, 2010).
- *Nonrepudiation*—protects against an individual's false denial of having performed a particular action and captures whether a user performed particular actions (i.e., sending or receiving a message) (NIST, 2013).

In general, these principles work together. Encryption provides confidentiality. Digital signatures and secure hashes ensure the integrity of messages sent and received. Redundancy of resources keeps the system available even under stress. Identities, certificates, passwords, and other mechanisms guarantee only authorized users access resources. Automatically collected records and logs may show which user accessed or modified specific parts of the system. When these logs are protected by some integrity mechanism, the result is a system with nonrepudiation.

2.2 PHYSICAL SECURITY AND PRIVACY

Physical protection aims to defend an area using the following approaches:

- *Deterrence*—prevent action via a credible threat of unacceptable counteraction and/or belief that the cost of action outweighs the perceived benefits (U.S. Department of Defense, 2016).
- *Detection*—the positive assessment that a specific object caused the alarm and/or the announcement of a potential malevolent act through alarms (U.S. Department of Energy, 2005).
- *Delay*—physical features, technical devices, security measures, or protective forces impede an adversary from accessing a protected asset or from completing a malevolent act (U.S. Department of Energy, 2005).
- *Response*—physical response to predetermined locations (based on the information provided by the intrusion detection systems) equipped with the force necessary to stop the advancement of the adversary.
- *Neutralize*—render enemy personnel or materiel incapable of interfering with a particular operation (U.S. Department of Defense, 2016).

Deterrence can be as innocuous as a sign indicating the presence of physical security components or a guard posted in a visible location to warn the potential adversary of the consequences of an attack. The next layer would be detection, usually accomplished with intrusion detection technologies, human watchers, or operational processes. Alarms may be employed to alert those protecting the asset (the trusted agents) or to scare off the attacker. Detection is effectively used in conjunction with delay. Barriers such as walls, deployed obstacles, storage containers, locks, and tamper-resistant devices take time for an adversary to penetrate, which provides delay (and some deterrence if they are visible). Automated access control systems provide log accesses to controlled areas as people pass through barriers and may gather identifying information from smart cards or other identifiers. Detection systems like motion detectors and cameras are deployed in and around barriers to quickly identify an abnormal event or adversarial action. The response to these types of events must be immediate and effective. Without a timely response, no threat can be completely neutralized. The response is initiated by the detection and must be performed quickly enough to engage the adversary prior to them traversing all of the barriers in their path. The responders require the capability to engage and neutralize the attackers for the overall protection strategy to be effective. If these elements are not properly utilized, even the most impenetrable defenses will eventually be defeated.

2.3 BLENDING INFORMATION AND PHYSICAL SECURITY AND PRIVACY

Security principles and controls in cyber security and physical security overlap but are not the same. Fig. 1 shows the cyber security principles and which physical security controls they would enable if translated to the physical domain. As is clear from the picture and will be illustrated in the examples that follow, authentication enables the most physical security controls. Unfortunately, authentication

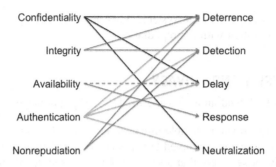

FIG. 1

Mapping of cyber security principles to the physical security controls they enable.

Implementation examples:	Confidentiality	Integrity	Availability	Authenticity	Nonrepudiation	Deterrence	Detection	Delay	Response	Neutralization
Barriers	+	+	-			++		+		+
Logs	-			+	++	+	+			
Alarms						+	++		+	
Encryption	++	+	-	+	+	+		+		+
Signatures		++		+	++			+		
Redundancy	-		++						-	
Identifiers	+			++	+	+	+	+		

FIG. 2

Mapping example security mechanisms (rows) to information security principles and physical security controls they enable (columns).

protocols are notoriously tricky to design and notedly absent in many CPS. We recommend following design principles such as those that have worked when implemented in the electric power grid (Khurana et al., 2010).

Fig. 2 shows an example of some security implementation mechanisms (the table rows) and the principles and controls to which they contribute (the columns). A "+" means the mechanism enables the principle or control. A "++" means that the mechanism is a primary means of obtaining the particular principle or control. A "−" means that implementing this mechanism may actually harm a particular security principle or control. For instance, barriers are a primary means of deterrence but actually may harm availability. These mappings show that availability and response are the least easy principles to implement via security mechanisms. We will reference Fig. 2 repeatedly throughout the chapter to recommend methods for example security problems.

3 DEFINING CPS

CPS is an umbrella term that includes systems of many other names including robotics, machine automation, industrial control systems (ICSs), process control systems, supervisory control and data acquisition systems, the industrial Internet, and the Internet of Things (IoT). These systems have different applications, architectures, and behaviors, but they all include a common attribute that makes them subsets of CPS: computerized components that interact with the physical world by sensing and controlling physical processes. CPS is the most general term and it implies communication may be happening among components of the system, but it does not clearly indicate whether the system connects to a network outside itself. *Embedded systems* is an older term for computational systems embedded into normal, "dumb" systems; however, embedded systems need not talk with each other or the larger Internet. The term *industrial Internet* connotes ICS and business-to-business linkages, but may not include consumer devices. Conversely, *IoT* has become the most popular term for CPS, but it evokes mostly commercial consumer devices (although infrastructural and business applications are included in the term's scope, NSTAC (2014)). We use CPS generally to mean any of these and use the individual terms when necessary for clarification.

3.1 SECURITY AND PRIVACY IN CPS

In this section we discuss the different application domains of CPS and the implications of failure in their security or privacy protections. The interconnectedness of CPS leads to interdependencies and system interactions that are not obvious to even careful inspection. The very nature of CPS affords both cyber and physical pathways for the adversary to obtain access to the system and greatly increases the adversary's options. These pathways may be fully protected in one domain or the other, but when these domains are joined into a single system, the only parts of the system that are truly protected are those in which both domains are protected. Meanwhile, defenses in either the cyber or physical component can be used to protect the other component in more ways than a pure cyber or physical system. For example, computerized skid detectors protect drivers from the physical danger of icy roads.

Security and privacy attack points in CPS may be at the interfaces between devices, on the devices themselves, in the infrastructure that supports them, from the Internet, and even from malicious users. Fig. 3 illustrates possible points of attack. Attackers may take advantage of the ambiguities of vulnerable communication protocols to mount an attack across an interface. They may exploit security flaws in weak implementations of application programming interfaces to compromise a component. Alternatively, they may take advantage of trust relationships between peer devices or between the devices and the infrastructures, clients, and users to whom they talk. Each of these vulnerability points must be covered by security protections or considered as potentially compromised system components from the perspective of other components.

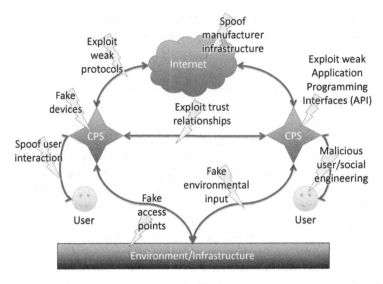

FIG. 3

Security attack points in CPS.

4 EXAMPLES OF SECURITY AND PRIVACY IN ACTION

Security and privacy in CPS are more complex than they appear. Until systems are analyzed holistically, privacy implications cannot be thoroughly understood. Part of the complexity of CPS is when they are invisibly connected to a larger network (which may in turn be connected to the Internet). The extent of the security and privacy boundaries for a device may suddenly become global in scope. Consumers are highly concerned about personal data being gathered about them and how it will be used (Groopman and Etlinger, 2015). Similarly, there is concern that the CPSs we depend on daily are plagued with security vulnerabilities. In this section, we present a series of examples to demonstrate how security and privacy are important to CPS and how difficult they are to ensure.

4.1 CYBER-PHYSICAL INFRASTRUCUTRE THREATS

The complex security implications of CPS were identified during a routine vulnerability assessment of a hydroelectric dam. The asset owners stated that their biggest fear was if both flood gates on the dam were opened at the same time, it would raise the water in the river below enough to flood half the town downstream. Two security surveys had been performed recently, one from a cyber security perspective and the other by a well-respected physical protection firm. Both assessed the dam as reasonably secure and noted the presence of a programmable logic controller (PLC) atop the dam. From a cyber perspective, the system was safe even though the PLC could be used to open both flood gates at the same time because a $10 padlock and a single tamper switch protected it. From the physical perspective the system seemed safe because, although these safeguards would be insufficient for a high-value target, the physical assessment did not consider the PLC's capabilities. However, the task time to cut the padlock off, defeat the single tamper switch, and connect a laptop to the PLC to override the security controls and open the gates was about 45 min while the quickest response would have taken approximately twice this time. This illustrates how the decades-old practice of assessing the security of CPS in domain-specific style provides an incomplete picture of the true security risks in holistic systems.

In this example, insufficient deterrence, detection, and authentication made the system vulnerable to an attack on the cyber system, potentially producing devastating physical effects. Referring to Fig. 2, we see ways to solve the lack of deterrence such as better physical barriers. Additionally, adding alarms would both increase deterrence and facilitate rapid detection. To enhance authentication, the system needs to require users to have unique identifiers and passwords so that even if one user plugged a laptop directly into the PLC, the attacker would not be able to use the system. Barriers and identifiers would also increase the delay time to use the system, giving authorities more time to react.

4.2 SMART CAR HACKING

In July of 2015, researchers Charlie Miller and Chris Valasek demonstrated to *Wired Magazine* how they could remotely hack into a Jeep Cherokee from 10 miles away while it was on the highway (Greenberg, 2015). By scanning the U.S. Sprint network for the car's Internet protocol address, they accessed the car's Internet-connected entertainment service. Unfortunately, this service was also connected to the car's controller area network (CAN), making it the only barrier between the 30–70 unprotected component system controllers and the external world. The researchers infected the service and overwrote the CAN-bus head node's firmware with a command processor that could issue commands to any system in the vehicle. The hackers could issue commands to disable the steering, abruptly engage the brakes, and even turn off the engine.

The dangerous violation of all five principles of cyber security caused the CPS to fail stunningly. From the physical side, security seemed fine, with physical locks to deter attackers, alarm systems to detect improper physical breach, and barriers to delay thieves trying to enter the vehicle. However, from the cyber side, the attackers easily identified vulnerable automobiles on the network due to a lack of confidentiality. Authentication was absent because no mechanism such as a unique login identifier and password was required to connect to the entertainment system. Checking if the new firmware was signed by a legitimate source could have enforced integrity. In a safe system (if such access could be obtained at all) the system would likely revert to a safe state or stop functioning rather than blindly accept an unverified update. System integrity was lacking because the head node could issue any command the attackers chose and no verification of commands (even ones that would induce dangerous situations) was performed. The hack made the system unavailable for use by the driver, a potentially deadly situation. Finally, nonrepudiation should have been used to record how the hack was accomplished, but the lack of logging or security identifiers made this impossible.

This example demonstrates how CPS can never be protected without enforcing the five principles of cyber security. Physical attacks were accomplished from within the physical protections by turning the cyber system against the physical side. No physical safeguards were in place to prevent unsafe acts such as violent turns of the steering wheel or braking at high speeds because the assumption was that anything operating from within the system's (weak) authentication process must be legitimate.

4.3 SMART HOME HACKING

Also in July 2015, CERT (2015) reported a textbook lack of security in home automation. Honeywell's Tuxedo Touch™ smart home controller can remotely lock/unlock the home and arm/disarm the security system. Honeywell connected the controllers to its Internet-accessible cloud servers, but the authentication mechanism was optional—it could be bypassed by ignoring the demand for a username and password as long as a legitimate user was logged in at the time. Thus, while the homeowner was logged in, an attacker could send commands in the user's name to turn on lights, operate appliances, set off alarms, view video camera feeds, unlock doors, and do anything else the system owner could do. The insecure cyber part of the CPS completely defeated the physical security offered by the system. These two vulnerabilities are officially called client-side authentication and cross-site request forgery, and they are both indexed in Mitre's Common Weakness Enumeration dictionary (available at https://cwe.mitre.org/).

Although the original design included a secure authentication system, the primary failing in this system was a lack of authentication (Bergstrom et al., 2001). Since all actions appeared to be legitimately performed by the system owner, no nonrepudiation existed. Without authentication, the system information could not be protected from falling into the wrong hands; confidentiality was broken and the integrity of the system could be violated. Furthermore, physical security controls of detection, delay, response, and neutralization were nullified because the attacker was treated as authorized. The only remaining security facets were availability and deterrence. This example illustrates how a serious deficit in one area of security can compromise most or all of the others. The system needed a strong authentication mechanism requiring full authorization every time a command was issued.

4.4 WEARABLE DEVICE VULNERABILITIES

Wearable devices may interact with collection points in stores, restaurants, along highways, or anywhere else we go, and these collection points may be nonobvious. Unclear controls and unexpected implications of sharing was the case with the infamous Fitbit sexual activity data-sharing scandal (Prasad et al., 2012). Users found that named categories of user-identifiable Fitbit data could be found via a simple web search. Some Fitbit users were surprised to find that this was public on the web and linked with identifying information, which is also a clear lack of confidentiality.

The problem was that the system designers wanted to maximize the benefits of information sharing but they did not make the implications clear to the users. Confidentiality and privacy breaches could have been avoided if the device had settings that by default did not share all categories of information and that notified users that they were sharing each class of information. Secondly, the system required no authentication to access the Fitbit information logs and made them publicly available. Thirdly, Fitbit linked the activities to individual identifiers that could easily be traced to their owners. This kind of embarrassment could have been avoided through use of private pseudonyms or anonymous sharing.

5 APPROACHES TO SECURE CPSs

Having completed an overview of security and privacy and the risks involved with CPS, we will now discuss principles for evaluating or designing CPS. While there are many general security and privacy practices (i.e., strong passwords), we focus on security mitigations and controls that are most pertinent to or have characteristics unique to CPS. We do not iterate classic cyber security literature. For those seeking instruction in the basics we suggest (Abadi and Needham, 1996).

5.1 SEGMENTATION

Segmentation applies the cyber-physical security control of deterrence by constructing a physical or logical barrier between groups of devices grouped according to communication, function, criticality, and risk. Segmentation in cyber systems may be accomplished through subnetting, encryption, virtual local area networks, access controls such as firewalls, access control lists, or software-defined networking. In general, computing assets that need to communicate with one another, that share the same risk profile, or that perform the same function should be put on a segment together. Devices that do not need frequent communication or that perform different kinds of functions should be separated. Critical functions should be split across separate machines if possible, making it difficult for an adverse circumstance to harm more than one critical function at a time.

Convenience may dictate that CPSs communicate over common IT networks or that multiple layers of functionality use the same infrastructure. However, when applying the cyber-physical security principle of segmentation, we recommend that barriers be erected between these layers and functions. Allowing unregulated access between segments with different risk profiles allows opportunities for less critical (and less well protected) functions to be used as a beachhead to attack more critical ones. The example of automotive hacking demonstrates how the less critical entertainment systems were exploited to access the critical real-time controls. Because the entertainment system is the only intermediary between the outside-world networks and the real-time, critical network, an adversary can send commands to the throttle, brake, etc., by compromising this system. Additionally, the CAN-bus protocol connecting the automotive subsystem controllers is designed for real-time communication, not security. CAN-bus has no authentication protocol, allowing any system on the segment to act at any privilege/priority level it chooses. Typical automotive network implementations have a single CAN-bus or separate busses organized by physical proximity of components. Assuming a CAN-bus is a requirement, segments should be separated by criticality, function, and risk rather than proximity. Separate segments may be joined, but security controls must be placed at the junction.

Connections between segments should employ the principles of least-privilege and need-to-know. Least-privilege provides client components or users general access to only the resources needed to fulfill their role. Need-to-know further restricts these privileges by allowing access to authorized resources only as needed to accomplish the current job. For example, a smartphone's word-processing application may need occasional access to the camera. Least-privilege would require the program to ask the user to grant this permission. Need-to-know would further prevent the application from using the camera while the user was typing, because it should not be taking pictures or video while the user is writing. Application firewalls and proxies are least-privilege measures to monitor and restrict the communication to only specific devices, protocols, and messages that are needed to communicate across the connection. State-aware protocols are need-to-know measures that prevent senseless command combinations like throwing a connected car's transmission into reverse while the car was moving forward rapidly.

Access time and user role are other dimensions of segmentation. Applications installed on and resources provided by devices in the network may be segmented. For example, iPhone applications must get permission to access the camera or other resources. However, access to the camera, once granted, endures beyond the transient need. Temporal segmentation, where access is granted only for a time, can improve security. Role-based access control is a form of segmentation that permits access to groups of functions needed to perform a particular job. If, for instance, one machine can be both a workstation and a server, the applications necessary for both of these tasks should be segmented, perhaps accessible only from separate user accounts. Specifically, the permissions required to operate these applications should be carefully monitored and restricted to only what is necessary for the particular function it must serve at that moment.

5.2 DEFENSE-IN-DEPTH

The principle of *defense-in-depth* prescribes layered defenses like a series of concentric walls protecting the vital "keep" of a castle. This principle implements the security controls of deterrence, detection, and delay to protect systems. For instance, a device that stores a private key should encrypt that sensitive information, restrict access to it via software barriers, and implement physical tamper-resistance. Defense-in-depth means that even if attackers break down the first protection, further layers will slow their advance. Coupling each barrier with alarms provides further deterrence and greater opportunity for detection.

5.3 DEFENSE-IN-BREADTH

If defense-in-depth is like protecting a single castle, *defense-in-breadth* is like coordinating the defenses of multiple castles. It implies the development of collaborative security where systems or components work together to defend each other so they will not be subject to a divide-and-conquer approach. Defense-in-breadth incorporates the concepts of least-privilege access and need-to-know.

The purpose of coordinating defenses is to limit the attack surface even though the number of devices may increase dramatically. Resilient edge-based defenses such as Digital Ants (Fink et al., 2014) can be used to coordinate defense. Smart devices with adaptive pattern recognition capabilities need the autonomy to detect attacks and respond collectively and globally. The intent is to prevent cascading failures in which an entire system is made vulnerable as a result of one poorly secured machine.

5.4 USER-CONFIGURABLE DATA COLLECTION/LOGGING

Data collection (especially data from personal CPS) can be very useful both for the user and for understanding dynamics and characteristics of groups. However, the utility of data collection must be considered in concert with preserving the privacy of the individual users. As with Fitbit's initial policy of collecting and sharing all data, users had a great utility to compare their fitness to the activities of the group. However, privacy controls were insufficient over the external visibility and identifiability of the data. When users discovered they could find out about the sexual activity recorded and unwittingly shared by others, the resulting debacle was very costly and embarrassing. One method of handling this problem would be to enforce stricter collection policies that are, by default, opt-in rather than opt-out. This will help better protect user privacy by allowing them to choose what information is shared. The default assumption must be that all of their data are private, so users must make a conscious decision to share their collected information. The data collection system must also make it clear to users exactly what is being shared and with whom. If Fitbit had explicitly listed for its users which items were being shared and with whom, they could have prevented the scandal. Such user-configurable privacy controls are applications of the principle of confidentiality.

5.5 PATTERN OBFUSCATION

One subtle way that CPS can be protected is by obfuscating the patterns of use. For example, ICS energy-draw patterns can imply the stage of an important process. Attackers could use this knowledge for reconnaissance or to cause damage to the system (Brownlee, 2015). Communication patterns in network traffic can also be mimicked by malicious entities so that intrusion detection systems are not alerted to unusual "conversations" between machines or to high throughput during odd hours. Even physical site visits to a remote ICS can form a pattern, which could give an attacker valuable information on when to attack a specific target. Obfuscation is a less obvious application of the principle of confidentiality.

In personal CPS, medical-related devices often publish information to doctors and the data may be aggregated *en masse* and posted to repositories. These repositories are useful for diagnosing conditions by comparing an individual to a population. Rather than posting exact data, the data can be resampled so that the collection is statistically identical but no longer individually identifiable. De-identification of medical and other sensitive personal data reduces legal liability in case of data theft.

5.6 END-TO-END SECURITY

End-to-end security refers to maintaining the security of data from transmission to reception and storage. Authentication, integrity, and encryption must be maintained at the application level throughout data communication between devices. As an example, a Fitbit stores data over a certain period of time that will then be uploaded via an Internet connection to the manufacturer's servers. In this example, the device, the connection method, and the final destination servers must be secure to provide end-to-end security. This can be accomplished by applying encryption on the device, to use a secure connection to transmit the data, and to ensure that the device company's servers are protected with a variety of virtual and physical methodologies.

5.7 TAMPER DETECTION/SECURITY

Deterrence and detection should be used to prevent the unauthorized manipulation of unmonitored equipment especially at remote or uncontrolled locations. This can be accomplished using strong locks, access locks that require specific authorization, security cameras, alarms, or any variety of other physical prevention and detection techniques. Additionally, authentication and nonrepudiation implemented via access logging can prevent unintentional access to the system and diagnose intrusions.

6 CONCLUSION

In this chapter we defined security and privacy and applied the classical definitions from both cyber and physical domains to the emerging domain of CPS. We illustrated the differing needs for security and privacy between infrastructural and personal CPS and demonstrated how connections between systems of both types imply the need for security and privacy throughout the entire infrastructure. Several examples demonstrated what happens when elements of security or privacy are neglected, and we discussed methods for ensuring that these systems are designed and used properly. Finally, we discussed the ongoing challenges consumers and the industry face as ubiquitous CPS becomes the new normal.

As the population of connected computerized devices like CPS continues growing exponentially, security and privacy must be taken seriously. Close attention must be paid to both the cyber and physical dimensions of CPS as well as their interplay. Although there are serious concerns, and the possibility of abuse and malice may be increased by the prevalence of these devices, both the costs and benefits must be considered when enacting design maxims or public regulations (Federal Trade Commission, 2015a,b; Wright, 2015). CPSs are here to stay and their presence will have an increasing impact on the daily lives of billions of people worldwide. As such, security and privacy concerns are incalculably important in the development, marketing, deployment, use, and obsolescence of CPS.

REFERENCES

Abadi, M., Needham, R., 1996. Prudent engineering practice for cryptographic protocols. IEEE Trans. Softw. Eng. 22 (1), 6–15. Available from: http://www.cse.chalmers.se/edu/year/2014/course/TDA602/prudent.pdf.

Barker, E., Smid, M., Branstad, D., Chokhani, S., 2013. A Framework for Designing Cryptographic Key Management Systems. National Institute of Standards and Technology, Gaithersburg, MD. NIST Special Publication 800-130.

Bergstrom, P., Driscoll, K., Kimball, J., 2001. Making home automation communications secure. Computer 34 (10), 50–56. Available from: http://ieeexplore.ieee.org/stamp/stamp.jsp?tp=&arnumber=955099&isnumber=20660.

Brownlee, L., 2015. The $11 trillion internet of things, big data and pattern of life (POL) analytics. Available from: http://www.forbes.com/sites/lisabrownlee/2015/07/10/the-11-trillion-internet-of-things-big-data-and-pattern-of-life-pol-analytics/.

CERT, 2015. Vulnerability note VU#857948—honeywell tuxedo touch controller contains multiple vulnerabilities. Available from: https://www.kb.cert.org/vuls/id/857948 (accessed 26.02.16).

Cherdantseva, Y., Hilton, J., 2013. A reference model of information assurance & security. In: 2013 Eighth International Conference on Availability, Reliability and Security (ARES), pp. 546–555. http://dx.doi.org/10.1109/ares.2013.72.

Committee on National Security Systems, 2010. National information assurance (IA) glossary. Available from: http://www.ncsc.gov/nittf/docs/CNSSI-4009_National_Information_Assurance.pdf.

Federal Trade Commission Staff Report, 2015a. Internet of things: privacy & security in a connected world. Available from: https://www.ftc.gov/system/files/documents/reports/federal-trade-commission-staff-report-november-2013-workshop-entitled-internet-things-privacy/150127iotrpt.pdf.

Federal Trade Commission Staff Report, 2015b. Careful connections: building security in the internet of things. Available from: https://www.ftc.gov/system/files/documents/plain-language/pdf0199-carefulconnections-buildingsecurityinternetofthings.pdf.

Fink, G., Haack, J., McKinnon, D., Fulp, E., 2014. Defense on the move: ant-based cyber defense. IEEE Secur. Priv. 12 (2), 36–43.

Greenberg, A., 2015. Hackers remotely kill a jeep on the highway—with me in it. Wired Magazine. Available from: http://www.wired.com/2015/07/hackers-remotely-kill-jeep-highway/; 2015.

Groopman, J., Etlinger, S., 2015. Consumer perceptions of privacy in the Internet of things: what brands can learn from a concerned citizenry. Altimeter Group, June 2015.

Khurana, H., Bobba, R., Yardley, T., Agarwal, P., Heine, E., 2010. Design principles for power grid cyber-infrastructure authentication protocols. In: 43rd Hawaii International Conference on System Sciences (HICSS), pp. 1–10.

Kissel, R. (Ed.), 2013. Glossary of Key Information Security Terms. National Institute of Standards and Technology, Gaithersburg, MD. NISTIR 7298 Revision 2.

NIST, 2013. Security and Privacy Controls for Federal Information Systems and Organizations. NIST, Gaithersburg, MD. Available from: http://dx.doi.org/10.6028/NIST.SP.800-53r4.

NSTAC (National Security Telecommunications Advisory Committee), 2014. NSTAC report to the President on the Internet of Things.

Pfleeger, C., Pfleeger, S.L., 2007. Security in Computing. Prentice-Hall, Boston, MA.

Prasad, A., Sorber, J., Stablein, T., Anthony, D., Kotz, D., 2012. Understanding sharing preferences and behavior for mHealth devices. In: Proceedings of the 2012 ACM Workshop on Privacy in the Electronic Society (WPES '12). ACM, New York, pp. 117–128.

U.S. Department of Defense, 2016. Department of defense dictionary of military and associated terms. Available from: http://www.dtic.mil/doctrine/new_pubs/jp1_02.pdf.

U.S. Department of Energy, 2005. Safeguards and Security Program Glossary—DOE M 470.4-7. U.S. DOE, Washington, DC. Available from: http://energy.gov/sites/prod/files/2013/05/f1/canceled7_SectionA_April16_2010.pdf.

Wright, J.D., Commissioner, Federal Trade Commission, 2015. How to regulate the internet of things without harming its future: some do's and don'ts. Available from: https://www.ftc.gov/system/files/documents/public_statements/644381/150521iotchamber.pdf.

PRINCIPLES

INTERFACING CYBER-PHYSICAL PRODUCTION SYSTEMS WITH HUMAN DECISION MAKERS

10

R. Müller, S. Narciss, L. Urbas

Technische Universität Dresden, Dresden, Germany

1 CYBER-PHYSICAL PRODUCTION SYSTEMS CHANGE HUMAN-MACHINE COOPERATION

Cyber-physical production systems (CPPS) are characterized by a combination of real (physical) objects and processes with information processing (virtual) objects and processes via open, partly global and continuously connected information networks (GMA 7.20, 2013). Within a CPPS, individual systems are linked according to current requirements, and connections are modified, terminated and re-instantiated at run time. As a consequence, all data and services can be made available wherever they are needed in the system. The connectivity of CPPS makes it possible to access a high amount of data stored in other systems and to exploit it to generate the necessary context information. On this basis, automated systems can adapt their behavior to the requirements of the current situation or process state (e.g., ISA 106, 2013), a task that has mainly been carried out by humans during the last 30 years of supervisory control (Sheridan, 1987).

Therefore, in CPPS a number of activities may not need human intervention anymore, such as control tasks including (re)configuration, parameterization, and optimization, as well as diagnosis, maintenance or transport. As a consequence, up to 50% of jobs in the manufacturing industries could be automated (Brynjolfsson and McAfee, 2011), particularly those involving well-defined repetitive procedures (Acemoglu and Autor, 2011). However, the remaining jobs in CPPS are expected to be reshaped and enriched, for instance by highly context-sensitive dispositive tasks (Hirsch-Kreinsen, 2014). While CPPS are expected to selforganize autonomously in routine and well-shaped non-routine situations, human operators will remain an essential part of the system to deal with unexpected events and incomplete data. Critically, in order to fulfill these functions operators need the skills to evaluate and plan whether and how CPPS components can be added, replaced, reconnected, adapted or maintained. The process of making such skilled and context-appropriate decisions consists of several challenging operations, such as information acquisition, information analysis, decision selection, action implementation (Parasuraman et al., 2000), and evaluation of action consequences. In CPPS, each of these phases can potentially be allocated to human decision makers, CPPS algorithms, or both (see Fig. 1).

Cyber-Physical Systems. http://dx.doi.org/10.1016/B978-0-12-803801-7.00010-9

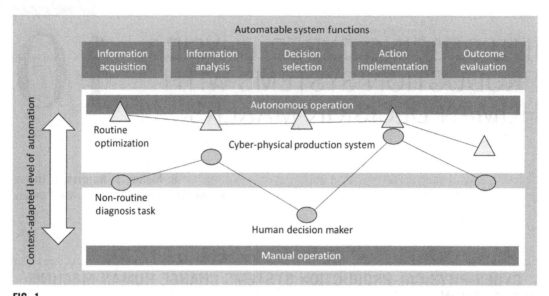

FIG. 1

Exemplary cooperation paths between CPPS and human decision makers during a problem-solving process.

If humans are to make complex decisions, their cognitive constraints need to be taken into account. For instance, humans are limited in their working memory capacity (Baddeley, 1986) and their ability to distribute attention across several aspects of the system, which often results in attentional tunneling (Endsley, 2013). Moreover, it is challenging for humans to evaluate the adequacy of their own mental models of the system and accommodate these models if necessary (Chinn and Brewer, 2001). Finally, decision-making can be affected by external and internal stressors, such as anxiety, time pressure, and mental workload. Any human-machine interface needs to be designed with these limitations in mind. However, in addition to such general concerns, the characteristics of CPPS pose two specific challenges for human decision-making: the system's *complexity* and its *flexibility*. To optimally support decisions, interface design should help operators deal with these challenges. Therefore the following sections will specify the cognitive tasks required to meet complexity and flexibility demands, and then use this as a starting point for defining interface design goals.

2 COMPLEXITY

The complexity of production systems necessarily increases as a function of smaller economic margins, faster and safer processes as well as higher demands on product quality and resource efficiency. Complexity arises when a system consists of numerous variables connected in ways that are at least partly intransparent. Additionally, progressions and cause-effect relationships tend to be nonlinear, the system dynamically changes even in the absence of operator interventions, and multiple goals need to be coordinated, which often are ill-defined or even contradictory (Dörner, 1989; Fischer et al., 2012). CPPS make this complexity manageable by means of context-aware automation, but the result is that systems get more complex internally. For instance, larger autonomous units decrease overt system complexity by reducing the number of components that operators need to be aware of. However, in the case of a malfunction such abstract system knowledge is insufficient, and instead operators need to understand the detailed setup and operation of the CPPS.

In this regard, CPPS come with a unique chance as they can provide access to the huge background knowledge stored in the digital twin of the physical systems (i.e., the cyber-part of a CPPS). However, operators need to receive this information in ways that appropriately support their problem-solving activities. Therefore psychological knowledge about the requirements of complex problem solving (Fischer et al., 2012) needs to form the theoretical basis for specific interface design goals. First, operators must be able to assess the problem structure by way of information reduction and synthesis, and build mental models of the system's problem space and goal states. Therefore CPPS interfaces should *support mental representations of the system's high-level constraints*. Second, different variables and their connections need to be understood at a level of detail that matches the requirements of a situation. Thus, *easy access to low-level features of the system* should be provided. Third, operators must be able to anticipate future developments while considering the system's current state and the results of previous actions, which calls for visualizations that *integrate the current system state with information about the past and future*. Fourth, an identification of the relevant variables is crucial for gathering the right information and planning appropriate actions, so the *relevance of information needs to be determined and displayed*. Finally, it is necessary for operators to monitor and evaluate the consequences of their actions, which can be supported by appropriate *feedback strategies*.

2.1 DESIGN GOAL C1: SUPPORT THE REPRESENTATION OF HIGH-LEVEL CONSTRAINTS

Systems can be described according to their constraints on multiple hierarchical levels of abstraction (Rasmussen, 1986), ranging from low-level aspects of their physical implementation to high-level aspects of their function, the production flow and the relations between subsystems (see Fig. 2). These high-level constraints need to be understood by operators so that they can form appropriate mental models of the system, which guide problem-solving activities. Support can be provided by strategies subsumed under the terms of *representation aiding* (Woods, 1991) and *ecological interface design* (Vicente and Rasmussen, 1992): Graphic displays are used to make the relations between variables directly perceivable. In complex process control environments, ecological interface design can improve performance (Jamieson, 2007) and situation awareness (Burns et al., 2008).

The integration of information can be supported by *configural displays* (Szalma, 2002; Bennett et al., 1993). They provide direct visual cues to the relationships between variables by mapping the values of different variables onto one geometric object or configuration to make an integrated variable emerge. For instance, in a configural display of a heat exchanger (Fig. 3, right), all diamonds would be aligned horizontally if the device was operated optimally. With deviations from horizontality being directly perceivable, this display reduces the mental effort that would be necessary to extract the same information from a standard display (Fig. 3, left). Thus, operators no longer need to integrate single measurements because their relations are displayed explicitly.

In CPPS, operators need to assess the connectivity and causal relationships between different subsystems of a plant. To facilitate this understanding of plant-wide dependencies, *intelligent topology analyzers* have been developed (Romero et al., 2014). They can merge structural models with

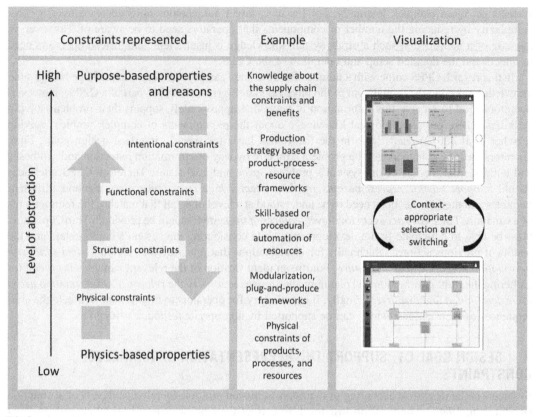

FIG. 2

Visualization of information in a CPPS according to its level of functional abstraction.

Adapted from Rasmussen, J., 1986. Information Processing and Human-Machine Interaction: An Approach to Cognitive Engineering.
North-Holland, New York, NY.

operational and process-related data, and transfer them to a visual interface. The next step is to visualize this information in ways that can easily be understood and navigated by human operators (Romero and Thornhill, 2014; Urbas et al., 2011). Moreover, fast changes of the plant topology in CPPS create new challenges such as the versioning of past topologies or integrating the partial models of distributed heterogeneous sources into an overall high-level topology model (Graube et al., 2011).

2.2 DESIGN GOAL C2: PROVIDE EASY ACCESS TO LOW-LEVEL FEATURES OF THE SYSTEM

During routine tasks in a CPPS, visualizations on high levels of abstractions usually are sufficient for operators to keep an overview. However, when there is a mismatch between their internal models and the external representations, operators should have the opportunity to *drill down to lower levels of abstraction*. The optimal level varies with the current task requirements. For instance, to represent the

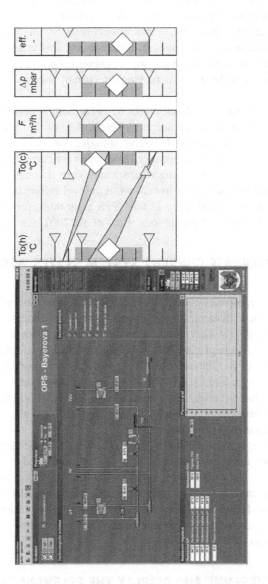

FIG. 3

Standard display *(left)* and configural display *(right)* of a heat exchanger.

From Urbas, L., Ziegler, J., Doherr, F., 2012. Produktergonomie in der Prozessautomatisierung. Z. Arbeitswissenschaft, 66, 169–182.

state and interaction of different concurrent processes in a plant, as it might be necessary in a diagnosis task, *animated functional mimic displays* (Bennett, 1992) can be used that convey information about the physical connections transporting matter or energy between components of a CPPS.

An even more detailed insight into the states of internal variables and programming logic is necessary when dealing with automation failures or when similar effects have different causes and thus call for different coping strategies. In these situations, *providing access to the raw data* enables operators to cross-check before acting. For instance, when a smart measurement instrument signals that maintenance is required, this may be due to systematic measurement bias or mechanical failure. To determine the cause, operators can inspect logic diagrams displaying the current values with the inputs and outputs of different function blocks.

In CPPS, motivating operators to cross-check might be even more beneficial than in traditional plants. This is because it allows them to gain a deeper understanding of the system and its behavior, which is challenging in these complex systems. However, a major drawback of cross-checking is its temporal cost, combined with the need to draw resources from other concurrent tasks. Therefore CPPS interfaces should support operators in determining whether and when it is necessary or beneficial to examine the raw data and when it is sufficient to rely on high-level information. One possibility is to provide information about the power and limits of the CPPS algorithms, for instance by displaying confidence intervals of state estimates and predictions (Barz et al., 2007, see Fig. 4).

2.3 DESIGN GOAL C3: INTEGRATE CURRENT SYSTEM STATES WITH THE PAST AND FUTURE

When planning interventions in complex systems, it is crucial to take the system's dynamic behavior into account, especially when these dynamics are nonlinear. The understanding of system dynamics can be supported by integrating information about the system's current state with information about past and predicted future states. An integration with past information can be achieved by visualizing the process and/or interaction history. Although *historic trend displays* do not tend to improve performance in routine control tasks (West and Clark, 1974), they can help operators to understand the development of system states (Nachreiner et al., 2006; Kindsmüller, 2006). This makes them valuable for CPPS, because it is exactly this understanding that is needed when dealing with unexpected events that are beyond the CPPS' capabilities to selforganize.

Besides integrating the present and past, operators need to be able to predict future system states and anticipate the consequences of their own actions. This can be supported by *predictive displays* (Wickens et al., 2000; Yin et al., 2014). Based on inference models, such displays provide a visualization of both the present state and the anticipated future state of a controlled process (see Fig. 4). Receiving a preview of the near-future parameter setup gives operators a chance to intervene proactively instead of merely reacting to changes. Predictive displays are expected to support proactive control even for non-routine tasks in CPPS.

2.4 DESIGN GOAL C4: DETERMINE AND DISPLAY THE RELEVANCE OF INFORMATION

The previous paragraphs emphasized the benefits of providing additional information, with the goal of enabling operators to gain a more complete understanding of the CPPS. However, at the same time this increased availability creates a serious risk of information overload. To manage the abundance of

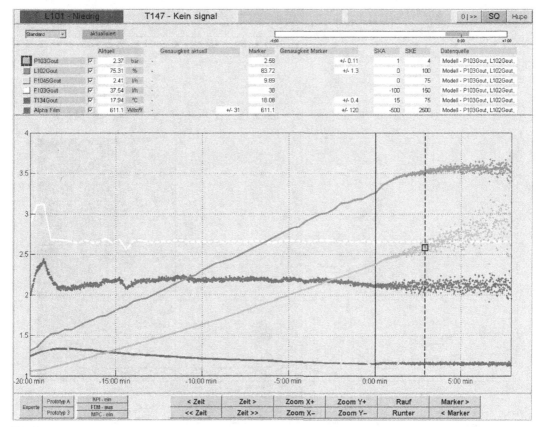

FIG. 4

Predictive multiline displays with two design alternatives for depicting the horizon-dependent confidence interval of the model-based predictions.

From Barz, A., Urbas, L., Wozny, G., 2007. Visualisierung modellbasierter Prozessgrößen im konventionellen Bedienen und Beobachten:
Nutzerzentrierte Darstellung im konventionellen Bedienen und Beobachten. P&A Kompendium 2007/2008: Das Referenzbuch für
Prozesstechnik und Automation. Publish-Industry Verlag, München.
(Continued)

information, strategies are needed to determine and display its relevance. Several statistical and model-based technologies are available to mathematically determine the situational relevance of information in CPPS, and the obtained relevance indicators can be used to filter, cue or alarm.

Filtering of data is appropriate when it keeps the structural logic of a display intact (e.g., in lists but not in mimics) and when there is no risk of depriving operators of necessary information. Obviously, the relevance algorithms would need to foresee all possible circumstances, which is not realistic in systems such as CPPS that are characterized by high numbers of unexpected events and non-routine tasks.

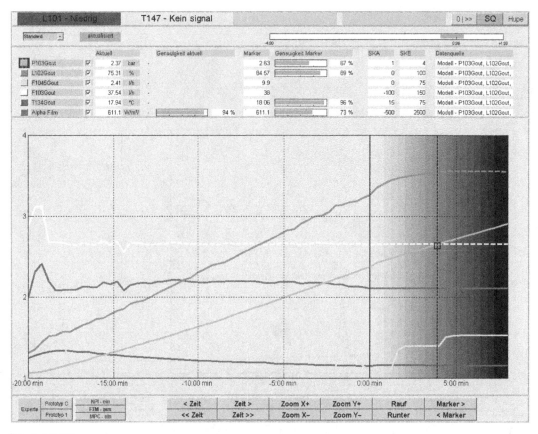

FIG. 4, CONT'D

To cope with uncertainty, *cueing strategies* such as information highlighting have been proposed (St. John et al., 2005; Yeh and Wickens, 2001). For instance, to indicate a potential problem the visual salience of a control display area can temporarily be increased by features such as color or shape (see VDI/VDE 3699-1, 2005, for a staged conspicuousness model for the process industries). However, such cueing can be problematic whenever the system is imperfect in determining the relevance of information, because operators are likely to miss relevant information when less relevant information is cued erroneously (Yeh et al., 1999). Therefore the visibility of presumably less relevant data should stay intact, for instance by variations of contrast, and the task of decluttering can be handed over to operators to guarantee that they remain aware of the information (Yeh and Wickens, 2001; Endsley et al., 2003).

As the detection of cues requires attentional resources, tasks that need constant attention or immediate action are better supported by *alarming* (EEMUA 191, 1999; VDI/VDE 3699-5, 2014). Current alarm strategies are based on binary on-off settings that result in conservative thresholds with a high number of false alarms. The latter reduce operators' trust in the automation (Lees and Lee, 2007) and thus they begin to ignore alarms, a phenomenon that is known as the *cry wolf effect* (Breznitz, 1984). To counteract this tendency, a promising strategy for dealing with uncertainty is to provide more graded indications of relevance. For instance, *likelihood alarms* have three or more stages, with each one indicating a different likelihood that the critical event is really present (Sorkin et al., 1988). Three-stage alarms can improve signal detection performance (i.e., more hits and fewer false alarms) and increase operators' trust relative to traditional two-state alarms (Ragsdale et al., 2012; Wiczorek and Manzey, 2014). Although operators still ignore some of the alarms, they now ignore those that are false while responding to correct alarms more often (Bustamante and Bliss, 2005). Alarm design, including support for cause-effect analysis, will gain even more importance in CPPS than in conventional systems, because CPPS are characterized by higher levels of automation in a wider range of situations.

2.5 DESIGN GOAL C5: DESIGN FEEDBACK STRATEGIES

Assessing the effects of control actions on system dynamics critically depends on feedback. In CPPS this topic requires special attention due to these systems' longer periods of full or partial automation in different contexts. Most importantly, the system must convey suitable *status and goal feedback*, informing operators about the progression of current processes and the achievement of goals. Based on regulatory paradigms from systems theory, feedback has to be considered as one of several basic components of a generic feedback loop (Flach et al., 2013; Narciss, 2013). Hence, to have an impact on performance, a feedback message has to provide information about the current state of a relevant control variable of the system for which clear standards or reference values must have been defined. Consequently, operators can detect discrepancies between the reference and current values, and derive control actions to reduce these discrepancies. However, this approach gets to its limits in non-routine situations where no standard is known.

Furthermore, the design of feedback depends on various characteristics of the person and task. A core issue of designing feedback strategies is to identify the conditions under which feedback will result in stable control (Flach et al., 2013). The interactive tutoring feedback (ITF) framework provides a basis for deducing design guidelines (Narciss, 2013). According to the ITF-framework, a *feedback strategy* can be considered as a coordinated plan which integrates clear and decisive statements regarding feedback conditions, feedback content, feedback scope and functions, feedback timing, as well as the form and modes of feedback presentation (Narciss, 2012). However, currently the design and implementation of elaborated feedback strategies is mostly restricted to routine tasks with a well-defined structure, and needs to be extended in order to cope with the non-routine tasks found in CPPS.

3 FLEXIBILITY

Flexible while resource-efficient production becomes more and more important as a consequence of an increasing individualization of products and a higher variety of resources such as bio-materials or renewable energy. The frequent changes in demands and environmental constraints make it necessary to quickly change goals, plans and tasks while working in a CPPS. In principle, the continuous integration with the digital plant provides all the technical information necessary for designing interfaces that support operators in performing these changes. However, interface design should also be grounded in psychological knowledge about the requirements for human action planning and control in flexible environments. First, operators need to be able to select relevant information, actions and goals from a large set of possibilities, and shield current plans from distraction by salient alternatives. To support such context-appropriate selection, the design of CPPS interfaces should *adjust information presentation to the current task context and avoid inappropriate transfer between tasks.* Second, operators need to coordinate different tasks or temporarily abandon them and then resume them after an interruption. This carries the risk of either performing incorrect actions after a task switch, or even forgetting to perform tasks altogether. To facilitate task switching and resumption, *prospective memory should be supported with external aids.* Third, to gather the information required for a given task, operators need to select visual presentations that are most helpful in the current production context, and overcome the tendency to perseverate. Thus, CPPS interfaces should *support switching between different forms of information presentation.* Finally, to face the frequent changes in CPPS setups and the resulting requirements for action selection, interfaces need to *provide guidance through tasks.*

3.1 DESIGN GOAL F1: CONTEXTUALIZE INFORMATION PRESENTATION

Due to the flexibility of CPPS, the current context given by products, processes, resources, and the associated tasks determines what actions and strategies are appropriate. For instance, consider the production of two products characterized by overlapping, bell-shaped effort-effect curves with different maxima. There is a process state region (between the maxima) where the correct way to influence the process depends on which of the two products is currently being produced, because within this region they require opposite control actions. Thus, in order to make appropriate decisions, operators need to know which production process is currently going on. To facilitate action selection in such situations, information presentation needs to be tailored to the task at hand. Furthermore, to prevent incorrect selections, operators must be aided in understanding that information which was highly relevant in another context can now be irrelevant, while previously irrelevant information has become relevant. Thus, unwanted carryover should be avoided.

CPPS frequently confront operators with new problems, and to solve new problems people usually rely on a transfer of knowledge from prior encounters with similar problems (Bassok, 1990; Kolodner, 1991). In this transfer, the understanding of structural correspondence is an important prerequisite for explanation and prediction (Lombrozo and Carey, 2006; Sloman, 2005). However, problem solvers often fail to spontaneously form structural analogies (Holyoak, 1985) and instead recall prior situations based on surface similarity (Ross, 1987). Therefore, operators should be aided in detecting and comparing the deep structure of problems. Support can be provided by various techniques, including explicit hints via *analogical comparisons* between domains (Gick and Holyoak, 1983; Goldwater and Markman, 2011), the *explication of causal patterns* (Goldwater and Gentner, 2015), pretraining with *schematic causal information* (Fernbach and Sloman, 2009) or the use of *simulations and animations* (Kubricht et al., 2015). To technically convey such information about structural and causal dependencies within a CPPS, a promising approach is offered by the interface design techniques discussed in Section 2.1, which aim at fostering an understanding of the system's high-level constraints. However, to meet the flexibility demands of CPPS, these techniques should be enriched by methods of supporting the direct comparison between different production contexts or CPPS configurations.

At the same time, the need to focus on a problem's deep structure should not be taken to indicate that surface features of information presentation are irrelevant with regard to the flexibility of CPPS. On the contrary, some physical interface features, such as the location of objects, can lead to an automatic retrieval of tasks that were associated with them in the past (Mayr and Bryck, 2007; Waszak et al., 2003), and surface similarity can trigger the retrieval of past problem-solving instances (Holyoak and Koh, 1987). Therefore different task contexts should be clearly differentiated in their visual appearance. This can be achieved by applying *univalent mappings* and *unequivocal visual structures*. For instance, moving the display elements for different tasks far apart (Wickens and Carswell, 1995) and presenting them in different colors (Wickens and Andre, 1990) can support focused attention while preventing distraction from currently irrelevant elements. Similarly, different pieces of equipment can be differentiated more clearly by labels with a large *Hamming distance* (Hamming, 1950) instead of technical abbreviations. The design of meaningful labels still is an open research question for distributed CPPS that compose highly flexible product, process, and resource frameworks. First approaches utilize the large body of terms and definitions that have been gathered in international standardization processes (Hadlich et al., 2010).

3.2 DESIGN GOAL F2: SUPPORT PROSPECTIVE MEMORY WITH EXTERNAL AIDS

CPPS pose high demands on operators' ability to coordinate different activities, switch between different tasks and remember that a particular task needs to be resumed after an interruption. This ability to remember intentions, also referred to as *prospective memory* (Kliegel et al., 2008), can account for numerous human errors in complex multicomponent tasks such as air traffic control or healthcare (Grundgeiger et al., 2014). In CPPS, prospective memory may be even more critical than in other automated systems, because CPPS are characterized by two determining factors of prospective memory performance: changes in context between different tasks, and the complexity of plans or goals that need to be maintained (Altmann and Trafton, 2002). Interface design can support prospective memory performance by outsourcing the control over intention retrieval to *external aids in the interface*. For instance, task resumption can be facilitated by keeping the task context visually present during the intermittent performance of other activities (Grundgeiger et al., 2013). Also, prospective memory is enhanced by specific cues to the identity of to-be-remembered tasks (Loft et al., 2013) and by salient reminders informing operators about the need to switch tasks. In conventional production systems, a typical measure is predefined checklists, for instance for electronic maintenance and supervision tasks. In CPPS, these checklists could be generated dynamically and adapted to the current context that is retrieved from digitized knowledge about the plant. Furthermore, checklists could be opened, modified or closed automatically whenever the technical system recognizes changes in task progression.

3.3 DESIGN GOAL F3: SUPPORT SWITCHING BETWEEN DIFFERENT FORMS OF INFORMATION PRESENTATION

To support the management of highly flexible CPPS, their information space is much larger than that of today's production systems. An appropriate reduction is achieved by means of displaying the relevance of information (see Section 2.4) and contextualizing information presentation (see Section 3.1). On the other hand, operators also need to get the possibility of exploring the large information space. To this end, information should be available on different levels of abstraction (see Sections 2.1 and 2.2). However, it is not enough to just add or subtract detail, but operators should be supported in switching between abstraction levels and changing conceptual perspective (cf. Rasmussen, 1986).

This is particularly relevant in flexible CPPS, because different tasks and production contexts call for different ways of mentally representing a problem, including different degrees of considering broad overviews versus exact parameter values. Some products are robust to variations in production parameters such as temperature or pressure, while others are highly sensitive to any deviations. Accordingly, for the first type of product it is sufficient to occasionally check high-level visualizations such as key performance indicators, while for the second type information should be available on a more detailed level, so that specific parameter values and settings can be inspected precisely, monitored in their development, and related to each other appropriately. To support operators in selecting an appropriate level of detail and in navigating between different views on the technical system, techniques of *semantic zooming* can be applied (Frisch et al., 2008; Schneider et al., 2014). At present, the capabilities of automation systems are quite restricted in this regard. It is good practice in current SCADA systems to support navigating between different views, and the basic methodologies for defining such high performance HMI are available (Hollifield et al., 2008). However, both the design methods and the integrated tool chains are not yet mature enough to support a cost-efficient engineering for CPPS (Urbas et al., 2012).

3.4 DESIGN GOAL F4: PROVIDE GUIDANCE THROUGH TASKS

In flexible CPPS with high amounts of non-routine tasks it is demanding to select which actions are to be performed at what time, in what order and in what manner. In some situations, valid information about task structure may already be available in the CPPS, so that it only needs to be communicated to the operator. In this case, the challenge is to provide step-by-step guidance, for instance by using the concept of *app-orchestration* (Pfeffer et al., 2013). Based on task models and app descriptions, it selects suitable apps from a predefined pool, adapts them to the task-specific needs, and connects them to an easily navigable app ensemble. Consequently, once the operator has finished a subtask (e.g., parametrizing a device), the orchestration immediately switches to the next app and automatically transfers all relevant data from previous processing steps.

However, a potential problem of such guidance systems is that operators are released from the need to think about the task structure and the next steps by themselves. This may discourage them from representing the task as a whole and instead make them conceive of it as a series of component operations. Thus, although receiving assistance can save time and improve performance, operators might not get sufficient opportunities to gain the expertise required for strategy selection and decision-making. As a consequence, dealing with unexpected events in a CPPS becomes difficult. The problem of too much assistance diminishing operators' competency has been termed the *assistance dilemma* (Koedinger and Aleven, 2007), and this dilemma needs to be considered in designing guidance systems that develop and maintain operators' situation awareness (Endsley, 1995).

For a considerable number of tasks in highly flexible CPPS, there is no predefined way of accomplishing them. In such situations, operators can be supported in selecting or developing appropriate strategies by *knowledge-based assistance systems*. For instance, case-based reasoning (CBR) systems (Watson and Marir, 1994) store past problem-solving episodes in relation to the particular production context. When the operator is confronted with a new problem, stored episodes are suggested as solutions, based on their match with the present problem description. First applications of CBR systems have been developed for industrial environments (Devaney and Cheetham, 2005). However, especially in CPPS that require sound inferences and creative solutions for new problems, CBR systems should mainly function as an extended memory while leaving the selection and combination of strategies to the operator (cf. Kolodner, 1991). Moreover, they should make it transparent what information their selection of cases is based on, which will support operators in evaluating the appropriateness of the suggested strategies.

4 DISCUSSION AND OUTLOOK

With the advent of CPPS, a clear trend toward higher automation in production systems is expected, which makes it possible to run more complex and flexible systems. This will add to the already observable trend to computerize routine cognitive tasks and thus modify the role of operators in manufacturing. Experience with computer-integrated manufacturing in the 1990s clearly indicates that despite the capability of full automation it is beneficial to keep humans in the loop. This is because humans are capable of coping with ambiguities, incomplete models and wrong data, they can supervise, optimize and innovate CPPS by asking the right questions, and they have the potential to responsibly deal with ill-defined or non-routine decision tasks. Therefore in CPPS the design of user interfaces is likely to become even more important than it is in today's supervisory control systems. The chapter selected several approaches from human factors research that might be helpful in tackling some of the design challenges imposed by CPPS. However, a good design of user interfaces for decision makers in CPPS needs to be complemented by structural considerations.

First, the present arguments assume that CPPS have access to a digital plant that collects all the information about the CPPS products, processes, and resources. As out-dated facts will result in wrong information and decisions the key to success is the proper *maintenance of the digital plant*. Therefore maintenance tasks that ensure a high fidelity and up-to-date digital plant need to become an integrated part of operators' daily work routines.

Second, operators and organizations need to understand that the *continuous development of competencies* will take a considerable amount of the daily work, for instance when implementing ubiquitously accessible simulator-based trainings. Besides acquiring process-oriented competencies such as decision-making, operators need to monitor and control how accurately they have mentally represented the CPPS. Without such metacognitive competencies, operators will not realize when their representations of the CPPS are erroneous or when it is advisable to use the assistance tools provided by CPPS interfaces.

Therefore human-machine cooperation in CPPS needs to become an interdisciplinary research endeavor, integrating various approaches from fields such as process control systems engineering, computer science, human factors, and instructional psychology.

REFERENCES

Acemoglu, D., Autor, D., 2011. Skills, tasks and technologies: implications for employment and earnings. In: Card, D., Ashenfelter, O. (Eds.), Handbook of Labor Economics. Elsevier, Amsterdam.
Altmann, E.M., Trafton, J.G., 2002. Memory for goals: an activation-based model. Cognit. Sci. 26, 39–83.
Baddeley, A.D., 1986. Working Memory. Clarendon Press, Oxford.

Barz, A., Urbas, L., Wozny, G., 2007. Visualisierung modellbasierter prozessgrössen im konventionellen bedienen und beobachten: nutzerzentrierte darstellung im konventionellen bedienen und beobachten. P&A Kompendium 2007/2008: Das Referenzbuch für Prozesstechnik und Automation. Publish-Industry Verlag, München.

Bassok, M., 1990. Transfer of domain-specific problem-solving procedures. J. Exp. Psychol. Learn. Mem. Cogn. 16, 522–533.

Bennett, K.B., 1992. The use of on-line guidance, graphic displays, and discovery learning to improve the effectiveness of simulation training. In: Regian, W., Shute, V. (Eds.), Cognitive Approaches to Automated Instruction. Erlbaum, Hillsdale, NJ.

Bennett, K.B., Toms, M.L., Woods, D.D., 1993. Emergent features and graphical elements: designing more effective configural displays. Hum. Factors 35, 71–97.

Breznitz, S., 1984. Cry-Wolf: The Psychology of False Alarms. Lawrence Erlbaum, Hillsdale, NJ.

Brynjolfsson, E., Mcafee, A., 2011. Race Against the Machine: How the Digital Revolution is Accelerating Innovation, Driving Productivity, and Irreversibly Transforming Employment and the Economy. Digital Frontier Press, Lexington, MA.

Burns, C.M., Skraaning, G.J., Jamieson, G.A., Lau, N., Kwok, J., Welch, R., Andresen, G., 2008. Evaluation of ecological interface design for nuclear process control: situation awareness effects. Hum. Factors 50, 663–679.

Bustamante, E.A., Bliss, J.P., 2005. Effects of workload and likelihood information on human response to alarm signals. In: Proceedings of the 13th International Symposium on Aviation Psychology, pp. 81–85.

Chinn, C.A., Brewer, W.F., 2001. Models of data: a theory of how people evaluate data. Cogn. Instr. 19, 323–393.

Devaney, M., Cheetham, B., 2005. Case-based reasoning for gas turbine diagnostics. In: The Eighteenth International FLAIRS Conference, Clearwater Beach, FL.

Dörner, D., 1989. Die Logik des Misslingens. Strategisches Denken in Komplexen Situationen. Rowohlt, Hamburg.

EEMUA 191, 1999. Alarm Systems—A Guide to Design, Management and Procurement. EEMUA, London.

Endsley, M.R., 1995. Toward a theory of situation awareness in dynamic systems. Hum. Factors 37, 32–64.

Endsley, M.R., 2013. Situation awareness. In: Lee, J.D., Kirlik, A. (Eds.), The Oxford Handbook of Cognitive Engineering. Oxford University Press, New York, NY.

Endsley, M.R., Bolte, B., Jones, D.G., 2003. Designing for Situation Awareness: An Approach to User Centered Design. Taylor & Francis, New York, NY.

Fernbach, P.M., Sloman, S.A., 2009. Causal learning with local computations. J. Exp. Psychol. Learn. Mem. Cogn. 35, 678–693.

Fischer, A., Greiff, S., Funke, J., 2012. The process of solving complex problems. J. Probl. Solving 4, 9–42.

Flach, J.M., Bennett, K.B., Jagacinski, R.J., Mulder, M., Van Paassen, R., 2013. The closed-loop dynamics of cognitive work. In: Lee, J.D., Kirlik, A. (Eds.), The Oxford Handbook of Cognitive Engineering. Oxford University Press, New York, NY.

Frisch, M., Dachselt, R., Brückmann, T., 2008. Towards seamless semantic zooming techniques for UML diagrams. In: SOFTVIS '08. HFES, Santa Monica, CA.

Gick, M.L., Holyoak, K.J., 1983. Schema induction and analogical transfer. Cogn. Psychol. 15, 1–38.

Gma 7.20, 2013. Cyber-Physical Systems: Chancen und Nutzen aus Sicht der Automation. GMA.

Goldwater, M.B., Gentner, D., 2015. On the acquisition of abstract knowledge: structural alignment and explication in learning causal system categories. Cognition 137, 137–153.

Goldwater, M.B., Markman, A.B., 2011. Categorizing entities by common role. Psychon. Bull. Rev. 18, 406–413.

Graube, M., Pfeffer, J., Ziegler, J., Urbas, L., 2011. Linked data as integrating technology for industrial data. In: Proceedings of 14th International Conference on Network-Based Information Systems (NBiS 2011).

Grundgeiger, T., Sanderson, P.M., Beltran Orihuela, C., Thompson, A., Macdougall, H.G., Nunnink, L., Venkatesh, B., 2013. Prospective memory in intensive care nursing: a representative and controlled patient simulator study. Ergonomics 56, 579–589.

Grundgeiger, T., Sanderson, P., Dismukes, R.K., 2014. Prospective memory in complex sociotechnical systems. Z. Psychol. 222, 100–109.

Hadlich, T., Mühlhause, M., Diedrich, C., 2010. Discovery and integration of information in a heterogeneous environment. In: IEEE International Conference on Emerging Technologies and Factory Automation. IEEE, Piscataway, NJ.

Hamming, R.W., 1950. Error detecting and error correcting codes. Bell Syst. Tech. J. 29, 147–160.

Hirsch-Kreinsen, H., 2014. Wandel von Produktionsarbeit—Industrie 4.0. WSI Mitt. 6, 421–429.

Hollifield, B., Oliver, D., Nimmo, I., Habibi, E., 2008. The High Performance HMI Handbook. PAS, Houston, TX.

Holyoak, K.J., 1985. The pragmatics of analogical transfer. In: Bower, G.H. (Ed.), The Psychology of Learning and Motivation. Academic Press, New York.

Holyoak, K.J., Koh, K., 1987. Surface and structural similarity in analogical transfer. Mem. Cognit. 15, 332–340.

Isa 106, 2013. Procedure automation for continuous process operations. Technical Report ISA-TR106.00.01-2013, ISA, North Carolina.

Jamieson, G.A., 2007. Ecological interface design for petrochemical process control: an empirical assessment. IEEE Trans. Syst. Man Cybern. 37, 906–920.

Kindsmüller, M.C., 2006. Trend-Literacy—Zur Interpretation von Kurvendarstellungen in der Prozessführung. Shaker Verlag, Aachen.

Kliegel, M., Mcdaniel, M.A., Einstein, G.O., 2008. Prospective Memory: Cognitive, Neuroscience, Developmental, and Applied Perspectives. Taylor & Francis Group/Lawrence Erlbaum Associates, New York, NY.

Koedinger, K.R., Aleven, V., 2007. Exploring the assistance dilemma in experiments with cognitive tutors. Educ. Psychol. Rev. 19, 239–264.

Kolodner, J., 1991. Improving human decision making through case-based decision aiding. AI Mag. 12, 52–68.

Kubricht, J.R., Lu, H., Holyoak, K.J., 2015. Animation facilitates source understanding and spontaneous analogical transfer. In: Dale, R., Jennings, C., Maglio, P., Matlock, T., Noelle, D., Warfaumont, A., Yoshimi, J. (Eds.), Proceedings of the 37th Annual Conference of the Cognitive Science Society. Cognitive Science Society, Austin, TX.

Lees, M.N., Lee, J.D., 2007. The influence of distraction and driving context on driver response to imperfect collision warning systems. Ergonomics 50, 1264–1286.

Loft, S., Smith, R.E., Remington, R.W., 2013. Minimizing the disruptive effects of prospective memory in simulated air traffic control. J. Exp. Psychol. Appl. 19, 254–265.

Lombrozo, T., Carey, S., 2006. Functional explanation and the function of explanation. Cognition 99, 167–204.

Mayr, U., Bryck, R.L., 2007. Outsourcing control to the environment: effects of stimulus/response locations on task selection. Psychol. Res. 71, 107–116.

Nachreiner, F., Nickel, P., Meyer, I., 2006. Human factors in process control systems: the design of human-machine interfaces. Saf. Sci. 44, 5–26.

Narciss, S., 2012. Feedback strategies. In: Seel, N. (Ed.), Encyclopedia of the Learning Sciences. Springer Science & Business Media, LLC, New York, NY.

Narciss, S., 2013. Designing and evaluating tutoring feedback strategies for digital learning environments on the basis of the interactive tutoring feedback model. Digit. Educ. Rev. 23, 7–26.

Parasuraman, R., Sheridan, T.B., Wickens, C., 2000. A model for types and levels of human interaction with automation. IEEE Trans. Syst. Man Cybern. 30, 286–297.

Pfeffer, J., Graube, M., Ziegler, J., Urbas, L., 2013. Vernetzte Apps für komplexe Aufgaben in der Industrie. atp Edition—Automatisierungstechnische Praxis, vol. 55, pp. 34–41.

Ragsdale, A., Lew, R., Dyre, B.P., Boring, R.L., 2012. Fault diagnosis with multi-state alarms in a nuclear power control simulator. In: Proceedings of the Annual Meeting of the Human Factors and Ergonomics Society, vol. 56, pp. 2167–2171.

Rasmussen, J., 1986. Information Processing and Human Machine Interaction: An Approach to Cognitive Engineering. North-Holland, New York, NY.

Romero, D.D., Thornhill, N.F., 2014. Integration, navigation and exploration of plant topology networks using the property-graph model. In: Proceedings of the IEEE/SICE International Symposium on System Integration, pp. 743–748.

Romero, D.D., Graven, T.-G., Thornhill, N.F., 2014. Towards the development of a tool for visualising plant-wide dependencies. In: Proceedings of the Emerging Technology and Factory Automation (ETFA), pp. 1–4.

Ross, B.H., 1987. This is like that: the use of earlier problems and the separation of similarity effects. J. Exp. Psychol. Learn. Mem. Cogn. 13, 629–639.

Schneider, F., Graube, M., Urbas, L., 2014. Integrierter Informationsraum: Responsive HMI für Leitwarte und Feld. VDI-Berichte 2222, Tagungsband USEWARE 2014.

Sheridan, T.B., 1987. Supervisory control. In: Salvendy, G. (Ed.), Handbook of Human Factors/Ergonomics. Wiley, New York, NY.

Sloman, S.A., 2005. Causal Models: How People Think About the World and Its Alternatives. Oxford University Press, Oxford.

Sorkin, R.D., Kantowitz, B.H., Kantowitz, S.C., 1988. Likelihood alarm displays. Hum. Factors 30, 445–459.

St. John, M., Smallman, H.S., Manes, D.I., Feher, B.A., Morrison, J.G., 2005. Heuristic automation for decluttering tactical displays. Hum. Factors 47, 509–525.

Szalma, J.L., 2002. Workload and stress of configural displays in vigilance tasks. In: Proceedings of the Human Factors and Ergonomics Society, vol. 46, pp. 1536–1540.

Urbas, L., Pfeffer, J., Ziegler, J., 2011. iLD-apps: usable mobile access to linked data clouds at the shop floor. In: Proceedings of the Workshop on Visual Interfaces to the Social and Semantic Web (VISSW 2011), Palo Alto, California.

Urbas, L., Obst, M., Stöss, M., 2012. Formal models for high performance HMI engineering. In: Troch, I., Breitenecker, F. (Eds.), Proceedings MathMod 2012.

VDI/VDE 3699-1, 2005. Process Control Using Display Screens—Principles. Beuth, Berlin.

VDI/VDE 3699-5, 2014. Process Control Using Display Screens—Alarms/Messages. Beuth, Berlin.

Vicente, K., Rasmussen, J., 1992. Ecological interface design: theoretical foundations. IEEE Trans. Syst. Man Cybern. 22, 1–18.

Waszak, F., Hommel, B., Allport, A., 2003. Task-switching and long-term priming: role of episodic stimulus-task bindings in task-shift costs. Cogn. Psychol. 46, 361–413.

Watson, I., Marir, F., 1994. Case-based reasoning: a review. Knowl. Eng. Rev. 9, 327–354.

West, B., Clark, J.A., 1974. Operator interaction with a computer controlled distillation column. In: Edwards, E., Lees, F.P. (Eds.), The Human Operator in Process Control. Taylor & Francis, London, UK.

Wickens, C.D., Andre, A.D., 1990. Proximity compatibility and information display: effects of color, space, and objectness on information integration. Hum. Factors 32, 61–77.

Wickens, C.D., Carswell, C.M., 1995. The proximity compatibility principle: its psychological foundation and relevance to display design. Hum. Factors 37, 473–494.

Wickens, C.D., Gempler, K., Morphew, M.E., 2000. Workload and reliability of predictor displays in aircraft traffic avoidance. Transport. Hum. Factors 2, 99–126.

Wiczorek, R., Manzey, D., 2014. Supporting attention allocation in multitask environments: effects of likelihood alarm systems on trust, behavior, and performance. Hum. Factors 56, 1209–1221.

Woods, D.D., 1991. Representation aiding: a ten year retrospective. In: Proceedings of the IEEE International Conference on Systems, Man, and Cybernetics, "Decision aiding for complex systems," vol. 2, pp. 1173–1176.

Yeh, M., Wickens, C.D., 2001. Attentional filtering in the design of electronic map displays: a comparison of color coding, intensity coding, and decluttering techniques. Hum. Factors 43, 543–562.

Yeh, M., Wickens, C.D., Seagull, F.J., 1999. Target cuing in visual search: the effects of conformality and display location on the allocation of visual attention. Hum. Factors 41, 524–542.

Yin, S., Wickens, C., Helander, M., Laberge, J.C., 2014. Predictive displays for a process-control schematic interface. Hum. Factors 57, 110–124.

SOCIAL NETWORK SIGNAL PROCESSING FOR CYBER-PHYSICAL SYSTEMS

11

T. Abdelzaher*, S. Wang*, P. Giridhar*, T.A. Amin*, P. Seetharamu*, H. Roy[†],
H. Wang*, L. Kaplan[†], E. Bowman[†], J. George[†]

University of Illinois at Urbana-Champaign, Champaign, IL, United States[*] *US Army Research*
Laboratory, Adelphi, MD, United States[†]

1 INTRODUCTION

We live in an exciting time, where the realm of cyber-physical systems is expanding from classical embedded applications, such as factory automation and process control, to new domains that increasingly feature humans in the loop. Examples of such new applications include disaster response, smart transportation, national security, and urban sustainability, to name a few. This chapter addresses the question of whether social networks, such as Twitter, can be explored as a novel sensing modality in such emerging cyber-physical applications in social spaces.

According to the United Nations, today, 54% of the world's population lives in cities. This percentage will increase to 66% by 2050 (Department of Economic and Social Affairs, United Nations, New York, 2014). Arguably, humans are some of the most versatile and widely deployed "sensors" in urban spaces. They are the owners and users of "smart things" on the Internet of Things; the survivors and first-responders in post-disaster operations; the commuters in intelligent transportation applications; and the witnesses of suspicious activity in national security scenarios. These individuals may relay important information such as real-time traffic conditions around recent accidents, locations of spreading post-disaster damage such as flooding or fires, real-time progress of unfolding unpredictable gatherings (e.g., demonstrations and protests), and unusual events that impact safety. Today, such observations are voluntarily shared on social media, leading to the prospect of exploiting social media as an observer of the physical world in future cyber-physical systems.

The complementary nature of human observations and physical sensor measurements offers compelling reasons for leveraging both. Consider an application such as disaster response. Measuring ground truth conditions in such an application is challenging. It requires damage assessment on a large scale. While specialized physical sensors can aid with damage assessment, for example, by detecting chemical leaks or reporting power disruptions, additional context can be reported by humans who are the disaster survivors or first responders. A sensor can report disconnection of a power line. Humans, on the other hand, can offer input that helps explain this disconnection, such as a fallen tree across the wires. The same wisdom applies to measuring traffic conditions in the city to help streamline

Cyber-Physical Systems. http://dx.doi.org/10.1016/B978-0-12-803801-7.00011-0

transportation. While traffic sensors might offer factual information about current traffic speed and backlog, human sources can offer additional context that helps explain the current conditions and anticipate their evolution. For example, they may report traffic accidents, road hazards, or maintenance problems that affect the flow of traffic. Taking such context into account can, in turn, help improve the accuracy of traffic forecasts. Indeed, exploiting humans as "intelligent sensors" in the smart city could significantly improve situation understanding. We therefore envision humans being an important "sensing subsystem" in future cyber-physical systems. They will contribute key measurements and complement traditional physical sensors by offering additional observations that help explain data, and that provide actionable information to improve planning and actuation.

It remains to conjecture on how human measurements can be seamlessly fused with more traditional sensing sources. Specialized applications might offer individual avenues for human feedback (e.g., via appropriate smartphone "apps" that interested users can download and use). This approach requires investment in publicity and proper incentives to recruit individuals to use these specialized applications (Gao et al., 2015; Koutsopoulos, 2013). Social networks offer a different avenue for data collection from human sources. Social sources already volunteer to post over 500 million tweets and over 80 million Instagram photos per day (Social Media Statistics, 2016), among other data. We show in this chapter that it is possible to exploit social networks as sensor networks to detect and track instances of physical events, offering input into cyber-physical systems. There is, of course, a mismatch in that social network data constitutes primarily unstructured text (e.g., Twitter) and images (e.g., Instagram), whereas feedback loops in cyber-physical systems need structured data. Mechanisms are needed to convert social network data into structured physical events and locations. Recent work has shown that events of interest (such as hazards or traffic accidents in the city) can be detected (Atefeh and Khreich, 2015), localized (Panteras et al., 2015), verified (Pasternack and Roth, 2011), and tracked over time (Guille and Favre, 2014), all in an automated fashion from social network data.

Public data uploaded to social networks can be thought of as a signal that reflects conditions of the physical world. The social signal in question acts as a novel *sensing modality*, not unlike acoustic, vibration, or magnetic sensing. Much the way physical objects induce distinguishable signals in their physical environment that can be detected by observing the physical medium, socially relevant events (such as car accidents, attacks, fires, natural disasters, anomalous events, parades, or protests) induce distinguishable signals in their social environment that can be detected by observing the social media. On physical media, signals from multiple objects are mixed as they impinge upon the sensor. These signals can be separated via an appropriate source separation or data association method, such as demultiplexing or band filtering. In social sensing, social signals emitted in response to events play the role of physical signals emitted from objects. A challenge lies in separating out the information bits about a single event from other information bits about other events. In that sense, the separation algorithm acts as a demultiplexer in electronic systems. The chapter describes recent progress in exploiting the social modality of sensing, motivated by the significant popularity and growth of modern social networks. Examples of extracting information from Twitter and Instagram are provided.

2 THE SOCIAL SIGNAL: A TWITTER STORY

If humans (on social media) were sensors, then their observations of physical events that surround them would be "measurements" of those events, and the chronological grouping of such observations that pertain to the same event instance over time would be the *observed trajectory* of the event instance in question. This trajectory evolves not only in physical space but also in some appropriately defined semantic space, such as the space of text embedding (Tang et al., 2015). This analogy gives rise to the prospect of representing the monitored environment as a set of event instances described by their trajectories and other metadata, with social networks being the instrument for event tracking. For the purposes of this paper, event tracking refers to reconstructing the evolution of event instances both in the physical world and in semantic space. An event instance can be broadly defined as an incident, independently observable by multiple humans within limited time and space. The idea that one can view Twitter users as sensors to detect and track events, such as earthquakes or protests, dates back several years (Sakaki et al., 2010).

Fig. 1 illustrates the analogy between event and target tracking. In physical spaces, objects such as tanks perturb the physical medium thereby eliciting a detectable physical response (such as acoustic waves or seismic vibrations). This response can be represented by changes in signal distribution in multiple signal frequency bands and can be detected by observing these deviations. Similarly, in social spaces, events such as accidents, protests, and confrontations perturb the surrounding social media thereby eliciting a detectable social response (such as the posting of tweets describing the event). This response can be represented by changes in signal distribution in multiple lexical frequency bands (i.e., frequency of occurrence of different keywords on the social medium) and can be detected by observing these deviations.

There is an important difference, however, between tracking tanks in physical media and tracking events in social media that is both a challenge and an opportunity. A tank (of a particular model) usually has a consistent signature in the physical medium that can be learned over time. Each instance of the tank will produce largely the same signature. This intuition was used in the past to identify different event *categories* on Twitter, such as earthquakes or typhoons using their characteristic signatures (i.e., keywords) (Sakaki et al., 2010). Beyond those category-wide keywords, events of the same category (e.g., different car accidents) may differ widely in the vocabulary used to describe them. Indeed, all accidents might share some category-related word (e.g., "accident" or some synonym). However,

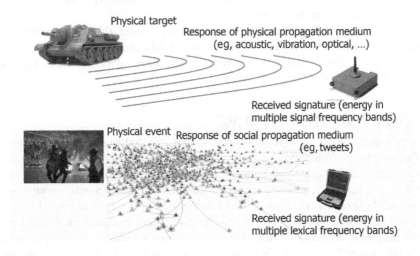

Physical target
Response of physical propagation medium
(eg, acoustic, vibration, optical, ...)

Received signature (energy in
multiple signal frequency bands)

Physical event Response of social propagation medium
(eg, tweets)

Received signature (energy in
multiple lexical frequency bands)

FIG. 1

The social sensing modality and its analogy with physical sensing.

references to individual accidents may be characterized by instance-specific words such as names of locations or landmarks where the accident occurred, references to the car types involved, or descriptions of their content. This makes it possible to demultiplex different instances. It also introduces a challenge, because those characteristic keywords of individual instances are not known in advance nor can they be easily learned through prior training because they are too specific to each instance. This challenge is reminiscent of the *data association problem* in target tracking (Schulz et al., 2001; Cox, 1993). It is also related to *stream clustering* literature in data mining (Aggarwal and Philip, 2006; Aggarwal, 2013), offering ideas for solving the problem.

Another problem to consider lies in the changing patterns of social media use. As social media users continue to increase in number and diversity, the lexical features of posted signals will change (e.g., special expressions, popular slang, and hashtags), reflecting the complex cultural terrain of the user population. Developing signal detection strategies that are agnostic to the specifics of the current use of the social media would thus be ideal.

2.1 PROBLEM FORMULATION

Consider discretizing time into slots. The monitored environment can be represented at any discrete time instant by a dynamically evolving set of ongoing event instances, where an event instance has a detection (or start) time and a finish time. Each event instance further belongs to a class or category label. For example, an event could be of class, say, a car accident, a parade, or a protest. Each event instance is further associated with a chronologically sorted list of all times-tamped tweets that describe it up to the current time, as well as a possibly evolving current location time-series. Given an event class, we may know a mobility model for the event. That is to say, given a current event location, we can determine the likelihood of future event location candidates. There are many scenarios in which this likelihood is easily computed. For example, events such as car accidents tend not to move. It is likely then that their predicted future locations coincide with their current ones. Other events, such as parades, may move slowly. Hence, likelihood of future locations can be inferred from the current direction and speed of motion. Finally, if we do not know the event mobility model, then future location likelihood is independent of current location.

The social medium is said to emit a signal. In the case of Twitter, this would be the set of tweets time-stamped in the current slot. The Twitter programming API can be used to collect tweets in real time as they are emitted. The challenge is to find event trajectories given this input. Fig. 2 presents a conceptual view of such an event tracking system.

2.2 THE SOCIAL SIGNAL FEATURE SPACE

Analogously to a multi-target tracking problem, consider the challenge of distinguishing different *instances* of an event type of interest. Taking an Occam's razor approach, we look for the simplest signal feature space in which individual event signatures can be independently detected and demultiplexed. The signal, in this case, is the frequency distribution of tokens (words) emitted on the social medium. Consider the lexicon of commonly used words in a language, such as English. Such a domain may contain around 10,000 words. We may want to distinguish thousands of concurrent event instances, each described by multiple characteristic words. In this case, the set of event instances populate the space of words rather densely. That is to say, there may be partial overlap between sets of words

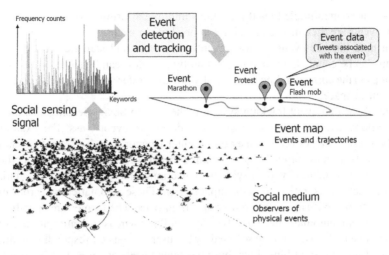

FIG. 2

An event tracking system.

commonly used in describing *different* event instances. This motivates considering higher powers of the signal; in this case, *word pairs*. In a language of 10,000 words, there are 100 million possible word pairs. This is several orders of magnitude larger than the number of event instances we might need to demultiplex within any given time slot. Consequently, within a given time slot, the set of word pairs that characterize ongoing event instances populate *very sparsely* the feature space of all possible word pairs. The probability of overlap (i.e., different event instances being characterized by the same word pair) is negligible. Accordingly, word pairs (or the "square" of the lexical signal) make good features for event detection. Two caveats accompany our observations:

- The validity of the sparsity observation in the feature space of keyword pairs hinges on the lack of strong correlations between the keywords used in the chosen pairs. If two words often come together as a single term, such as "Dodgers Stadium" or "Angela Merkel", the term should be considered as a single keyword. Accordingly, some preprocessing is needed to lump generally co-occurring words into a single one, which is a technique well-studied in phrase and frequent pattern mining (El-Kishky et al., 2014; Han et al., 2007). Applying ideas from these techniques, we remove from consideration keywords pairs, where the individual keywords co-occur with a much higher probability than the product of the probabilities of their occurrence individually. With that preprocessing, we ensure lack of strong correlations between keywords in a pair.
- Sparsity ensures that if two event instances are different, their discriminative keyword pairs are different with high probability. The inverse is not always true. Given two different discriminative keyword pairs, they may or may not be of two different event instances. This will be the case, for example, if the event instance has more than two high frequency keywords, allowing for multiple alternative subsets of two keywords to uniquely characterize the event. Such subsets would have to be consolidated.

An event tracking system therefore trivially operates as follows. As input tweets arrive, they are inspected for the discriminative keyword pairs they contain and placed into the corresponding bins.

A tweet may be placed in multiple bins if it contains multiple discriminative keyword pairs. Bins containing similar tweets are then merged resulting in one bin per event. Once the input has been demultiplexed into a data bin per event, various event parameters and statistics can be computed. For example, events can be counted. Their start and end times-tamps can be trivially found by inspecting tweet times-tamps in the corresponding bins. Finally, location references in tweets inside a bin can help localize the event in space.

An example of such an event tracking system is the Apollo Social Sensing Toolkit (Wang et al., 2014). Apollo allows users to enter keywords that define an event class. For example, to collect information about civil unrest events, a user might enter keywords such as "protest" and "confrontation". Apollo collects from the input Twitter stream all tweets containing the user-specified keywords. It then performs event detection by identifying new discriminative keywords that are indicative of the occurrence of distinct events, reflected in the input stream. These automatically learned event-specific keywords are not to be confused with the user-specified ones. The former are specific to individual event instances and are automatically identified by Apollo, whereas the latter are used to define the event class of interest to the user and are selected by the user. Tweets corresponding to different events are then separated into different bins. Each bin represents tweets about a specific event (e.g., tweets about a specific protest) in the collected stream, and is associated with its own event-specific learned keywords. These functions are described below.

2.3 EVENT DETECTION

When a new event occurs, it changes the distribution of keywords emitted on the social medium. In our feature space of word pairs, this causes new signal spikes. A suitable measure for detecting such spikes is *information gain*. When a new event occurs, keyword pairs characteristic to that event will be present disproportionately in the current window compared to the previous one. It is possible to compute information gain of a keyword pair in a window as the amount of information gained in distinguishing this window from previous windows if we were told whether or not the given keyword pair occurred in that window. Clearly, pairs that occur more disproportionately in the current window offer more information gain. This idea was first proposed in Giridhar et al. (2014) to distinguish different traffic accidents.

Discriminative keyword pairs are those that have a sufficiently high information gain, as determined by a threshold selected depending on the granularity of events we wish to observe. Selecting this threshold is akin to calibrating the sensitivity of a physical sensor. A bin corresponding to each newly detected discriminative keyword pair is then created, enumerating tweets that contain that pair. Note that, an event may give rise to more than one discriminative keyword pair. Therefore, we consolidate different bins when there is substantial overlap in the tweets they contain. A distance metric is needed between the content of different bins based on the statistical distribution of words in the bins. The distance between two bins decides if they need to be consolidated. Examples of common distance metrics between statistical distributions of words are the *Jaccard distance*, the *term frequency difference ratio*, the *cosine similarity distance*, and the *Kullback-Leibler (KL) divergence*. Of these, prior work has shown that the Jaccard distance performs consistently well. It is also the simplest metric, since it is the only one of the aforementioned four that does not consider word frequency. The reason, we believe, is that the individual bins are small enough that it is inaccurate to estimate the true probability of each keyword solely based on its frequency of occurrence in the bin. Hence, metrics that depend on probability distributions, approximated as word counts, are less accurate.

2.4 EVENT LOCALIZATION

Prior work addressed the challenge of identifying location references in tweets. For example, a reference to a street name in a tweet can be identified and treated as a *location tag*. To localize events, an event location is decided by a majority vote of the location tags of tweets in the event bin. If none of the tweets contains a location reference, then the bin is not localized. Table 1 summarizes the fraction of localized bins in four datasets collected in 2015, labeled by (i) *Disaster*, (ii) *Protest*, (iii) *Traffic*, and (iv) *War*.

- *Disaster*: A dataset collected using keywords "disaster", "humanitarian", and "earthquake".
- *Protest*: A dataset collected using keywords "protest" and "confrontation".
- *Traffic*: The keywords used here are "traffic", "accident", and "stuck".
- *War*: The keywords used here are "rebels", "attack", and "bombing".

Note that, a location was computed for more than half of the events. This is a positive result considering that, on Twitter, usually fewer than 2% of the tweets are geotagged. Table 2 summarizes the data collected.

Next, we demonstrate the accuracy of event demultiplexing by comparing different schemes for separating multiple concurrent events. The comparison results are shown in Table 3. The table compares different algorithms for associating tweets with distinct events. The key aspect of each algorithm

Table 1 Summary of Localization Ratio of Bins

Dataset	Disaster	Protest	Traffic	War
Localization ratio	80.96% (642/793)	79.94% (562/703)	52.89% (997/1885)	83.81% (528/630)

Table 2 Summary of Twitter Datasets

Event Type	Start Time (UTC)	End Time (UTC)	# Tweets Collected	# Events Detected	Locations Identified
Disaster	Apr. 19 19:41:08 2015	May 02 16:58:47 2015	319,181	121	104 countries or cities
Protest	Apr. 01 06:03:51 2015	May 04 06:50:40 2015	329,336	178	129 countries or cities
Traffic	Apr. 19 12:12:50 2015	May 02 17:05:59 2015	480,717	392	307 countries or cities
War	Apr. 19 19:45:17 2015	May 02 17:03:21 2015	391,711	103	95 countries or cities

Table 3 Comparison Between Distance-Based and Location-Based Clustering Schemes

Scheme	Disaster (Error)	Protest (Error)	Traffic (Error)	War (Error)
Jaccard	**19.24%**	12.33%	**0.83%**	**4.12%**
Location	33.07%	**9.12%**	2.11%	18.02%

lies in how it performs tweet clustering (with the understanding that different clusters will be associated with different events). Specifically, we compare location-based clustering to clustering based on lexical features (bins consolidated using the Jaccard distance). In all cases, the unit to cluster is a bin of tweets with the same discriminative keyword pair. The table shows the percentage of incorrectly grouped bin pairs (i.e., bins grouped together that are not about the same event and bins about the same event that are not grouped together). We call this percentage, *consolidation error*.

From Table 3, we observe that for Disaster, War, and Traffic, Jaccard-distance-based clustering performs better than its location-based counterpart (with a lower consolidation error), whereas in contrast, for the Protest dataset, location-based clustering works better. The result can be explained by the nature of each event type. Namely, events that are well aligned with city boundaries are most suitable for localization, since city-level features are easiest to extract. Smaller-scale events as well as events of global interest are harder to merge correctly based on location.

2.5 EVENT TRACKING

Event tracking extends the event detection algorithm in a straightforward manner by applying bin consolidation across successive time slots with the purpose of monitoring the evolution of an event over time. Thus, after consolidating the bins in the current time slot (based on Jaccard distance), we consolidate the bins within the two successive most recent slots. An open question is how to assess the accuracy of tracking. In physical target tracking, a target has a unique location over time, making semantics of tracking unambiguous. In event tracking, multiple conditions in the environment can be associated with the event being monitored. Consequently, it becomes more challenging to determine what an ideal measurement of an event trajectory should include.

To facilitate consolidation of descriptions (i.e., bins) pertaining to the same event, we use a sliding window approach where successive slots overlap in time. To study the performance of this lexically inspired tracking system, it is interesting to compare it to two baselines:

1. *HashTag*: The events are detected and tracked purely by using the hashtags in the tweet texts. When two tweets have some common hashtag, they are treated as the same event.
2. *Clustering*: We choose Aggarwal and Yu's stream clustering algorithm (Aggarwal and Philip, 2006), where each tweet is represented by a high-dimensional word-frequency vector and clustering is performed based on Cosine similarity. A weight is defined for each tweet that decays over time. Clusters with small weights are removed.

Table 4 summarizes the accuracy comparison between the different algorithms for event tracking, where accuracy is defined as 1—consolidation error (as defined in the previous section). The table demonstrates that high accuracy is possible.

Table 4 Event Tracking Accuracy Comparison

Solution	Disaster	Protest	Traffic	War
HashTag	59.75%	47.41%	60.03%	42.17%
Clustering	79.13%	77.52%	87.62%	81.20%
Lexical	85.74%	92.09%	98.42%	95.61%

To give an example of the content of an event-bin, in Table 5, we show tweets that were automatically associated with the same bin (using the algorithms described in this chapter) in the aftermath of the Nepal Earthquake, which occurred on April 24, 2015. Tragically, this earthquake resulted in the death of more than 8000 people in Nepal. The social media analysis algorithms, described in this chapter, detected this event due to the rise of tweets with new high-information-gain keyword pairs on the social medium. Table 5 shows detected keyword pairs and example tweets from their clusters.

From the table, we observe that at the beginning of the earthquake, people focused more on the earthquake itself, using keyword pairs such as "earthquake" and "death" in tweets. As the earthquake developed, people focused on the relief efforts, using text with "donations" and "humanitarian" as keywords. Later people focused on survivors from Nepal, using keyword pairs such as "survivor" and "hospital". Neither the original occurrence of the event nor any of the above keyword pairs was known to our algorithms in advance. They were detected automatically and associated with the same event based on discussed distance metrics. The example shows the capability of tracking real-world events.

Table 5 Nepal Earthquake Tracking Summary

Date	New Detected Keywords	Sample Tweets
04/25/2015	nepal, earthquake, death	Powerful magnitude-7.8 earthquake that rocked Nepal triggered an avalanche on Mount Everest http://t.co/MULEuWhx3Q http://t.co/QeRKg8QgYp
		RT @BBCBreaking: At least 876 killed in Nepal #earthquake; deaths also reported in India, Tibet & Bangladesh http://t.co/3BTo9l1QZ4 http://u2026
04/26/2015	help, nepalearthquake	RT @cnnbrk: At least 2,263 people have died in Nepal from massive #NepalEarthquake and aftershocks, official says. http://t.co/hCyjO7YyS7
		Anyone with information about my son Joseph Patrick please help #NepalEarthquake #Pray4Joe http://t.co/X2Kn7mOtRO http://t.cou2026
04/27/2015	surges, devastation, drone, thankyoupm, donations	Nepal #earthquake: Death toll surges to 3,218; four aftershocks felt in last 12 hours http://t.co/Njvru9k2kQ
		@cnni: New drone footage shows the extent of devastation from the #NepalEarthquake: http://t.co/7PiPjayQZ1 https://t.co/phIGRkYoZQ
		#ThankYouPM for massive rescue and relief operation by India in Nepal after #earthquake
		Nepal Earthquake: Facebook to Match Donations Made for Victims http://t.co/aLooadYNxjFreeSubmission http://t.co/J90dT2qnXb
04/28/2015	salute2indianforces, koirala, sanjay, humanitarian	Thank you very much Indin Forces for being with us.It means alot…. #Salute2IndianForces
		CNN's Dr. Sanjay Gupta performs surgery on girl in Nepal: CNN's Dr. Sanjay Gupta performed a life-saving…http://t.co/4EtmH28EwC#tcot

Continued

Table 5 Nepal Earthquake Tracking Summary—cont'd

Date	New Detected Keywords	Sample Tweets
04/29/2015	survivor, hours, hospital, field, miracle	Live: Nepal earthquake kills 4,352, PM Sushil Koirala says death toll could reach 10,000: A high-intensity ear…http://t.co/A68VtR6hWK
		Nepal earthquake survivor drank urine while trapped for 82 hours http://t.co/v9DHM5Jhnf#worldnews
		That is amazing, Nepal Army rescued a 4-month kid alive after 22 hours! ::http://t.co/KzJPJeZDCx https://t.co/HvTkvS0Ba0via@sharethis
04/30/2015	pakistan, serves, masala, teenage, lydia	RT @haaretzcom: Nepal earthquake updates / Israeli field hospital opens, to treat 200 people per day http://t.co/PMwRlRT6YO http://t.co/s9i
		Pakistan serves 'beef masala' to earthquake-hit Nepal via /r/worldnews http://t.co/GoFJO09mJP
		Teenage boy pulled out of rubble alive five days after Nepal earthquake http://t.co/0kiAigYE7M#telegraph#news
		Lydia Ko donating earnings to Nepal relief effort: The 18-year-old Ko, ranked No. 1 in the world, successfully…http://t.co/2nquCITqJa

3 EVENT DETECTION AND TRACKING WITH INSTAGRAM

While the above discussion focused on Twitter, event detection and tracking is also possible with Instagram; the popular picture sharing network. Unlike the case with Twitter, a considerable percentage of Instagram content is geotagged making event detection and tracking easier. Instagram is popularized due to the proliferation of picture-taking devices (e.g., over 2 billion smart phone users at present). At the time of writing, Instagram has more than 400 million users, who collectively upload 80 million pictures a day (Social Media Statistics, 2016). This is up from 300 million, 150 million, and 30 million users in 2014, 2013, and 2012, respectively. Based on a sample of images we collected that are publicly viewable, more than 15% contain location metadata, making it meaningful (given the large total volume) to consider Instagram as a tool for event detection and tracking.

Detecting user-specified types of events based on pictures calls for a capability to associate the pictures with specific event keywords (e.g., ones entered by the user in a search query). Fortunately, Instagram users frequently associate customized metadata with images to identify what an image represents. Specifically, Instagram allows users to *tag* images they upload and also associate a spatial location based on the GPS. Our experiments show that 80% of Instagram images are tagged with at least one keyword, and 19% have a nonempty comments field. This makes it possible to search Instagram images for those matching event-specific keywords. The results are clustered by location to identify the number and location of matching events.

The analogy between Instagram and target tracking is straightforward. Each image (matching event-specific keywords) can be considered as the location of a binary proximity sensor that detected the event. Prior literature in networked sensing describes multiple algorithms for target detection and

Table 6 Average Error in Location Estimation (miles)

Dataset	DBScan	KMeans	Distance Based
Taylor Swift Concerts	4.42	0.027	0.037
Paris Attacks	1.489	0.37	0.28
Maroon V Concerts	1.414	0.120	0.126
Marathons	2.07	1.95	0.57
Tornadoes	10.88	5.433	8.04

tracking using proximity sensors. These algorithms can be applied with little or no modifications to recognize the occurrence of events (and identify their locations) from the distribution of concentrations of images that contain event-relevant tags. In the simplest implementation, event location is simply the centroid of locations of images in the cluster corresponding to this event. The cluster itself is found using a standard clustering algorithm, such as K-Means or DBScan. Therefore, a query for an event such as "*#ParisAttack*" or "*#Marathon*" will retrieve pictures with annotations matching the query. The pictures from each time window are clustered, from which events matching the query can be detected, localized, and tracked over time.

Table 6 shows the location error (in miles) in estimating the location of multiple events, listed in the table. The error is computed by clustering locations of related images using each of the three clustering algorithms, then computing the distance between the cluster centroid and true event location. In most cases, a localization accuracy below 0.5 miles is achieved. In some cases, finer-grained localization is possible. Of the event types considered, the one with the largest error was localization of tornadoes. We believe there are two reasons for this phenomenon. First, unlike the other events considered, tornadoes struck less populated areas. Hence, the density of human observers was notably lower, leading to fewer observations and reduced accuracy. Second, observers of tornadoes likely maintained a safe observation distance, making the location of "sensors" (the individuals taking photos) less of an indicator of the actual location of the tornado. In contrast, for example, concert events were attended by many individuals who were in close proximity to the performer, leading to errors as small as 60–100 m, which is the size of a concert venue.

4 VERACITY OF MEDIA POSTS

The above discussion focused on event detection, localization, and tracking. A related important problem is veracity analysis. An analogy is appropriate here with communication across noisy channels. Much like receiving signals over a noisy channel, social media descriptions of events may include distortion and noise. It is appropriate then to develop ground truth estimators that attempt to reconstruct the "original signal" (i.e., the ground truth in the physical world) from the received distorted version shared on social media.

There are two approaches for ground truth estimation in current research literature. The first direction attempts to model how humans evaluate the credibility of data. Machine learning approaches are used to train classifiers based on many data samples judged (i.e., labeled) by a human observer as credible or not. The goal is to train the classifier to distinguish credible information (Sikdar et al., 2014).

These classifiers consider a large number of features pertaining to content, sources, frequency of occurrence, and context, and determine which features are more predictive of human credibility judgment.

The second approach to ground truth estimation uses purely statistical techniques borrowed from data fusion literature, such as maximum likelihood estimation (Eliason, 1993). These techniques do not interpret content semantics. Rather, they consider the statistical properties of content dissemination. Modeling data sources as unreliable sensors, the question addressed by statistical analysis techniques is to determine the probability that an observation is true, given the set of sources that agree or disagree with it (Wang et al., 2012, 2014). In this approach, each tweet is associated with a binary variable that indicates whether the tweet is true or false. We conceptualize the content of the tweet as a *claim* made by a group of sources. The claim can be true or false. Thus, the group of sources who mentioned the tweet are thought of as "binary sensors" that report the corresponding claim. The goal is to correctly guess the binary truth value for each claim. Clearly, if credibility of sources was known, this would be an easy problem. The challenge lies in that source credibility is generally not known and individual claims are not easy to verify. Hence, one has to jointly estimate both the reliability of sources and the truth values of their claims. A recent book describes advances in such joint estimation (Wang et al., 2015).

Note that, for this approach to work, one needs to determine not only the reliability of individual sources but also their dependencies. Such dependencies arise, for example, when one source may be influenced by another, which may lead it to agree with (e.g., re-post) a piece of data without independently verifying its correctness. From a reliability perspective, dependencies among sources lead to the possibility of correlated errors. In order to correctly quantify error correlations, it becomes necessary to accurately model human behavior as an imperfect information relay.

A recent human subjects study (Roy et al., 2016) explored social factors influencing online information transmission. The study seeks an understanding of how human biases and trust relations impact information sharing. Not surprisingly, the study shows that the probability of information forwarding depends significantly on trust relations and biases. Accounting for these relations is therefore key in modeling error correlations, an issue that impacts ground truth estimation. The study offers a foundation for future models of communities as noisy information filters. The amplification and/or attenuation properties of both the signal and the noise through such a filter depend on community-specific trust relations and beliefs. This allows the formulation of statistical ground-truth estimation problems that account for the filter's transfer function. Conversely, given this transfer function, it allows the estimation of trust relations and biases from a community's information propagation patterns. These problems offer great opportunities for future research at the interface of cyber-physical systems and social science.

5 CONCLUSIONS

In this chapter, we argued that the social modality of sensing is not unlike other sensing modalities, such as magnetic, acoustic, seismic, or proximity sensing. A useful practice is to transform the signal received from the environment into an appropriate spatial or frequency domain, and then perform signal processing on that domain. This chapter described an exercise in applying the above approach to Twitter text and Instagram image metadata. It regarded physical events in a social environment as stimuli that cause the social medium to react. It then represented the signal emitted on the social medium by distributions in a lexical frequency domain (in the case of Twitter) or physical space (in the case of Instagram). Signals (or appropriate features thereof) were then clustered in that domain to zoom in on individual events. Evaluation results showed that the approach is successful at detecting and tracking physical events. Issues of content reliability were also touched upon. The success of the approach suggests the utility of social network as sensing sources, offering not only event statistics, but also associated text that provides additional context on each event, and quantifies reliability of different claims.

ACKNOWLEDGMENTS

Research reported in this chapter was sponsored in part by the U.S. Army Research Laboratory and was accomplished under Cooperative Agreement W911NF-09-2-0053, DTRA grant HDTRA1-10-1-0120, and NSF grant CNS 13-29886, and supported in part by an appointment to the Student Research Participation Program at the U.S. Army Research Laboratory administered by the Oak Ridge Institute for Science and Education through an interagency agreement between the U.S. Department of Energy and USARL. The views and conclusions contained in this document are those of the authors and should not be interpreted as representing the official policies, either expressed or implied, of the Army Research Laboratory or the U.S. Government. The U.S. Government is authorized to reproduce and distribute reprints for Government purposes notwithstanding any copyright notation here on.

REFERENCES

Aggarwal, C.C., 2013. A survey of stream clustering algorithms.

Aggarwal, C.C., Philip, S.Y., 2006. A framework for clustering massive text and categorical data streams. In: SDM. SIAM, pp. 479–483.

Atefeh, F., Khreich, W., 2015. A survey of techniques for event detection in Twitter. Comput. Intell. 31 (1), 132–164.

Cox, I.J., 1993. A review of statistical data association techniques for motion correspondence. Int. J. Comput. Vis. 10 (1), 53–66.

Department of Economic and Social Affairs, United Nations, New York, 2014. World Urbanization Prospects, The 2014 Revision.

El-Kishky, A., Song, Y., Wang, C., Voss, C.R., Han, J., 2014. Scalable topical phrase mining from text corpora. Proc. VLDB Endow. 8 (3), 305–316.

Eliason, S., 1993. Maximum Likelihood Estimation. Logic and Practice. Quantitative Applications in the Social Sciences, 96Sage Publications, Beverly Hills, CA.

Gao, H., Liu, C.H., Wang, W., Zhao, J., Song, Z., Su, X., Crowcroft, J., Leung, K.K., 2015. A survey of incentive mechanisms for participatory sensing. IEEE Commun. Surv. Tutorials 17 (2), 918–943.

Giridhar, P., Amin, M.T., Abdelzaher, T., Kaplan, L., George, J., Ganti, R., 2014. Clarisense: clarifying sensor anomalies using social network feeds. In: PERCOM Workshops. IEEE.

Guille, A., Favre, C., 2014. Mention-anomaly-based event detection and tracking in Twitter. In: 2014 IEEE/ACM International Conference on Advances in Social Networks Analysis and Mining (ASONAM), August 2014. pp. 375–382.

Han, J., Cheng, H., Xin, D., Yan, X., 2007. Frequent pattern mining: current status and future directions. Data Min. Knowl. Disc. 15, 55–86.

Koutsopoulos, I., 2013. Optimal incentive-driven design of participatory sensing systems. In: INFOCOM, 2013 Proceedings IEEE, April 2013.pp. 1402–1410.

Panteras, G., Wise, S., Lu, X., Croitoru, A., Crooks, A., Stefanidis, A., 2015. Triangulating social multimedia content for event localization using Flickr and Twitter. Trans. GIS 19 (5), 694–715.

Pasternack, J., Roth, D., 2011. Generalized fact-finding. In: Proceedings of the 20th International Conference Companion on World Wide Web, WWW '11. ACM, New York, NY, USA, pp. 99–100.

Roy, H., Bowman, E.K., Kase, S.E., Abdelzaher, T., 2016. Investigating social bias in information transmission: Experimental design and preliminary analyses. In: Proceedings from ICCRTS: 21st International Command & Control Research & Technology Symposium.

Sakaki, T., Okazaki, M., Matsuo, Y., 2010. Earthquake shakes Twitter users: real-time event detection by social sensors. In: WWW.

Schulz, D., Burgard, W., Fox, D., Cremers, A.B., 2001. Tracking multiple moving targets with a mobile robot using particle filters and statistical data association. In: IEEE ICRA. IEEE.

Sikdar, S., Adal, S., Amin, M., Abdelzaher, T., Chan, K., Cho, J.H., Kang, B., O'Donovan, J., 2014. Finding true and credible information on Twitter. In: 2014 17th International Conference on Information Fusion (FUSION), July 2014. pp. 1–8.

Social Media Statistics, 2016. https://www.brandwatch.com/2016/03/96-amazing-social-media-statistics-and-facts-for-2016/.

Tang, J., Qu, M., Mei, Q., 2015. Pte: predictive text embedding through large-scale heterogeneous text networks. In: Proceedings of the 21th ACM SIGKDD International Conference on Knowledge Discovery and Data Mining, KDD '15. ACM, New York, NY, USA, pp. 1165–1174.

Wang, D., Abdelzaher, T., Kaplan, L., 2015. Social Sensing: Building Reliable Systems on Unreliable Data. Elsevier Science & Technology, Waltham, MA.

Wang, D., Amin, T., Li, S., Kaplan, T.A.L., Pan, S.G.C., Liu, H., Aggrawal, C., Ganti, R., Wang, X., Mohapatra, P., Szymanski, B., Le, H., 2014. Humans as sensors: an estimation theoretic perspective. In: IPSN.

Wang, D., Kaplan, L., Le, H., Abdelzaher, T., 2012. On truth discovery in social sensing: a maximum likelihood estimation approach. In: IPSN.

VALIDATION, VERIFICATION, AND FORMAL METHODS FOR CYBER-PHYSICAL SYSTEMS

12

P. Bagade, A. Banerjee, S.K.S. Gupta

Arizona State University, Tempe, AZ, United States

1 INTRODUCTION

Cyber-physical systems (CPSs) are complex networks of computing systems each interacting with their physical environment through exchange of information and energy (Gupta et al., 2013). By definition, complex systems exhibit emergent behavior, which is only observed when the networked system is considered as a whole, and cannot be estimated by mere composition of individual component behavior. Aggregate effects are the manifestations of emergent behavior through macroscopic properties, which affect controller actions, resource consumption, and operational characteristics. Design and implementation of CPSs to manage and control aggregate effects, either to limit their harmful consequences or to utilize their benefits, is essential.

One of the key techniques used to design CPSs is formal modeling and analysis. Formal models are mathematical constructs that can be used to specify behavior of a complex system in terms of discrete or continuous variation of its system variables. Using formal models, a CPS can be designed using three steps: (a) formal modeling: the CPS behavior can be extracted in terms of a number of continuous or discrete system variables; (b) verification: the models can be simulated over time or space or theoretically analyzed to determine whether the model meets the requirements proposed by the designer, also known as verification; and (c) validation: the models can be implemented and evaluated, either in real-life deployment or in some virtual environment emulating a real-life deployment, to check whether the implementation meets the requirements, also known as validation. In this book chapter, we give a brief overview of the formal models, verification techniques, and validation methodologies that are most recently proposed to control aggregate effects in CPSs. Let's begin by discussing some CPS examples, and the inefficiencies caused by aggregate effect agnostic controllers. We will then discuss the challenges of incorporating aggregate effects modeling in the three essential steps of CPS design.

1.1 EXAMPLES OF AGGREGATE EFFECT AGNOSTIC CPS CONTROLLERS

Aggregate effects have been recently observed in CPSs such as medical devices (Trachette and Helen, 1999), autonomous vehicles (Kornienko et al., 2004), and smart grids (Lu et al., 2013). Controllers that are agnostic of aggregate effects can induce hazardous system configurations. To illustrate this issue let

Cyber-Physical Systems. http://dx.doi.org/10.1016/B978-0-12-803801-7.00012-2

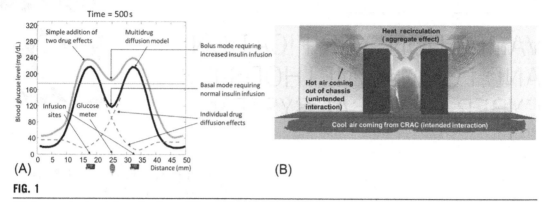

FIG. 1

(A) Aggregate drug interaction in multidrug artificial pancreas, and (B) hot spots created due to aggregate heat recirculation in data center.

us consider a practical example of multidrug infusion system. As shown in Fig. 1A, there are two infusion sites from which insulin is delivered. The glucosemeter is situated in between the two infusion sites. If the blood glucose level due to insulin infusion goes above 180 mg/dL, the controller infuses bolus insulin (a high dosage) and immediately brings down the glucose level. The controller is proactive and uses a model of glucose insulin interaction to estimate when such bolus is required. If the model of aggregate effect simply adds up the individual drug infusion effects, the controller will end up infusing more insulin than required. A more accurate model of aggregate effect of drug interaction dynamics (Fig. 1A) shows that the blood glucose level at the glucosemeter site does not go beyond the threshold. In another example involving data center cooling control systems, a controller might want to estimate hot spot locations and distribute cooling accordingly. As shown in Fig. 1B, the hot spots in the data center depend on air flow patterns that can only be analyzed if the data center is considered as a single macroscopic system. A wrong estimation of hot spots may either lead to over-provisioned cooling (Jonas et al., 2007; Mukherjee et al., 2009), or thermal runaway of servers (Tang et al., 2006).

1.2 CHALLENGES OF INCORPORATING AGGREGATE EFFECTS IN MODEL CHECKING

Model checking using hybrid systems require the development of a reachability analysis technique that computed the evolution of the CPS over time. The problem of reachability analysis is undecidable for even simple linear systems. Incorporation of models of aggregate effects is further expected to exacerbate the problem. Hence, it is essential to analyze whether impact of aggregate effects is significant enough to incorporate in model checking.

1.2.1 Challenge 1: Analyzing the impact of aggregate effects

Mathematical characterization of aggregate behavior is nontrivial due to several issues. The stochastic nature of the interacting agents such as computing systems and physical environment plays an important part in causing aggregate behavior. Several researchers have considered information theoretic explanation to predict emergence, which can be used in case of aggregate effects for CPSs. The conclusion is when randomness within individual components is reduced and stochastic interaction between different components is increased, aggregate effects can be observed. Although such efforts provide a means to formally express which system designs may result in aggregate effects, they fall short in estimating its magnitude.

1.2.2 Challenge 2: Complexity of model checking

It has been observed that models of aggregate effects in CPS has four characteristics (Jonas et al., 2007; Mukherjee et al., 2009; Tang et al., 2006): (a) the effect of individual actions of computing units on the physical system, (b) a change in individual action due to information obtained from other computing units in the network, (c) randomness in the environment, and (d) the invocation of new physical processes due to the simultaneous action of the networked computing units.

The effect of an individual interaction of a computing unit with its environment is often nonlinear and spatio-temporal in nature. Table 1 gives an overview of the type of nonlinearity in different CPS examples, and the available solutions to mathematically characterize them. The interactions being time variant exacerbates the problem. Modeling the individual cyber-physical interaction requires a hybrid approach in order to address the discrete computational and continuous physical processes as noted by several researchers (Fränzle et al., 2014). Hence, a hybrid automata is often used, which represents the computational processes as discrete states and the physical processes as differential equations in each state. However, most researchers consider only linear temporal dynamics for the ease of analysis of the models. There has been a spatio-temporal hybrid automata (STHA) model proposed to consider both spatial and temporal dynamics of CPSs (Banerjee and Gupta, 2013). However, the analysis of STHA is limited for one space dimensions and linear partial differential equations (Banerjee and Gupta, 2013). There has been very limited work on nonlinear hybrid automata with time variant dynamics and almost no work on nonlinear STHA, which is required to accurately model aggregate effects in CPSs.

Traditional composition semantics may not characterize aggregate effects: The basic understanding from the research on emergence is that in complex systems such as CPSs, the individual computing unit exhibits stochastic behavior while they interact with each other in a stochastic manner. The information content of a stochastic process is computed in terms of the entropy (information content (Johnson et al., 2013)) of the corresponding data generated by the process. When the entropy of information transferred between different computing units becomes greater than the entropy of the individual units in a CPS, emergent behavior can potentially occur. In such cases, microscopic views, where the CPS is a composition of individual computing units and macroscopic views, where the CPS network is an entirely new system, differ in their observed behavior.

The difference in behavior of microscopic and macroscopic views of a CPS caused due to increase in entropy of transferred information and decrease in entropy in individual processes is called emergence.

The mathematical implication of such a definition is that aggregate effects cannot be characterized or modeled by composing models of individual components of the CPS. The CPS has to be considered as a single system and not a collection of individual computing units. Such requirements may often increase the complexity of the system, for example, it may require new states in the system in addition to the Cartesian combination of state space of the individual components. Hence approaches such as the one used for modeling network of medical devices (Arney et al., 2010) or networks of autonomous vehicles (Wongpiromsarn et al., 2012a) may not be accurate enough. Evidently they have not been able to characterize events such as drug interaction or selforganization.

Table 1 Nonlinearity, Dimensionality, Stochastic Nature in CPSs and Available Solution Methods

Linearity	Parameter Type	Independent Dimensions	Type of State Transition	Solution Approach	Analysis Type	Example
Linear	Fixed	1	Linear	Fixed point computation	Approximate reachability analysis using zonotopes (Girard and Guernic, 2008)	UAV trajectory estimation (Mukherjee and Gupta, 2009), steam boiler controller (Henzinger and Wong-toi, 1996)
Linear	Fixed	1	Linear	Computation with finite error	Time bound reachability analysis (Kim, 2012)	Sensor networks (Kim, 2012)
Linear	Fixed	1	Nonlinear	Computation with finite error	Symbolic reachability analysis (Jha et al., 2007)	Air traffic alert and collision avoidance system (Jha et al., 2007)
Nonlinear (multiplicative)	Fixed	1	Linear	Estimation with finite error	Piecewise linearization approach (Han and Krogh, 2006)	Electric power networks (Han and Krogh, 2006)
Nonlinear (multiplicative)	Stochastic	1	Linear	Estimation with finite error	Conservative linearization (Asarin et al., 2003)	Bergman's minimal model for insulin glucose interaction (Wang et al., 2010)
Nonlinear (power)	Fixed	1	Nonlinear	Numerical methods	HySon (Bouissou et al., 2012)	Aviation modeling (Bouissou et al., 2012)
Nonlinear (time variant, multiplicative)	Fixed/random	≥2	No solution	No solution	No solution	Diffusion equation, Navier stokes equation

1.2.3 Challenge 3: Safety verified CPS controller may not be implementable in practice

The safety verified controllers often assume an error free sensor, zero delay sensing and actuation, infinite sampling frequency, and capability to estimate numbers with infinite precision. These controller configurations may not satisfy the requirements when implemented using sensing, actuation, and computation platforms that have delays, limited sampling frequency and finite precision. A major impetus in CPS research can be to synthesize implementable controllers from the safety verified hybrid models, which maintain the safety properties in practice.

2 FORMAL REPRESENTATION OF CPSs

Mathematical characterization of aggregate behavior is nontrivial due to their stochastic nature, nonlinearity, time variance, etc. A few important formal representation methods are discussed here.

2.1 STATE-BASED MODELING (HYBRID SYSTEMS AND TEMPORAL LOGIC)

State-base models are mathematical constructs that represent the behavior of the CPS in terms of discrete modes of controller software and continuous state variables of physical system. In each discrete mode a certain control decision is evaluated based on the values of certain continuous state variables in CPSs. The variation in state variables is evaluated using differential equations which represent the behavior of CPS. Hybrid automata are popularly used to represent the discrete computing models and continuous variables in a single mathematical construct. Hybrid automaton also enables transitions between discrete modes, which are decided by associated guard conditions on state variables. Tools such as Shift (Deshpande et al., 1997), Ptolemy (Eker et al., 2003), and BAND-Aide (Banerjee et al., 2010) are used for CPS hybrid automata modeling.

Consider a cooling control system (CRAC), which receives hot air as input and provides cold air as output. The CRAC operates in two modes $\{m_1, m_2\}$ (Fig. 2) and each mode extracts a constant amount of power $P_{ac}(m_i)$ from the incoming air at temperature T_{in}. The outgoing cold air at temperature T_{out} of the CRAC gets heated by aggregate effects of heat sources in a room generating power P_{gen}. The CRAC changes mode based on a threshold on the inlet temperature T_{thresh}. The CRAC switches modes

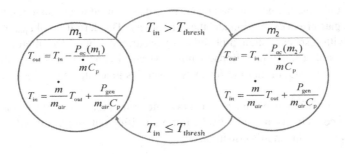

FIG. 2

Hybrid automata model of a cooling control system with two states, each having different dynamics and state transitions governed by constraints on the inlet temperature.

depending on threshold crossing conditions of inlet temperature. The dynamic equations governing the inlet and outlet temperatures of the CRAC at a given mode m_i are given by Eq. (1)

$$T_{out} = T_{in} - \frac{P_{ac}(m_i)}{\dot{m}C_p},$$

$$\dot{T}_{in} = \frac{\dot{m}}{m_{air}}T_{out} + \frac{P_{gen}}{m_{air}C_p}. \tag{1}$$

where $\dot{m}$ is the mass flow rate from the CRAC, m_{air} is the mass of air inside the room, and C_p is the specific heat of air. Given this CRAC system, its hybrid model is as shown in Fig. 2. The hybrid model of CRAC consists of two modes $L = \{m_1, m_2\}$ and the initial mode can be m_1. The continuous state dynamics can be seen in Eq. (1), where for each mode the power drawn by the CRAC $P_{ac}(m_i)$ changes.

There are several variants of hybrid automata. The most commonly used are timed automata (Jiang et al., 2015) and linear hybrid automata. A timed automata assumes that the dynamic equations can only be of the form $\dot{t} = $ constant. A linear hybrid automata assumes that the dynamic equations can only be of the form of linear first-order differential equations. The models provide limited support to represent non-linear and spatio-temporal nature of aggregate effects in CPSes.

Temporal logic formula: It represents the system property w.r.t. time which will be true for infinite state sequence of the system or until other condition becomes true for that system. Temporal logic formula is used to represent guard and invariant conditions of the hybrid automata (Clarke et al., 1983). It is also used to specify CPS properties such as safety property can defined as, the value P should be always less than its threshold value P_{th}, G ($P \leq P_{th}$).

2.2 INFORMATION THEORY

The theory of emergence provides a mathematical tool that can potentially be used to detect and characterize aggregate effects. Several researchers have considered information theoretic explanation to predict emergence, which can be used in case of aggregate effects for CPSs (Chan et al., 2010; Grosu et al., 2009; Johnson et al., 2013; Jiao and Torquato, 2011; Prokopenko et al., 2009). It has been observed that the relative impact of emergence with respect to system complexity can be expressed in terms of the efficiency of prediction e_{pred}. Given an abstraction of the CPS, e_{pred} is defined as the ratio of how much information on aggregate effect can be predicted to the statistical complexity of the abstraction. Given two abstractions the parameter e_{pred} can be used to evaluate the trade-off between the impact of aggregate effect estimation and the complexity. The definition of e_{pred} only distinguishes aggregate effects with respect to their probability of occurrence. In CPSs, often rare events cause a long-lasting and possibly catastrophic impact on the system. Examples include nocturnal hypoglycemia, potentially causing death or thermal runaway of servers in a data center due to delay in response from cooling systems.

At the core of information theoretic approach lays the estimation of entropy of information content used by a given process (Johnson et al., 2013). For a specific continuous variable V_1, the entropy is given by the well-known Shannon's equation:

$$H(V_1) = \sum_{\forall x \in \mathbb{R}} P(v_i = x) \log (P(v_i = x)) \tag{2}$$

where $P(v_i = x)$ denotes the probability that the variable v_i takes the value x.

According to the definition, the emergence in a CPS is observed when the total entropy of individual processes is less than the total entropy of data transferred between processes by a critical value. For example, let us assume that the insulin infusion rate only takes three values: a basal rate, a correction bolus rate and a braking rate. Let us consider that the infusion pump is in basal mode for 80% of the time and in correction and braking modes 10% of the time each. Hence the entropy of each of the infusion process individually is: $(0.8 \log(0.8) + 0.1 \log(0.1) + 0.1 \log(0.1)) = 0.92$. Thus together with the two processes of insulin and glucagon infusion, the total entropy is around 1.84 bits. However, if we consider $G(t)$, the data transferred between the processes, it can assume a wide range of values as obtained from simultaneously solving drug diffusion equation and Bergman minimal model, which represents insulin interaction with glucose (Andersen and Højbjerre, 2002). Even if we restrict $G(t)$ to four equally opportune values its entropy is 2 bits, which is 0.16 bits greater than the individual processes combined. Thus, emergent behavior can be expected from such as system. Indeed multichannel infusion pumps are observed to control hypoglycemia better than individual application of two hormones one by one (Herrero et al., 2013).

2.3 MULTIAGENT REPRESENTATION

One of the popular ways of representing networked CPSs is by using the concept of multiagent systems. The collaboration between multiple agents resulting in group behavior fits nicely with the concept of CPS aggregate effects, which are also the result of communication between multiple agents. There have been several successful efforts in employing the concepts of multiagent systems in CPS design and analysis. The research is mostly applied towards robotics (Bhatia et al., 2011; Wongpiromsarn et al., 2012b), unmanned aerial vehicles (UAVs) (Luxhøj and Öztekin, 2009), or driverless cars (Google self-driving car project, 2015). The core idea is representing the CPS as a set of entities. These entities interact using some form of control message exchange. Based on the control messages, a controller in each entity computes a decision in order to minimize a global objective. For better understanding of this topic, let us take the example of collision avoidance in an UAV environment. The entities here are the UAVs. Each UAV has: (a) speed, (b) position sensor, (c) a mechanism to communicate with other UAVs, and (d) a control algorithm that takes the speed and position of other UAVs and decides on the next speed or target position, based on certain global objective. The global objective can be real-time requirements, such as avoidance of collision, or offline properties, such as reconnaissance of maximum targets with the least fuel cost.

Once such an objective is formulated, the idea is to perform a distributed optimization to find out a control strategy. This control strategy consists of a set of conditions on the speed and position of other UAVs and an equation to derive the speed and position of each UAV over time in response to the conditions. Typically such optimizations are nonlinear and techniques such as Lyapunov optimization, Lyapunov barrier functions are obtained to find out a control strategy. The principal drawbacks of such techniques are that Lyapunov functions or barrier functions are very difficult to derive from even a simplified representation of the CPS design problem. There are no automated techniques to derive Lyapunov functions since the physical dynamics of CPSs may be represented using various types of differential equations. Another drawback is that control Lyapunov techniques often do not take into account the sensing and actuating delays associated with a practical system. Hence the control strategy derived in theory may not work in practice.

3 VERIFICATION

The aggregate effects in mission critical CPSs modeled using techniques in Section 2 are required to be analyzed to ensure their trustworthy operation. Some of the important CPS verification techniques are discussed below:

3.1 MODEL CHECKING

To verify the correctness of CPSs with aggregate effects, model checking on CPS properties specified using temporal logic formulas or first-order logic formulas is performed (Kalajdzic et al., 2015; Clarke and Zuliani, 2011) as discussed in Section 2.1. This method checks if the CPS model satisfies the specified property. Semantically, if M is the CPS model and φ is the property, model checking methods verifies whether M entails φ (Clarke et al., 1983). Model checking for CPSs uses reachability analysis technique to see if the specified property holds for all system reachable states. Such exhaustive exploration of state space can be computationally intensive and time consuming. Automated techniques can be used for CPS model checking (Frehse, 2005). The state-of-art of currently available reachability analysis techniques are discussed in the following section.

3.1.1 Reachability analysis

In order to estimate the aggregate effects in CPSs accurately, the nonlinear nature of the system interactions should be considered in the analysis. Table 2 classifies the research on computing reach set for nonlinear system. The approach of linearization of the nonlinear equations has been widely proposed (Bagade et al., 2013; Asarin et al., 2003, 2007). In each linear piece, the available zonotope-based solutions are directly applicable. These approximation errors are nonlinear in terms of the size of the pieces and may even increase exponentially with respect to the image computation time step. To improve the linearization error, pieces of linear regions are considered to overlap (Dang et al., 2009, 2010). This method increases the number of state variables, increasing the computation complexity. The reach set approximation error still increases in a nonlinear fashion with increase in size of linearization pieces. HyperTech (Henzinger et al., 2000) proposes the use of interval numerical method to approximate the reach set for nonlinear hybrid system. It uses rectangular shape to represent the nonrectangular region shape, resulting in the increase in wrapping error. This error keeps on growing with each iteration due to the accumulation of errors from the previous steps. To avoid these linearization errors, recently few methods have been proposed which compute reachable sets of nonlinear system without linearizing them. One of the methods proposed polynomial Zonotope to capture the nonlinearity of the system (Althoff, 2013). This approach models error while converting nonlinear system equations into polynomial differential equations by adding uncertainty. Similar approaches have been attempted in Dang (2006), where Bernstein polynomial or Bezier curves are used. Each of these approaches require division of the initial set into simplices, the image for each of these simplices can then be represented as a combination of Bernstein polynomial or Bezier or Taylor surfaces. However, the overapproximation error increases as a quadratic function of the size of these simplices. The verification of the parameterized hybrid systems using first-order discrete dynamics logic has been proposed in (Platzer, 2007) linear systems only. The use of satisfiability modulo theory is proposed to encode discontinuous invariant conditions for nonlinear hybrid systems (Cimatti et al., 2012). However, the reachability analysis for these systems is not specified.

Table 2 Available Solutions for Reachability Analysis of Nonlinear Hybrid Systems

Technique	Type of Nonlinearity	Error Bound	Example Used for Modeling
Partitioning of state space (Asarin et al., 2007)	Second order differential equation	$O(h^2)$, h is simplex size	Van der Pol oscillator, Biquad lowpass filter
Piecewise linear (Asarin et al., 2003; Lee et al., 2015)	Lipschitz continuous	$O(h^3)$	Van der Pol oscillator
Linear interpolation (Dang et al., 2010)	Lipschitz continuous	$O(h^3)$	Biochemical network
Interval numerical method (Henzinger et al., 2000)	All types of nonlinear dynamics	Error varies with different nonlinear functions	Thermostat with delay, two-tank system, air traffic conflict resolution
Taylor series expansion (Benvenuti et al., 2012), Bernstein polynomial or Bezier curves (Dang, 2006)	Second order differential equation	$O(h^2)$	Van der Pol oscillator, Biquad lowpass filter
Use of polynomial zonotopes (Althoff, 2013)	Differential equations which can be represented as polynomials	$O(h^2)$	Van der Pol oscillator, Biological aging model
Verifying hybrid systems with parameters using first-order logic (Platzer, 2007)	Linear dynamics, no reachability for nonlinear	Not applicable	Train control system
Encoding hybrid systems (Cimatti et al., 2012)	Polynomial dynamics with discontinuous invariant	Not applicable	Braking control system of trains
Proposed approach with EB splines	Compactly supported continuous and smooth functions such as diffusion dynamics	$O(\tau+h)$, τ is time discretization step and h is set discretization step	Brusselator, artificial pancreas

3.2 THEOREM PROVING

Theorem proving is widely being used for CPSs verification, which provides mathematical reasoning on the correctness of system properties (Platzer and Quesel, 2008; Banerjee and Gupta, 2013; Ábrahám-Mumm et al., 2001; Manna and Sipma, 1998; Ouimet and Lundqvist, 2007). Unlike model checking, theorem proving takes less time as it reasons about the state space using system constraints only, not on all states on state space. However, fully automated techniques are less popular for theorem proving as automated generated proofs can be long and difficult to understand (Ouimet and Lundqvist, 2007). KeYmaera (Platzer and Quesel, 2008) theorem prover uses an automated prover, real quantifier elimination and symbolic computations in computer algebra systems for hybrid system verification. The STHA (Banerjee and Gupta, 2013) tool takes into account spatial and temporal variation of continuous dynamics of physical system are proved properties for CPSs. Further, tools such as PVS (Manna and Sipma, 1998) and STeP (Ábrahám-Mumm et al., 2001) use only invariant conditions for theorem proving, not the whole system logic. Thus, these solutions cannot be used to estimate aggregate effects accurately.

3.3 SIMULATION

Simulations of various network components and their interaction with physical world are widely used in designing wireless networks. Thus, they can be readily used to consider aggregate effects in CPSs. WCPS (Li et al., 2013) focuses on simulating wireless civil infrastructural control systems and the effects of network delays and data loss on control. Truetime (Cervin et al., 2003) is a Matlab-based simulator used for simulating the effect of network performance on continuous control systems. Further, to capture the wireless network characteristic more accurately, NCSWT (Eyisi et al., 2012) and PiccSIM (Kohtamaki et al., 2009) integrate control system simulators with NS-2 for wireless network simulation tools such as WCPS, GISOO (Aminian et al., 2013) use TOSSIM (Levis et al., 2003) and COOJA (Osterlind et al., 2006) respectively to have more realistic wireless network models. Other network simulators model both packet level and continuous traffic flows in the network (Ahn and Danzig, 1996; Kesidis et al., 1996; Kiddle et al., 2003; Melamed et al., 2004; Liu, 2007; Banerjee et al., 2013). However, a common disadvantage of these simulators is that they do not consider events from continuous systems. These events can change parameters of the control system and hence change the nature of continuous dynamics that govern the system variables. Further, they do not consider any form of time refinement to accurately estimate event timings and hence take the fixed time step approach towards simulation which can result in approximate estimation of aggregate effects.

3.4 SYMBOLIC EXECUTION

Symbolic execution is a technique to analyze CPS code and automatically generate test case inputs that might cause errors in the software. In this technique, the variables in a program are represented using symbols and the steps of the program are executed to track the values of the variable in terms of expressions on the symbols. For any branch statement, the symbolic equation is compared with a threshold to generate input data that can result in changes in program sequence. This methodology can be used to generate test cases for CPS software. Such test cases can then be simulated using physical system simulators to analyze the effects of different inputs on the CPS system. This technique has been proposed in many recent works (Sasnauskas et al., 2011; Majumdar, 2012). The work in (Păsăreanu and Rungta, 2010) proposes Symbolic PathFinder (SPF) that has interfaced symbolic execution with a JAVA compiler to symbolically analyze Java bytecode in complex CPS. Symbolic execution to verify aggregate effects can be complex as the conditions for change in variable values are dependent on various computing units in the CPS.

4 VALIDATION

Validation of a system provides whether the requirements set forth in specification phase are satisfied by the implementation. In this section, we discuss validation techniques for CPSs.

4.1 EXPERIMENTAL VALIDATION

Experimental validation of safety-critical systems such as medical devices, air-crafts are necessary to ensure their trustworthiness. Such validation might take a long time due to the availability of resources, time required to complete the experiments in real-life scenarios, etc. Thus, experiments are required to be designed using scientific methods such as design of experiments to improve efficiency of obtaining results. Experimental validation for medical devices is performed using clinical studies. Such trails are an essential part of medical devices approval as mandated by Food and Drug Administration. As humans are involved in clinical studies, approval from Institutional Review Board is required. One example of such study is clinical trials performed on GeMREM (Nabar et al., 2011) model, a generative model for resource efficient patient monitoring using ECG sensor. Clinical studies for medical treatments such as multidrug chemotherapy involve further challenges in ensuring patient safety from aggregate effects of different injection sites.

4.2 EMULATION

Hardware emulation of physical system gives in depth feedback on CPS design and thus can be effectively used to emulate aggregate effects in CPSs. The process is usually orders of magnitude faster than the simulation method. However, it is challenging due to implementation of differential equations governing physical processes in the CPS. Towards this effect, CyberHeart (The Human body, 2015), a virtual platform to emulate patient-specific computational models of heart has been proposed. It can accelerate medical device validation and speed up their development process. Generic infusion pump project (2011) focuses on safety-assured model-based development of infusion pump. It executes implementation of formally verified infusion pump design on a virtual layer that emulates infusion pump operations. Emulation platforms for CPSs such as Smart grids are extensively researched (Kinsy et al., 2011; Mallouhi et al., 2011; Dondossola et al., 2009; Genge et al., 2012). MARTHA (Kinsy et al., 2011) uses power electronic circuit compilation for faster emulation as compared to traditional methods using field programmable gate arrays. Testbeds such as TASSCS (Mallouhi et al., 2011), CRUTIAL (Dondossola et al., 2009) and (Genge et al., 2012) are proposed to analyze security in smart grids. The AirStar testbed (Jordan and Bailey, 2008) is developed by NASA Langley to test new aircraft technologies, which cannot be tested using actual flights. STHA model (Banerjee and Gupta, 2013) for skin temperature rise due to sensor heating has been validated with hardware implementation of a physical system. Fig. 3 shows the register grid to emulate models of human physiology. The human skin temperature change model is represented using Finite Difference Time Domain form

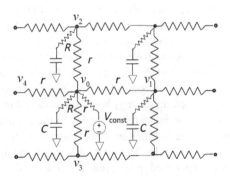

FIG. 3

Register grid representing analog behavior of physical system.

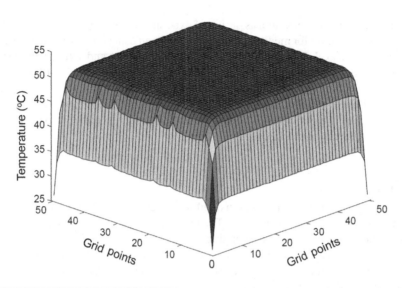

FIG. 4

50 × 50 resistor grid circuit output plot.

of Penne's bioheat equation (Tang et al., 2005). Fig. 4 shows the result of simulating the 50×50 register grid network in Cadence. The plot verifies the simulation result with the actual physical behavior of human physiology according to Penne's Bio Heat equation.

4.3 CONTROLLER SYNTHESIS

The safety verified controllers often assume an error free sensor, zero delay sensing and actuation, infinite sampling frequency, and capability to estimate numbers with infinite precision. These controller configurations may not satisfy the requirements when implemented using sensing, actuation, and computation platforms that have delays, limited sampling frequency, and finite precision. A major impetus in the CPS research is to synthesize implementable controllers from the safety verified hybrid models, which maintain the safety properties in practice. Researchers have proposed various methods for CPS controller synthesis (De Wulf et al., 2004; Di Guglielmo et al., 2013; Bouyer et al., 2011; Dreossi and Dang, 2014; Huang et al., 2015). The Almost-ASAP (De Wulf et al., 2004) semantics forces the controller to take action within bounded delay, $\Delta > 0$, which can be obtained using reachability analysis technique. However, the problem of computing Δ is undecidable. For automating the synthesis of controller parameters, bounded model checking strategy has been proposed in the literature (Di Guglielmo et al., 2013). Further, to consider sensor and actuation delay in the controller, time automaton is synthesized into an equivalent robust time automaton while preserving the original timing constraints within a desired delay (Bouyer et al., 2011). The controller synthesis approach has also been used for finding parameters satisfying specified properties of biological models (Dreossi and Dang, 2014) and for security attack detection (Huang et al., 2015). Although these approaches consider the limitations of implementing timing automaton in a real-life setting, they assume linear dynamics for the physical systems. However, most of the physical systems have nonlinear dynamics, where these methods can add errors in parameter synthesis. Further, analyzing aggregate effects can help to estimate physical dynamics more accurately.

5 CONCLUSIONS

This book chapter discusses the usage of formal modeling and analysis on CPSs that exhibit aggregate effects under networked operation of its subsystem. The formal models can be used for theoretical verification with respect to requirements. A requirement verified model can be converted to an implemented CPS, either manually or through automated controller generators. Such implementations can be validated to match the model properties through experiments in real-life deployment or virtual emulated environments. The main aim of such techniques is to design CPSs that can control aggregate effects in an efficient, safe, and cost-effective manner. The book chapter only introduces the various available techniques and discusses these methods through concrete CPS examples. The reader is encouraged to read further on using the references to gain better understanding of the topics.

ACKNOWLEDGMENTS

The authors would like to thank National Science Foundation, CNS grant #1218505, IIS grant #1116385 and National Institutes of Health #EB019202 for supporting their work.

REFERENCES

Ábrahám-Mumm, E., Hannemann, U., Steffen, M., 2001. Verification of hybrid systems: formalization and proof rules in PVS. In: Proceedings, Seventh IEEE International Conference on Engineering of Complex Computer Systems, 2001. IEEE, pp. 48–57.

Ahn, J.S., Danzig, P.B., 1996. Packet network simulation: speedup and accuracy versus timing granularity. IEEE/ACM Trans. Netw. 4 (5), 743–757.

Althoff, M., 2013. Reachability analysis of nonlinear systems using conservative polynomialization and non-convex sets. In: Proceedings of the 16th International Conference on Hybrid Systems: Computation and Control. ACM, pp. 173–182.

Aminian, B., Araujo, J., Johansson, M., Johansson, K.H., 2013. GISOO: a virtual testbed for wireless cyber-physical systems. In: Industrial Electronics Society, IECON 2013–39th Annual Conference of the IEEE. IEEE.

Andersen, K.E., Højbjerre, M., 2002. A Bayesian approach to Bergman's minimal model. Insulin 50 (100), 200.

Arney, D., Pajic, M., Goldman, J.M., Lee, I., Mangharam, R., Sokolsky, O., 2010. Toward patient safety in closed-loop medical device systems. In: Proceedings of the 1st ACM/IEEE International Conference on Cyber-Physical Systems. ACM, pp. 139–148.

Asarin, E., Dang, T., Girard, A., 2003. Reachability analysis of nonlinear systems using conservative approximation. In: Maler, O., Pnueli, A. (Eds.), Hybrid Systems: Computation and Control, LNCS 2623. Springer-Verlag, Berlin, pp. 20–35.

Asarin, E., Dang, T., Girard, A., 2007. Hybridization methods for the analysis of nonlinear systems. Acta Informatica 43 (7), 451–476.

Bagade, P., Banerjee, A., Gupta, S.K.S., 2013. Safety assurance of medical cyber-physical systems using hybrid automata: a case study on analgesic infusion pump. In: Medical CPS Workshop.

Banerjee, A., Gupta, S.K.S., 2013. Spatio-temporal hybrid automata for safe cyber-physical systems: a medical case study. In: 2013 ACM/IEEE International Conference on Cyber-Physical Systems (ICCPS). IEEE, pp. 71–80.

Banerjee, A., Banerjee, J., Varsamopoulos, G., Abbasi, Z., Gupta, S.K., 2013. Hybrid simulator for cyber-physical energy systems. In: 2013 Workshop on Modeling and Simulation of Cyber-Physical Energy Systems (MSCPES). IEEE, pp. 1–6.

Banerjee, A., Kandula, S., Mukherjee, T., Gupta, S.K., 2010. BAND-AiDe: A tool for cyber-physical oriented analysis and design of body area networks and devices. ACM TECS, Wireless Health Systems Special Issue.

Benvenuti, L., Bresolin, D., Collins, P., Ferrari, A., Geretti, L., Villa, T., 2012. Ariadne: Dominance checking of nonlinear hybrid automata using reachability analysis. In: Proceedings of the 6th International Conference on Reachability Problems. RP'12. Springer-Verlag, Berlin, Heidelberg, pp. 79–91.

Bhatia, A., Maly, M.R., Kavraki, L.E., Vardi, M.Y., 2011. Motion planning with complex goals. IEEE Robot. Autom. Mag. 18 (3), 55–64.

Bouissou, O., Mimram, S., Chapoutot, A., 2012. Hyson: set-based simulation of hybrid systems. In: 23rd IEEE International Symposium on Rapid System Prototyping (RSP), pp. 79–85.

Bouyer, P., Larsen, K.G., Markey, N., Sankur, O., Thrane, C., 2011. Timed automata can always be made implementable. In: CONCUR 2011—Concurrency Theory. Springer, Berlin, pp. 76–91.

Cervin, A., Henriksson, D., Lincoln, B., Eker, J., Årzén, K.-E., 2003. How does control timing affect performance? IEEE Control Syst. Mag. 23 (3), 16–30.

Chan, W.K.V., Son, Y.-J., Macal, C.M., 2010. Agent-based simulation tutorial-simulation of emergent behavior and differences between agent-based simulation and discrete-event simulation. In: Proceedings of the Winter Simulation Conference. Winter Simulation Conference, pp. 135–150.

Cimatti, A., Mover, S., Tonetta, S., 2012. A quantifier-free SMT encoding of non-linear hybrid automata. In: Formal Methods in Computer-Aided Design (FMCAD), 2012. IEEE, pp. 187–195.

Clarke, E.M., Zuliani, P., 2011. Statistical model checking for cyber-physical systems. Automated Technology for Verification and Analysis. Springer, Berlin. pp. 1–12.

Clarke, E.M., Emerson, E.A., Sistla, A.P., 1983. Automatic verification of finite state concurrent system using temporal logic specifications: a practical approach. In: Proceedings of the 10th ACM SIGACT-SIGPLAN Symposium on Principles of Programming Languages. ACM, pp. 117–126.

Dang, T., 2006. Approximate reachability computation for polynomial systems. In: Hespanha, J., Tiwari, A. (Eds.), Hybrid Systems: Computation and Control. Vol. 3927 of Lecture Notes in Computer Science, Springer, Berlin, Heidelberg, pp. 138–152.

Dang, T., Le Guernic, C., Maler, O., 2009. Computing reachable states for nonlinear biological models. Computational Methods in Systems Biology. Springer, Berlin. pp. 126–141.

Dang, T., Maler, O., Testylier, R., 2010. Accurate hybridization of nonlinear systems. In: Proceedings of the 13th ACM International Conference on Hybrid Systems: Computation and Control. ACM, pp. 11–20.

De Wulf, M., Doyen, L., Raskin, J.-F., 2004. Almost asap semantics: from timed models to timed implementations. Hybrid Systems: Computation and Control. Springer, Berlin. pp. 296–310.

Deshpande, A., Göllü, A., Varaiya, P., 1997. SHIFT: A Formalism and a Programming Language for Dynamic Networks of Hybrid Automata. Springer, Berlin.

Di Guglielmo, L., Seshia, S., Villa, T., et al., 2013. Synthesis of implementable control strategies for lazy linear hybrid automata. In: 2013 Federated Conference on Computer Science and Information Systems (FedCSIS). IEEE, pp. 1381–1388.

Dondossola, G., Deconinck, G., Garrone, F., Beitollahi, H., 2009. Testbeds for assessing critical scenarios in power control systems. Critical Information Infrastructure Security. Springer, Berlin. pp. 223–234.

Dreossi, T., Dang, T., 2014. Parameter synthesis for polynomial biological models. In: Proceedings of the 17th International Conference on Hybrid Systems: Computation and Control. ACM, pp. 233–242.

Eker, J., Janneck, J.W., Lee, E., Liu, J., Liu, X., Ludvig, J., Neuendorffer, S., Sachs, S., Xiong, Y., et al., 2003. Taming heterogeneity—the ptolemy approach. Proc. IEEE 91 (1), 127–144.

Eyisi, E., Bai, J., Riley, D., Weng, J., Yan, W., Xue, Y., Koutsoukos, X., Sztipanovits, J., 2012. NCSWT: an integrated modeling and simulation tool for networked control systems. Simul. Model. Pract. Theory 27, 90–111.

Fränzle, M., Lygeros, J., Haddad, M., Mertikopoulos, P., Lesage, J.R.C., Lennartson, B., Mitra, S., Stouraitis, T., Theodoridis, S., Chen, Y., et al., 2014. On communications, control, and signal processing (ISCCSP 2014). IEEE Control Syst. Mag.

Frehse, G., 2005. PHAVer: algorithmic verification of hybrid systems past hytech. Hybrid Systems: Computation and Control. Springer, Berlin, pp. 258–273.

Generic infusion pump project, 2011. http://rtg.cis.upenn.edu/medical/gpca/gpca.html (accessed 08.31.15).

Genge, B., Siaterlis, C., Fovino, I.N., Masera, M., 2012. A cyber-physical experimentation environment for the security analysis of networked industrial control systems. Comput. Electr. Eng. 38 (5), 1146–1161.

Girard, A., Guernic, C., 2008. Zonotope/hyperplane intersection for hybrid systems reachability analysis. In: Proceedings of the 11th International Workshop on Hybrid Systems: Computation and Control. HSCC'08. Springer-Verlag, Berlin, Heidelberg, pp. 215–228.

Google self-driving car project, 2015. http://www.google.com/selfdrivingcar/ (accessed 08.31.15).

Grosu, R., Smolka, S.A., Corradini, F., Wasilewska, A., Entcheva, E., Bartocci, E., 2009. Learning and detecting emergent behavior in networks of cardiac myocytes. Commun. ACM 52 (3), 97–105.

Gupta, S.K.S., Mukherjee, T., Venkatasubramanian, K.K., 2013. Body Area Networks: Safety, Security, and Sustainability. Cambridge University Press, Cambridge.

Han, Z., Krogh, B., 2006. Reachability analysis of nonlinear systems using trajectory piecewise linearized models. In: American Control Conference, pp. 1505–1510.

Henzinger, T.A., Wong-toi, H., 1996. Using hytech to synthesize control parameters for a steam boiler. In: Formal Methods for Industrial Applications: Specifying and Programming the Steam Boiler Control, LNCS 1165. Springer-Verlag, Berlin, pp. 265–282.

Henzinger, T.A., Horowitz, B., Majumdar, R., Wong-Toi, H., 2000. Beyond hytech: hybrid systems analysis using interval numerical methods. Hybrid Systems: Computation and Control. Springer, Berlin. pp. 130–144.

Herrero, P., Georgiou, P., Oliver, N., Reddy, M., Johnston, D., Toumazou, C., 2013. A composite model of glucagon-glucose dynamics for in silico testing of bihormonal glucose controllers. J. Diabetes Sci. Technol. 7 (4), 941–951.

Huang, Z., Wang, Y., Mitra, S., Dullerud, G., 2015. Controller synthesis for linear time-varying systems with adversaries. arXiv preprint arXiv:1501.04925.

Jha, S., Brady, B.A., Seshia, S.A., 2007. Symbolic reachability analysis of lazy linear hybrid automata. In: Proceedings of the 5th International Conference on Formal Modeling and Analysis of Timed Systems. FORMATS'07. Springer-Verlag, Berlin, Heidelberg, pp. 241–256.

Jiang, Y., Zhang, H., Li, Z., Deng, Y., Song, X., Gu, M., Sun, J., 2015. Design and optimization of multiclocked embedded systems using formal techniques. IEEE Trans. Ind. Electron. 62 (2), 1270–1278.

Jiao, Y., Torquato, S., 2011. Emergent behaviors from a cellular automaton model for invasive tumor growth in heterogeneous microenvironments. PLoS Comput. Biol. 7 (12), e1002314.

Johnson, J.J., Tolk, A., Sousa-Poza, A., 2013. A theory of emergence and entropy in systems of systems. Procedia Comput. Sci. 20, 283–289.

Jonas, M., Varsamopoulos, G., Gupta, S.K., 2007. On developing a fast, cost-effective and non-invasive method to derive data center thermal maps. In: 2007 IEEE International Conference on Cluster Computing. IEEE, pp. 474–475.

Jordan, T.L., Bailey, R.M., 2008. NASA Langley's AirSTAR testbed: a subscale flight test capability for flight dynamics and control system experiments. In: AIAA Guidance, Navigation and Control Conference and Exhibit, pp. 18–21.

Kalajdzic, K., Jegourel, C., Bartocci, E., Legay, A., Smolka, S.A., Grosu, R., 2015. Model checking as control: feedback control for statistical model checking of cyber-physical systems. arXiv preprint arXiv:1504.06660.

Kesidis, G., Singh, A., Cheung, D., Kwok, W., 1996. Feasibility of fluid event-driven simulation for atm networks. Global Telecommunications Conference, Communications: The Key to Global Prosperity, vol. 3. pp. 2013–2017.

Kiddle, C., Simmonds, R., Williamson, C., Unger, B., 2003. Hybrid packet/fluid flow network simulation. In: Proceedings of the Seventeenth Workshop on Parallel and Distributed Simulation, 2003 (PADS 2003). IEEE, pp. 143–152.

Kim, K.-D., 2012. PhD Dissertation Middleware and Control of Cyber-Physical Systems: Temporal Guarantees and Hybrid System Analysis.

Kinsy, M., Khan, O., Celanovic, I., Majstorovic, D., Celanovic, N., Devadas, S., 2011. Time-predictable computer architecture for cyber-physical systems: digital emulation of power electronics systems. In: 2011 IEEE 32nd Real-Time Systems Symposium (RTSS). IEEE, pp. 305–316.

Kohtamaki, T., Pohjola, M., Brand, J., Eriksson, L.M., 2009. PiccSIM toolchain-design, simulation and automatic implementation of wireless networked control systems. In: ICNSC'09, International Conference on Networking, Sensing and Control, 2009. IEEE, pp. 49–54.

Kornienko, S., Kornienko, O., Levi, P., 2004. Generation of desired emergent behavior in swarm of micro-robots. ECAI, vol. 16. p. 239.

Lee, H.-S.L., Althoff, M., Hoelldampf, S., Olbrich, M., Barke, E., 2015. Automated generation of hybrid system models for reachability analysis of nonlinear analog circuits. In: 20th Asia and South Pacific-Design Automation Conference (ASP-DAC), pp. 725–730.

Levis, P., Lee, N., Welsh, M., Culler, D., 2003. TOSSIM: accurate and scalable simulation of entire TinyOS applications. In: Proceedings of the 1st International Conference on Embedded Networked Sensor Systems. ACM, pp. 126–137.

Li, B., Sun, Z., Mechitov, K., Hackmann, G., Lu, C., Dyke, S.J., Agha, G., Spencer Jr., B.F., 2013. Realistic case studies of wireless structural control. In: Proceedings of the ACM/IEEE 4th International Conference on Cyber-Physical Systems. ACM, pp. 179–188.

Liu, J., 2007. Parallel simulation of hybrid network traffic models. In: PADS'07, 21st International Workshop on Principles of Advanced and Distributed Simulation, 2007. IEEE, pp. 141–151.

Lu, X., Wang, W., Ma, J., Sun, L., 2013. Domino of the smart grid: an empirical study of system behaviors in the interdependent network architecture. In: 2013 IEEE International Conference on Smart Grid Communications (SmartGridComm). IEEE, pp. 612–617.

Luxhøj, J.T., Öztekin, A., 2009. A regulatory-based approach to safety analysis of unmanned aircraft systems. Engineering Psychology and Cognitive Ergonomics. Springer, Berlin. pp. 564–573.

Majumdar, R., Saha, I., Shashidhar, K.C., Wang, Z., 2012. CLSE: Closed-loop symbolic execution. In: NASA Formal Methods Symposium. Springer, Berlin, Heidelberg, pp. 356–370.

Mallouhi, M., Al-Nashif, Y., Cox, D., Chadaga, T., Hariri, S., 2011. A testbed for analyzing security of SCADA control systems (TASSCS). In: Innovative Smart Grid Technologies (ISGT), 2011 IEEE PES. IEEE, pp. 1–7.

Manna, Z., Sipma, H.B., 1998. Deductive verification of hybrid systems using step. Hybrid Systems: Computation and Control. Springer, Berlin. pp. 305–318.

Melamed, B., Pan, S., Wardi, Y., 2004. HNS: a streamlined hybrid network simulator. ACM Trans. Model. Comput. Simul. 14 (3), 251–277.

Mukherjee, T., Gupta, S.K.S., 2009. MCMA + CRET: a mixed criticality management architecture for maximizing mission efficacy and tool for expediting certification of UAVs. In: IEEE Workshop on Mixed Criticality: Roadmap to Evolving UAV Certification, CPSWeek'09.

Mukherjee, T., Banerjee, A., Varsamopoulos, G., Gupta, S.K.S., Rungta, S., 2009. Spatio-temporal thermal-aware job scheduling to minimize energy consumption in virtualized heterogeneous data centers. Comput. Netw. 53 (17), 2888–2904.

Nabar, S., Banerjee, A., Gupta, S.K.S., Poovendran, R., 2011. GeM-REM: generative model-driven resource efficient ECG monitoring in body sensor networks. In: 2011 International Conference on Body Sensor Networks (BSN). IEEE, pp. 1–6.

Osterlind, F., Dunkels, A., Eriksson, J., Finne, N., Voigt, T., 2006. Cross-level sensor network simulation with COOJA. In: Proceedings 31st IEEE Conference on Local Computer Networks.

Ouimet, M., Lundqvist, K., 2007. Formal Software Verification: Model Checking and Theorem Proving. Embedded Systems Laboratory, MIT.

Păsăreanu, C.S., Rungta, N., 2010. Symbolic pathfinder: symbolic execution of java bytecode. In: Proceedings of the IEEE/ACM International Conference on Automated Software Engineering. ASE'10. ACM, New York, NY, pp. 179–180.

Platzer, A., 2007. Differential dynamic logic for verifying parametric hybrid systems. Automated Reasoning with Analytic Tableaux and Related Methods. Springer, Berlin. pp. 216–232.

Platzer, A., Quesel, J.-D., 2008. KeYmaera: a hybrid theorem prover for hybrid systems (system description). Automated Reasoning. Springer, Berlin. pp. 171–178.

Prokopenko, M., Boschetti, F., Ryan, A.J., 2009. An information-theoretic primer on complexity, self-organization, and emergence. Complexity 15 (1), 11–28.

Sasnauskas, R., Dustmann, O., Kaminski, B., Wehrle, K., Weise, C., Kowalewski, S., 2011. Scalable symbolic execution of distributed systems. In: 2011 31st International Conference on Distributed Computing Systems (ICDCS), pp. 333–342.

Tang, Q., Tummala, N., Gupta, S.K.S., Schwiebert, L., 2005. Communication scheduling to minimize thermal effects of implanted biosensor networks in homogeneous tissue. IEEE Trans. Biomed. Eng. 52 (7), 1285–1294.

Tang, Q., Gupta, S.K., Stanzione, D., Cayton, P., 2006. Thermal-aware task scheduling to minimize energy usage of blade server based datacenters. In: 2nd IEEE International Symposium on Dependable, Autonomic and Secure Computing. IEEE, pp. 195–202.

The Human Body, 2015. Exploring a New Frontier of Cyber-Physical Systems: The Human Body. https://www.nsf.gov/mobile/news/news_summ.jsp?cntn_id=135105&org=NSF&from=news (accessed 08.31.15).

Trachette, L..J., Helen, M.B., 1999. A mathematical model to study the effects of drug resistance and vasculature on the response of solid tumors to chemotherapy. Math. Biosci. 164, 17–38 (1085).

Wang, Y., Eskridge, K.M., Galecki, A.T., 2010. Unknown title. J. Health 2 (3), 188–194.

Wongpiromsarn, T., Mitra, S., Lamperski, A., Murray, R.M., 2012a. Verification of periodically controlled hybrid systems: application to an autonomous vehicle. ACM Trans. Embed. Comput. Syst. 11 (S2), 53.

Wongpiromsarn, T., Topcu, U., Murray, R.M., 2012b. Receding horizon temporal logic planning. IEEE Trans. Autom. Control 57 (11), 2817–2830.

BENCHMARKING OF CYBER-PHYSICAL SYSTEMS IN INDUSTRIAL ROBOTICS

THE ROBOCUP LOGISTICS LEAGUE AS A CPS BENCHMARK BLUEPRINT

13

T. Niemueller*, G. Lakemeyer*, S. Reuter[†], S. Jeschke[†], A. Ferrein[‡]

Knowledge-Based Systems Group, RWTH Aachen University, Aachen, Germany[*] Institute Cluster IMA/ZLW & IfU, RWTH Aachen University, Aachen, Germany[†] Mobile Autonomous Systems & Cognitive Robotics Science, Aachen University of Applied Sciences, Aachen, Germany[‡]

1 INTRODUCTION

Manufacturing industries are on the brink of widely accepting a new paradigm for organizing production by introducing perceiving, active, context-aware, and autonomous systems. This is often referred to as Industry 4.0 (Kagermann et al., 2013), a move from static process chains toward more automation and autonomy. The corner stones for this paradigm shift are *smart factories*, which are context-aware facilities in which manufacturing steps are considered as services that can be combined efficiently in (almost) arbitrary ways allowing for the production of various product types and variants cost-effectively even in small lot sizes, rather than the more traditional chains, which produce only a small number of product types at high volumes. These smart factories are made possible by means of *cyber-physical systems (CPSs)*, which combine computational with sensing and actuation capabilities to directly observe and influence the production process in a more decentralized fashion. They can communicate vertically within the overall business processes (e.g., to retrieve information like the priority of an order) and horizontally with other machines on the factory floor. Such smart factories require a capable logistics system, which efficiently maintains the material flow and highly parallel execution of individual productions. Autonomous mobile robots are a natural choice for this task, as they can apply multi-robot coordination to the emerging problem of *supply chain optimization (SCO)* and even take over some parts of the machine handling. (Supply chains describe logistic networks that comprise interlinked logistic actors (Niemueller et al., 2013c), in this chapter, we focus on intra-logistics for the material flow of a smart factory by a group of autonomous mobile robots.) To develop such systems, a *testbed* is required to ensure the fitness of a system for a specific task. Furthermore, *benchmarks* are needed that allow the quantification of the performance by a given set of aspects to compare systems with each other.

Cyber-Physical Systems. http://dx.doi.org/10.1016/B978-0-12-803801-7.00013-4

193

In this chapter, we will outline a method for designing such benchmarks by identifying *key performance indicators (KPI)* from the underlying industrial context and using such metrics for the evaluation of integrated systems within an environment of comprehensible and manageable size. We focus in particular on multi-robot systems as one of the most challenging CPS categories. As a specific example, we introduce the *RoboCup Logistics League (RCLL)*, which models the task of in-factory logistics and is inspired by actual industrial contexts. We describe how the analysis of data recorded during actual test runs (competitive games) lead to the better understanding and identification of relevant KPIs that will hopefully allow the quantification of the overall system performance by fine-grained criteria in the future. This specific instance of the workflow to create a *holistic benchmark* that is a benchmark that combines the assessment of an integrated (multi-robot) system with a domain-specific set of KPIs to evaluate the performance, can serve as a blueprint to create benchmarks also for other industry-inspired scenarios.

The chapter is structured as follows. In Section 2 we define the terminology and the example logistics scenario. Following on, in Section 3, we introduce the RoboCup Logistics League as an example of the logistic robot scenario. We describe general criteria for a benchmark in Section 4 and other scenarios that exist today. In Section 5 we present the RCLL as a benchmark and our quantitative data analysis. KPIs and their exemplary application to the RCLL are described in Section 6 before we discuss the transferability of our approach to other scenarios in Section 7 and conclude in Section 8.

2 CYBER-PHYSICAL SYSTEMS IN A SMART FACTORY

CPSs integrate computation and physical processes (Lee, 2008). The term has undergone several development stages starting with simple identification devices such as RFID (radio frequency identification, passive information retrieval via wireless communication) chips, followed by devices with sensors and actuators that required centralized processing and instruction. The latest generation—considered as the canonical CPS (Hermann et al., 2015)—are embedded systems that extend this by integrating computation and can be networked. In the context of this paper, we limit the meaning to the most complex version of such CPSs, that is *autonomous mobile robots*. Such systems have a variety of sensors and actuators to perceive, manipulate, move within the environment, and coordinate to form a multi-robot system. A *smart factory* (SF) is a context-aware production facility that considers, e.g., object positions or machine status, to assist in the execution of manufacturing tasks (Lucke et al., 2008). It can draw information from the physical environment or a virtual model, e.g., a process simulation, order, or product specification. It is designed to cope with the challenges that arise from the desire to produce highly customized goods, which result in the proliferation of variants (Lucke et al., 2008) and therefore smaller lot sizes. A CPS, in the sense of an autonomous mobile robot, requires a *task-level executive*, which is the highest level of decision-making software component that determines the tasks and steps necessary to achieve its intended goal, and that executes and monitors the required behaviors. Typical approaches can be roughly divided in three categories (Niemueller et al., 2015b): state machine based controllers like SMACH (Bohren and Cousins, 2010) or XABSL (Loetzsch et al., 2006), reasoning systems from Procedural Reasoning Systems (Ingrand et al., 1996) or rule-based agents (Niemueller et al., 2013b) to more formal approaches like GOLOG (Levesque et al., 1997), and finally automated planning systems with varying complexity and modeling requirements, for example based on PDDL (McDermott et al., 1998) and its various extensions. In this sense, the task-level executive is what makes a CPS an *intelligent system*.

For the remainder of this chapter, we envision a SF that offers various manufacturing services, i.e., a number of machines that can refine, assemble, or modify a workpiece toward a final product. Given a product specification, the SF determines the required processing steps and involved machines. Such a SF rather offers a number of special production technologies than just production types (Hermann et al., 2015) and requires more complex logistics for efficient parallel production. The challenge then becomes developing a multi-robot system capable of efficiently determining the assignment of robots to logistics tasks, and machine handling. The robots must coordinate in order to avoid resource conflicts (e.g., two workpieces requiring processing at the same machine) and cease the opportunity for cooperation (e.g., to meet a tight schedule for a high value product). This entails the studying of various methods for the task-level executive and coordination, and to balance the *trade-off between centralization and distribution*.

3 THE ROBOCUP LOGISTICS LEAGUE

RoboCup (Kitano et al., 1997) is an international initiative to foster research in the field of robotics and artificial intelligence. Besides robotic soccer, RoboCup also features application-oriented leagues that serve as common testbeds to compare research results. Among these, the industry-oriented RoboCup Logistics League (RCLL) (RoboCup Logistics website: http://www.robocup-logistics.org) tackles the problem of production logistics in a smart factory. Groups of three robots have to plan, execute, and optimize the material flow and deliver products according to dynamic orders in a simplified factory. The challenge consists of creating and adjusting a production plan and coordinate the group (Niemueller et al., 2013a).

The RCLL competition takes place on a field of 12 m × 6 m partially surrounded by walls (Fig. 1). Two teams of three robots each are playing at the same time. The game is controlled by the *referee box (refbox)* (Niemueller et al., 2013c), a central software component for sending orders to the robots, controlling the machines and collecting teams' points. Additionally, log messages and game reports are sent to the refbox, which allows for detailed game analyses and benchmarks. It is also used in a simulation of the RCLL (Zwilling et al., 2014). After the game is started, no manual interference is allowed, robots receive instructions only from the refbox and must act fully autonomously. The robots must plan and coordinate their actions to efficiently fulfill their mission (cf. Niemueller et al. (2015b) for a characterization of the RCLL as a planning domain). The robots communicate among each other and with the refbox through WiFi. Communication delays and interruptions are common and must be handled gracefully.

Each team has an exclusive set of six machines of four different types of modular production system (MPS) stations. The refbox assigns a zone of 2 m × 1.5 m to each station (position and orientation are randomized within the zone). Each station accepts input on one side and provides processed workpieces on the opposite side. Machines have markers that uniquely identify the station and side. A signal light indicates the current status, such as "ready", "producing", or "out-of-order".

FIG. 1

Teams Carologistics (robots with additional laptop) and Solidus (pink parts) during the RCLL finals at RoboCup 2015.

The game is split in two major phases. In the *exploration phase* robots need to roam the environment and discover machines assigned to their teams. For such machines, it must then report the marker ID, position zone, and light signal code. For correct reports, robots are awarded points, for incorrect reports negative points are scored.

In the *production phase*, the goal is to fulfill orders according to a randomized schedule by refining raw material through several processing steps and eventually delivering it. An order states the product to be produced, defined as a workpiece consisting of a cup-like cylindrical colored base, zero, or up to three colored rings, and a cap. An exemplary production chain for a product of the highest complexity is shown in Fig. 2. All stations require communication through the refbox to prepare and parametrize them for the following production step. Using different types of machines, the robots have to assemble different final products.

The robot used in the competition is the Robotino 3 by Festo. It features a round base with a holonomic drive, infrared short-distance sensors, and a webcam. Teams are allowed to add additional sensors and computing devices as well as a gripper for workpiece handling (Fig. 3).

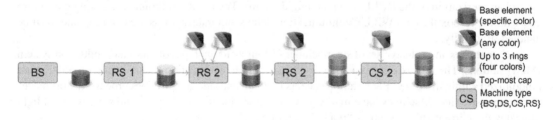

FIG. 2

Steps for the production of a high complexity product in RCLL 2015.

FIG. 3

Robot approaching a ring station in production.

4 ROBOT BENCHMARKING

In the chosen example, the goal is to evaluate a robotics scenario in an industrial context. Therefore we now describe aspects and criteria of a robotics benchmarks in general, verify that our specific scenario indeed constitutes a suitable benchmark, and mention some other existing robot benchmarks.

4.1 ROBOT BENCHMARK FEATURES

The key question is whether a scenario fulfills some general criteria to consider it a proper robot benchmark like (cf. Dillmann, 2004): (1) the robot needs to perform a real mission; (2) the benchmark must be accepted in the field; (3) the task has a precise definition; (4) repeatability, independence, unambiguity of the test, and (5) collection of ground-truth data.

Robotic systems are integrated from a large number of hardware and software components. Other questions that arise are whether specific or smaller groups of components (modules) can and should be tested, whether the analysis of the full system is of interest, or to aim for a combination of both (Amigoni et al., 2013). In the context of a multi-robot system another important dimension is the overall fleet-level performance, i.e., how does the robot group perform as a whole.

A final relevant question is whether to run fully reproducible tests, or whether to test the flexibility of the approach. In the former case, robots would be presented with the same environment, task, and parametrization every time. This especially allows the benchmarking of the optimization capabilities of a system or module, i.e., how well can it adapt to the environment over time and what is the best- or worst-case performance (depending on the chosen scenario). If testing for flexibility, those three elements could be varied (within a certain domain-specific range), for example, introducing uncertainty in the task structure (varying, e.g., time parameters when something must be started or completed), or other agents. This in particular allows the testing of the robustness and average performance of systems and modules.

A crucial part for a benchmark are metrics, which allow to quantification of the performance and comparison of different methods and systems. These metrics are often neglected when benchmarks are described, as it can be hard to find the proper method to calculate a meaningful and decisive value. The metrics are also not intrinsic to an environment or system, but they rather describe the criteria (optimization, robustness, etc.) relevant for the benchmark operator.

4.2 RCLL AS A BENCHMARK

First, we verify the general requirements outlined by Dillmann (2004): (1) the robots' task is to maintain and optimize the material flow in the factory and they thus accomplish a real mission; (2) the scenario has been accepted as a full RoboCup league with participants (in the RCLL alone) from all over the globe, thus we can state that the benchmark has been accepted in the field; (3) the rulebook provides a detailed specification (RCLL Technical Committee, 2015), furthermore the referee box provides accountable agency to the environment (and simulation), the task therefore has a precise definition; (4) the scenario can be run many times, either with the same or with a randomized parametrization, it is independent from the observer and the results are unambiguous; and (5) the referee box logs messages sent over the network, game reports, and updates to its fact base comprising information about the game available and relevant for the refbox, additionally a prototype was developed using field-mounted sensors for product tracking (Niemueller et al., 2013c), thus ground-truth data are available.

An important aspect that can be tested is the *system-level performance of a single robot system*, i.e., what difference does it make to add or remove a robot. Necessarily, this also includes *module-level aspects*, e.g., how good are the path planning or collision avoidance capabilities of the robot while being deployed in a real task. But most importantly, the *fleet-level* performance is crucial, which is determined in particular by the effectiveness of the coordination strategy and cooperation capabilities. Unlike approaches like RoCKIn (see Section 4.3), however, the performance is evaluated within an industry-inspired scenario, rather than crafting the scenario such that the benchmark requirements are met and potentially diverting much farther from an actual industrial scenario.

The referee box allows the scenario to be run either with the same environment and parametrization, or randomizing machine positions and order time windows each time. This way, optimization for specific scenarios can be evaluated for best and worst case runs, or the average performance and robustness without bias can be evaluated.

The RCLL's metric is implicitly given by its grading scheme. Complex production steps and delivery are awarded with points and thus drive the performance. We will have a close look at the evaluation criteria and ideas for improvement in Section 5.

4.3 OTHER TESTBED AND BENCHMARK SCENARIOS

The RoCKIn (RoCKIn website: http://rockinrobotchallenge.eu/) project is another recent and prominent benchmarking approach with an industrial aspect. There, the goal is to combine system-level and module-level benchmarking results into a single system (Amigoni et al., 2013). While there are key differences between the RCLL and the RoCKIn@Work approach (e.g., single vs. multi-robot, module-level aspects vs. focus on overall integrated system evaluation, scenario design vs. solution design based on industry-inspired scenario), we deem compatible elements in both approaches (Niemueller et al., 2015a). An obstacle in this regard might be the fact that RoCKIn@Work focuses on multiple short-term tasks, while the RCLL strives for long-term autonomy by using a single scenario for enduring test runs. Other interesting benchmark scenarios making use of robotic competitions are, for instance, the European Land Robot Trials (ELROB) (Schneider et al., 2010), and RoboCup@Home (Iocchi et al., 2015).

5 A BENCHMARK FOR LOGISTIC ROBOTS IN A SMART FACTORY

In this section, we combine our general considerations of CPS in smart factories (Section 2) and its special instance RCLL with the criteria from Section 4 to form a relevant benchmark.

5.1 RCLL 2014 EVALUATION

Recently, we presented a data analysis based on 75 GB of refbox data organized using MongoDB (Niemueller et al., 2012) of the RoboCup competition 2014 (Niemueller et al., 2015c). As an example, we focus on the RoboCup 2014 final of the RCLL between the Carologistics (cyan) and the BBUnits (magenta), which ended with a score of 165 to 124 (video of the final is available at https://youtu.be/_iesqH6bNsY). While the overall theme of the game was the same in 2014, the game was played on a more constrained field with simpler machines compared to 2015, as shown in Fig. 4. Products were

FIG. 4

Carologistics (Robotino 2 with laptops) and BavarianBendingUnits (Robotino 3) during the RCLL finals at RoboCup 2014.

represented by pucks moved on the ground and placed at machines. Each team had 16 machines of seven different types and there were only three different kinds of products. We briefly introduce the figures that make up our analysis before we draw conclusions based on our observations.

5.1.1 Analysis—Description

Fig. 5 shows the machines (M1–M24) grouped per team above the horizontal game-time axis. Each row expresses the machine's state over the course of the game. Gray means it is currently idle. Green means that it is actively processing (busy) or blocked while waiting for the next input. After a work order has been completed, the machine waits for the product to be picked up (orange/waiting). The machine can be down for maintenance for a limited time (dark red/down). Sometimes it is used imprecisely (yellow/imprecise), that is, the product is not placed properly under the RFID device. The row "Deliveries"

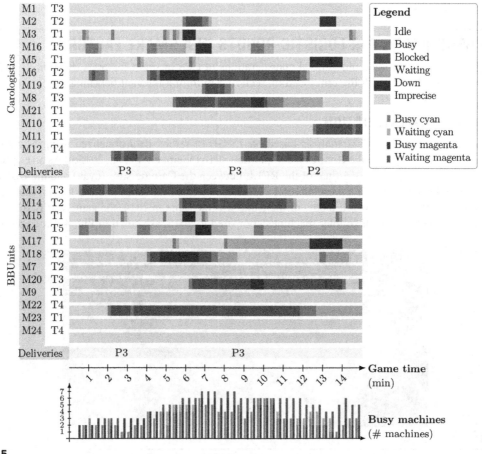

FIG. 5

Machine states over the course of the final game at RoboCup 2014. The lower graph shows the occupied machines per 20 second time block.

shows products that are delivered at a specific time. Below the time axis, Fig. 5 shows the busy machines over time. Each entry consists of a cyan (light grey) and magenta (dark grey) column and represents a 20 second period. The height of each column shows the number of machines that are producing (bright team color) or waiting for the product to be retrieved (dark team color).

Fig. 6 shows the adherence to the delivery schedule. The horizontal axis denotes the game-time and each row a particular order. Recall that an order is defined by a product type and amount (P1–P3) and the start and end time for the delivery. In the diagram, red boxes (unfulfilled) indicate orders that have been missed. Light green boxes (completed) show fulfilled orders, while dark green (partial) orders indicate that only some of the total amount of requested products has been delivered. Red dots indicate the actual time of delivery. Robots send periodic beacon signals that may include the position of the robot. This allows for analyzing the typical paths and hotspots on the field. As one example, Fig. 7 visualizes the data of a game in simulation by the Carologistics team as a 2D histogram (heatmap). Regions of the map without overlay have not been visited by any robot. Red (medium grey) areas indicate frequent operation of any of the robots in that area fading to blue (dark grey) according to the given scale for less frequent travel.

5.1.2 Analysis—Conclusions

Fig. 5 shows that the production strategies of both teams differ significantly. The cyan team (Carologistics) is goal oriented toward moving workpieces fast through the production steps, indicated by shorter busy and blocked times (green/busy) as well as fewer and very short periods where a machine is waiting for the output to be picked up (orange/waiting). The magenta team (BBUnits), on the other hand, has on average more machines occupied (bar chart at the bottom) resulting in longer times that workpieces remain in a machine (dark green/blocked areas when waiting for the next input workpiece and orange/waiting areas when waiting for output to be picked up).

Already in anticipation of the next section, we recognize that the cyan team achieved a significantly lower throughput time, that is the time that a workpiece requires from the first to the last processing step. The magenta team, however, had a higher number of machines in use at a given time. Only the lower throughput time enabled the cyan team to deliver the P2 product (Fig. 6). Additionally, it allowed

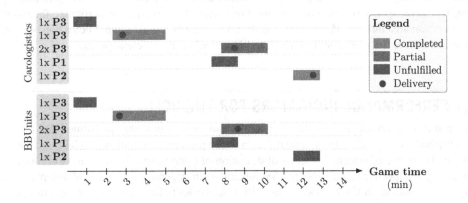

FIG. 6

Adherence to delivery schedule (finals RoboCup 2014).

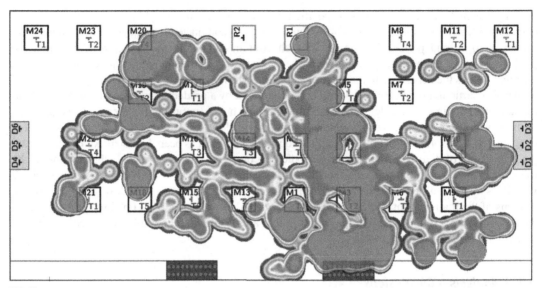

FIG. 7

2D histogram (heatmap) for the positions of all three robots of one team combined on the playing field over the course of a full game in simulation. Red (medium grey inner areas) indicates areas of frequent travel fading according to the scale to blue (dark grey outer boarders) for less time spent at a position.

for more recycling of material (recyclable material is only available after a more complex machine has completed a work order).

The major observation here is that the grading implicitly favors low throughput production—which was not intended. Recognizing this lead to reasoning about alternative and more fine-grained evaluation metrics that we introduce in the next section.

A secondary observation based on Fig. 7 is the existence of areas that are visited frequently. One such place, naturally, is the blue (dark grey) insertion area, where robots need to go to pick up raw material. The heatmap also reveals corridors used frequently for traveling. In particular, the route from the blue insertion area through M3 and M6 and then M2 and M4 is used often. Such a hotspot can slow down the team more than using sub-optimal, but less frequently traveled routes.

6 KEY PERFORMANCE INDICATORS FOR THE RCLL

Performance evaluation requires metrics to quantify how well a task was performed and to compare different methods and systems. In industrial contexts key performance indicators (KPI) are used to analyze the performance of systems in terms of deviation of operational planning and execution. These metrics are highly specific to the problem domain and even depend on the particular department performing the evaluation. In the following, we will describe KPIs for logistics in production and their adaptation and application in the RCLL, as an example of how existing well-known KPIs can be transferred to a related robotics scenario.

6.1 KEY PERFORMANCE INDICATORS IN LOGISTICS

High efficiency of logistics systems can be achieved by maximizing performance, or by minimizing costs (Nyhuis and Wiendahl, 2006). Using KPIs, an operationalization of the logistic objective is possible. Fig. 8 shows exemplary indicators assigned to such objectives. As an example, we describe two of the most important logistic KPIs: *throughput time* (the time span for an order to be created) and *utilization* (of production machines) in more detail and point out some intricacy known as the scheduling dilemma of logistics. For a more detailed discussion of KPIs for logistic processes relevant for the RCLL we refer to Niemueller et al. (2015c).

The *throughput time (TTP)* of a process is defined as the duration from the start to the end of the processing of an order (Lödding, 2012): $TTP = T_{proc.\ end} - T_{proc.\ start}$. As an example, Fig. 9 shows the throughput of a product of type P2, which consists of a raw product S0 and two intermediate products S1 and S2. The intermediate products have to be manufactured in advance in order to assemble the product P2. The manufacturing of an intermediate product S2 consists of two operations on the machines T1 and T2, as well as the time span needed for transportation and waiting times. While the final

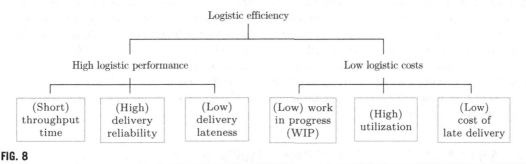

FIG. 8

Exemplary key performance indicator within production logistics.

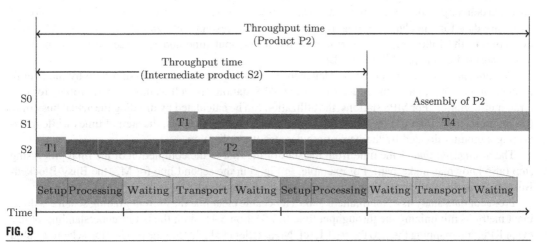

FIG. 9

Throughput time components.

assembly of product P2 can only be done after all necessary intermediate products are ready, the manufacturing of the intermediate products S1 and S2 can be *parallelized*. Thus, the resulting throughput time of product P2 is formed by the throughput time of the intermediate product S2 and the final assembly.

The *utilization U* is the usage ratio of production resources such as production machines or transportation entities and the operation period. Thus, the utilization in terms of a production machine describes the ratio of a machine's operating time and a duration of the desired reference period (Lödding, 2012): $U = \sum T_{\text{operations}} / T_{\text{duration of reference period}} \times 100\%$.

Effective logistic processes strive for the maximization of logistic performance while minimizing costs. In terms of the introduced KPIs, this demands for short throughput times and high utilization of logistic resources. To maximize *throughput time* of an order, the order has to be processed as soon as it arrives by triggering the manufacturing of all necessary intermediate products, the transport among the production machines as well as the assembly of the final product. For an undelayed flow of the order through the production facility, all production and logistic resources should be idle and waiting for processing the active order. In contrast, a high *utilization of resources* demands continuous input and output of orders to minimize waiting and idle-times. This can only be achieved by continuously feeding production machines in order to prevent shortages in the material flow. But this will slow down the throughput time, as orders have to wait to be processed. Hence, high resource utilization and short throughput times cannot be achieved together. This conflict between the objectives of *logistic performance* and *logistic costs* is called the *scheduling dilemma of logistics* (Nyhuis and Wiendahl, 2006). As one strives to maximize logistic performance by decreasing throughput time, logistic costs increase as this can only be achieved by decreasing the utilization of production resources. It is up to the benchmark operator to balance this trade-off.

6.2 APPLICATION OF KEY PERFORMANCE INDICATORS IN THE RCLL

The aim of RCLL is to simulate a realistic, yet simplified, production environment. With a limited set of resources (mobile transportation robots and stationary production machines) the teams have to produce and deliver products according to a dynamic production schedule. By means of the referee box all necessary data for calculating throughput time as well as resource utilization (and further KPIs) are available. In the following we describe how the throughput time and resource utilization of KPIs can be adapted and applied in the RCLL.

The *utilization U* of a resource is calculated by dividing its actual processing time by the overall evaluation time (e.g., game time). Resources are MPS stations as well as three mobile robots for material transportation. For MPS stations, the utilization can be calculated by dividing the actual busy time (bright green/busy areas in Fig. 5) by the overall game time. For robots, the sum of times while transporting a product divided by the game time denotes the utilization.

The *throughput time* is the time from the start of the first to the completion of the final processing step (also known as makespan). For example, in Fig. 5 in the second line for M2, the Busy-Blocked-Busy cycle is the throughput time for a production of 84 seconds.

From our data analysis of tournament games it became clear that the current grading scheme's dominant metric is minimizing the throughput time (cf. Section 5.1). With the better understanding of relevant KPIs and mapping them to the RCLL (cf. Niemueller et al., 2015c) the production schedule could now be adjusted and a more fine-grained grading scheme developed, e.g., to award certain best-in-class capabilities. For example, to simulate large volume production (which still can benefit from dynamic autonomous logistics, for example for variant production, where robots are used only for some processing steps), a standing order for a certain product would benefit from the maximization of machine utilization.

7 TRANSFERRING HOLISTIC BENCHMARKS TO OTHER DOMAINS

A holistic robot benchmark is based on an industry-inspired scenario at a comprehensible scale where robotic task performance is evaluated with existing or slightly adapted KPIs. When introducing robotics in an industrial scenario, the goals are usually efficiency and/or flexibility. In the context of smart factories, additionally more process autonomy is desired. To verify whether these goals have been reached, valid metrics are necessary to quantify the impact. Several factors must be considered, e.g., whether the overall factory performs better with the help of robots, or whether more robots increase performance or lead to unexpected bottlenecks. Following the example of the RCLL, we have presented an idea how this can be accomplished for autonomous in-factory logistics. This can be extended into a blueprint for generally applying robotic benchmarks in industrial scenarios:

1. Take the relevant scenario and scale it down to a comprehensible size introducing robots (or CPS in the more general sense) in certain areas.
2. Integrate the system on a small scale and verify its feasibility. A simulation environment adequately representing the environment can be a viable option.
3. Apply existing or slightly adapted KPIs relevant to the scenario that have been used before.
4. If the results are satisfactory, scale up the approach monitoring the development of the KPIs.

The latter step seems especially important to small and medium enterprises, for which the introduction of complex technologies can pose a high risk. Starting out at a smaller scale, or in simulation, with continuous evaluation, allows for an informed decision of the viability of the project early in the process.

8 CONCLUSION

We expect autonomous mobile robots to play an important role in smart factories in the context of Industry 4.0. Mobile robots are complex CPSs that combine sensors, actuators as well as computing devices that enable them to reason in response and interact with the environment. The ability of these systems to reason in response to their environment and to select a corresponding behavior poses new challenges for evaluation and certification. On the one hand, the autonomy of these systems expands their operational envelop and make them suitable for a broad area of applications. On the other hand, companies selling mobile robots have to give reliable predictions of the systems behavior. Therefore extensive tests have to be conducted. As the systems behavior evolves in combination with other autonomous systems, a test of all possible situations becomes infeasible.

In this paper, we have introduced the concept of an holistic benchmark, which combines an environment and task inspired by an industrial context with selected KPIs, which are metrics relevant for the specific scenario.

As a specific example for such a logistics benchmark scenario we described the RCLL. It has a proper balance of medium complexity, comprehensible size, and yet the necessary challenges to yield meaningful results. Based on the analysis of data recorded through several RCLL competitions and simulations we explained the insight that the current grading scheme prefers low throughput production over a high machine utilization. These are two of several KPIs that we have identified as being relevant and interesting, and could be used for further develop the league as a benchmark. It shows that it is important to choose both the scenario and metrics with respect to the intended industrial context.

The general idea of a holistic benchmark and its example implementation for the RCLL can serve as a blueprint for further testbeds and benchmarks.

ACKNOWLEDGMENTS

T. Niemueller was supported by the German National Science Foundation (DFG) research unit *FOR 1513* on Hybrid Reasoning for Intelligent Systems (http://www.hybrid-reasoning.org).

REFERENCES

Amigoni, F., Bonarini, A., Fontana, G., Matteucci, M., Schiaffonati, V., 2013. Benchmarking through competitions. In: European Robotics Forum—Workshop on Robot Competitions: Benchmarking, Technology Transfer, and Education.
Bohren, J., Cousins, S., 2010. The SMACH high-level executive. IEEE Robot. Autom. Mag. 17(4).
Dillmann, R., 2004. Benchmarks for Robotics Research. EURON. http://www.cas.kth.se/euron/euron-deliverables/ka1-10-benchmarking.pdf Technical report.
Hermann, M., Pentek, T., Otto, B., 2015. Design Principles for Industrie 4.0 Scenarios: A Literature Review. Working Paper 01/2015. Technische Universität Dortmund.
Ingrand, F., Chatila, R., Alami, R., Robert, F., 1996. PRS: a high level supervision and control language for autonomous mobile robots. In: IEEE International Conference on Robotics and Automation (ICRA). vol. 1.
Iocchi, L., Holz, D., Ruiz-del Solar, J., Sugiura, K., van der Zant, T., 2015. Robocup@home: analysis and results of evolving competitions for domestic and service robots. Artif. Intell. 229, 258–281.
Kagermann, H., Wahlster, W., Helbig, J., 2013. Recommendations for implementing the strategic initiative INDUSTRIE 4.0. Final Report. Platform Industrie 4.0.
Kitano, H., Asada, M., Kuniyoshi, Y., Noda, I., Osawa, E., 1997. Robocup: The Robot World Cup Initiative. In: 1st International Conference on Autonomous Agents.
Lee, E.A., 2008. Cyber Physical Systems: Design Challenges. In: 11th IEEE International Symposium on Object Oriented Real-Time Distributed Computing (ISORC-08).
Levesque, H.J., Reiter, R., Lespérance, Y., Lin, F., Scherl, R.B., 1997. Golog: a logic programming language for dynamic domains. J. Logic Program. 31(1–3).
Lödding, H., 2012. Handbook of Manufacturing Control: Fundamentals, Description, Configuration. Springer Science & Business Media, New York/Berlin.
Loetzsch, M., Risler, M., Jungel, M., 2006. XABSL—a pragmatic approach to behavior engineering. In: IEEE/RSJ International Conference on Intelligent Robots and Systems (IROS).
Lucke, D., Constantinescu, C., Westkämper, E., 2008. Smart factory—a step towards the next generation of manufacturing. In: Manufacturing Systems and Technologies for the New Frontier, The 41st CIRP Conference on Manufacturing Systems.

McDermott, D., Ghallab, M., Howe, A., Knoblock, C., Ram, A., Veloso, M., Weld, D., Wilkins, D., 1998. PDDL—The Planning Domain Definition Language. AIPS-98 Planning Competition Committee Technical report.

Niemueller, T., Lakemeyer, G., Srinivasa, S.S., 2012. A generic robot database and its application in fault analysis and performance evaluation. In: IEEE/RSJ International Conference on Intelligent Robots and Systems (IROS).

Niemueller, T., Ewert, D., Reuter, S., Ferrein, A., Jeschke, S., Lakemeyer, G., 2013. RoboCup Logistics League Sponsored by Festo: a competitive factory automation testbed. In: RoboCup Symposium 2013.

Niemueller, T., Lakemeyer, G., Ferrein, A., 2013b, Incremental task-level reasoning in a competitive factory automation scenario. In: AAAI Spring Symposium 2013—Designing Intelligent Robots: Reintegrating AI.

Niemueller, T., Lakemeyer, G., Ferrein, A., Reuter, S., Ewert, D., Jeschke, S., Pensky, D., Karras, U., 2013c, Proposal for advancements to the LLSF in 2014 and beyond. In: ICAR—1st Workshop on Developments in RoboCup Leagues.

Niemueller, T., Ferrein, A., Reuter, S., Jeschke, S., Lakemeyer, G., 2015a, The RoboCup Logistics League as a holistic multi-robot smart factory benchmark. In: Open Forum on Evaluation of Results, Replication of Experiments and Benchmarking in Robotics Research at IROS 2015.

Niemueller, T., Lakemeyer, G., Ferrein, A., 2015b, The RoboCup Logistics League as a benchmark for planning in robotics. In: WS on Planning and Robotics (PlanRob) at International Conference on Automated Planning and Scheduling (ICAPS).

Niemueller, T., Reuter, S., Ferrein, A., Jeschke, S., Lakemeyer, G., 2015c, Evaluation of the RoboCup Logistics League and derived criteria for future competitions. In: RoboCup Symposium.

Nyhuis, P., Wiendahl, H.P., 2006. Logistic production operating curves-basic model of the theory of logistic operating curves. CIRP Ann. Manuf. Technol. 55(1).

RCLL Technical Committee, 2015. The RoboCup Logistics League Rulebook for 2015. RoboCup Logistics League Technical report.

Schneider, F.E., Wildermuth, D., Brüggemann, B., Röhling, T., 2010. European Land Robot Trial (ELROB) towards a realistic benchmark for outdoor robotics. AT&P J. Plus 2, 97–102.

Zwilling, F., Niemueller, T., Lakemeyer, G., 2014. Simulation for the RoboCup Logistics League with real-world environment agency and multi-level abstraction. In: RoboCup Symposium.

LOCALIZATION IN CYBER-PHYSICAL SYSTEMS

14

G. Liu*, Q. Chen*, Y. Liu†, Q. Yang*

Montana State University, Bozeman, MT, United States* Southeast University, Nanjing, China†

1 INTRODUCTION

Due to the success of cyber-physical systems (CPSs) (Lee, 2008), radio-frequency identification (RFID) technology has attracted a lot of attention from both industry and academia, e.g., smart home, Industry 4.0, object identification, and indoor localization (Finkenzeller and Waddington, 1999; Nazari Shirehjini et al., 2012; Almaaitah et al., 2010).

An RFID system consists of several readers and RFID tags. RFID tags are attached to objects of interest, and send back reply signal when they receive query signals from readers. RFID tags can be categorized as proactive and passive tags. The proactive tags are equipped with a battery and will transmit the reply signal back to the readers once it receives the query signals. The passive tags, on the other hand, are battery free and transmit the reply signal by backscattering the energy of the query signal from the reader. Although RFID technology was first proposed for identifying objects, it is now being used in localizing objects.

In early CPSs (Lymberopoulos et al., 2015), e.g., wireless sensor networks (Kuruoglu et al., 2009; Moore et al., 2004), wireless localization methods can be categorized into two groups: range-based and range-free approaches (Zhao et al., 2013).

The goal of range-based approaches is to obtain an accurate position for a target. Range-based approaches use wireless devices to measure the wireless signals transmitted from a target. Based upon the received signals, range-based approaches use signal processing techniques to obtain the angle of arrival (AoA) (Niculescu and Nath, 2003a; Xiong and Jamieson, 2013), time difference of arrival (Peng et al., 2007), and time of arrival (Hofmann-Wellenhof et al., 2012) of the received signals. Then, the Euclidean distances from the receiving devices to the target can be calculated. Finally, the target's location is computed by analyzing the geometric relation between the target and these devices.

In contrast to range-based approaches, the goal of the range-free approach, such as Niculescu and Nath (2003b), Ni et al. (2004), Wang and Katabi (2013), and Zhong and He (2009), is to find the proximity area where a target is located. The requirement of range-free approaches is that there must be anchor nodes with known positions around the target. Range-free approaches employ indirect measures, such as RSS (Ni et al., 2004), hop count (Niculescu and Nath, 2003b), and signal profile similarity Wang and Katabi (2013), to quantify the distances between anchor nodes and the target. A closer measure indicates a closer distance between an anchor node and the target. Then, the anchor node(s) with the closest measure will be selected. Finally, the target's location is estimated as the proximity area nearby the selected anchor nodes.

Although the principle of RFID localization system is similar to above-mentioned approaches, it differs in practical implementations and high accuracy considering the characteristics of a RFID localization system. First, the powerful signal processing capability of a RFID reader enables it to carry out complicated analysis on received signals. Therefore, in a RFID localization system, readers are able to estimate the relative positions of a target tag to the readers by analyzing the backscattered radio frequency (RF) signals from the tag. An RFID tag that passively reflects received signals, however, does not have the signal processing capability. Since most of the existing RFID localization systems employ passive tags; we will use the term tag for simplicity throughout the rest of the chapter. Second, an RFID localization system often consists of several readers and large amount of tags because readers are much more expensive than tags. What's more, RFID tags are energy efficient, especially for those passive RFID tags that work without any battery. Therefore it is possible to deploy huge amount of tags in an interested area. Third, the communication range of a regular RFID localization system is 10–20 m (Yang et al., 2014), so RFID localization systems achieve a fine-grained accuracy (error of 7 mm) (Yang et al., 2014) in an indoor environment, even in the nonline-of-sight (NLOS) scenarios (error of 11.2 cm by Wang and Katabi, 2013). Such high localization accuracy is not achievable for other CPS systems. Meanwhile, due to the above differences, the techniques used in a RFID localization system can hardly be adopted by other CPS systems.

Because of the aforementioned advantages, RFID localization is widely applied in indoor environments. It was firstly used in robotic navigation applications (Tsukiyama, 2003; Han et al., 2007; Kulyukin et al., 2004), then extended to CPS applications. For example, RFID localization is used to identify misplaced items (Liu et al., 2013; Wang and Katabi, 2013), locate the position and orientation of interested items in a NLOS environment (Wang and Katabi, 2013), and track mobile baggage (Yang et al., 2014).

As illustrated in Fig. 1, RFID localization can be better understood from two perspectives: RF signal processing and accuracy improvement. RF signal processing accounts for analyzing the backscattered RF signals and calculating the position of the target tag. It can be categorized as received signal strength (RSS)-based, phase-based, and multipath profile-based. In addition to analyzing the processed RF signals, various accuracy improvement approaches are employed improve the localization accuracy in practical implementation.

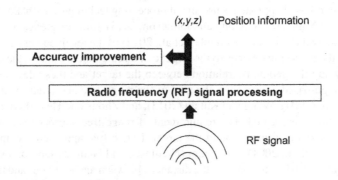

FIG. 1

Architecture of RFID localization.

Despite the variety of RFID localization approaches, there remain many research challenges in RFID localization. For example, RSS-based approaches can be easily implemented but with low accuracy. Phase-based and multipath profile-based approaches achieve higher accuracy but are time-consuming and poorly scalable.

In this chapter, we first present a review on existing RFID localization technologies, and then identify their limitations and potential improvements. The focus of this chapter will be RF signal processing in RFID localization. In addition, we will introduce various accuracy improvement techniques within different RF signal processing schemes.

2 RFID LOCALIZATION MODES

An RFID localization system can work in three different modes: (1) tag localization without reference tags, (2) tag localization with reference tags, and (3) reader localization, as illustrated in Fig. 2.

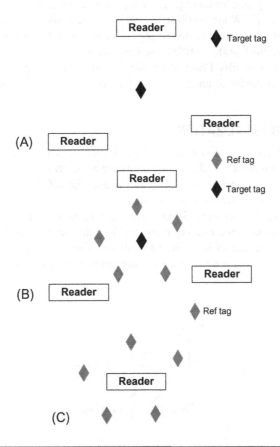

FIG. 2

Different working modes of a RFID localization system: (A) tag localization without reference tags, (B) tag localization with reference tags, and (C) reader localization.

In the first mode shown in Fig. 2A, RFID tags are attached to a target object. Localization of the target is realized by analyzing the RF signals received at the reader. In the second mode shown in Fig. 2B, extra reference RFID tags whose locations are known are deployed to assist in the localization of the target tag, such as in Ni et al. (2004). Comparing the RFID signals backscattered from the reference and the target tag, the environmental impact of the signals such as multipath effect can be eliminated. In the third mode shown in Fig. 2C, RFID tags are treat as references and the RFID reader becomes the target, such as (Kulyukin et al., 2004; Tsukiyama, 2003; Park and Lee, 2013). By communicating with these tags, the reader is able to localize itself.

3 PRINCIPLE OF RFID LOCALIZATION

In this section, we will introduce the principles of different RFID localization approaches. Existing RFID localization approaches can be categorized into three groups: RSS based (e.g., Ni et al., 2004; Bekkali et al., 2007), phase based (e.g., Li et al., 2009; Yang et al., 2014), and multipath profile based (e.g., Wang et al., 2013; Wang and Katabi, 2013). In these approaches, a reader calculates the location of the target tag(s) by analyzing the strengths, phases and multipath profiles of the received signals backscattered from the tag(s). An RSS-based approach can be easily implemented, but does not provide accurate localization results. Phase and multipath profile-based approaches offer higher accuracy but require additional hardware and/or software to perform complex signal processing tasks.

3.1 RSS-BASED RFID LOCALIZATION

The strength of a received RF signal is widely used in localizing wireless devices that send or backscatter the signal, e.g., in Bluetooth, WiFi, and cellular networks. RSS can be easily measured and usually is accessible from the physical layer. Nearly all commercial off-the-shell (COTS) RFID readers can provide RSS information in received RF signals.

The fundamental assumption of an RSS-based approach is that the strength of the received signals at a reader decreases as its distance to a tag increases, i.e., the shorter the distance, the stronger the signal strength. That is to say, RSS is an indicator of the distance between a reader and a tag. As shown in Fig. 3, the distance between a reader and a tag can be estimated based on the RSS of received signals.

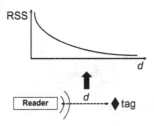

FIG. 3

Principle of RSS-based localization approaches.

The RSS of backscattered signals measured at a reader can be expressed as:

$$\text{RSS}[dB] = P_L(d_0)[dB] - 10n\log_{10}\left(\frac{d}{d_0}\right) + S,$$

where d is the distance between the tag and reader, $P_L(d_0)$ is the path loss along a reference distance d_0, which is usually set as 1 m, n is an empirical constant depending on the signal propagation environment, S is the shadowing factor—a random variable follows the Gaussian distribution. From this equation, we can get the target tag's distance d to the reader as:

$$d = d_0 \times 10^{\frac{P_L(d_0)-\text{RSS}+S}{10n}}.$$

Integrating the distances calculated from multiple readers, the location of the target tag can be calculated.

RSS-based localization approach can be easily implemented on COTS RFID readers given the RSS of received RF signals are provided. In Hightower et al. (2000), an empirical equation is derived to reveal the relationship between distance and RSS. In Ni et al. (2004), differences of RSS from reference tags are captured to estimate the relative distances between a reader and tags. Since RSS can be significantly affected by the communication environment, e.g., multipath effects and reflections, it is often inaccurate in estimating the distance between a reader and tags.

3.2 PHASE-BASED RFID LOCALIZATION

Currently, many COTS RFID readers provide not only RSS but also phase information of received RF signals. Phase information can be used to compute the target tag's location by calculating the distances between the target tag and different readers or a reader's different antennas. Since phase information is resilient to the effects of various communication environments, it offers higher accurate localization results than RSS-based approaches.

To obtain the phase information of backscattered RF signals, received signals are first converted to baseband signals. The baseband signal can be represented by the in-phase (I) and quadrature (Q) components. Ideally, the phase of a received signal can be computed by

$$\varphi = \arctan\frac{Q(t)}{I(t)}$$

The phase offset of a received RF signal on a reader includes three components: (1) the phase offset φ_{prop} generated by the round-trip signal propagation between the reader and the tag, (2) the phase offset φ_{reader} introduced by the RFID reader when it processes the signal, and (3) the phase offset φ_{tag} caused by the tag when it backscatters the signal. Thus, the total phase offset of the received RF signal on the reader is

$$\varphi = \varphi_{\text{prop}} + \varphi_{\text{reader}} + \varphi_{\text{tag}} \tag{1}$$

where φ_{prop} can be used to calculate the distance from the reader and tag(s). The relationship between φ_{prop} and the distance d can be expressed as:

$$\varphi_{\text{prop}} = 2d \times k = 2d \times \frac{2\pi}{\lambda} = 2d \times \frac{2\pi f}{c} \tag{2}$$

where $2d$ is the distance that the RF signal traverses, λ is the signal's wave length, f is the signal's frequency, c is the speed of light. The above equation indicates that phase offset increases linearly as the distance increases, however, it returns to zero periodically. In fact, phase information can be used to calculate not only the target tag's location but also its velocity if it is on a moving object. Next, we will introduce several phase-based RFID localization methods.

3.2.1 Time domain phase difference of arrival (TD-PDOA)

Phase difference of RF signals received at different times can be used to estimate the velocity of a moving tag/target (Yanakiev et al., 2007; Nikitin et al., 2010). As shown in Fig. 4, a reader receives two signals at times t_2 and t_1 when the distances from the tag to the reader are d_1 and d_2, respectively. We use V_r to denote the velocity on the direction from the reader to the tag, and V_t for the velocity orthogonal to V_r. Then, V_r can be expressed as the derivative of the distance to time:

$$V_r = \lim_{\Delta t \to 0} \frac{\Delta d}{\Delta t} = \lim_{\Delta t \to 0} \frac{d_2 - d_1}{t_2 - t_1}.$$

Multiplying $2k$ on both sides, we have

$$V_r \times 2k = \lim_{\Delta t \to 0} \frac{2k \times d_2 - 2k \times d_1}{t_2 - t_1}.$$

According to Eq. (1), the above equation can be rewritten as:

$$V_r \times 2k = \lim_{\Delta t \to 0} \frac{\left(\varphi_{prop2} + \varphi_{reader} + \varphi_{tag}\right) - \left(\varphi_{prop1} + \varphi_{reader} + \varphi_{tag}\right)}{t_2 - t_1}.$$

Since the phase offsets φ_{reader} and φ_{tag} are the same for signals received at times t_2 and t_1, the radial velocity V_r can be computed from

$$V_r = \lim_{\Delta t \to 0} \frac{\varphi_{prop2} - \varphi_{prop1}}{t_2 - t_1} \times \frac{1}{2k} = \frac{\partial \varphi}{\partial t} \times \frac{c}{4\pi f}.$$

The TD-PDOA method can only estimate the radial velocity V_r, and has no way to find the tangent velocity V_t, thus limits its application in tracking moving tags.

3.2.2 Frequency domain phase difference of arrival (FD-PDOA)

Phase difference of RF signals received at different frequencies can be used to estimate the distance between a reader and a tag (Li et al., 2009; Nikitin et al., 2010). As illustrated in Fig. 5, a reader transmits RF signals on two different frequencies f_1 and f_2.

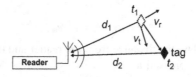

FIG. 4

Phase information is used to estimate the velocity of a moving tag.

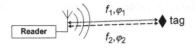

FIG. 5

Phase information is used to estimate the distance from a reader to tag.

Since different frequency means different wavelength, the phases of received signals backscattered from a tag will be different on the reader. Therefore the distance d can be calculated by comparing the phase difference in received signals. According to Eq. (2), we have

$$\varphi_{\text{prop1}} = d \times \frac{4\pi f_1}{c},$$

$$\varphi_{\text{prop2}} = d \times \frac{4\pi f_2}{c},$$

where d is the distance between the tag and RFID reader. Subtracting the above equations, we have

$$\varphi_{\text{prop2}} - \varphi_{\text{prop1}} = \varphi_2 - \varphi_1 = d \times \frac{4\pi}{c}(f_2 - f_1).$$

It is correct because the phase offsets φ_{reader} and φ_{tag} are the same in φ_{prop1} and φ_{prop2}. Therefore, the distance d from the reader to the tag can be calculated as:

$$d = \frac{c}{4\pi} \times \frac{\varphi_2 - \varphi_1}{f_2 - f_1} = \frac{c}{4\pi}\frac{\partial \varphi}{\partial f}.$$

To apply the FD-PDOA method, a reader should be capable to quickly adjust its transmission frequencies, which is not available on COTS RFID readers.

3.2.3 Space domain phase difference of arrival (SD-PDOA)

Phase difference of received signals on different antennas can be used to estimate the direction of arrival of received signals (Azzouzi et al., 2011; Nikitin et al., 2010). As illustrated in Fig. 6, a reader transmits an RF signal to a target tag. We also assume that there are at least two receiving antennas on the reader. In fact, the reader can use an antenna array to achieve higher localization accuracy.

The relationship between the signal's direction θ, also called arrival of angel (AoA), and the distance difference that the signal travels to reach antennas Rx1 and Rx2 is

$$\sin\theta \approx \frac{d_2 - d_1}{a}, \tag{3}$$

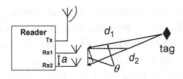

FIG. 6

Phase information is used to estimate the direction of a tag.

where a is the distance between these two antennas, d_1 and d_2 are the distances from the tag to antenna Rx1 and Rx2, respectively. The phase changes from the tag to the reader's antennas are

$$\varphi_{prop1} = kd_1,$$

$$\varphi_{prop2} = kd_2,$$

where $k = 2\pi f/c$ is the wave vector that is proportional to the frequency. Rewriting these equations, we obtain the distances from the tag to the antenna Rx1 and Rx2 as:

$$d_1 = \frac{\varphi_{prop1}}{k},$$

$$d_2 = \frac{\varphi_{prop2}}{k}.$$

Subtracting these two equations, we have

$$d_2 - d_1 = \frac{\varphi_{prop2} - \varphi_{prop1}}{k} = \frac{\varphi_2 - \varphi_1}{k} = \frac{c}{2\pi f}(\varphi_2 - \varphi_1). \tag{4}$$

Substituting Eq. (4) into Eq. (3), we get

$$\sin\theta \approx \frac{c}{2\pi f}\frac{(\varphi_2 - \varphi_1)}{a}.$$

Finally, the signal's AOA can be calculated by the following equation:

$$\theta = \arcsin\left[\frac{c}{2\pi f}\frac{(\varphi_2 - \varphi_1)}{a}\right].$$

The SD-PDOA method can accurately locate the direction of a tag. Compared to other phase-based methods, SD-PDOA requires an RFID reader to be equipped with multiple antennas. However, SD-PDOA cannot handle the NLOS scenarios.

3.2.4 Hologram
Instead of using only one reader, the holographic localization method (Parr et al., 2013; Yang et al., 2014) makes use of multiple RFID readers to calculate the location of a tag by analyzing the phase information in received signals on all readers.

The phase information of a received RF signal at a certain reader can be expressed as:

$$\varphi = \left(\frac{2\pi}{\lambda}2d\right) \bmod 2\pi = \left(\frac{4\pi}{\lambda}\sqrt{(x_0 - x)^2 + (y_0 - y)^2}\right) \bmod 2\pi, \tag{5}$$

where (x_0, y_0) and (x, y) denote the coordinates of the reader and the tag λ is the wavelength of the carrier signal and d is the distance from the reader to the tag. According to Eq. (5), a phase value measured by a reader implies multiple possible distances from itself to the tag. In particular, the possible locations inferred by a reader lay on multiple concentric circles around it as shown in Fig. 7.

Based on different phase values measured at several readers, an image consisting of multiple overlapped circles can be constructed. By splitting the image into squares, a hologram is formed in the way that any square is assigned a weight that is proportional to the number circles crossing this square. The weight of a square actually represents the likelihood of the tag being located there.

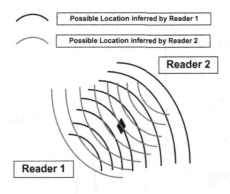

FIG. 7

Possible locations inferred by different readers.

The hologram can therefore be viewed as a likelihood distribution of the target tag's possible positions. In a hologram, the weight of a square located at (x_i, y_i) can be expressed as:

$$A(x_i, y_i) = \left| \sum_{m=0}^{n} A_m e^{j\left(\frac{4\pi d_{im}}{\lambda} - \varphi_m\right)} \right|,$$

where d_{im} denotes the distance between the square and a reader m, φ_m denotes the phase value of the signal received by reader m, and A_m is the strength of the signal received at reader m.

In summary, TD-PDOA, FD-PDOA, and SD-PDOA archive high accuracy in localization; however, TD-PDOA, FD-PDOA are too complex to be implemented on COTS RFID readers. TD-PDOA requires an RFID reader to perform multiple phase measurements within a short period of time (e.g., 1 ns), while FD-PDOA requires an RFID reader to transmit signals of multiple frequencies. Although SD-PDOA is applicable on COTS RFID readers, it cannot handle NLOS scenarios. In the end, the hologram technique can be implemented on any COTS RFID reader that provides phase information and achieves higher accuracy (Yang et al., 2014) than other phase-based approaches; it, however, requires many readers being deployed in the target area, i.e., it is not always applicable in practical real-world applications. Worse still, the hologram technique cannot handle NLOS scenario.

3.3 MULTIPATH PROFILE-BASED RFID LOCALIZATION

RSS- and phase-based localization techniques only work in the line-of-sight scenarios, where no obstacles exist between readers and tags. In many RFID localization applications, however, NLOS situations are expected. To realize RFID localization in an NLOS environment where signal reflections and multipath effects exist, multipath profile-based approaches (Wang et al., 2013; Wang and Katabi, 2013) make use of signals' profiles, such as multipath profiles, to estimate the location of RFID tags.

The outline of a multipath profile-based RFID localization approach (Wang and Katabi, 2013) can be seen in Fig. 8, where the third localization mode (tag localization with reference tags) is used. In Fig. 8, an RFID reader (with one antenna) moves along a track and keeps transmitting and receiving RF signals to emulate an antenna array. When a backscattered signal is received, the reader conducts a high

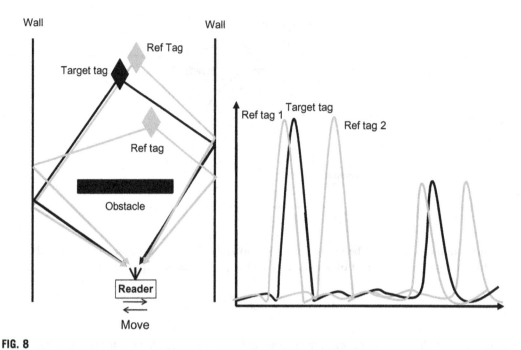

FIG. 8

Principle of a multipath profile-based RFID localization.

speed omnidirectional scan to measure the signal strength of signals from various directions. This is achieved by the reader continuously adjusting its phase controller.

For a signal reflected by a certain tag, its signal strength on direction with angle θ can be expressed as:

$$B(\theta) = \left| \sum_{k=0}^{K-1} w(k,\theta) \cdot s(t_k) \right|^2$$

$$w(k,\theta) = e^{-j\frac{2\pi}{\lambda}t_k v \cos\theta},$$

(6)

where t_k is the kth time instance, $s(t_k)$ is the received signal at t_k, v is the antenna's moving speed, $w(k,\theta)$ is the complex weight assigned to $s(t_k)$ when the antenna scanning is steered to the direction θ, λ is the wavelength of the carrier. Eq. (6) indicates that at time t_k, the received signal $s(t_k)$ can be considered as the signal received at the antenna located at position $t_k v$.

With the emulated antenna array, the reader can measure the signal strength on different directions, where a profile of the received signals' strength on various directions can be plotted, as shown in Fig. 8. The spikes on the left indicate the reader strong signals reflected off the left wall, and those on the right correspond to signals reflected off the right wall. We can see that the tags close to each other will have similar multipath profiles. Therefore, it is possible to find the reference tag that is the closest to the target tag by comparing their multipath profiles. The position of the reference tag is then considered the target tag's location.

In summary, the multipath profile-based approach outperforms the RSS and phase-based approaches in NLOS scenarios. However, it requires an RFID reader to quickly scan the signals received from various directions. This is not applicable to most COTS RFID readers. In addition, it requires a large amount of reference tags to be deployed.

4 CASE STUDIES

In this section, we introduce three classic RFID localization systems as case studies to further explain how the RSS, phase and multipath profile-based approaches are implemented in real applications. We also use Table 1 to present a comparison among several popular RFID localization systems.

4.1 LANDMARC

LANDMARC is one of the earliest RFID localization system (Ni et al., 2004), which adopts an RSS-based scheme and thus can be easily implemented. Most importantly, it inspired the research community to use RFIDs to localize objects rather than to identify objects. As seen in Fig. 9, LANDMARC in the second mode with reference tags assistance in localizing the target tag.

Readers in LANDMARC do not measure the exact RSS of signals backscattered from tags. Instead, it uses eight levels to represent the RSS of received signals, and each level corresponds to a distance. The LANDMARC system consists of two networks, a RFID network that tracks tags, and a wireless network that forwards tracking results to the Internet. Based on the RSS of signals backscattered from the reference and target tags, k nearest reference tags (to the target tag) can be identified. The location of the target tag can be expressed as:

$$(x, y) = \sum_{i=1}^{k} w_i(x_i, y_i),$$

where w_i is a weight assigned to each identified reference tag. The weight is quantified by the RSS difference between the reference tag i and the target tag. The smaller the difference, the larger the weight is. Experiment results show that LANDMARC's localization error is around 1 m.

Table 1 A Comparison of Several RFID Localization Systems

System	Deployment	Technique	Accuracy (m)
SportOn, Hightower et al. (2000)	Readers	RSS	3
LANDMARC, Ni et al. (2004)	Readers/ref tags	RSS	1
Tagoram, Yang et al. (2014)	Readers	Phase	0.007
New Measurement, Nikitin et al. (2010)	Readers	Phase	0.21
RF-Compass, Wang et al. (2013)	Readers	Multipath profile	0.03
PinIt, Wang and Katabi (2013)	Readers/ref tags	Multipath profile	0.11

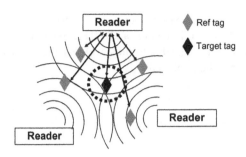

FIG. 9

Illustration of the LANDMARC RFID localization system.

4.2 TAGORAM

Tagoram is a phase-based localization scheme that employs the hologram technique (Yang et al., 2014). It achieves high localization accuracy (with an error of 7 mm) in an indoor environment. It also works well in a fast-changing environment, i.e., it can be used to track moving objects. Tagoram is tested not only in a laboratory environment, but also in real experiments—baggage tracking in three airports. Both laboratory and real-world experiments show that Tagoram is an efficient and accurate RFID localization method.

As shown in Fig. 10, two readers in the Tagoram system are placed beside a target tag. When the tag is moving, e.g., on a conveyor, the readers receive signals from different directions. If we look at the system from the tag's perspective, there are more than two virtual readers deployed along the path along which the tag is moving. With several readers in place, a hologram can be drawn based on the phase information contained in received signals on virtual readers. The phase information measured on each reader can be represented as:

$$\theta = \left(\frac{2\pi}{\lambda} \times 2d + \theta_t + \theta_r + \theta_{\text{tag}} \right) \bmod 2\pi$$

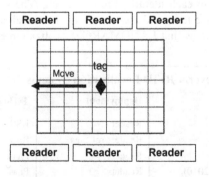

FIG. 10

Principle of the Tagoram system where the boxes with solid lines are RFID readers and those with the dashed lines are virtual readers.

where λ is the wave length, d is the distance between the tag and reader, $\theta_t + \theta_r + \theta_{\text{tag}}$ is called the diversity term that captures the signal's phase rotations caused by the reader's transmitter and receiver circuits and the tag's reflection characteristic.

To eliminate the impact of the diversity term, Tagoram makes use of an augmented differential hologram where the phase values measured on each reader is replaced by the phase difference between readers. In this way, the phase offsets generated by the tag and readers can be avoided because the offsets are the same in every measurement. The average localization error of Tagoram is 7 mm, i.e., the most accurate result in existing RFID localization systems.

4.3 PinIT

PinIt is a multipath profile-based localization system that achieves accurate localization in a NLOS environment (Wang and Katabi, 2013). It is by far the most accurate RFID localization system in NLOS environments. As shown in Fig. 3, the PinIt uses reference tags to assist in the localization of a target tag. Based on multipath profiles of target and reference tags, PinIt locates the target tag by finding its k closest reference tags.

In the PinIt system, a reader moves along a line to emulate an antenna array. By continuously querying the (target and reference) tags, multipath profiles composed of signal strengths in all possible directions of tags are constructed on the reader. The similarity of the multipath profiles between different tags represents the physical closeness between them. Since the absolute multipath profile of a tag is mixed with phase changes caused by signal reflections, the dynamic time warping (DTW) technique is used to compare the similarity of two profiles. Instead of comparing absolute values of two profiles, DTW first finds the best alignment between them and then measures the shift. Only the reference tags with small multipath profile shifts from the target tag are considered in the localization algorithm.

To efficiently find the position of a target tag located within a large amount of reference tags, PinIt uses a hierarchical approach to avoid querying a large amount of reference tags. As shown in Fig. 11, PinIt first queries the reference tags that are within a large (1 m) range and identifies a coarse-grained region where the target tag may be located. Then, PinIt queries tags in this region and gradually reduces the region until the requested resolution (e.g., 15 cm) is reached. Based on the experiments conducted in a library, PinIt is able to provide accurate localization with a mean error of 11.2 cm.

5 POTENTIAL RESEARCH ISSUES

Through the literature review, we note that the accuracy issue of RFID localization has been well addressed. The newest the RFID localization system (i.e., Tagoram) can achieve high localization accuracy with error less than 7 mm. The next step is to apply existing or newly developed RFID localization systems into real-world applications. The following trade-off issues need to be considered when a RFID localization system is deployed in practice.

5.1 ACCURACY VS. DEPLOYMENT

An ideal RFID localization system should be easily implemented and deployed with high accuracy. Although some RFID localization schemes (e.g., PinIt) provide high accuracy, large amounts of

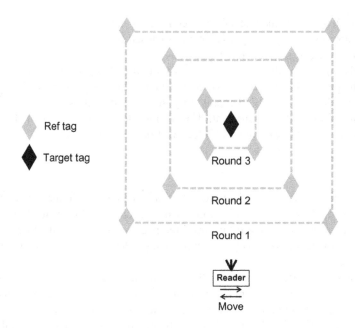

FIG. 11

Illustration of the hierarchical localization method in the PinIt system.

reference tags are needed. The higher the accuracy, the more reference tags would be needed. Even though passive RFID tags are cheap and small, it is either inconvenient or impractical to deploy tons of reference tags with known locations before conducting the localization task. In addition, reference tags could be easily blocked by metal materials, e.g., aluminum foil used in food and pharmaceutical packaging. Reference tags can also be torn off or destroyed by persons with benign or malicious purpose. Therefore, how to achieve high localization accuracy with a limited number of reference tags remains a problem.

5.2 ACCURACY VS. COMPLEXITY

Although phase- and multipath profile-based RFID localization approaches could achieve very accurate localization results without the help of reference tags, they typically require more computational, hardware, and/or software resources on a reader to process RF signals and perform mathematical calculations. For time-critical applications or recourse-limited devices, these methods seem to be inefficient and impractical. Therefore, how to achieve high accurate localization with reasonable and affordable resources should be a potential research issue deserved to be followed up.

5.3 ACCURACY VS. PORTABILITY

Due to the rapid development of the Internet of things (IoTs), an explosion of various IoT applications can be expected in the near future. That is to say, diverse system requirements are inevitable in RFID localization systems. On the other hand, it is impossible to design a one-size-fits-all localization system that addresses all the requirements from various applications. Therefore, an ideal RFID localization system should be adjustable according to the requirements posed by different applications. In other words, how to make a RFID localization system portable to various applications is another open yet challenging issue.

6 CONCLUSION

In this chapter, we investigate existing RFID localization technologies. We give a detailed introduction on the principles of various RFID localization approaches and follow case studies on three classic approaches. We also identify the advantages and drawbacks of each approach. In the end, we provide a discussion on the trade-off issues in a RFID localization system, which may lead to future research activities in the RFID localization domain.

REFERENCES

Almaaitah, A., Ali, K., Hassanein, H.S., Ibnkahla, M., 2010. 3D passive tag localization schemes for indoor RFID applications. In: 2010 IEEE International Conference on Communications (ICC). IEEE, Cape Town, South Africa, pp. 1–5.

Azzouzi, S., Cremer, M., Dettmar, U., Kronberger, R., Knie, T., 2011. New measurement results for the localization of UHF RFID transponders using an angle of arrival (AoA) approach. In: 2011 IEEE International Conference on RFID (RFID). IEEE, Orlando, FL, pp. 91–97.

Bekkali, A., Sanson, H., Matsumoto, M., 2007. RFID indoor positioning based on probabilistic RFID map and kalman filtering. In: Third IEEE International Conference on Wireless and Mobile Computing, Networking and Communications, 2007 (WiMOB 2007). IEEE, White Plains, NY, p. 21.

Finkenzeller, K., Waddington, R., 1999. RFID Handbook: Radio-Frequency Identification Fundamentals and Applications. Wiley, New York, NY.

Han, S., Lim, H., Lee, J., 2007. An efficient localization scheme for a differential-driving mobile robot based on RFID system. IEEE Trans. Ind. Electron. 54 (6), 3362–3369.

Hightower, J., Want, R., Borriello, G., 2000. SpotON: An indoor 3D location sensing technology based on RF signal strength. UW CSE 00-02-02. University of Washington, Department of Computer Science and Engineering. Seattle, WA, p. 1.

Hofmann-Wellenhof, B., Lichtenegger, H., Collins, J., 2012. Global Positioning System: Theory and Practice. Springer Science & Business Media.

Kulyukin, V., Gharpure, C., Nicholson, J., Pavithran, S., 2004. RFID in robot-assisted indoor navigation for the visually impaired. In: Proceedings. 2004 IEEE/RSJ International Conference on Intelligent Robots and Systems, 2004 (IROS 2004). vol. 2. IEEE, Sendai, Japan, pp. 1979–1984.

Kuruoglu, G.S., Erol, M., Oktug, S., 2009. Localization in wireless sensor networks with range measurement errors. In: Fifth Advanced International Conference on Telecommunications, 2009. AICT '09, pp. 261–266.

Lee, E.A., 2008, May. Cyber physical systems: Design challenges. In: Object Oriented Real-Time Distributed Computing (ISORC), 2008 11th IEEE International Symposium on. IEEE, Orlando, FL, pp. 363–369.

Li, X., Zhang, Y., Amin, M.G., 2009, April. Multifrequency-based range estimation of RFID tags. In: RFID, 2009 IEEE International Conference on. IEEE, Orlando, FL, pp. 147–154.

Liu, V., Parks, A., Talla, V., Gollakota, S., Wetherall, D., Smith, J.R., 2013. Ambient backscatter: wireless communication out of thin air. In: Proceedings of the ACM SIGCOMM 2013 Conference on SIGCOMM. ACM, Hongkong, China, pp. 39–50.

Lymberopoulos, D., Liu, J., Yang, X., Choudhury, R.R., Sen, S., Handziski, V., 2015. Microsoft indoor localization competition: experiences and lessons learned. GetMobile: Mobile Comp. and Comm. 18 (4), 24–31.

Moore, D., Leonard, J., Rus, D., Teller, S., 2004. Robust distributed network localization with noisy range measurements. In: Proceedings of the 2nd International Conference on Embedded Networked Sensor Systems. SenSys'04. ACM, New York, NY, pp. 50–61.

Nazari Shirehjini, A., Yassine, A., Shirmohammadi, S., 2012. An RFID -based position and orientation measurement system for mobile objects in intelligent environments. IEEE Trans. Instrum. Meas. 61 (6), 1664–1675.

Ni, L.M., Liu, Y., Lau, Y.C., Patil, A.P., 2004. LANDMARC: indoor location sensing using active RFID. Wirel. Netw. 10 (6), 701–710.

Niculescu, D., Nath, B., 2003a, April. Ad hoc positioning system (APS) using AOA. In: INFOCOM 2003. Twenty-Second Annual Joint Conference of the IEEE Computer and Communications. vol. 3. IEEE Societies, San Francisco, CA, pp. 1734–1743.

Niculescu, D., Nath, B., 2003b. DV based positioning in ad hoc networks. Telecommun. Syst. 22 (1–4), 267–280.

Nikitin, P.V., Martinez, R., Ramamurthy, S., Leland, H., Spiess, G., Rao, K.V.S., 2010. Phase based spatial identification of UHF RFID tags. In: 2010 IEEE International Conference on RFID, pp. 102–109.

Park, S., Lee, H., 2013. Self-recognition of vehicle position using RFID passive RFID tags. IEEE Trans. Ind. Electron. 60 (1), 226–234.

Parr, A., Miesen, R., Vossiek, M., 2013. Inverse SAR approach for localization of moving RFID tags. In: 2013 IEEE International Conference on RFID (RFID). IEEE, Penang, Malaysia, pp. 104–109.

Peng, C., Shen, G., Zhang, Y., Li, Y., Tan, K., 2007. Beepbeep: a high accuracy acoustic ranging system using cots mobile devices. In: Proceedings of the 5th International Conference on Embedded Networked Sensor Systems. SenSys '07. ACM, New York, NY, pp. 1–14.

Tsukiyama, T., 2003, June. Navigation system for mobile robots using RFID tags. In: Proceedings of the International Conference on Advanced Robotics (ICAR). Taipei, Taiwan.

Wang, J., Katabi, D., 2013. Dude, where's my card?: RFID positioning that works with multipath and non-line of sight. In: Proceedings of the ACM SIGCOMM 2013 Conference on SIGCOMM. ACM, Hongkong, pp. 51–62.

Wang, J., Adib, F., Knepper, R., Katabi, D., Rus, D., 2013. RF-Compass: robot object manipulation using RFIDs. In: Proceedings of the 19th Annual International Conference on Mobile Computing & Networking. ACM, Miami, FL, pp. 3–14.

Xiong, J., Jamieson, K., 2013. Arraytrack: a fine-grained indoor location system. In: Presented as Part of the 10th USENIX Symposium on Networked Systems Design and Implementation (NSDI 13). USENIX, Lombard, IL, pp. 71–84.

Yanakiev, B., Eggers, P., Pedersen, G.F., Larsen, T., 2007, September. Assessment of the physical interface of UHF passive tags for localization. In: Proc. 1st Int EURASIP RFID Workshop. Vienna, Austria, pp. 1–4.

Yang, L., Chen, Y., Li, X.-Y., Xiao, C., Li, M., Liu, Y., 2014. Tagoram: real-time tracking of mobile RFID tags to high precision using cots devices. In: Proceedings of the 20th Annual International Conference on Mobile Computing and Networking. MobiCom '14. ACM, New York, NY, pp. 237–248.

Zhao, J., Xi, W., He, Y., Liu, Y., Li, X.-Y., Mo, L., Yang, Z., 2013. Localization of wireless sensor networks in the wild: Pursuit of ranging quality. IEEE/ACM Trans. Netw. 21 (1), 311–323.

Zhong, Z., He, T., 2009. Achieving range-free localization beyond connectivity. In: Proceedings of the 7th ACM Conference on Embedded Networked Sensor Systems. SenSys '09. ACM, New York, NY, pp. 281–294.

GREEN CYBER-PHYSICAL SYSTEMS

15

C. Estevez, J. Wu
University of Chile, Santiago, Chile

1 ROLE OF GREEN CYBER-PHYSICAL SYSTEMS

Environmental issues are constantly gaining more attention as the general public becomes more aware of the consequences of environment degradation. As a result, various initiatives have been taken to counteract the effects induced by our own technology. Green topics are inherently multi-disciplinary, which include energy efficiency, energy self-sustainability, environmental care, and others. As discussed and defined in Wu et al. (2015), green topics include not only energy efficiency issues, but a much broader range of topics that address the well-being of the environment and the impact on the health of future organisms.

Cyber-physical systems (CPSs) represent a huge breakthrough in the evolution of computerized interconnectivity. This vast pervasive network not only includes traditional computers, but also a whole new breed of smart autonomous devices that seamlessly interconnect via machine-to-machine (M2M) communications to provide a whole new set of benefits, including many types of services. The range of locations is basically unbound, as CPS devices can be in vehicles, buildings, habitats, humans, utility grids, containers, garments, smart phones, plants, even underwater (as discussed ahead). These may impact environmental and operational states. Since CPS may ubiquitously exist, they may have great potential to aid the green objectives. The creation of the term of CPS is often attributed to Dr. Helen Gill from National Science Foundation in the year 2006. However, it has been found that the term "Cyber-Physical" was used much earlier. For example, Yacov Y. Haimes used the term of "Cyber-Physical" in his publication in 2002, which is quite close to the time the term of Internet of Things (IoT) was created (1999). Although both IoT and CPS are intended to increase the connections for the cyberspace and the physical world, IoT focuses more on networking, while CPSs are more relevant to information exchange and feedback (Ma, 2011).

The role of green CPS is to distribute and utilize all the sensed data in a manner that every device or technological entity that can benefit, from an environment standpoint, will receive it. The data are translated into information, which leads to smarter and greener decisions. Any ICT aspect that contributes to this system falls within the scope of Green CPS. Solutions may be anywhere from practical to theoretical, proof-of-concept to fully functional, and layer-specific to whole systems. Regardless of the solution, every contribution is important and this chapter focuses on discussing the most recent advances, which may encourage more relevant investigations in the future and the development of applications to support a sustainable world.

Cyber-Physical Systems. http://dx.doi.org/10.1016/B978-0-12-803801-7.00015-8

225

2 GREEN CPS RECENT ADVANCES

For the convenience of discussions, we classify the relevant works as the following categories: energy efficiency, self-sustainability, and environmental issues (see Fig. 1).

2.1 ENERGY EFFICIENCY

Energy efficiency is one of the broadest topics covered in the green spectrum. Energy conservation indirectly contributes to the reduction of the carbon fingerprint. In many cases there are additional advantages such as cost reduction, greater device independence, among others. Since the energy efficiency topics are quite broad, the relevant subsection is divided into the following topics: energy management, M2M and smart city, protocols, and scheduling.

2.1.1 Energy management

Energy management is an important subject. The efficiency of the management determines greatly the energy efficiency of the entire system. Here we discuss some practical and theoretical works.

The work in Ilic et al. (2010) proposes modeling energy systems as cyber-based physical systems, introduces a novel cyber-physical dynamical model whose mathematical description depends on the cyber technologies supporting the physical system, and shows how the proposed cyber-physical model is used to develop interactive protocols between the controllers embedded within the system and the network operator. One interesting aspect that is worth pointing out is that, among the cyber information feedback considered in this work, the price was included. For example, if the cost of operating a

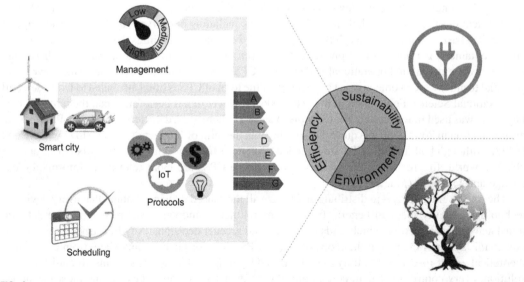

FIG. 1

Topics discussed are divided into: energy efficiency, self-sustainability, and environmental issues.

specific local device is greater than the cost of using external resources the device could decide to shut down. The proposed dynamic model greatly depends on the cyber technologies supporting the physical system. It has been shown here that the inclusion of these models results in a cyber-based model that lends itself to distributed sensing and actuation within a complex system.

A major issue in green CPS is how to design an energy management strategy for ensured sustainability of energy computing while the computation processes and physical processes continue to interact. One approach is to model the energy management strategy as the balance between the power produced by renewable resources and power demand of the workload (Ge et al., 2014). The dynamic power balance strategy (DPBS) can coordinate the work mode of workload and batteries ensuring the sustainability of energy computing for the energy management system of a specific device. The solution takes the runtime environment of DPBS and simplifies it into a supervisory control system that takes the output power of the renewable energy as an input parameter, and then models the DPBS using a finite state machine (FSM) based on the workload's power demand model and the battery's SOC model.

The work Fuhong et al. (2014) discussed a trade-off problem between bandwidth and energy. To manage the energy resources, a cooperative differential game model was proposed. The game model was based on the work Miao et al. (2010, 2012), where the layers are the attributes of interest, in this case bandwidth and energy (i.e., only two players). The steps to find the fair resource allocation solution are: (1) computing the optimal cost of grand coalition, the feedback Nash equilibrium, and the optimal cost for intermediate coalitions; (2) calculating Shapley value and allocating the total cooperative payoff; and (3) allocating over time of each player Shapley value. With the aid of these steps, it is possible to find the optimal point that minimizes the resource consumption. The results reflect that as the bandwidth increases the energy decreases, and vice-versa.

Considering physical sensors with certain sensing capabilities in an IoT sensory environment, in this work Liu et al. (2014) it was proposed to design an efficient energy management framework to control the duty cycles of these sensors under quality-of-information (QoI) expectations in a multitask-oriented environment. The QoI is characterized by a number of attributes including accuracy, latency, and physical context (specifically, sensor coverage). There are various technical challenges in sensor energy and data quality management. A major one that drives this work involves the large-scale management of heterogeneous devices that are expected to populate IoT systems. To address this challenge, an energy management service (and supporting algorithms) that is transparent to and compatible with any lower layer protocols and over-arching applications was designed, while providing long-term energy-efficiency under the satisfactory QoI constraints. To achieve this design it was necessary to introduce the concept of sensor-to-task relevancy to explicitly consider the sensing capabilities offered by a sensor to the applications and QoI requirements of the task. Then, a generic information fusion function was presented to compute the critical covering set of any given task in selecting the sensors to service a task over time. Finally, a runtime energy management framework based on the previous design elements was proposed to control the duty cycles of a sensor, where the control decision was made optimally considering the long-term task usage statistics and the service delay of each task serves as the constraint. To demonstrate the effectiveness of the management system, a scenario with 15 sensors and four tasks was created where the sensors can cover anywhere between one to four tasks, the goal was to distribute the energy consumption as evenly as possible. Results have shown that the sensors that covered four tasks where able to delegate tasks efficiently to those that covered less tasks, hence having significant energy savings.

Looking at an applied case, the work in Pan et al. (2015) raised an interesting question. Even with feedback control of a particular amenity, for example heating, that is set to turn off if the temperature reaches a certain value. Is this energy efficient if there is no one in the room, house, our building? This work incorporates a location-based feedback that controls the energy modes of various appliances, putting them into energy-saving mode when a user is leaving the building/home/office, and pre-heats/cools if the user is approaching. Using a LEED-gold-certificated green office building, a unique IoT experimental testbed was built for the energy efficiency and building intelligence data collection. For 1 year the building was monitored. The results have shown that, due to the centralized and static building controls, the actual running of green buildings may not be energy efficient even though they may be "green" by design. An IoT framework with smart location-based automated and networked energy control, which uses smartphone platform and cloud computing technologies to enable multi-scale energy proportionality including building-, user-, and organizational-level energy proportionality demonstrates good energy efficiency. It was shown in a proof-of-concept IoT network and control system prototype that carries out real-world experiments that demonstrated the effectiveness of the proposed solution. The broad application of the proposed solution has not only led to significant economic benefits in terms of energy saving, home/office network intelligence improvement, but also bought about huge social implications in terms of global sustainability.

2.1.2 Machine-to-machine and smart city

The first set of topics presented are related to M2M and smart city with the advantages of implementing Green IoT under various scenarios. From a telecommunication operator standpoint energy-saving solutions can bring many benefits. This work Lin et al. (2011) described energy-saving solutions for telecommunication operators using the IoT. There is a significant number of equipment in a telecommunication operator enterprise, therefore reducing the energy consumption would have a beneficial impact on the environment and also result in a reduction in energy costs. The solution can be described to have two parts: (1) monitoring—a real-time comprehensive monitoring system to collect energy consumption information focused mainly on the base stations, and (2) control—based on the energy state of the equipment (mainly base stations) there is a linkage between the monitoring devices and the power management system. The objective is straightforward: to make intelligent decisions based on the environment state. The proposed role of the CPS is classified into three categories:

1. Energy measurements: the measurements can not only provide real-time feedback and control, but also gather statistics and therefore make smarter long-term decisions; like a preemptive load balancing.
2. Remote monitoring and control: through the use of remote monitoring devices, connected to the operator via the cellular network, the equipment condition can be monitored (such as temperature, humidity, light, and so on) to obtain feedback that could affect how the power consumptions are distributed.
3. Intelligent linkage: by intelligently linking the environment information provided by the IoT to the management system, the operator's equipment can perform optimally and safely via controlling the environment system, without overdoing it, with the necessary exact amount of energy.

With these tools it is possible to reduce unnecessary energy consumption of equipment.

The work Sundar Prasad and Kumar (2012) raised an interesting topic on the energy consumption of the sensors in IoT. Some research works analyzed the benefits of energy savings by discussing the energy consumption with and without the presence of IoT sensors, but, when discussing the benefits of having IoT, the energy required to run the IoT network itself is often not mentioned, which usually consists of a massive amount of low-powered devices. The proposed solution for the case of a massive distributed IoT sensor network, is to have a coordinated effort, although relevant strategies have been also discussed in Chao et al. (2011), which is succeeded by Chao and Wu (2015). The key to the energy efficiency in Sundar Prasad and Kumar (2012) is the scheduling algorithm, which required the time to be slotted and the nodes decide to be active or not based on the information gathered from its neighbors, which are constantly transmitting their activity decision. The successful synchronization allows for the nodes to consume energy only when necessary, hence saving this valuable resource. This solution, however, assumes a very specific architecture, that is, massive distributed wireless nodes with overlapping cells, which may not be the case for all (or even most) types of networks.

Moving on to a bigger picture, we will discuss the potential energy savings in an IoT-enabled smart city (Sanchez et al., 2014). Improving the efficiency of city services and facilitating a more sustainable development are some of the main drivers of the smart city concept. Here a case was presented in which the city of Santander, Spain, used IoT devices to sense the environment and make smarter decisions. In this trial, the indicators of interest where energy related and the sensing environment consisted of sensors that detect the presence of people and cars, and the controlled infrastructure is the street lighting. The lighting is programmed to dim when no activity is detected, and the intensity of the light varies depending on various factors, such as the detection of a pedestrian, a car, rain, and other. Sensors not only detect presence, but also light intensity, such that the street lighting can be adjusted under cloudy conditions, even during the day, if necessary. This last factor increases the energy consumption during the day (if dark conditions are met), nevertheless this energy is minimal compared to the energy consumed during the night. The tests reported in this work were performed during the night, in clear and rainy conditions. As expected the energy savings were significant, the energy consumed in the IoT-enabled system is approximately one-third of the energy consumed before the trial. Additionally it has been reported that the CO_2 emissions was lowered.

Smart buildings play an important role in smart cities. In Jinlong et al. (2013), a framework to simulate CPS was proposed with the target application of energy savings in buildings. The framework was designed with the following characteristics: distributed, heterogeneous, autonomous, real-time processing, and mass data processing. The framework organized in separate agents and the characteristics mentioned above apply to the manner in which the agents communicate. In the provided simulation scenario, the agents include temperature, humidity, supplied air temperature, supplied air humidity, and others factors. It can be seen that the proposed framework is able to build a system where all the agents collaborate to find a suitable balance.

2.1.3 Protocols

The topic of protocol and scheduling is critical at the lower levels, in both communication and computing. This work is generally more theoretical and simulation based, as the work presented here shows. Here two scheduling and two protocol topics are discussed.

In Estevez and Kailas (2012) a process-stacking multiplexing access algorithm was designed for single channel operation. The concept is simple and intuitive, but its implementation is not trivial. The key to stacking single channel events is to operate while simultaneously obtaining and handling a posteriori time-frame information of scheduled events. This information is used to shift a global time pointer that the wireless access point manages and uses to synchronize all serviced nodes. This scheme does not work under a time-frequency division multiple access (TDMA-FDMA) hybrid technique as its predecessor technique (Estevez et al., 2011), so it can only be compared with TDMA. In this scheme, the slot duration is variable and the base station is given the capability to interlace data with processes. This allows for the same algorithm that is scheduling the frames to schedule idle time to mimic the same energy self-sustainability capability of its preceding work. The protocol reads the frame size ahead of time and allows the protocol to compute the exact amount of time required to transmit, once the transmission in complete, which included the size of the next frame in the header, the base station replies with the next time to send. The node then sleeps until the specified time, saving valuable energy. This work is being continuously improved upon and is on-going (Estevez and Azurdia, 2015).

Unlike existing power-saving protocols of human-to-human (H2H) networks, Chao et al. (2011) emphasized the importance of designing a dedicated M2M power saving mechanism. Energy efficiency is an important trait that is gained from a more innate M2M network. Unlike H2H terminals, a lot of machines related to specific M2M applications are fixed and not expected to move randomly, and, further, the M2M communications timings are also less critical than those of H2H communications. The solution consists on improving power savings for both M2M devices and network operations. Removal of unnecessary M2M activities will be beneficial to conserve the power of M2M devices. On the operation side, improvements in operations and optimized signaling flow were also proposed, which may help reduce the power consumption of M2M devices indirectly. Even though there were no exact computations of energy savings, the work does present a significant reduction in communications activities due to the mechanism migration from H2H to M2M.

IoT envisions the notion of ubiquitous connectivity of everything. However, the current research and development works have been mostly restricted to scalar sensor data based IoT systems, thus leaving a gap to benefit from services and application enabled by the Internet of Multimedia Things (IoMT) (Alvi et al., 2015). Recently, IETF ROLL working group standardized an IPv6 routing protocol for low-power and lossy networks (RPL) for resource constrained devices. RPL builds a tree-like network topology based on some network metric optimization using RPL objective functions (Alvi et al., 2015). Previous RPL implementations for scalar sensor data communication are not feasible for IoMT, since multimedia traffic pose distinct network requirements. The goal of this work is to design an enhanced version of RPL for IoMT in which the sensed information is essentially provided by the multimedia devices. The proposed RPL implementation incorporates application specific quality of service (QoS) and energy requirements. The performance is highly dependent on the link quality. If the link is poor, the effective throughput would drop and the amount of packet delivered per unit time would decrease, in which case the energy must be increased. If the throughput requirement is being met, the energy can be reduced to a level where the QoS requirements are still met. The proposed protocol, referred to as Green-RPL, is compared to two techniques (ETX and OF0). Results have shown that it is able to outperform OF0 in energy efficiency and throughput; nevertheless, when compared to ETX, in outperforms it in terms of throughput but not energy efficiency.

The work Zhou et al. (2015) mainly has shown that the IoT can reach places that many would not imagine. As with the previous case, an energy efficient protocol was discussed, nevertheless this protocol was optimized for underwater operation. For the reader that is unfamiliar with underwater wireless sensor networks (UWSN), this is a field mainly to explore the vast ocean space. This work proposed an Enhanced Channel-aware Routing Protocol (ECARP), which is an improved version of the existing CARP protocol. The main contribution of this work over its predecessor is the improved energy savings. It is able to achieve this in two ways:

(1) Improved data transmission: when the sensor measurement does not change (within a certain threshold) the protocol transmits a much smaller packet that only informs that the value remains the same rather than retransmitting the whole data packet.
(2) Improved routing: rather than constantly sending routing control messages to determine the best path, a memory is implemented so that it selects only from previously successful paths, reducing the amount of signaling between nodes.

Results show that for the simulation parameters used that ECARP uses 25% of the energy used by CARP to perform the same tasks.

2.1.4 Scheduling
In another relevant work Liang et al. (2013), energy efficiency was addressed in these types on networks, more specifically long-term evolution-advanced (LTE-A). For IoT applications, continuous low-rate streaming data may be reported from devices over a long period of time, imposing stringent requirements on power saving. To manage power consumption, 3GPP LTE-A has defined the discontinuous reception (DRX) and transmission (DTX) mechanisms to allow devices to turn off their radio interfaces and enter sleep mode in various patterns. Existing literature has paid much attention to the evaluation of the performance of DRX/DTX; however, research is on-going into how to tune DRX/DTX parameters to optimize energy cost. In Liang et al. (2013), the DRX/DTX optimization is addressed by maximizing the sleep periods of devices while guaranteeing the QoS, specifically on the aspects of traffic throughput, packet delay, and packet loss rate in IoT applications. Efficient schemes to optimize DRX/DTX parameters and schedule packets at the base station are proposed. Simulation results have shown that these schemes can guarantee the aforementioned QoS attributes while saving the energy of the device.

Another scheduling work was described in Abdullah and Yang (2013). Here the network is more generic than the previous case. It is somewhat comparable to a sensor network. The nodes in this topology have different hierarchies, basically sensor nodes and, what are referred to as, brokers. The brokers act as an intermediary between the sensor nodes and the sink (edge node). The authors described that, in earlier works, the scheduling had performed solely based on expiration times, but claimed to improve the performance by taking into consideration the overall IoT system efficiency. Additionally, the routing algorithm is also an energy saving feature of the proposed system. From the results obtained, it can be observed that the proposed scheduling and routing algorithms provide better energy efficiency and response times. It should be mentioned that improving the response time is what results in savings in energy.

2.2 SELF-SUSTAINABILITY
Energy sustainability is well established as one of the prominent enabling technologies for the pervasive development of the IoT. Energy harvesting is one of the main contributors to energy independence, also commonly referred as energy self-sustainability. When true independence is reached, the only obstacle remaining is battery life; nevertheless, in theory, devices could potentially be self-sustainable for many years.

In Meng and Jin (2011) a simple wireless node was built using a solar panel energy harvesting source, lithium batteries for stable power, various sensors, and a processor to program the energy management. The processor used was the S3C6410 ARM11, which is a low-cost, low-power, high-performance microprocessor. Some external interfaces used in this testbed are radio-frequency identification (RFID) readers, infrared sensors, environmental sensors, and multi-channel sensors. The wired communication interfaces supported include RS485, RS232, Universal Serial Bus (USB), and Ethernet, and the wireless communication interfaces supported are composed of GSM (global system for mobile communications), GPRS (general packet radio service), CDMA (code division multiple access), Zigbee, WiFi, and Bluetooth. For the test performed in this work only Ethernet and Zigbee where used. For the energy collection and management system a solar battery board is used alongside a 25 F super capacitor, which collects the energy generated by the solar panel. For a stable energy source, lithium batteries are used. For the test, the terminal carries out a regular transmission. The transmission lasts 3 seconds and it is repeated every 5 minutes. Using these experimental setup parameters the terminal is able to maintain self-sustainability for a long period of time (not specified).

This work Roselli et al. (2014) summarized various energy harvesting methods oriented for IoT, which implicitly means that the energy collectors are generally small in size. One of the energy harvesting devices is called the hybrid solar-rectenna. It is a device that is equipped with a small solar panel and an antenna that acts much like a rectifying circuit, converting an analog AC (alternate current) signal into a DC (direct current) signal. Once in DC form it can be used to power small low-power devices. The energy harvested is also combined with the energy harvested from the solar panel, making this little device more convenient. Another prototype worth mentioning is an energy harvesting system composed mainly of piezoelectric parts that can be installed in shoes, which is able to collect energy from the user when walking. The prototypes mentioned here can be inserted in clothing and optimistically power e-Health sensors and actuators, such as blood pressure sensors, insulin pumps, electrocardiogram (ECG) sensors, electromyography (EMG) sensors, motion sensors, etc.

The building blocks for the IoT are the sensors that, by incorporating energy harvesting devices, allow for independent nodes increasing the pervasiveness. Many types of energy harvesting devices are being developed; however, there is still only a limited understanding of the properties of various energy sources and their impact on energy harvesting adaptive algorithms. In Gorlatova et al. (2015), the work focused on characterizing the kinetic (motion) energy that can be harvested by a wireless node with an IoT form factor and on developing energy allocation algorithms for such nodes. This work also described some methods for estimating harvested energy from acceleration traces, and studied energy generation processes associated with day-long human routines, such as walking, running, cycling, etc. These observations provided insights into the design of motion energy harvesters, IoT nodes, and energy harvesting adaptive algorithms. Data were collected using accelerometers and estimating the energy harvested, but there do not seem to be a direct measurement. The estimated energy harvested walking is in the order of 200 µW and running is 800 µW. These are significant power values that could easily activate low-power body sensor networks.

Energy self-sustainability is a very desirable attribute in IoT sensing devices, arguably it is an indispensable attribute. In the work Kamalinejad et al. (2015), a wireless energy harvesting system for IoT (WEH-IoT) is described. In the proposed system, the antenna is able to not only receive data, but, when not in use, it is able to absorb the EM (electromagnetic) radiation to charge the device, or at the very least extend the time between charges. The system proposed to have a rectifying component near the antenna connection that converts AC (more specifically radio frequency) to DC power. Once in the

DC domain it can be added to the power supply. This harvesting system can extend the life-time of the battery, particularly when the energy consumption is relatively low.

2.3 ENVIRONMENTAL ISSUES

Pollution is the consequence of existing technologies, energy inefficient systems have a greater toll on the environment. The solutions discussed here are aimed mainly at monitoring pollution: the control aspect is not yet automated. It can be said that the feedback loop is completed by humans.

In Tao et al. (2014), it has claimed that to properly implement an energy-saving and emission-reduction (ESER) policy, it is fundamental to perform an effective quantitative evaluation of ESER. Current ESER evaluation technology is isolated from the enterprise information systems, such as research planning, product data management, and customer relationship management. To address this problem Tao et al. (2014) proposed a novel method for ESER life-cycle assessment based on an IoT system and the bill of material (BOM). Some issues that the current ESER evaluation system encounters are: (1) the data required in existing assessment methods are collected manually, and the assessment relies primarily on existing data—it is required to have real-time and dynamic requirements, which are not met; (2) current studies focus on environmental and process industries, and treats these entities as a unit—the data yields information regarding the entity as a whole, but no details are collected at different levels or stages of production incapacitating the ability to optimize these further; (3) ESER assessment tools are independent from the existing enterprise information systems, hindering the integration and interaction of these; and (4) it uses client/server software architecture, which can be arduous to upgrade and maintain. The solutions proposed are obtained in a straightforward manner from the weaknesses, and these are: (1) meet real-time and dynamic requirements by using IoT, (2) study BOM-based evaluation theories/methods for a multi-structure and multi-level ESER evaluation, (3) have the ESER evaluation done in existing enterprise information systems, and (4) switch to a browser/client software architecture, which is operating-system independent, easily upgradable, and has lower maintenance costs.

Another work related to air pollution is discussed in Xuyao et al. (2013). The online automatic monitoring system of malodor pollution, which can perform real-time, automatic monitoring and risk assessment, was designed to improve the handling ability of malodor emergency alerts. In this work, remote sensing monitoring odorous pollutants diffusion, assessment and pre-warning for emergencies were studied utilizing the ArcGIS database with the real-time data monitoring of malodor, meteorological parameters, and other tools. The data center of the monitoring system was structured on a Hadoop cloud computing platform. Relying on the environmental protection supervision, cooperative supervision is achieved, which was supported by IoT for large-scale monitoring networks. A test trial was developed in Dagang Industrial Park of the Tianjin Binhai New Area. The system has proved that it directly enriches the environmental IoT enhancing the malodor monitoring level that, consequently, improves the environmental supervision capability in China.

Yet another example of environmental monitoring can be seen in Pokric et al. (2014). Here a solution, called ekoNET, was developed for real-time monitoring of air pollutants and other weather indicators, such as temperature, pressure, and humidity. ekoNET is based on low-cost sensors, which are simple to deploy, use, and maintain. This initiative was intended to be used in the context of the IoT-domain of Smart Cities. ekoNET devices are equipped with GPS and GPRS communication capabilities, as well as carbon dioxide (CO_2), ozone (O_3), nitrogen monoxide boron (NO-B_4), nitrogen dioxide boron (NO_2-B_4), and carbon monoxide boron (CO-B_4) concentration sensors. The ekoNET system can help target the location of environmental air-related issues that may arise in a smart city.

We close the chapter with a very important topic for the environment, which is water sustainability. Water is the lifeblood of the planet hence plays a vital role in the proper functioning of the Earth's ecosystems and human activities. Water is used in agriculture, manufacturing, transportation, and energy production. Key technologies, such as sensing technology, wireless communications, hydrodynamic modeling, and data analysis, enable intelligently networked water-related CPSs with embedded sensors, processors, and actuators that can interact with the water environment. In this work it is mentioned that energy harvesting is possible using biological energy via microbial fuel cells (MFCs) that generate electricity through electrochemical reactions with a type of common and safe bacteria (manganese oxidizing microorganisms) ubiquitous in water. This will allow continuous sensing and greater (or complete) independence from batteries. There are a few challenges surrounding water sensing but an important issue is hydrodynamic modeling. With so many type of water systems, such as coastal marine and fresh water, which have irregular geometries, make fine-scale modeling difficult. Computing technology advances can diminish this issue. As better tools become available, CPS will become an important contributor to water sustainability.

Finally, it is appropriate to conclude this chapter with a publication Airehrour et al. (2015) that discusses if energy efficient standards are enough to counteract the increasing carbon footprint that is being generated by our own technology. This is an important environmental issue. The take-away point of this publication is that promoting energy efficiency, without curbing overall energy consumption, may not necessarily reduce the carbon footprint. The reason been that energy-efficient technology will lower the energy consumption price per unit potentially causing the corollary effect of increasing the amount of equipment. The cost of the equipment is a one-time investment, but the cost of energy is a recurrent toll. The desire to grow can balance our ethical concerns.

3 SUMMARY

Table 1 organizes the work discussed by topic of interest. The first four subtopics fall within the energy efficiency main topic. Self-sustainability and environmental issues are not categorized further. The table summarizes well all the recent research work presented in this chapter.

It has been shown here that CPS unquestionably has the potential to generate a great positive impact on the care of the environment. In a broad sense, green CPS can improve energy efficiency and help reduce environmental pollution. With the aid of energy harvesting, CPS systems can become independent and scatter further into more inaccessible locations helping in this way to monitor greater portions of our environment. Exciting topics include automated city lighting, device activation using user-location-based criteria, and energy independent nodes. Promising energy efficient protocols and scheduling techniques point toward even greater energy savings in the future. Energy harvesting is making energy independence a reality and pollution control is becoming smarter and more pervasive. The chapter closes with an interesting publication discussing how a corollary effect caused by the proliferation of energy-efficient systems can potentially increase the carbon footprint. It is the expectation and desire of this chapter to shine some light on use of environmentally friendly technology and techniques, as well as to promote an ethical conscience in the use of these.

Table 1 Discussed Work Organized by Topic

Topic	Recent Work
Energy management	Modeling (Ilic et al., 2010), strategy (Ge et al., 2014), energy-bandwidth tradeoff (Fuhong et al., 2014), modeling (Miao et al., 2010), modeling (Miao et al., 2012), quality-of-information (Liu et al., 2014), framework (Pan et al., 2015)
M2M and smart city	Telecom operators (Lin et al., 2011), Internet of Things (IoT) (Sundar Prasad and Kumar, 2012), machine-to-machine (M2M) cellular networks (Chao et al., 2011; Chao and Wu, 2015), smart city trial (Sanchez et al., 2014), in-building (Jinlong et al., 2013)
Protocols	Time-frequency multiplexing (Estevez and Kailas, 2012), variable-time slot, time-frequency division multiple access TDMA (Estevez et al., 2011; Estevez and Azurdia, 2015), M2M synchronization (Chao et al., 2011), routing (Alvi et al., 2015), routing (Zhou et al., 2015)
Scheduling	QoS consideration (Liang et al., 2013), IoT message scheduling (Abdullah and Yang, 2013)
Self-sustainability	IoT design (Meng and Jin, 2011), energy harvesting (Roselli et al., 2014), kinetic energy harvesting (Gorlatova et al., 2015), wireless energy harvesting (Kamalinejad et al., 2015)
Environmental issues	Emission reduction (Tao et al., 2014), malodor pollution monitoring (Xuyao et al., 2013), environmental context-awareness (Pokric et al., 2014), carbon footprint (Airehrour et al., 2015)

ACKNOWLEDGMENTS

This work is partially funded by project FONDECYT 11121655.

REFERENCES

Abdullah, S., Yang, K., 2013. An energy-efficient message scheduling algorithm in Internet of Things environment. In: Proceedings of 2013 9th International Wireless Communications and Mobile Computing Conference (IWCMC), July 2013, pp. 311–316. http://dx.doi.org/10.1109/IWCMC.2013.6583578.

Airehrour, D., Gutierrez, J., Liu, W., Wu, J., 2015. When Internet raised to the things power: are energy efficiency standards sufficient to curb carbon footprints? In: 2015 IEEE Globecom Workshops (GC Wkshps), December 2015, pp. 1–6. http://dx.doi.org/10.1109/GLOCOMW.2015.7413978.

Alvi, S.A., Shah, G.A., Mahmood, W., 2015. Energy efficient green routing protocol for Internet of multimedia things. In: 2015 IEEE Tenth International Conference on Intelligent Sensors, Sensor Networks and Information Processing (ISSNIP), April 2015, pp. 1–6. http://dx.doi.org/10.1109/ISSNIP.2015.7106958.

Chao, H., Wu, J., 2015. Machine-to-machine (M2M) communications, architecture, performance and applications. In: Anton-Haro, C., Dohler, M. (Eds.), Optimizing Power Saving in Cellular Networks for Machine-to-Machine (m2m) Communications. Woodhead Publishing Ltd, UK.

Chao, H., Chen, Y., Wu, J., 2011. Power saving for machine to machine communications in cellular networks. In: Proceedings of 2011 IEEE GLOBECOM International Workshop on Machine-to-Machine Communications, December 2011, pp. 389–393. http://dx.doi.org/10.1109/GLOCOMW.2011.6162477.

Estevez, C., Azurdia, C., 2015. Bottom-layer solutions for 60 GHz millimeter-wave wireless networks: modulation and multiplexing access techniques. Telecommun. Syst., 1–17.

Estevez, C., Kailas, A., 2012. Energy-efficient process-stacking multiplexing access for 60-GHz mm-wave wireless personal area networks. In: Engineering in Medicine and Biology Society (EMBC), 2012 Annual International Conference of the IEEE, August 2012, pp. 2084–2087. http://dx.doi.org/10.1109/EMBC.2012.6346370.

Estevez, C., Wei, J., Kailas, A., Fuentealba, D., Chang, G.K., 2011. Very-high-throughput millimeter-wave system oriented for health monitoring applications. In: 2011 13th IEEE International Conference on e-Health Networking Applications and Services (Healthcom), June, pp. 229–232. http://dx.doi.org/10.1109/HEALTH. 2011.6026753.

Fuhong, L., Qian, L., Xianwei, Z., Yueyun, C., Daochao, H., 2014. Cooperative differential game for model energy-bandwidth efficiency tradeoff in the Internet of Things. China Commun. 11 (1), 92–102. http://dx. doi.org/10.1109/CC.2014. 6821311.

Ge, Y., Dong, Y., Zhao, H., 2014. An energy management strategy for energy-sustainable cyber-physical system. In: 2014 9th International Conference on Computer Science & Education (ICCSE). IEEE, pp. 413–418.

Gorlatova, M., Sarik, J., Grebla, G., Cong, M., Kymissis, I., Zussman, G., 2015. Movers and shakers: kinetic energy harvesting for the Internet of Things. IEEE J. Sel. Areas Commun. 33 (8), 1624–1639. http://dx.doi.org/ 10.1109/JSAC.2015. 2391690.

Ilic, M.D., Xie, L., Khan, U.A., Moura, J.M., 2010. Modeling of future cyber-physical energy systems for distributed sensing and control. IEEE Trans. Syst. Man Cybern. Part A: Syst. Humans 4 (40), 825–838.

Jinlong, W., Qianchuan, Z., Yin, Z., 2013. An effective framework to simulate the cyber-physical systems with application to the building and energy saving. In: Control Conference (CCC), 2013 32nd Chinese. IEEE, pp. 8637–8641.

Kamalinejad, P., Mahapatra, C., Sheng, Z., Mirabbasi, S., Leung, V.C.M., Guan, Y.L., 2015. Wireless energy harvesting for the Internet of Things. IEEE Commun. Mag. 53 (6), 102–108. http://dx.doi.org/10.1109/ MCOM.2015.7120024.

Liang, J.M., Chen, J.J., Cheng, H.H., Tseng, Y.C., 2013. An energy-efficient sleep scheduling with QoS consideration in 3GPP LTE-advanced networks for Internet of Things. IEEE J. Emerging Sel. Top. Circuits Syst. 3 (1), 13–22. http://dx.doi.org/10.1109/JETCAS.2013.2243631.

Lin, C., Wang, J., He, Q., 2011. Internet of Things technology for energy saving of telecom operators. In: Proceeding of IET International Conference on Communication Technology and Application (ICCTA 2011), October 2011, pp. 614–618. http://dx.doi.org/10.1049/cp.2011. 0741.

Liu, C.H., Fan, J., Branch, J.W., Leung, K.K., 2014. Toward QoI and energy-efficiency in Internet-of-Things sensory environments. IEEE Trans. Emerg. Top. Comput. 2 (4), 473–487. http://dx.doi.org/10.1109/ TETC.2014.2364915.

Ma, H.D., 2011. Internet of Things: objectives and scientific challenges. J. Comput. Sci. Technol. 26 (6), 919–924.

Meng, Q., Jin, J., 2011. The terminal design of the energy self-sufficiency Internet of Things. In: Proceeding of 2011 International Conference on Control, Automation and Systems Engineering (CASE), July 2011, pp. 1–5. http://dx.doi.org/10.1109/ICCASE.2011.5997619.

Miao, X.N., Zhou, X.W., Wu, H.Y., 2010. A cooperative differential game model based on transmission rate in wireless networks. Oper. Res. Lett. 38 (4), 292–295.

Miao, X.N., Zhou, X.W., Wu, H.Y., 2012. A cooperative differential game model based on network throughput and energy efficiency in wireless networks. Optim. Lett. 6 (1), 55–68.

Pan, J., Jain, R., Paul, S., Vu, T., Saifullah, A., Sha, M., 2015. A Internet of Things framework for smart energy in buildings: designs, prototype, and experiments. IEEE Internet Things J. 2 (6), 527–537. http://dx.doi.org/ 10.1109/JIOT. 2015.2413397.

Pokric, B., Krco, S., Drajic, D., Pokric, M., Jokic, I., Stojanovic, M.J., 2014. ekoNET—environmental monitoring using low-cost sensors for detecting gases, particulate matter, and meteorological parameters. In: Proceeding of Innovative Mobile and Internet Services in Ubiquitous Computing (IMIS), July 2014, pp. 421–426. http:// dx.doi.org/10.1109/IMIS.2014. 57.

Roselli, L., Borges Carvalho, N., Alimenti, F., Mezzanotte, P., Orecchini, G., Virili, M., Mariotti, C., Goncalves, R., Pinho, P., 2014. Smart surfaces: large area electronics systems for Internet of Things enabled by energy harvesting. Proc. IEEE 102 (11), 1723–1746. http://dx.doi.org/10.1109/JPROC.2014.2357493.

Sanchez, L., Elicegui, I., Cuesta, J., Munoz, L., 2014. On the energy savings achieved through an Internet of Things enabled smart city trial. In: Proceedings of 2014 IEEE International Conference on Communications (ICC), June 2014, pp. 3836–3841. http://dx.doi.org/10.1109/ICC.2014.6883919.

Sundar Prasad, S., Kumar, C., 2012. An energy efficient and reliable Internet of Things. In: Proceeding of 2012 International Conference on Communication, Information Computing Technology (ICCICT), October 2012. pp. 1–4. http://dx.doi.org/10.1109/ICCICT. 2012.6398115.

Tao, F., Zuo, Y., Xu, L.D., Lv, L., Zhang, L., 2014. Internet of Things and BOM-based life cycle assessment of energy-saving and emission-reduction of products. IEEE Trans. Ind. Inf. 10 (2), 1252–1261. http://dx.doi.org/10.1109/TII.2014.2306771.

Wu, J., Thompson, J., Zhang, H., Kilper, D.C., 2015. Green communications and computing networks [series editorial]. IEEE Commun. Mag. 53 (11), 148–149.

Xuyao, Y., Hui, Y., Kexin, X., Yijin, S., Jinhang, L., 2013. Design and implementation of on-line automatic monitoring system for malodor pollution incidents based on ArcGIS. In: Proceedings of 2013 IEEE 11th International Conference on Electronic Measurement Instruments (ICEMI), vol. 2, pp. 524–528. http://dx.doi.org/10.1109/ICEMI.2013.6743121.

Zhou, Z., Yao, B., Xing, R., Shu, L., Bu, S., 2015. E-CARP: an energy efficient routing protocol for UWSNs in the Internet of underwater things. IEEE Sens. J. 16 (11), 4072–4082. http://dx.doi.org/10.1109/JSEN.2015.2437904.

WIRELESS RECHARGEABLE SENSOR NETWORKS FOR CYBER-PHYSICAL SYSTEMS

C. Lin*,†, B. Xue*,†, F. Xia*,†, G. Wu*,†, J. Deng‡

Dalian University of Technology, Dalian, China Key Laboratory for Ubiquitous Network and Service Software of
Liaoning Province, Dalian, China† University of North Carolina at Greensboro, Greensboro, NC, United States‡*

1 WRSNs AND RELATED APPLICATIONS

Wireless sensor networks (WSNs) are composed of a large number of inexpensive micro-sensor nodes, which can be deployed in the specific area to form a multihop and selforganized network for monitoring. The sensors are used to identify and extract useful information from the environment. Functionality of WSNs includes environmental sensing, information collecting, and processing perceived objects.

With the option to mount various types of sensors ranging from temperature, magnetic, pressure, acoustic sensors to more complex gyroscopes, imaging, infrared, video sensors, a WSN provides a convenient way to access information in the physical world (Akyildiz et al., 2002; Sohrabi et al., 2000). It starts by finding an increasing number of applications from those within our daily lives to many mission-critical tasks. The past decade has witnessed the widespread adoption of WSNs in a variety of fields, including environmental monitoring, ecosystem surveillance, physical hazards prevention, and daily activity recognition. Most applications of WSNs, such as structural monitoring for the Golden Gate Bridge, agricultural rain-fed farming decisions, and forest fire detection, require a long-lived WSN with prolonged network lifetimes.

However, sensor nodes are typically supplied with batteries that can only store a limited amount of energy, which has been the biggest impediment. Due to the limited energy capacity of the battery used at each sensor node, WSNs can only remain operational for a limited amount of time. The network performance is thus limiting the survival time of network and it is one of the most important factors that hinder large-scale network deployment.

The increasing demand for more complex sensors leads to higher energy consumption on the part of sensor nodes. To this end, energy conservation has been one of the primary focuses in WSN research in the past decade. Since replacing the sensor's battery is infeasible or risky in many applications (Werner-Allen et al., 2006; Yu et al., 2005), most of the research aims to reduce energy consumption. For a single node, duty cycling is one of the most effective methods to save energy, but it reduces the usefulness of the node (Ye et al., 2002).

Cyber-Physical Systems. http://dx.doi.org/10.1016/B978-0-12-803801-7.00016-X

In recent years, advances in energy-harvesting have enabled sensor nodes with renewable energy sources (like rechargeable batteries) (Kansal and Srivastava, 2005) to be deployed in the region of interest for long-term monitoring and surveillance purposes. The novel concept of wireless rechargeable sensor network (WRSN) has been proposed to address the limited network lifetime issue of WSNs. In a WRSN, sensor nodes utilize the energy-harvesting technology to convert harvested energy, such as solar, wind, and radio frequency energies, into direct current to replenish their power supplies so that the WRSN can operate sustainably.

Although recharging allows a sensor to prolong its lifetime by gathering energy from natural sources, it is often a very slow process, highly dependent on random environmental factors like the intensity of sunlight. Typically, the average rate of recharging would be much less than the average energy discharge rate during the active (i.e., sensing and transmission) period (Dai et al., 2013a). As a result, a sensor would need to spend most of its lifetime in the off state, when it is not sensing, but only recharging, requiring an unreasonably large number of sensors to be deployed. Therefore, finding an easy and reliable way to replenish sensor's battery begins to attract more attention within the sensor network research community.

Fortunately, breakthroughs in wireless charging technology have opened up a new dimension to the power sensor nodes in terms of distance without any wires or plugs. Pioneered by Tesla (1914) a century ago, it is only recently that wireless charging enjoys so much popularity after the experimental realization by Kurs et al. (2007). It has shown that a total of 60 W of energy can be transferred between two magnetically coupled coils over an air gap of 2 m with 40% efficiency. The experimental prototype was soon extended to power multiple devices (Kurs et al., 2010). In the meantime, the fast development of mobile devices and stagnant battery technology delivered the impetus to drive wireless charging technology into commercialization and many products of close-distance over-the-air charging are now available. For example, the charging pad called "Powermat©" can recharge multiple cell phones and PDAs (Personal Digital Assistants) simultaneously by simply putting them on the pad (Powermat, 2015).

However, wirelessly charging sensors for the entire network in a single charging station is not practical due to its sometimes large size (Xie et al., 2012b). Many pieces of research have proposed using mobile chargers or Wireless Charging Vehicles (WCVs) in WRSNs. Compared to energy-harvesting sensor nodes, wireless rechargeable sensor nodes can be smaller, enabling a large-range of applications, such as embedded infrastructure sensing and human activity recognition. A typical WRSN with wireless energy transfer technology includes several key network components: WCVs, base station (BS), and sensor nodes in various states, as illustrated in Fig. 1.

Since WCV is an excellent option for charging all the nodes in WRSNs, interesting research results have been published. In 2011, Yang and Wang (2015) proposed the prototype of WRSNs, which laid a foundation for the application. Some important progress has been made, mainly related to the periodic charging method, collaborative charging method and system performance assessment, and so on, which is discussed in Section 3.

2 CHARGING MODEL

WCVs, BS, and rechargeable sensor nodes are important endpoints of energy flow in WRSN. In the process of energy flow, the energy of sensor nodes with depleting level is replenished, in order to maintain the normal operation of the network. Therefore, the attributes and behavior of energy flow's endpoints have a great impact on charging efficiency, and then become a decisive factor affecting the lifetime of WRSNs.

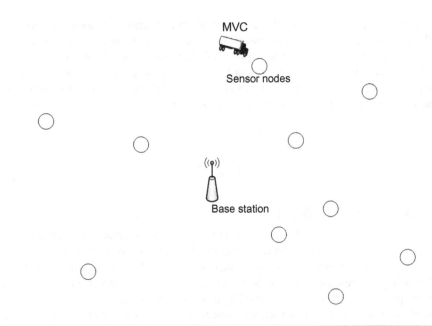

FIG. 1

A paradigm of wireless rechargeable sensor networks.

On the other hand, the charging problem of WRSNs can be simplified to the problem of how to effectively schedule a WCV to serve every sensor node in the WRSN. Therefore, we are required to construct a charging mode firstly by assuming the attributes and formulating the behavior of the WCV, BS, and sensor nodes. Hence, the abstracted problem of energy charging scheduling will be made concrete.

In WRSNs, completion of a charging task relies on the cooperation of sensor nodes and WCVs, so the description about the behavior of sensor nodes and WCVs can be converted to the description of the charging task.

2.1 CHARGING MODEL FOR SENSOR NODES

WRSNs are composed of a large number of cheap micro-sensor nodes, so we described the network by the number and distribution of sensor nodes. The network model is usually denoted as a set of sensor nodes $V = \{v_1, v_2, ..., v_m\}$, which are distributed randomly in a two-dimensional (2D) area, and the location of a node v_i is denoted as (x_i, y_i).

Since the battery capacity of sensor nodes is the main constraint of a WRSN's lifetime, we are required to construct the energy consumption model of sensor nodes. Sometimes, we formulate principles about energy consumption from the theoretical aspects as Eq. (1; Yang and Wang, 2015):

$$E(T) \leq R(T) + E_0 \tag{1}$$

where T is the time duration, $E(T)$ is the total energy consumption of the network in time duration T, $R(T)$ is the total energy replenished into the network by the charging vehicles in T, and E_0 is the initial energy of all the nodes (Yang and Wang, 2015).

In addition, we can even use these attributes with some restrictions to calculate information that makes the charging problem formalization easier (Jiang et al., 2014). For example, for the ith sensor, its power consumption per unit time (working power) P_i depends on the rate of event generating f_i and the unit energy consumption e_i. Therefore, P_i can be calculated using Eq. (2). Moreover, some underlying characteristics of the charging requests can also be modeled. For example, the arrival rate of the request λ can be estimated as Eq. (3) with threshold factor α.

$$P_i = f_i \times e_i \tag{2}$$

$$\lambda = \frac{\sum_{i=1}^{m} P_i}{(1 - \alpha) \times W} \tag{3}$$

2.2 CHARGING MODEL FOR WCVs

As the executor of charging task, WCV's behavior and attributes are the main factors that determine the charging efficiency. Battery capacity and traveling speed should be assumed firstly according to the needs of charging model, which are the basic attributes of WCVs. For example, in Zhang et al. (2015), mobile chargers are assumed to be homogeneous. For every charger, the battery capacity is P, the traveling speed is V, and energy consumed by traveling one unit distance is C, moreover, energy consumption of WCV are also taken into account. Due to the needs of scheduling algorithm and charging model, we can add some more descriptions about WCVs, such as the number of WCVs in WRSN with multiple WCVs. In Dai et al. (2013c), they considered the minimum of WCVs problem in a general 2D network so as to keep the network running forever.

The main task of a WCV is to choose a charging request and charging for the corresponding sensor node. Since scheduling charging requests is one of the main tasks of WCVs, we generally use queue theory to describe charging requests. In a WSRNs, a WCV maintains a requesting queue that gathers the real-time charging requests from the sensor when their residual energy falls below a certain threshold, and we use $M/G/1$ queue of Kendall's notation (Jiang et al., 2014) for deeply analyzing the characteristics of the requests.

In a WRSN with multiple mobile chargers, another focus of the charging model is to determine the number of the WCVs. A promising method is to group sensor nodes into slices and assign one WCV for each slice (Madhja et al., 2015). Madhja et al. divided the network into M equal sized slices, one for each WCV. Thus, every WCV is responsible for charging nodes that belong to its slice. We denote D_i as the set of sensor nodes that belong to slice j, i.e., to the jth WCV's group, as shown in Fig. 2.

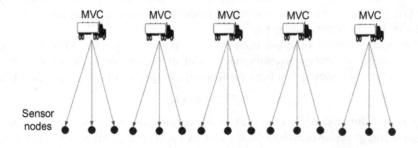

FIG. 2

Hierarchical collaborative charging schemes (Madhja et al., 2015).

2.3 CHARGING MODEL FOR CHARGING TASK

The charging model for the charging task contains the main information reflecting the state of a sensor node, which is the foundation of the scheduling decision. The WCV chooses the charging task with the highest priority by comparing and judging the information that each charging task contains. Therefore, there are various types of task model. A typical task model is a tuples containing main information scheduling algorithm needs, with some restrictions. For example, in Li and He (2010), most tasks of an embedded soft real-time multitask system are periodic tasks and every periodic task τ_i is described with a 7-tuple:

$$\tau_i = (S_i, WCET_i, AET_i, BCET_i, D_i, T_i, P_i) \tag{4}$$

where $BCET_i \leq AET_i \leq WCET_i \leq D_i$ and $D_i = T_i$. S_i is a point in time at which a real-time task becomes ready to or is activated to execute, $BCET_i$ is the best case execution time, $WCET_i$ is the worst case execution time, AET_i is the actual execution time which is varying during every request in an uncertain environment, D_i is the relative deadline of task τ_i, P_i is the relative criticality of task τ_i, and T_i is the period of task τ_i.

In particular, we can even use task model to describe the scheduling behavior of WCVs by planning a path. For example (Dai et al., 2013a): we denote the path of the WCV as $P = \left(\pi_0, \pi_1, ..., \pi_{|V_s|}, \pi_{|V_s|+1} \right)$, where $\pi_0 = \pi_{|V_s|+1} = BS$ and $\{\pi_i\}_{i=1}^{|V_s|} = V_s$. Denote t_{ij} as the time required for the WCV to move between sensor v_i and v_j.

In conclusion, the charging model is a set of descriptions for sensor nodes, WCVs and charging task. In addition, we are required to construct mathematical formulas and a function model according to the needs of the scheduling algorithm.

3 CLASSIC CHARGING SCHEMES

In WRSNs, one or more WCVs are responsible for replenishing energies for all rechargeable sensors, in order to solve the limitation of energy. Therefore, it is necessary for WCVs to charge nodes before they exhaust their battery power, otherwise, the dead nodes can no longer be replenished with energy by WCV and continue to operate. This issue can be solved by developing appropriate scheduling schemes. Hence, scheduling strategies for charging have become a prominent issue.

Charging schemes can be divided into two categories: deterministic methods (Xie et al., 2012a,b, 2013a,b; Shi et al., 2011; Fu et al., 2013) and nondeterministic methods (He et al., 2013). In deterministic methods, charging for individual nodes is carried out in a periodic and deterministic manner. For example, in Shi et al. (2011), they assumed that the WCV starts from the service station, visits each sensor node once in a cycle and ends at the service station. Further, data flow routing in the network is invariant with time, with both routing and flow rates being part of the optimization problem, which are shown in Fig. 3.

However, deterministic charging methods usually require explicit system information, such as exact node location, channel status, and so on, which are difficult or sometimes even impossible to obtain in practical applications.

Nondeterministic methods (He et al., 2013) are usually on-demand, in which sensors send their energy charging requests to WCVs when their energy levels run below a threshold. Upon the reception of a request, a WCV will immediately rearrange the order of recorded charging tasks, select a candidate sensor, and proceed. As a traditional and classic on-demand charging scheme, Nearest-Job-Next with Preemption (NJNP) (He et al., 2013) allows the mobile charger to switch to a spatially closer target node if the new requesting node is closer to the mobile charger. More specifically, under NJNP, each

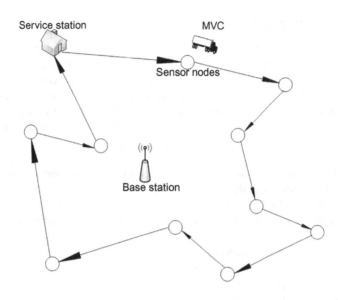

FIG. 3

Deterministic charging methods.

charging completion of nodes and the arrival of new charging requests trigger the re-selection of the next to-be-charged node, and the mobile charger selects the spatially closest requesting node at that time as the next node to charge.

In order to understand the research status at home and abroad of charging schemes in WRSN, we generally introduce these classic charging schemes from three aspects: periodic charging methods, collaborative charging mechanism, and performance evaluation.

Periodic charging method (Xie et al., 2012a,b, 2013a,b; Shi et al., 2011; Fu et al., 2013): are combined with node distribution model and energy consumption model. A charging problem can be converted to the traveling salesman problem and calculate a Hamilton loop as WCVs charging trajectory. A periodic charging method is mainly divided into a multimode charging method and a single-node charging method. In multinode charging method, WCV can charge the battery for multiple nodes within the changing scope of WCV at the same time, greatly improving the charging efficiency (Shi et al., 2011; Xie et al., 2012a). On the contrary, WCV can only charge one node at a time in single-node charging, which is relatively inefficient (Xie et al., 2012b).

On the basis of a multinode charging method, Xie and other researchers study the path planning problem where WCVs are regarded as the mobile BS (Xie et al., 2013a,b) by establishing smallest enclosing disk (SED) (Welzl, 1991), and an SED covers all the sensor nodes in the network. With the node position as the center of a circle and the charge loss rate as the radius, they set up a concentric circle structure and regarded the overlap part of the concentric circles as the resting spot of WCV. Similarly, Fu et al. (2013) propose the discretization of wireless charging planning theory, which sets up the concentric circles structure in the SED and searches of the resting spot of WCV from the overlapping area. However, in Dai et al. (2013c), the researchers point out that the amount of calculation based on SED is large and it is not suitable for large-scale WSNs.

Collaborative charging mechanism: He et al. and others proposed that the periodic charging problems have limitations, such as deterministic, cyclical. The nondeterministic factors of the network will cause immeasurable effects for both of energy demand and energy supply (He et al., 2013). Therefore, we need to adopt a cooperative charging mechanism (Guo et al., 2013), in order to meet the network's heterogeneous needs of nondeterministic factors, dynamic topology and node properties. This paper proposed a WCV dynamic path programming method: NJNP and an evaluation system from aspects of throughput and charging delay. In the WCV charging schedule and routing strategy, they proposed a WCV charging scheduling method based on a tree structure (He et al., 2014), which reduces the charging consumption and charging delay. Li et al. (2011) proposed a J-RoC method combining routing protocol with charging policy. The WCV will update global energy status information and then schedule to charge. At the same time, nodes use rechargeable awareness routing protocol to select a path of low energy consumption for transmission. In that way, the energy consumption will be balanced and the lifetime of the network will be prolonged.

Although a collaborative charging method can effectively solve the impact of uncertainty factors in WRSNs, it still neglects the reliability of charging demand information and real-time transmission requirements. Failure or delay of charging demand information may lead to disability for WCV to arrive at charging position before it is out of energy. It will influence the reliability of the network. So the real-time and reliability for demanding information transmission need to be focused along with hybrid scheduling problem of demand information and gathering information.

Performance evaluation: Jiang et al. (2011) and Cheng et al. (2013) analyzed the optimization scheduling problem of WRSN under the condition of random events in detail. Dai et al. (2013a,b) established a performance evaluation standard on the basis of network quality of monitoring (QoM) and optimized system performance from the aspects of WCV behavior, data transmission protocol, collaborative control, and so on. Madhja et al. (2013) put forward the charging decision problem and proved its complexity. They lucubrated how to weigh the path of WCV, charging decision of WCV and charging the amount of WCV to optimize the performance of the system.

In our latest work (Lin et al., 2015), we proposed a double warning thresholds with double preemption charging scheme, in which double warning thresholds are used when residual energy levels of sensor nodes fall below certain thresholds. Besides that, we also proposed a task merging and splitting charging strategy HCCA (Hierarchical Clustering Charging Algorithm) and HCCA-TS (Hierarchical Clustering Charging Algorithm based on Task Splitting) for large scaled WRSNs (Lin et al., 2016). Simulation results indicate that HCCA can enhance the performance in terms of reducing charging times, journey time and average charging time simultaneously. Moreover, HCCA-TS can further improve the performance of HCCA.

The above-mentioned optimizing methods can all improve the functioning of a network, but the parameter combination of a system needs to be continuously optimized to enhance the system's performance, because that the charging process in massive WRSN is a lengthy process. In addition, in recent years, researchers have studied multiWCV collaborative charging mechanism in WRSN (Madhja et al., 2013). Zhang et al. (2012, 2015) design multiWCV collaborative charging mechanism, where WCV can transmit energy to each other, and schedule the charging order to optimize the charging efficiency of the system.

4 CHARGING SCHEMES FOR STOCHASTIC EVENT CAPTURE

Due to the huge number of potential applications of WRSN, many approaches have been proposed to extend the network lifetime. The energy constraint problem in WRSN can be overcome by breakthroughs in wireless power charging technologies, and there have emerged a number of applications employing wireless power charging technology to enhance event monitoring performance, such as wireless identification and sensing platform applied in fields ranging from individual activity recognition to large-scale urban sensing and structural health monitoring applications, where a civil structure is instrumental with sensor nodes capable of being powered solely on energy transmitted to the sensor node wirelessly by a mobile helicopter.

Moreover, many efforts focus on issues of stochastic event capture, which is a fundamental problem in WSN design and concerns the scheduling of sensors' duty cycles to maximize their ability to capture interesting events of a probabilistic nature. An abundance of research results in charging schemes to maximize QoM for stochastic event capture have been made, because QoM defined as the average information gained per event by the network is a basic performance evaluation criteria.

Existing practical approaches with wireless power charging technologies adopted to address the energy constraint in the WSNs focus on using one single or multiple WCVs to move around the sensors and charge them in turn during the travel schedule for applications such as routing and data gathering. Jiang et al. (2011) are the first to exploit wireless power charging by WCVs for efficient stochastic event capture, motivated by these applications, which jointly determine the WCV's movement and sensor activation schedules to maximize the QoM. On the basis of their efforts to pave the way to the maximum QoM charging and scheduling problem, Dai et al. (2013a) studied its relaxed version, and a solution approached by transforming this version into a submodular function maximization problem, under the condition that event utility function is concave. Consequently, a move closer to solving the original problem has been made by considering the travel time of the WCV.

Next, we will introduce the two scheduling schemes in detail, which lay a foundation for the study of a charging scheme for stochastic event capture in WRSN.

Jiang et al. first derived a wireless recharging model based on experimental data as shown in Fig. 4 and then formulated the QoM problem.

They considered N sensor nodes, indexed from 1 to N, placed at N points of interest (PoIs) in a 2D lane as shown. A reader can travel to visit the WCV tags at PoIs along a closed curve (Ω). In addition, they make simplifying assumptions that each sensor can only monitor one PoI, the charging time for each sensor is identical, the event staying time follows exponential distribution, and all sensors follow a simple periodic schedule, sensors monitor PoIs for q time of every p time.

After a series of analyses of the periodic scheduling behavior of WCV, they defined the QoM in the WRSNs as: $\mathrm{QoM} = \lim\limits_{T \to \infty} \sum_{i=1}^{N} c_i / \sum_{i=1}^{N} m_i$, where assuming during time T, there are m_i stochastic

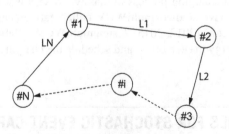

FIG. 4

Wireless recharging model based on QoM optimization.

events occurring at PoI i, and WCV tag i captures c_i events. Then the problem can be formulated as how to schedule the movement of the reader and the operation of WCV tags to maximize QoM.

It is of great interest to jointly investigate how the reader plans its path to provide sufficient energy and how the WCV tags should duty cycle themselves to maximize the QoM. There are mainly three typical cases: QoM in independent aggressive wake-up, QoM in joint aggressive wake-up, and QoM in joint periodic wake-up. Therefore, they study the optimal scheduling problem in these three cases and formulate the corresponding theorem in order to get maximum value of QoM.

Dai et al. (2013a) relaxed some assumptions in Jiang et al. (2011), and considered the scenario in which a WCV periodically travels within the sensor network and recharges a selected group of sensors wirelessly to enable them for the task of stochastic event capture. They analyzed the QoM of stochastic event capture, which considers the possibility that the same PoI may be monitored by multiple sensors. They formulate the MQCSP (Maximum QoM Charging and Scheduling Problem) and R-MQCSP (Relaxed version of MQCSP) problems, and show that both of them are NP-hard (Non-deterministic Polynomial). Then, R-MQCSP is reformulated as a monotone submodular function maximization problem under a sufficient special condition. This reformulation of R-MQCSP allows an algorithm that achieves a one-sixth approximation for the QoM maximization. At last, they also discussed a special case where the active time slots constraint, which refers to the constraint on the number of active time slots in a sensor's schedule caused by small battery capacity or a long period of charging process, can be lifted. It allows a better and unified algorithm achieving a series of approximation factors (up to $1 - 1/e$) under different parameter settings. On the basis of their R-MQCSP results, they proposed an approximation algorithm for MQCSP, which considers the WCV's travel time. Besides, they also proposed an approximation algorithm for the special case without the active time slots constraint.

5 MULTIPLE-WCV COLLABORATIVE CHARGING SCHEMES

As an excellent option for charging all the nodes in the network, using a single mobile charger (WCV) traveling through the network fields to recharge every sensor node has been proposed in many recent works. These algorithms work effectively in small-scale networks. However, most of them assume that a mobile charger has a sufficient amount of battery to cover the entire WSN and make a round trip back to the BS. In large-scale networks these algorithms do not work efficiently, especially when the amount of energy the WCV can provide is limited. Therefore, more and more research is committed to studying multiple-WCV collaborative charging.

In collaborative mobile charging, multiple mobile chargers work together in order to accomplish a given set of objectives. These objectives include charging sensors at different frequencies with a minimum number of mobile chargers and reaching the farthest sensor for a given set of mobile chargers, subject to various constraints, including speed and energy limits of mobile chargers.

Next, we will introduce three scheduling schemes in detail, which lay a foundation for the study of multiple-WCV collaborative charging in WRSN.

Zhang et al. (2012) are the first to introduce a novel collaborative mobile charging paradigm, which allows energy transfer between mobile chargers. They propose a scheduling algorithm *PushWait*, which is proven to be optimal and can cover a one-dimensional (1D) WSN of any length to demonstrate the advantages of this novel paradigm in coverage and energy efficiency. It is the first to consider the collaborative mobile charging paradigm. For the nonuniform case, which is conjectured to be NP-hard, they first presented two observations from space and time aspects to remove some impossible scheduling choices. Then, they proposed a heuristic algorithm, *ClusterCharging(β)*, in which sensors are grouped into clusters.

Later, Zhang et al. (2015) developed a set of scheduling algorithms for five different scenarios ranging from the simplest one to the hardest one, in an effort to provide some potential guidelines for the future design of charging scheduling.

First of all, they derived a wireless recharging model of N stationary sensor nodes distributed over a 2D area, where the location of the ith node s_i is denoted as (x_i, y_i) with the battery capacity b_i and energy consuming rate r_i. The recharging cycle of a sensor is defined as the time period that the sensor with a full battery can survive without being charged: $\tau'_i = b_i / r_i$. Wireless charging efficiency between a charger and a sensor node is defined as η_i.

Then they defined three conditions: (K1) all sensor nodes are distributed along a 1D line. Without loss of generality, they let $\forall 1 \leq i \leq N, y_i = 0$. In (K2) all sensor nodes have the same recharging cycle τ, i.e., $\forall 1 \leq i \leq N, \tau'_i = \tau$. In (K3) wireless energy transfer has no energy loss, i.e., $\eta_1 = \eta_2 = 1$. They defined $\overline{Kj}$ to indicate that Kj does not hold, and $j \in \{1, 2, 3\}$ to discuss for five conditions from the simplest one to the hardest one: $K1K2K3$, $K1K2\overline{K3}$, $K1\overline{K2}K3$, $K1\overline{K2}\overline{K3}$, $\overline{K1K2K3}$. Then, they proposed five different collaborative mobile charging paradigms for different scenarios, for example, *PushWait* for $K1K2K3$, η*PushWait* for $K1K2\overline{K3}$, *ClusterCharging(β)* for $K1\overline{K2}K3$, η*ClusterCharging(β)* for $K1\overline{K2}\overline{K3}$, and $H\eta$*ClusterCharging(β)* for $\overline{K1K2K3}$.

When applying collaborative charging, another problem that we cannot ignore is that of determining the minimum number of WCVs.

Dai et al. (2013c) are the first to consider the minimum mobile chargers (Min-MCP) problem in general 2D WRSNs mainly based on the classical results of distance constrained vehicle routing problem (DVRP), i.e., how to find the minimum number of energy-constrained WCVs and their recharging routes given a 2D WRSN, so as to keep the network running forever. They proposed approximation algorithms to address Min-MCP by proving for any $\varepsilon > 0$, there is no $(2 - \varepsilon)$ approximation algorithm for DVRP on a general metric space, which is the best result as far as we know. Based on this result, they proved that Min-MCP is NP-hard, and its inapproximability bound is the same as that of DVRP and proposed approximation algorithms to address Min-MCP.

6 CONCLUSION

In this chapter, we firstly introduce the structure and function of WRSNs, which leads to the broad application prospects of WRSN. To solve the performance bottleneck problem, we discuss ways to replenish energy for sensor nodes. Existing energy replenishing technologies are energy-harvesting technology and wireless charging technology.

Then, we emphatically introduce charging schemes, where WCVs are used to charge sensor nodes with wireless charging technology, because of its obvious superiority compared to energy harvesting technology. We can understand the essence of charging schemes by studying the charging model. Moreover, we can also learn the development status of charging schemes by summarizing the classic charging schemes.

Finally, we discuss two typical categories of charging schemes: charging schemes for stochastic event capture and multiple-WCV collaborative charging schemes. We introduce the necessity of studying the two typical categories of charging schemes and several scheduling schemes laying a foundation for the study.

REFERENCES

Akyildiz, I.F., Su, W., Sankarasubramaniam, Y., et al., 2002. Wireless sensor networks: a survey. Comput. Netw. 38, 393–422.

Cheng, P., He, S., Jiang, F., et al., 2013. Optimal scheduling for quality of monitoring in wireless rechargeable sensor networks. IEEE Trans. Wirel. Commun. 12, 3072–3084.

Dai, H., Jiang, L., Wu, X., et al., 2013a. Near optimal charging and scheduling scheme for stochastic event capture with rechargeable sensors. In: 2013 IEEE 10th International Conference on Mobile Ad-Hoc and Sensor Systems (MASS). IEEE, Hangzhou, China, pp. 10–18.

Dai, H., Wu, X., Xu, L., et al., 2013b. Practical scheduling for stochastic event capture in wireless rechargeable sensor networks. In: 2013 IEEE Wireless Communications and Networking Conference (WCNC). IEEE, Shanghai, China, pp. 986–991.

Dai, H., Wu, X., Xu, L., et al., 2013c. Using minimum mobile chargers to keep large-scale wireless rechargeable sensor networks running forever. In: 2013 22nd International Conference on Computer Communications and Networks (ICCCN). IEEE, Nassau, Bahamas, pp. 1–7.

Fu, L., Cheng, P., Gu, Y., et al., 2013. Minimizing charging delay in wireless rechargeable sensor networks. In: 2013 IEEE International Conference on Computer Communications (INFOCOM 2013). IEEE, Torino, Italy, pp. 2922–2930.

Guo, S., Wang, C., Yang, Y., 2013. Mobile data gathering with wireless energy replenishment in rechargeable sensor networks. In: 2013 IEEE International Conference on Computer Communications (INFOCOM 2013). IEEE, Torino, Italy, pp. 1932–1940.

He, L., Gu, Y., Pan, J., et al., 2013. On-demand charging in wireless sensor networks: theories and applications. In: 2013 IEEE 10th International Conference on Mobile Ad-Hoc and Sensor Systems (MASS). IEEE, Hangzhou, China, pp. 28–36.

He, L., Cheng, P., Gu, Y., et al., 2014. Mobile-to-mobile energy replenishment in mission-critical robotic sensor networks. In: 2014 IEEE International Conference on Computer Communications (INFOCOM 2014). IEEE, Toronto, ON, Canada, pp. 1195–1203.

Jiang, F., He, S., Cheng, P., et al., 2011. On optimal scheduling in wireless rechargeable sensor networks for stochastic event capture. In: 2011 IEEE 8th International Conference on Mobile Adhoc and Sensor Systems (MASS). IEEE, Valencia, Spain, pp. 69–74.

Jiang, L., Dai, H., Wu, X., et al., 2014. On-demand mobile charger scheduling for effective coverage in wireless rechargeable sensor networks. In: Stojmenovic, I. (Ed.), Mobile and Ubiquitous Systems: Computing, Networking, and Services. Springer, Tokyo, Japan, pp. 732–736.

Kansal, A., Hsu, J., Srivastava, M., Raqhunathan, V., 2006. Harvesting aware power management for sensor networks. 43rd ACM/IEEE Design Automation Conference, San Francisco, CA, pp. 651–656.

Kurs, A., Karalis, A., Moffatt, R., et al., 2007. Wireless power transfer via strongly coupled magnetic resonances. Science 317, 83–86.

Kurs, A., Moffatt, R., Soljačić, M., 2010. Simultaneous mid-range power transfer to multiple devices. Appl. Phys. Lett. 96, 044102.

Li, X., He, X., 2010. The improved EDF scheduling algorithm for embedded real-time system in the uncertain environment. In: 2010 3rd International Conference on Advanced Computer Theory and Engineering (ICACTE). IEEE, Chengdu, China. V4-563-V564-566.

Li, Z., Peng, Y., Zhang, W., et al., 2011. J-RoC: a joint routing and charging scheme to prolong sensor network lifetime. In: 2011 19th IEEE International Conference on Network Protocols (ICNP). IEEE, Vancouver, Canada, pp. 373–382.

Lin, C., Xue, B., Wang, Z., et al., 2015. DWDP: a double warning thresholds with double preemptive scheduling scheme for wireless rechargeable sensor networks. In: 17th IEEE International Conference on High Performance Computing and Communications (HPCC 2015). IEEE, New York, NY, pp. 503–508.

Lin, C., Wu, G., Obaidat, M.S., et al., 2016. Clustering and splitting charging algorithms for large scaled wireless rechargeable sensor networks. J. Syst. Softw. 113, 381–394.

Madhja, A., Nikoletseas, S., Raptis, T.P., 2013. Efficient, distributed coordination of multiple mobile chargers in sensor networks. In: Proceedings of the 16th ACM International Conference on Modeling, Analysis & Simulation of Wireless and Mobile Systems. ACM, Barcelona, Spain, pp. 101–108.

Madhja, A., Nikoletseas, S., Raptis, T.P., et al., 2015. Hierarchical, collaborative wireless charging in sensor networks. In: Proceedings of the IEEE Wireless Communications and Networking Conference (WCNC). IEEE, New Orleans, USA.

Powermat. Available at: http://www.powermat.com.

Shi, Y., Xie, L., Hou, Y.T., et al., 2011. On renewable sensor networks with wireless energy transfer. In: 2011 IEEE International Conference on Computer Communications (INFOCOM 2011). IEEE, Shanghai, China, pp. 1350–1358.

Sohrabi, K., Gao, J., Ailawadhi, V., et al., 2000. Protocols for self-organization of a wireless sensor network. IEEE Pers. Commun. 7, 16–27.

Tesla N. (1914) Apparatus for transmitting electrical energy. Google Patents.

Welzl, E., 1991. Smallest Enclosing Disks (Balls and Ellipsoids). Springer, Graz, Austria.

Werner-Allen, G., Lorincz, K., Ruiz, M., et al., 2006. Deploying a wireless sensor network on an active volcano. IEEE Internet Comput. 10, 18–25.

Xie, L., Shi, Y., Hou, Y.T., et al., 2012a. On renewable sensor networks with wireless energy transfer: the multi-node case. In: 2012 9th Annual IEEE Communications Society Conference on Sensor, Mesh and Ad Hoc Communications and Networks (SECON). IEEE, Seoul, Korea, pp. 10–18.

Xie, L., Shi, Y., Hou, Y.T., et al., 2012b. Making sensor networks immortal: an energy-renewal approach with wireless power transfer. IEEE/ACM Trans. Netw. 20, 1748–1761.

Xie, L., Ycb, Shi, Hou, Y.T., et al., 2013a. Bundling mobile base station and wireless energy transfer: modeling and optimization. In: 2013 IEEE International Conference on Computer Communications (INFOCOM 2013). IEEE, Turin, Italy, pp. 1636–1644.

Xie, L., Shi, Y., Hou, Y.T., et al., 2013b. On traveling path and related problems for a mobile station in a rechargeable sensor network. In: Proceedings of the Fourteenth ACM International Symposium on Mobile Ad Hoc Networking and Computing. ACM, Bangalore, India, pp. 109–118.

Yang, Y., Wang, C., 2015. Wireless Rechargeable Sensor Networks. Springer, London, UK.

Ye, W., Heidemann, J., Estrin, D., 2002. An energy-efficient MAC protocol for wireless sensor networks. In: 2002 IEEE International Conference on Computer Communications (INFOCOM 2002). IEEE, Orlando, FL, USA, pp. 1567–1576.

Yu, L., Wang, N., Meng, X., 2005. Real-time forest fire detection with wireless sensor networks. In: Proceedings of 2005 International Conference on Wireless Communications, Networking and Mobile Computing. IEEE, Las Vegas, Nevada, USA, pp. 1214–1217.

Zhang, S., Wu, J., Lu, S., 2012. Collaborative mobile charging for sensor networks. In: 2012 IEEE 9th International Conference on Mobile Adhoc and Sensor Systems (MASS). IEEE, Las Vegas, Nevada, USA, pp. 84–92.

Zhang, S., Wu, J., Lu, S., 2015. Collaborative mobile charging. IEEE Trans. Comput. 64, 654–667.

APPLYING BEHAVIORAL GAME THEORY TO CYBER-PHYSICAL SYSTEMS PROTECTION PLANNING

17

S.T. Hamman*,†, K.M. Hopkinson*, L.A. McCarty†

The Air Force Institute of Technology, WPAFB, OH, United States * *Cedarville University, Cedarville, OH, United States* †

1 INTRODUCTION

As civilization enters the fourth industrial revolution (Industry 4.0), cyber-physical systems (CPSs) will play an increasingly expanding role in society (Jazdi, 2014; Zhou et al., 2015). Because society's dependence on CPSs is directly proportional to its attractiveness to terrorists and other adversaries who are motivated to inflict maximum harm on their enemies (Brown et al., 2006), protection planning is a vital aspect of the ongoing operation of any real-world CPS.

Large-scale CPSs pose some unique security challenges because they may be geographically dispersed and located in remote areas where it is difficult to provide physical security (Neuman, 2009). Providing adequate protection resources in such contexts is infeasible due to the large attack surface and the limited availability of man hours and money. Therefore, large-scale CPS protection planning is necessarily an exercise in the allocation of scarce protection resources (Fletcher and Liu, 2011).

Not only are protection resources scarce, but they must be allocated in light of the fact that adversaries are strategic actors. The U.S. Office of Homeland Security (2002) warn that adversaries (e.g., terrorists, enemy nation states, etc.) perform reconnaissance and undertake intensive planning before making an attack. Any CPS protection scheme that does not take into account attack scenarios waged by intelligent adversaries is naïve and inadequate.

The research presented in this chapter lays the foundation for approaching the challenge of CPS protection planning in view of these realities. Our approach is to model the protection scenario as a newly formulated security game based on the Colonel Blotto (CB) game from game theory. A security game has been defined as a "game-theoretic model that captures essential characteristics of resource allocation decision making" (Fielder et al., 2014, p. 18). We then "solve" the game by applying the concept of level-k reasoning from behavioral game theory. Behavioral game theory allows us to model the strategic nature of intelligent adversaries and provides insights into anticipating and countering their most likely actions. The goal of our approach is to neither over- nor under-protect the critical sites in a CPS infrastructure, thereby achieving the biggest "bang for the buck."

Furthermore, we eliminate much of the human element in protection planning by leveraging a mathematical programming solver to determine level-k solutions to an integer linear program (ILP) which models the security CB game. The solver outputs a precise allocation of protection resources across critical sites. The ILP is applicable to any size CPS and any amount of protection resources.

In order to provide a clear illustration of how the methodology can be applied to a real-world CPS, we show how it would allocate protection resources across a notional smart grid special protection system (SPS).

Lastly, we validate our approach by conducting an attack competition with human subjects based on the parameters of the SPS. We believe that using human subjects is the only legitimate way to validate the effectiveness of an approach that has as its goal the countering of attacks waged by intelligent adversaries. Other ways to measure effectiveness, including performing computer simulations, generating random attacks, or even constructing mathematically rigorous models, fall short because it cannot be convincingly demonstrated that they fully capture the intelligence and strategic nature of human beings.

2 RELATED WORK

Our work is related to other research efforts that attempt to allocate scarce protection resources effectively over CPSs. Feng et al. (2013) promote building attack trees to find the most damaging attack paths, thereby identifying where protection resources are needed most. Similarly, Fletcher and Liu (2011) recognize the impossibility of protecting every aspect of a CPS infrastructure. They introduce an integrated methodology to prioritize security requirements with the goal of ensuring that the most important tasks are addressed first, instead of proceeding in an ad hoc manner, such as "easiest first" or "least expensive first."

Other work has been done regarding the use of game theory in CPS protection planning. Cardenas et al. (2009) emphasize the need for estimation algorithms that capture realistic attack models, and they suggest that game theoretic techniques for modeling rational adversaries may be useful for this task. He et al. (2012) use game theory to model the probabilities of successful attacks as a function of the number of components that are attacked and defended. Ma et al. (2011) attempt to find the Nash Equilibria for a game theoretic formulation of a CPS security scenario, and they distinguish between the degradability and the survivability of CPSs after attacks. Vigo et al. (2013) find an optimum solution to a CPS security game by utilizing linear programming. Backhaus et al. (2013) incorporate human decision making into a model of defending SCADA control systems by including one level of level-k reasoning.

Although not in the context of CPSs, Tambe (2012) cite successful, real-world implementations of security systems that rely on computer-generated solutions to security games. Their algorithms are currently being used by the Los Angeles International Airport and the U.S. Coast Guard, among others, to derive inspection schedules.

3 APPROACH

When trying to defend a large-scale distributed CPS, protection planners are faced with the dilemma of allocating limited protection resources (e.g., man hours and money) as efficiently as possible over multiple vulnerable sites. If the sites are not all equally valuable, it does not make sense to allocate the resources evenly over the sites (the *equal allocation strategy*). A much better strategy would be to allocate the resources proportionately according to the relative values of the sites (the *proportional allocation strategy*). However, these two approaches, like any plain optimization formulation, fail to capture the effects of the attacker behavior in the model (Alpcan and Basar, 2011). Consequently, these two natural approaches to the scarce resource allocation problem are inadequate.

Our approach to large-scale CPS protection planning is to try and anticipate how an adversary would allocate his attack resources and then to deploy defensive resources accordingly. To accomplish this, we leverage the concept of level-k reasoning from behavioral game theory.

3.1 LEVEL-K REASONING

When engaging in a strategic contest, the first step a person typically employs in formulating a strategy is to make an educated guess as to what his opponent will do. From that point, he can arrange his strategy to beat his opponent's putative strategy. Of course, he may realize his opponent is also rational and is likely following a similar procedure. This might lead him to try and beat the strategy that he imagines his opponent is going to use to try and beat his initial strategy. This type of back-and-forth reasoning could theoretically continue indefinitely. Behavioral game theorists, who have extensively studied the dynamics of human beings engaged in strategic interactions, call this thought process level-k reasoning (Camerer, 2003).

In the concept of level-k reasoning, the natural, instinctual strategy is denoted as the level-0 (L0) strategy, the first logical extension of it as the L1 strategy, then L2, and so on. To summarize, the Lk type implicitly assumes his opponent is an L$(k-1)$ type.

Over decades and in many contexts, behavioral game theorists have empirically studied how many levels deep people typically descend into the level-k reasoning process. One noteworthy attempt to isolate the level-k reasoning process from confounding variables is the 11–20 money request game (Arad and Rubinstein, 2012a). In this game, two participants, independently of one another, are asked to choose an amount of money between $11 and $20, and they are told they will be given whatever amount of money they choose. Additionally, they are told that they will earn a $20 bonus if they choose exactly $1 less than the other participant.

The L0 strategy in this game is to ask for $20—it is the instinctual starting point since it is the highest amount of money available. From there, the L1 strategy is to ask for $19 in anticipation of the other participant asking for $20, because this will result in the $20 bonus. The L2 strategy is to ask for $18, and the L3 strategy is to ask for $17. Around 80% of the subjects in a study conducted with 108 participants chose between $17 and $20, and the authors demonstrate that these choices represent between three and zero levels of level-k reasoning, respectively.

This leaves the other 20% of participants who chose between $11 and $16. Did they employ more in-depth reasoning than the other 80% of participants? Based on ex post interviews with the subjects, the answer to this question is a definitive, "No." Indeed, behavioral game theory researchers have repeatedly demonstrated that humans rarely (if ever) continue to four or more levels of reasoning, either because they do not believe their opponents will continue that far, or because they stop when they reach the limit of their mental capacity (Arad and Rubinstein, 2012a). People who do not employ level-k reasoning typically describe their strategies as being based on "gut instincts," guesses, or intuition. These strategies require very little time to formulate relative to level-k reasoning strategies and typically perform poorly in strategic contests.

Numerous level-k reasoning studies have been conducted on vastly different pools of people and the results are similar to those of the 11–20 money request game. Researchers have concluded that the majority of people, no matter what their country of origin, level of intelligence, profession, ethnicity, gender, etc., employ between zero to three levels of reasoning (Camerer, 2003). Our approach is to leverage this basic trait of human nature to derive efficient protection resource allocations. However, before we can apply level-k reasoning to the scarce resource allocation problem, we need a formal model of the problem, which the CB game from game theory provides us.

3.2 THE CB GAME

The CB game (Gross and Wagner, 1950) was devised to capture the strategic dynamics inherent in allocating scarce resources over multiple sites. In the canonical CB game, two colonels, A and B, compete over K independent battlefields of total aggregate value U. The colonels control M and N soldiers, respectively, and they must distribute them over the K battlefields independently of one another (in the cover of darkness as the eponymous construction goes). Their choices are revealed simultaneously (continuing the illustration, at dawn), and whichever colonel has allocated the most soldiers to a particular battlefield wins that battlefield. Each colonel's goal is to maximize his own utility.

There are many variations of the basic CB game. The colonels may have the same amount of soldiers or different amounts. The values of the battlefields may be homogenous or heterogeneous. The colonels may agree or disagree on the values of the battlefields. There are also different ways to resolve ties, including denoting a default winner in all cases, splitting the utility between the colonels, or not awarding the utility to either colonel.

Arad and Rubinstein (2012b) have demonstrated that when people play the CB game, they exhibit level-k reasoning. Therefore, this strategic model provides us with a sound basis for applying level-k reasoning to CPS protection planning.

We model CPS protection planning as a specific type of CB game where the defender and the attacker are the two colonels, protection and attack resources are the soldiers, and CPS critical sites are the battlefields. In order to capture the nuanced dynamics of CPS protection planning, we select the following game variations:

- We assign both the defender and the attacker the same number of soldiers, making them equally matched. This makes sense because both the defender and the attacker would naturally allocate resources in proportion to the size and value of the CPS infrastructure.
- We assign both the defender and the attacker 100 soldiers. This choice allows us to easily identify the proportion of resources allocated to each site (i.e., each soldier is 1% of a colonel's budget).
- We assume that the battlefield values will be heterogeneous. It is likely that, based on their location in the infrastructure and their differing responsibilities, the critical sites of any large-scale distributed CPS will have different amounts of utility.
- We assume that the attacker and the defender will value the battlefields symmetrically. Because our model is predicated on a well-planned attack, we assume that the attacker will conduct substantial reconnaissance and will have accurate values for the sites.
- In the case of a tie, we split the battlefield utility evenly.

In addition to these variations we must make a tweak to the canonical CB game to transform it into a security game. In protection planning, unlike in the war version of the game, the critical sites are not neutral ground—they are all owned by the defender by default. In the classic CB game, in the scenario where neither the defender nor the attacker allocates resources to a particular battlefield, the battlefield's value goes unawarded. However, in the same situation in a security context, the defender would win that site because he owns all of the sites to begin with. Therefore, in what we term the *security CB game*, unattacked battlefields are automatically awarded to the defending colonel, even if those sites are not protected.

3.3 CALCULATING LEVEL-*K* STRATEGIES

We devised an ILP to model the security CB game. A mathematical programming solver (e.g., CPLEX) is able to efficiently calculate the best responses to any set of opponent strategies. By bootstrapping the model with the L0 strategy, the ILP can calculate strategies at any depth of level-*k* reasoning by first computing the best response to the L0 strategy, (i.e., the L1 strategy) and then the best response to that strategy, and so on, until the desired level is reached.

The L0 strategy in the CB game is the proportional allocation strategy, as demonstrated by Arad and Rubinstein (2012b). The proportional allocation strategy for a colonel with N total of soldiers, and a game with set K of battlefields and U total utility is calculated as follows:

$$n_j = \frac{u_j}{U \times N}, \forall j \in K \tag{1}$$

Even though this approach allows one to calculate any level of level-*k* reasoning strategy, which level should CPS protection planners select? The answer is that it depends on what level the attackers select. Based on the findings from behavioral game theory described earlier, it can be assumed that they will use between zero and three levels of level-*k* reasoning.

This might make it appear like the L4 strategy would be the best choice, but the evidence from the experiment we conducted with human subjects (detailed below) strongly supports the choice of L3 as the correct defensive strategy for the security CB game. This makes sense because as strategies become more distant from lower-level strategies, they overprotect some sites at the expense of others (i.e., they "overthink" the problem). Therefore, the best place to compete is as near as possible to the majority of the anticipated attacks. The choice of L3 as the best strategy is also consistent with the results of behavioral game theory competitions (e.g., Arad and Rubinstein, 2009).

The ILP is as follows:

Let P be the set of possible attack strategies and K be the set of battlefields.

Let N be the number of soldiers the attacker and defender each have.

Let a_{pj} be the number of soldiers placed at battlefield j in attack strategy p, for $1 \le j \le |K|, 1 \le p \le |P|$.

Let z_{jn} be a decision variable where:

$$z_{jn} = \begin{cases} 1, & \text{defender places exactly } n \text{ soldiers at battlefield } j \\ 0, & \text{otherwise} \end{cases} \quad \text{for } 1 \le j \le |K|, 0 \le n \le N.$$

Let ε_{pjn} be an indicator variable that is calculated off-line for each $1 \le p \le |P|, 1 \le j \le |K|, 0 \le n \le N$.

$$\text{For } n \ne 0, \varepsilon_{pjn} = \begin{cases} 1, n > a_{pj} \\ 0, n = a_{pj}, \text{ so a tie results in 0 points.} \\ -1, n < a_{pj} \end{cases}$$

$$\text{For } n = 0, \varepsilon_{pj0} = \begin{cases} 1, 0 = a_{pj} \\ -1, 0 < a_{pj} \end{cases}, \text{ so a 0-0 score results in a win for the defender.}$$

Let u_j represent the utility of battlefield j for $1 \leq j \leq |K|$. A model to find an optimal strategy for the defender becomes:

$$\text{maximize}: \sum_{p=1}^{|P|} \sum_{j=1}^{|K|} \sum_{n=0}^{N} u_j \varepsilon_{pjn} z_{jn} \qquad (2)$$

$$\text{subject to}: \sum_{j=1}^{|K|} \sum_{n=0}^{N} n \cdot z_{jn} = N$$

$$\sum_{n=0}^{N} z_{jn} = 1 \quad \forall j$$

$$z_{jn} \in \{0, 1\} \quad \forall j, n$$

The objective function is the number of wins, minus the number of losses, where ties count as 0 for $n \neq 0$. The first constraint enforces the rule that the defender and attacker each have only N soldiers. The second constraint ensures that the defense cannot place two different amounts of soldiers at one battlefield.

---

4 ILLUSTRATION

In order to illustrate how our approach would allocate scarce protection resources in a real security setting, in this section we detail a specific large-scale distributed CPS and a realistic, although hypothetical, attack scenario.

4.1 ATTACKING THE SMART GRID

Lloyds (2015) imagines a realistic power grid attack scenario conducted by a highly motivated and capable adversary. The report describes a meticulously planned cyber attack involving considerable reconnaissance and effort that hinges on planting malware in smart grid safety control systems. In the scenario proposed, the malware lies dormant until activated during a peak period of electricity demand, at which point it attacks grid components in a coordinated manner and eventually triggers a *cascading blackout*. A cascading blackout is an "avalanche" of power outages that spreads rapidly and uncontrollably over a vast region. To prevent cascading outages from occurring, *load shedding*, the practice of taking "blocks of customers off-line in order to prevent a total collapse of the electric system" (U.S. Department of Energy, 2004, p. 7), is typically performed.

Our attack scenario is based on Lloyds (2015), and is oriented around the SPS described in Ross et al. (2013), which is representative of a general, large-scale CPS. As the power grid evolves into the smart grid, SPSs will be increasingly relied upon to help maintain grid stability. SPSs are CPSs made up of communicating nodes located at key points in the grid that automatically detect and correct power imbalances in a coordinated manner.

The SPS outlined by Ross et al. (2013) is distributed over 30 power distribution substations. The goal of an adversary in our hypothetical attack scenario is to infiltrate the nodes of the SPS and cause them to ignore load-shedding commands on demand. Ironically, it is by keeping customers *online* that the adversary hopes to maximize damage. If the adversary can cause a major source of

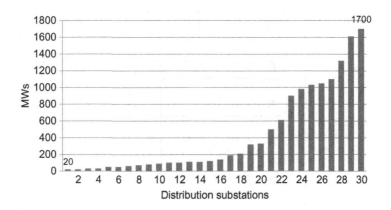

FIG. 1

The distribution of 13,030 MWs over the 30 distribution substations in Ross et al. (2013).

power generation to go offline, perhaps through a physical attack on a key generator, and then prevent mitigating load shedding from taking place, he may be able to create enough of a power imbalance to trigger a cascading blackout.

Because each node manages a different number of megawatts (MWs) (i.e., some distribution substations may be located in urban areas and others in rural areas), they are not all equally attractive to the adversary. His goal is not to obtain control over as many of the 30 nodes as possible, but to gain control over as many MWs as possible. It is assumed that the adversary, having done considerable reconnaissance, knows the average number of MWs flowing through each distribution substation.

It is also assumed that the adversary does not know how protection resources have been allocated to the nodes. His goal is to allocate more attack resources to particular nodes than the defender has allocated to protecting them. As demonstrated earlier in this chapter, based on findings from behavioral game theory, it is highly likely that the attacker will start with the proportional allocation strategy and then employ between zero and three levels of level-k reasoning to allocate his attack resources. This assumption is put to the test in the competition detailed below.

With the infrastructure detailed, we have identified the two parameters needed by the mathematical programming solver to calculate level-k resource allocations from the ILP: the number of critical sites (30, based on the 30 distribution substations) and their values (the average number of MWs they control, which we round to the nearest 10). These data were compiled from the model power grid in Ross et al. (2013) and are shown graphically in Fig. 1. The entire power system is comprised of 13,030 MWs, spread out over 30 substations, ranging from 20 to 1,700 MWs each.

Table 1 shows the CPLEX calculated allocations for each of the L0 through L5 strategies. It illustrates that as level-k reasoning increases, the trend is to devote more resources to the largest sites at the expense of the lesser sites.

5 VALIDATION

To test our approach, we conducted an experiment with human subjects, which we believe provides the most accurate validation possible (in an experimental setting) of the effectiveness of our approach.

Table 1 The L0 Through L5 Allocations of the 100 Defensive Units Across the 30 Sites

ID	1–7	8–15	16	17	18	19	20	21	22	23	24	25	26	27	28	29	30
L0	0	1	1	1	2	2	3	4	5	7	8	8	8	8	10	12	13
L1	0	0	0	0	0	3	4	5	6	8	9	9	9	9	11	13	14
L2	0	0	0	0	0	4	0	6	0	9	10	10	10	10	12	14	15
L3	0	0	0	0	2	0	0	0	0	10	11	11	11	11	13	15	16
L4	0	0	0	0	1	0	2	0	2	0	12	12	12	12	14	16	17
L5	0	0	1	2	0	2	2	2	2	0	0	13	13	13	15	17	18

As level-k reasoning increases, the trend is to devote more resources to the largest sites at the expense of the lesser sites. The experimental results confirm that the L3 strategy achieves the best balance across all sites, neither over- nor under-allocating resources.

5.1 EXPERIMENTAL DETAILS

We solicited volunteers from among the engineering and computer science majors at a private Midwestern university. Ninety-two human subjects participated. They competed for gift cards and were given a week to compile their submissions, which incentivized thoughtful participation. The participants were analytically minded, with an average ACT Math score of 29.83, which exceeds the 93rd percentile. We believe they are a fair representative sample of intelligent and motivated people in general.

The participants were provided with a prompt based on the SPS defined earlier, outlining the 30 critical sites and their values. There were asked to compete in two different competitions, one as defenders of the infrastructure and one as attackers. (The competitions were subtly different due to the nuance in the security CB game where the defenders own all of the sites by default.) They were tasked with allocating 100 indivisible units of resources across the 30 sites with the goal of winning as much utility as possible. They were clearly informed of all of the specific dynamics of the game, including the equally matched opponent, the rule of the defender winning unattacked sites by default, and the splitting of utility in case of a tie.

The attack competition served as the basis to measure the defense submissions against. The defense submissions were scored by matching each of them against all of the attack submission in head-to-head contests. The total amount of utility won over all of the head-to-head contests served as the ranking criteria.

The aggregated attack strategies the subjects submitted are shown in Fig. 2.

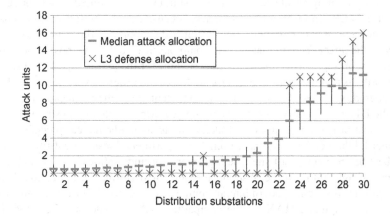

FIG. 2

The aggregate human attack strategies ($n=92$). The median attack allocations approximately align with the proportional allocation strategy (compare with Fig. 1). The vertical bars denote the middle 50% of attack allocations. The L3 defense allocations defeat approximately 75% of the attack strategies in all nine sites where resources are allocated, which includes the eight most valuable sites. This is a remarkably efficient allocation of defensive resources.

5.2 EXPERIMENTAL RESULTS

Fig. 3 illustrates the results of the defense competition, including the scores of all ninety-two human participants and our computer-generated L0 to L5 strategies. The L3 strategy performed remarkably well, finishing the best of any of the level-k strategies in 3rd place.

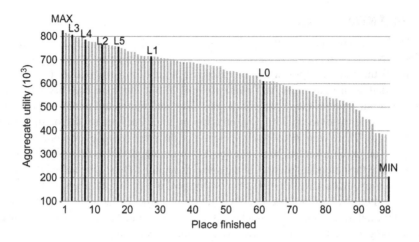

FIG. 3

The results of the defense competition ($n=98$). The L3 strategy achieved 32% more aggregate utility than L0 (i.e., the proportional allocation strategy) and only 2% less than the maximum possible utility.

The L0 strategy finished in a tie for 61st place (three of the human competitors also submitted the L0 strategy). This is an alarming result because it is the natural way that limited protection resources would be allocated across a CPS infrastructure in practice, because it is the most "commonsense" approach. However, it is not a *strategic* approach. The reason it did poorly in the competition is presumably because many of the attackers anticipated this "commonsense" defensive posture and allocated their attack resources accordingly. This result provides support for the notion that the L0 strategy is naïve.

Also noteworthy is that the first three level-k strategies did progressively better, peaking at L3, and then began to degrade with the L4 strategy. This is consistent with behavioral game theory findings, and suggests that the L4 and L5 strategies are over-optimized. Meanwhile the L3 strategy competed in the same space as most of the competitors, and likely due to its mathematical precision, out-performed most of the other L3 thinkers in the competition. The experimental results confirm that the L3 strategy achieves the best balance across all sites, neither over- nor under-allocating resources.

To help gain insight into a more absolute measure of performance (as opposed to the relative performance against this specific pool of human competitors), we used CPLEX to solve for the optimal defensive strategy (designated MAX) in the competition. MAX could only be determined *after* the competition because it requires perfect knowledge of all of the attack strategies. (No participant, including us, was privy to the specific attack strategies during the competition.) MAX represents the maximum achievable amount of utility.

We also calculated the minimum achievable amount of utility (designated MIN) by entering a strategy with zero resources allocated to all of the sites. MIN achieved some utility because it won sites where no attack resources were allocated.

In this competition, the total number of possible allocations of 100 units of resources over 30 sites equals:

$$\binom{100+30-1}{30-1} \approx 6 \times 10^{28} \tag{3}$$

Table 2 Competition Result Details for Select Defense Strategies ($n = 98$)		
ID	**Place in Competition**	**Percentage of All Possible Strategies Outperformed (%)**
L0	T61	82.2
L1	26	97.3
L2	12	99.2
L3	3	99.7
L4	7	99.5
L5	17	99.0
Human-best	1	99.8
Human-worst	98	10.5
MAX	N/A	100.0
MIN	N/A	0.0

The level-k strategies peak at L3, which finished 3rd place in the competition against 92 human competitors, and outperformed approximately 99.7% of all possible strategies against this set of human attackers.

This is an enormous number of strategies, of which our competition submissions represents only a minute subset. We believe that if all possible defensive submissions were to be entered into the competition, they would be roughly normally distributed between MIN and MAX. Based on this assumption, and by setting MIN and MAX each three standard deviations on either side of the mean, we were able to use the cumulative distribution function for the standard normal distribution to calculate an absolute percentage score for every strategy in the competition. These values are summarized in Table 2.

The L3 strategy performed better than 99.7% of all possible defensive allocations against this pool of attackers. For comparison purposes, L0 performed better than 82.2%, and the last place human finisher in the competition performed better than only 10.5%. On average, the defense strategies submitted into our competition outperformed 89% of all possible defensive allocations, which is not surprising since they were created by humans. They would naturally outperform the random (i.e., unintelligent) strategies that make up the vast majority of all possible strategies.

6 CONCLUSION

In conclusion, in this chapter we argue that any promising approach to CPS protection planning *must* take into account the strategic nature of powerful and highly motivated adversaries. Failure to do so is naïve and unlikely to be effective. To accomplish this, we leverage behavioral game theory, which is a field that has extensively studied and documented how human beings behave in strategic scenarios. To make concrete the connection between CPS security and behavioral game theory, we introduce the security CB game, which is a rigorous model of the scarce resource allocation problem inherent in defending any large-scale CPS from attack.

Furthermore, by using CPLEX in conjunction with an ILP, we automate the allocation computations. This is helpful because it provides a mathematically sound basis for resolving the trade-offs and difficult decisions inherent in the scarce protective resource allocation problem.

Most importantly, we demonstrate that it is possible to do much better than the obvious, straightforward approach of allocating scarce protection resources proportionately across a CPS infrastructure. The proportional allocation strategy is highly unlikely to effectively counteract an attacker's resource allocations, as demonstrated by the attack competition we conducted, because it is naïve and not strategic. On the other hand, the computer-generated L3 strategy performed very well in the competition, demonstrating the validity of our overall approach.

The findings from this work are intended to be the *beginning* of CPS protection planning, not the end. In other words, any serious attempt to protect a large-scale CPS will involve the strategic allocation of scarce resources as a starting point. The security CB game and the ILP provide protection planners with data to help them make the difficult decisions inherent in this task. Because of its general nature, our approach is applicable to a wide variety of CPSs situated in various contexts, but it does not provide protection planners with the implementation details they ultimately need.

Future work could explore how the allocation of man hours and money can be made concrete in a specific CPS infrastructure. For example, log auditing and network monitoring are two tasks where man hours are invariably in limited supply. How, therefore, should security personnel divide up these man hours amongst the many nodes that make up large-scale CPSs? We argue that, as opposed to a proportional allocation of man hours, they should choose the mathematically optimized L3 strategy. In another example, Neuman (2009) notes that security in CPS must be accomplished in part by making the nodes resilient because they are inherently vulnerable. But resilience is inherently a matter of degree. The security CB game and L3 reasoning can help determine where "extra resilience" should be placed and exactly how much is necessary. For example, if the budget is available to purchase additional security measures (e.g., biometric authentication, surveillance cameras, antitamper hardware, etc.) for geographically disparate sites in a CPS, our methodology can shed light on how one should choose which sites to upgrade and by how much.

DISCLAIMER

The views expressed in this paper are those of the authors and do not reflect the official policy or position of the United States Air Force, the Department of Defense, or the U.S. Government.

REFERENCES

Alpcan, T., Basar, T., 2011. A Decision and Game-Theoretic Approach. Cambridge University Press, Cambridge.

Arad, A., Rubinstein, A., 2009. Let game theory be? Calcalist 11 May. Available from: Tel Aviv University (08.02.16).

Arad, A., Rubinstein, A., 2012a. The 11–20 money request game: a level-k reasoning study. Am. Econ. Rev. 102 (7), 3561–3573.

Arad, A., Rubinstein, A., 2012b. Multi-dimensional iterative reasoning in action: the case of the Colonel Blotto game. J. Econ. Behav. Organ. 84 (2), 571–585.

Backhaus, S., Bent, R., Bono, J., Lee, R., Tracey, B., Wolpert, D., Xie, D., Yildiz, Y., 2013. Cyber-physical security: a game theory model of humans interacting over control systems. IEEE Trans. Smart Grid 4 (4), 2320–2327.

Brown, G., Carlyle, M., Salmeron, J., Wood, K., 2006. Defending critical infrastructure. Interfaces 36 (6), 530–544.

Camerer, C., 2003. Behavioral Game Theory: Experiments in Strategic Interaction. Princeton University Press, Princeton.

Cardenas, A., Amin, S., Sinopoli, B., Giani, A., Perrig, A., Sastry, S., 2009. Challenges for securing cyber physical systems. In: Workshop on Future Directions in Cyber-Physical Systems Security. Available from: Berkeley's Center for Hybrid and Embedded Software Systems (23 January 2016).

Feng, X., Lu, T., Guo, X., Liu, J., Peng, Y., Gao, Y., 2013. Security analysis on cyber-physical systems using attack tree. In: Ninth International Conference on Intelligent Information Hiding and Multimedia Signal Processing, pp. 429–432. Available from: IEEE Portal: Xplore Digital Library (31.01.16).

Fielder, A., Panaousis, E., Malacaria, P., Hankin, C., Smeraldi, F., 2014. Game theory meets information security management. In: Proceedings of the Twenty-Ninth IFIP TC 11 International Conference, pp. 15–29. Available from: Springer Link (31.01.16).

Fletcher, K.K., Liu, X., 2011. Security requirements analysis, specification, prioritization, and policy development in cyber-physical systems. In: Fifth International Conference on Secure Software Integration and Reliability Improvement Companion, pp. 106–113. Available from: IEEE Portal: Xplore Digital Library (31.01.16).

Gross, O., Wagner, R., 1950. A continuous Colonel Blotto game. Available from: RAND Corporation (27.08.15).

He, F., Zhuang, J., Rao, N., 2012. Game-theoretic analysis of attack and defense in cyber-physical network infrastructures. In: Proceedings of the Industrial and Systems Engineering Research Conference, pp. 1–8. Available from: University at Buffalo (31.01.16).

Jazdi, N., 2014. Cyber physical systems in the context of Industry 4.0. In: 2014 IEEE International Conference on Automation, Quality and Testing, Robotics, pp. 1–4. Available from: IEEE Portal: Xplore Digital Library (27.01.16).

Lloyds, 2015. Business blackout: the insurance implications of a cyber attack on the U.S. power grid. Available from: Lloyds (27.08.15).

Ma, C.Y.T., Rao, N.S.V., Yau, D.K.Y., 2011. A game theoretic study of attack and defense in cyber-physical systems'. In: IEEE Conference on Computer Communications Workshops, pp. 708–713. Available from: IEEE Portal: Xplore Digital Library (23.01.16).

Neuman, C., 2009. Challenges in security for cyber-physical systems. In: DHS Workshop on Future Directions in Cyber-Physical Systems Security, pp. 1–4. Available from: Rutgers' Center for Information Management, Integration, and Connectivity (31.01.16).

Ross, K.J., Hopkinson, K.M., Pachter, M., 2013. Using a distributed agent-based communication enabled special protection system to enhance smart grid security. IEEE Trans. Smart Grid 4 (2), 1216–1224.

Tambe, M., 2012. Security and Game Theory: Algorithms, Deployed Systems, Lessons Learned. Cambridge University Press, New York.

U.S. Department of Energy, 2004. Final report on the August 14, 2003 blackout in the United States and Canada: causes and recommendations. Available from: U.S. Department of Energy (27.08.15).

U.S. Office of Homeland Security, 2002. National strategy for homeland security. Available from: Office of Homeland Security (27.08.15).

Vigo, R., Bruni, A., Yuksel, E., 2013. Security games for cyber-physical systems. In: Proceedings of the Eighteenth Nordic Conference, pp. 17–32. Available from: Springer Link (31.01.16).

Zhou, K., Liu, T., Zhou, L., 2015. Industry 4.0: towards future industrial opportunities and challenges. In: Twelfth International Conference on Fuzzy Systems and Knowledge Discovery, pp. 2147–2151. Available from: IEEE Portal: Xplore Digital Library (23.01.16).

ABOUT THE AUTHORS

Seth T. Hamman received the B.A. degree in Religion from Duke University, Durham, NC, in 2002 and the M.S. degree in Computer Science from Yale University, New Haven, CT, in 2011. He is currently pursuing a Ph.D. in Computer Science at the Air Force Institute of Technology (AFIT), Wright-

Patterson AFB, OH. He is an Assistant Professor of Computer Science in the School of Engineering and Computer Science at Cedarville University, in Cedarville, OH. His research interests are in the areas of cybersecurity and education.

Kenneth M. Hopkinson received the B.S. degree from Rensselaer Polytechnic Institute, Troy, NY, in 1997 and the M.S. and Ph.D. degrees from Cornell University, Ithaca, NY, in 2002 and 2004 respectively, all in Computer Science. He is a Professor of Computer Science at the Air Force Institute of Technology (AFIT), Wright-Patterson AFB, OH. His research interests are in the areas of simulation, networking, and distributed systems.

Lindsey A. McCarty received the B.S. degree in Mathematics from Cedarville University, Cedarville, OH, in 2006, and the M.S. and Ph.D. degrees from the University of Michigan, Ann Arbor, MI, in 2008 and 2012, respectively, in Applied and Interdisciplinary Mathematics. She is an Assistant Professor of Mathematics at Cedarville University. Her research interests include scheduling, large-scale optimization, and interdisciplinary math.

PROBABILISTIC GRAPHICAL MODELING OF DISTRIBUTED CYBER-PHYSICAL SYSTEMS

18

S. Sarkar*, Z. Jiang*, A. Akintayo*, S. Krishnamurthy[†], A. Tewari[‡]

Iowa State University, Ames, IA, United States[*] Sony Computer Entertainment America, San Diego, CA, United States[†]
ExxonMobil Research & Engineering Company, Annadale, NJ, United States[‡]

1 INTRODUCTION

Cyber-physical systems (CPSs) combine entities in the physical world with the computing and communication capabilities for monitoring and control (Fig. 1). With the advent ubiquitous sensing, advanced computation and strong connectivity, modern CPSs such as transportation networks (Work and Bayen, 2008), power grids (Sridhar et al., 2012), power plants (Kesler, 2011) and integrated buildings (Kleissl and Agarwal, 2010) have shown tremendous potential in terms of increased efficiency, robustness and resilience—a fact that is reflected in the great interest in CPS research shown by the U.S. government (Kramer, 2014), the EU (Fitzgerald et al., 2013), other countries around the world (NIST, 2013) and industry in general (Bradley et al., 2013). However, to realize such potential, effective and efficient modeling and analysis tools have to be developed for CPSs that are scalable, robust, flexible and adaptive. In general, state-of-the-art CPS modeling tools are heavily dependent on domain knowledge and first principles and hence lack in scalability, robustness and flexibility due to large variations in system configuration and complex subsystem interactions (Krishnamurthy et al., 2014). Data-driven approaches can potentially alleviate some of these critical issues in CPS modeling (Freeman et al., 2013) with various applications such as detecting faults and cyber-attacks as shown for power systems in (Kosut et al., 2010). However, a major challenge for data-driven approaches is to assimilate heterogeneous data types, such as discrete, event-driven data generated by the cyber space and continuous, temporal data generated by the physical space (Liu et al., 2016).

In this context, this chapter discusses some recent developments in the data-driven modeling paradigm for distributed complex CPSs using machine learning techniques, such as probabilistic graphical models (PGMs) (e.g., symbolic dynamic filtering and Bayesian networks). The potential applications of PGMs can be monitoring and diagnostics, event propagation and impact analysis, cross-application information exchange, supervisory control and seamless human-machine interaction. We will present various key aspects of this modeling approach, namely data preparation, model learning and decision-making with two energy systems application—integrated buildings and wind energy farms.

The chapter is organized in four sections including the present one. Section 2 gives a brief introduction to PGMs, a class of machine learning models extremely relevant for CPS modeling. Among

Cyber-Physical Systems. http://dx.doi.org/10.1016/B978-0-12-803801-7.00018-3

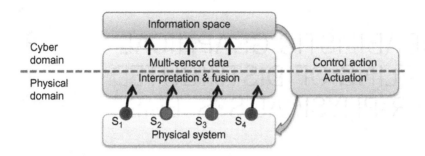

FIG. 1

CPS: combing physical space with information space via computing, communication and control.

Sarkar, S., 2011. Autonomous Perception and Decision Making in Cyber-Physical Systems. PhD thesis, The Pennsylvania State University Graduate School, State College, PA 16802.

various PGM techniques, this section focuses on Bayesian networks and a special type of Markov network called Symbolic Dynamic Filtering as Section 3 presents different CPS applications of these tools. Section 3 is distributed in three subsections providing three CPS modeling use cases—spatial interconnectivity modeling, multiscale temporal dynamics modeling and finally spatiotemporal causality modeling. The chapter is summarized and concluded with future research directions in Section 4.

2 PROBABILISTIC GRAPHICAL MODELS

PGMs have been widely used in various machine learning applications as they provide a succinct mechanism to model the joint distribution of statistically dependent random variables. The main benefit of PGMs is the ability to represent the joint distribution as a graph, which allows one to draw inferences about the underlying system without even knowing the parametric form of the model. Typically, a random variable is denoted as a node in the graph, while statistical dependencies are represented as edges (undirected or directed) between nodes. In this section, we briefly introduce two types of PGMs, namely Bayesian networks and a special type of Markov network called Symbolic Dynamic Filtering.

2.1 BAYESIAN NETWORK

Bayesian network is a type of PGM that allows one to capture causal information (cause and effect) using directed edges (Kohler and Friedman, 2009; Gershman and Blei, 2012). Each node defines a conditional distribution of itself, given the parent nodes. The directionality of the edges is such that no directed cycles are included in the overall graph. Hence, Bayesian networks are considered as directed acyclic graphs (DAGs). The overall joint distribution of the network is computed as a product of the conditional distribution defined by every node in the framework.

Learning and inference are the two main problems associated with Bayesian networks. The former involves learning the structure (the DAG) and the parameters of the conditional probability distributions that best explain the given (training) data. Bayesian networks support both supervised (i.e., with labeled training data) and unsupervised (i.e., without target labels) learning. The second problem of inference pertains to finding probabilistic answers to user defined queries. For example, a user may seek the joint distribution of a subset of random variables given the observed values of another disjoint subset of random variables. Since, Bayesian networks only encode node-based conditional probabilities, finding answers to such queries is not straightforward. However, efficient algorithms exist that allow one to find the exact answer to an arbitrary query using a secondary structure (such as junction tree) and a message-passing architecture (Kohler and Friedman, 2009). Anomaly detection and root-cause isolation in CPSs can both be interpreted as inference problems.

In the CPS community, graphical security models are already quite popular for visually representing and analyzing vulnerabilities in a system. Threat trees and Bayesian networks are two of the well-known graphical formalisms for security modeling (Kordy et al., 2013; Laboratory, 2013). Bayesian networks are versatile as they can be constructed from attack models and domain knowledge, or learned from data. On the other hand, attack graphs model how multiple vulnerabilities can be combined to result in an attack. Bayesian attack graphs combine attack graphs with computational procedures of Bayesian networks (Liu and Man, 2005). Wang et al. (2008) proposed a probabilistic security metric for nodes in an attack graph and provided an algorithm for computing this metric in an attack graph. Frigault and Wang (2008) provided a method to assign conditional probability to nodes in a Bayesian attack graph based on common vulnerability scoring system (CVSS) scores and used that to calculate security metrics. They later extended their work to dynamic Bayesian networks to account for the evolving nature of vulnerabilities and availabilities of software patches (Frigault et al., 2008). Likewise, Houmb et al. quantified security risk level from CVSS estimates of frequency and impact using Bayesian networks (Houmb et al., 2010). A Bayesian network modeling approach for separating different sources of uncertainty, such as uncertainty in attacker actions and attack success, for real-time security analysis is described in Xie et al. (2010). Feng and Xie (2012) provided an algorithm for merging expert knowledge and information stored in databases into a single Bayesian network. PGMs have also been successfully used for root-cause analysis in different domains. For instance, Bayesian networks have been used for fault isolation in electrical power system (Choi et al., 2011), automotive systems (Jansson, 2004), telecommunication networks (Velasco, 2012) and manufacturing processes (Velasco, 2012). However, the majority of the earlier work mentioned above deals with purely cyber security issues or only physical fault scenarios. It also mostly relies on labeled training data for nominal and all possible anomalous conditions. In Section 3, we will present novel ideas for using Bayesian network models for anomaly detection and root-cause analysis in CPSs with unlabeled data.

2.2 SYMBOLIC DYNAMIC FILTERING (SDF): A SPECIAL TYPE OF MARKOV NETWORK

Markov models are a class of statistical models that have a wide range of applications, such as speech recognition and natural language processing (Cardie and Mooney, 2006; Daw and Finney, 2003). These models are shown to be efficient in learning probabilistic dependencies among random variables in both directed and undirected fashion. Hidden Markov models (HMMs) have been particularly useful for learning temporal dynamics of an underlying process (Rabiner, 1994). Several modifications for HMMs have been proposed, such as IHMM (Beal et al., 2002) integrated several parameters to three hyper-parameters to model countably infinite hidden state sequence and IHHMM (Heller et al., 2009) extended it to infinite number of hierarchical levels and (Wakabayashi and Miura, 2012) applied a forward-backward algorithm to reduce model complexity through the order of operations.

However, Markov models with hidden states often rely on iterative learning algorithms that may be computationally intensive. To alleviate such issues, symbolic dynamic filtering (SDF) was recently proposed (Ray, 2004; Rajagopalan and Ray, 2006) based on the concepts of symbolic dynamics and probabilistic finite state automata (PFSA). Several improvements related to coarse graining of continuous variables (Sarkar et al., 2013), state splitting and merging techniques for PFSA (Mukherjee and Ray, 2014), efficient inference algorithms (Sarkar et al., 2009) and hierarchical model learning (Akintayo and Sarkar, 2015) have been proposed over the last decade within the SDF framework. SDF-based tools have been successfully used for modeling/analysis of various complex CPSs such as nuclear power plants (Jin et al., 2011), coal-gasification systems (Chakraborty et al., 2008; Chakraborty et al., 2012) and gas turbine engines (Sarkar et al., 2008). Therefore, SDF is suitable for capturing subsystems' behaviors, as well as their interactions in the form of spatiotemporal features (Jiang and Sarkar, 2015). Real-time inference using SDF has also been shown to be more efficient compared to other competitive feature extraction techniques (Rao et al., 2009).

The core steps of SDF for extracting useful features from dynamical systems as delineated in literature (Sarkar et al., 2012) (shown in Fig. 2) are summarized as data preprocessing, symbol sequence generation via partitioning, encoding the symbol sequence with a PFSA and finally extracting the salient features from the PFSA for decision-making. Other similar feature extraction methods

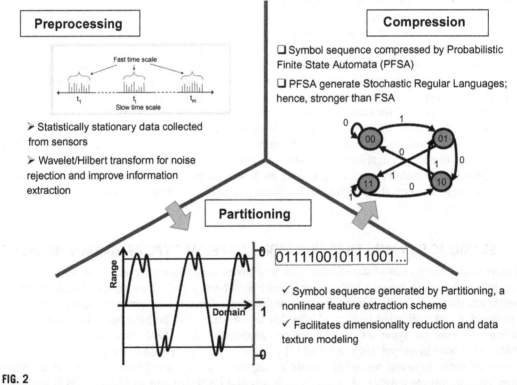

FIG. 2

Basics of the symbolic dynamic filtering (SDF) scheme.

include PCA (Jollife, 1989), multidimensional scaling (MDS) (Cox and Cox, 1994) and locally linear embedding (LLE) (Saul and Rowels, 2000). It is therefore a useful tool for online pattern discovery in addition to its suitability for offline parametric identification.

As mentioned above, the key aspect of symbolization involves partitioning that discretizes continuous data with a trade-off between information loss and model complexity. Note that this process enables a uniform abstraction for both continuous and discrete data obtained from a CPS. Furthermore, this resolves critical scaling issues and improves robustness to noise and spurious disturbances. Uniform partitioning, maximum entropy partitioning (Chau, 2001), SFNNP (Buhl and Kennel, 2005), as well as partitioning to optimize class separability (Sarkar et al., 2012), are among some of the most useful partitioning methods. In the recent study, a multivariate discretization scheme called the maximally bijective discretization (MBD) (Sarkar et al., 2013) was proposed for modeling complex CPSs via maximally preserving the input-output data relationship for a dynamical system.

Data discretization is followed by the symbolization process and construction of PFSA models, namely the D-Markov machines. Symbol sequences generated from observation using partitioning are modeled with a D-Markov machine to capture the temporal characteristics embedded in the symbol sequence. An extension called xD-Markov machine was also proposed (Sarkar et al., 2014) to capture the causal dependencies between multiple symbol sequences. Such models were shown to be sufficient to capture a fairly general class of causal dependencies while correlation-based analysis fails to perform the task (Chattopadhyay, 2014). Details of construction of both machine types may be found in Sarkar et al. (2014).

In the following section, we will present some new developments for SDF-based CPS modeling schemes, both for identification of hierarchical temporal dynamics (Akintayo and Sarkar, 2015) and as spatiotemporal cause-effect relationships (Jiang and Sarkar, 2015).

3 PROBABILISTIC GRAPHICAL MODELING APPLIED TO CPSs

Many complexities in CPS modeling arise from the spatial distributedness, multiscale nature of the temporal dynamics as well as complex spatiotemporal interactions. This section presents use cases addressing all of these three aspects and demonstrating efficacies of PGMs for various CPS applications.

3.1 BAYESIAN NETWORK MODELING OF SPATIALLY DISTRIBUTED CPS

This subsection presents a CPS use case based on large building systems, where detection of an anomaly and root-cause analysis are extremely difficult due to the spatial distributedness and complex subsystem interactions. For example, a large commercial building can be thought of as a typical distributed CPS involving various physical subsystems (e.g., heating, ventilation, and air conditioning (HVAC) systems, lighting and security systems) with computation, communication and control enabling optimal operation. Detection of cyber-attacks, physical faults and root-cause analysis are critical for such a building system for improving energy efficiency, maintaining safety and reducing operation cost.

The building system considered here is instrumented with networked sensors and actuators to control the HVAC system. The sensors and actuators exchange information with the building automation system through BacNet, a data communication protocol (ASHRAE, 2014; Roth et al., 2005). The networked building system described above is vulnerable to different types of cyber-attacks, such as a

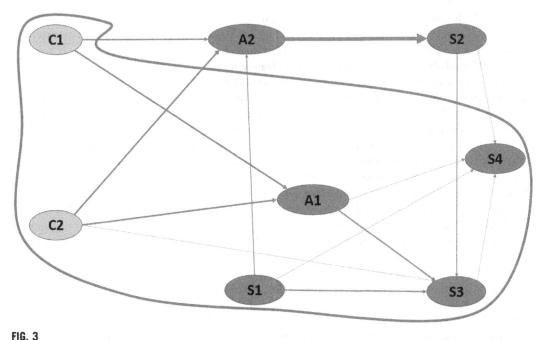

FIG. 3

Bayesian network for building HVAC system. The thickness of the arrows represents the strength of influence from the parent node to the child node and the *(green in color version)* encircled region shows the Markov blanket of node A2.

denial of service (DoS) attack. The physical faults in a building HVAC system can occur in the form of malfunctioning actuators, e.g., leaky water valves and stuck air dampers. In this work, we primarily focus on data integrity attacks that are launched from BacNet. As shown in Fig. 3, data are collected for two cyber variables (C1 and C2), two actuator variables (A1 and A2) for controlling heating and air flow to a building zone respectively and four sensor variables (S1, S2, S3, and S4) for monitoring zone parameters such as temperature and relative humidity. C1 is a BacNet issued identifier that uniquely identifies a sensor or actuator in the building system within the BacNet network and C2 identifies the BacNet operation performed on the actuator or sensor.

3.1.1 Bayesian network-based learning and inference

The cyber data collection is event-driven and physical data are collected every second and recorded in the building automation system logs. Details on synchronizing cyber and physical data can be found in Krishnamurthy et al. (2014) and Sarkar (2011). A Bayesian network was learnt (both structure and parameters) using nominal operation data via off-the-shelf algorithms. The resulting Bayesian network characterizes the normal operation and, hence, is capable of detecting anomalies as low probability events and with the detected anomaly, the root-cause isolation can be conducted correspondingly. Fig. 3 shows a graphical representation of the learnt model where the thickness of an edge between a pair of nodes signifies the (directed) strength of influence between the nodes.

For inference, we define anomaly score (AS) with respect to random variable sets $X(t)$ (target set) and $Y(t)$ (evidence set), as given in Eq. (1), which quantifies the degree of deviation of an observed state from its most likely state. We use this metric for the purpose of anomaly detection and root cause isolation as described in the sequel:

$$AS(X(t), Y(t)) = -\log \frac{P_{X|Y(t)}(X(t))}{\max\left(P_{X|Y(t)}\right)} \qquad (1)$$

In Eq. (1), $P_{X|Y}(X(t))$ is the posterior probability of the target variable state actually observed at an instant t, and $\max\left(P_{X|Y(t)}\right)$ is the probability of the most probable target state, given the evidence set $Y(t)$. Therefore the lower the probability of the observed target variable state, the higher the value of the anomaly score. The posterior distribution is obtained using the Junction Tree algorithm (Kohler and Friedman, 2009), which enables efficient computation of arbitrary joint posteriors in Bayesian networks. Note that the posterior distribution in Eq. (1) is a conditional distribution. Given a target node, our scalable solution limits the search for the root-cause to the nodes present in the Markov blanket of the target node where the Markov blanket of a node is defined as a set containing its parent nodes, children nodes and all the other parent nodes of its children. For example, in Fig. 3 the (green in color version) encircled region shows the Markov blanket of A2.

Upon anomaly detection, the root-cause isolation step aims at ranking the candidate nodes or "suspects" by how likely they are to be the cause of the detected anomaly and then designating the node that is the most likely as the root-cause of the anomaly. We define a metric called root-cause potential (RCP) to generate this rank order. We denote the state of the target T at an instant t by $T(t)$. $RCP_{T(t)}(N_i)$ quantifies the likelihood that node N_i is the root cause of this anomaly. Then the node with the highest value of RCP among the candidate nodes is treated as the most likely root-cause of the detected anomaly. Three algorithms for root-cause isolation are formulated based on the choice evidence set. While detail algorithms can be found in Krishnamurthy et al. (2014), they are briefly illustrated in Fig. 4. These algorithms

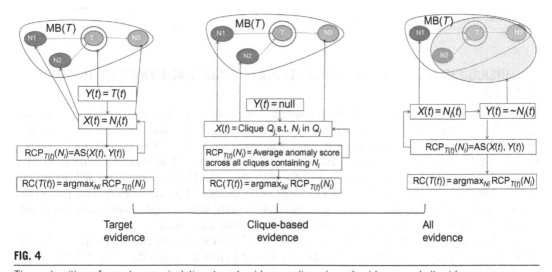

FIG. 4

Three algorithms for root-cause isolation: target evidence, clique-based evidence and all evidence.

essentially define the RCP metric in different ways based on a spectrum of evidence set. On one end only the target node (used for anomaly detection) is used as evidence, on the other end of the spectrum lies the case with all nodes (except the node under investigation) used as evidence. In the middle, a clique evidence-based algorithm is formulated that achieves both high performance as well as computational advantages.

3.1.2 Results

The results in Table 1 show that the target evidence algorithm has low computational complexity and that its accuracy is also relatively poor. The all-evidence algorithm can be more accurate, although there may be a significant chance of over-fitting. The clique-based algorithm has the lowest computational overhead and the best root-cause accuracy for all the anomalies as it exploits a natural partitioning of the graphical model using cliques and examines individual cliques in an isolated manner to discover local anomalies. Interested readers are encouraged to go to Krishnamurthy et al. (2014) for further details.

Table 1 Comparison of Root-Cause Isolation Algorithms in Terms of Root-Cause Accuracy and Computational Speed

RCP Algorithm	Root-Cause Accuracy				Computational Time per Test Record (ms)
	Cyber A1 (%)	Cyber A2 (%)	Physical A1 (%)	Physical A2 (%)	
Target evidence	2	0	49	84	90
Clique evidence	67.3	94	77	99	84
All evidence	56	80	77	99	379

A1, *Actuator 1*; A2, *Actuator 2*; ms, *milliseconds.*

3.2 HIERARCHICAL LEARNING OF MULTISCALE TEMPORAL DYNAMICS IN CPS WITH SDF

Complex CPSs often show many different dynamic behaviors due to change in operating conditions, system health characteristics and environmental disturbances. While most of these characteristics change over different slow-time scale epochs, it is important to be able to capture such dynamic behaviors from fast-time scale time-series data. More importantly, it may not be feasible to have labeled datasets for all possible conditions of a real-life CPS. Therefore unsupervised algorithms are required that may not even have the knowledge of a number of possible behaviors a priori (i.e., nonparametric). In this context, we present a hierarchical SDF scheme in this subsection that is able to learn multiscale temporal dynamics from a streaming time-series data.

As obtaining ground truth can be tricky for real CPSs, a simulated chaotic nonlinear duffing system (given by Eq. (2)) with known behaviors is used to illustrate the efficacies of the proposed algorithm:

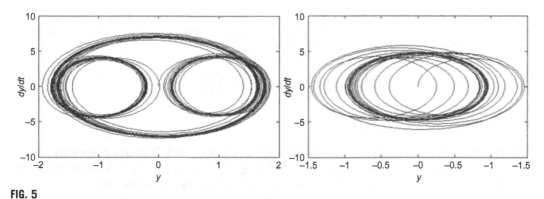

FIG. 5

Phase-space plot of duffing system with: $\beta=0$ *(left)* and $\beta=0.40$ *(right)*.

$$\frac{d^2x(t)}{dt^2} + \beta\frac{dx}{dt} + \alpha_1 x(t) + \lambda x(t)^3 = A\cos(\omega t) \tag{2}$$

where $A=22.0$ is the amplitude of the forcing function, $\omega=5.0\,\text{rad/s}$ is its frequency, $\alpha_1=1.0$ is the excitation harmonics and spring stiffness, $\lambda=1.0$. However, the dynamics of the system are known to change by varying the damping factor, β, which shows the behavior before bifurcation with values $<\sim0.28$ and after bifurcation with values $>\sim0.28$ as shown in Fig. 5 (Rao et al., 2009). For the validation results, two types of dynamics are considered—one before bifurcation ($\beta=0$) and one after ($\beta=0.40$).

3.2.1 SDF based inference

In order to learn a hierarchical SDF, we first introduce a classification process within the SDF framework. Let K classes model K different quasi-stationary behaviors identified in the data and are denoted by C^1, C^2, ..., C^K over the same alphabet, Σ. Each class, C^i is modeled by an ergodic (equivalently irreducible) PFSA $G_i = \left(Q^i, \Sigma, \delta^i, \Pi^i\right)$, where $i=1, 2,..., K$. Also for each class, C^i let a symbol string $S^i \triangleq s_1^i s_2^i,...,s_N^i$ be already identified from the streaming data. The state transition function and the set of states, Q of the D-Markov machine are fixed by choosing an appropriate depth, D and let the (probability) matrix be denoted by $\Pi_{mn}^i \geq 0$ and $\sum_{n=1}^{|\Sigma|} \Pi_{mn}^i = 1$. The a priori probability density function, $f_{\Pi_m^i | S^i}$ of the random row-vector Π_m^i, conditioned on a symbol string, S^i follows a Dirichlet distribution (Ferguson, 1973; Sethuraman, 1994) as described below:

$$f_{\Pi_m^i | S^i}\left(\theta_m^i | S^i\right) = \frac{1}{B\left(\alpha_m^i\right)} \prod_{n=1}^{|\Sigma|} \left(\theta_{mn}^i\right)^{\alpha_{mn}^i - 1} \tag{3}$$

where θ_m^i is a realization of the random vector, Π_m^i and $B\left(\alpha_m^i\right)$ is the normalizing constant. Furthermore,

$$\alpha_{mn}^i = N_{mn}^i + 1 \tag{4}$$

where N_m^i is the number of times the symbol σ_n in S^i is emanated from the state q_m.

By the Markov property of the PFSA G^i, the $(1 \times |\Sigma|)$ row-vectors, $\{\Pi_m^i\}$, $m = 1, \ldots, |\Sigma|$ are statistically independent of each other. Therefore the a priori joint density $f_{\Pi^i|S^i}$ for the probability morph matrix Π^i conditioned on the symbol string S^i, is given as:

$$f_{\Pi^i|S^i}\left(\theta^i|S^i\right) = \prod_{m=1}^{|Q|}\left(N_m^i + |\Sigma| - 1\right)! \prod_{n=1}^{|\Sigma|} \frac{\left(\theta_m^i\right)^{N_{mn}^i}}{\left(N_{mn}^i\right)!} \tag{5}$$

where $\theta^i = \left[\left(\theta_1^i\right)^T \left(\theta_2^i\right)^T \ldots \left(\theta_{|Q|}^i\right)^T\right] \in [0, 1]^{|Q| \times |\Sigma|}$

Then, let a new slow time epoch be constituted by a symbol string $\hat{S}$. Now, the probability that the symbol string belonging to a particular class of PFSA $(Q, \Sigma, \delta, \Pi^i)$ is a product of independent multinomial distribution (Wilks, 1963) given that the exact morph matrix Π^i is known:

$$Pr\left(\hat{S}|Q, \delta, \Pi^i\right) = \prod_{m=1}^{|Q|}\left(\hat{N}_m\right)! \prod_{n=1}^{|\Sigma|} \frac{\left(\Pi_{mn}^i\right)^{\hat{N}_{mn}}}{\left(\hat{N}_{mn}\right)!} \tag{6}$$
$$\triangleq Pr\left(\hat{S}|\Pi^i\right) \text{ as } Q \text{ and } \delta \text{ are kept invariant}$$

$\hat{N}_{mn}^i$ is the number of times the symbol σ_n is emanated from the state $q_m \in Q$ in the symbol string $\hat{S}$ and $\hat{N}_m \triangleq \sum_{n=1}^{|\Sigma|} \hat{N}_{mn}$ and is similar to N_{mn}^i in defined for S^i. Combining Eqs. (3) and (5) we obtain the conditional probability of the test data behavior given model in training data:

$$Pr\left(\hat{S}|S^i\right) = \prod_{m=1}^{|Q|} \frac{\left(\hat{N}_m\right)!\left(N_m^i + |\Sigma| - 1\right)!}{\left(\hat{N}_m + N_m^i + |\Sigma| - 1\right)!} \times \prod_{n=1}^{|\Sigma|} \frac{\left(\hat{N}_{mn} + N_{mn}^i\right)!}{\left(\hat{N}_{mn}\right)!\left(N_{mn}^i\right)!} \tag{7}$$

where Stirling's approximation formula, $\log(n!) = n \log n - n$ (Pathria, 1996) is helpful by considering $\log\left(Pr\left(\hat{S}|S^i\right)\right)$ with statistically many samples for both N^i and $\hat{N}$. Detail derivations can be found in Akintayo and Sarkar (2015), Sarkar et al. (2009), and Sarkar et al. (2011).

3.2.2 Hierarchical SDF learning

The idea of online learning of various dynamical characteristics from time-series data begins with learning a model with the first slow time epoch of data. Following that, the process involves performing inference on a new slow time epoch data $\hat{S}_{\tau_j}$ (at current epoch τ_j) as described above either to classify it as one of the existing classes C^i (in the set of existing classes, $C = \{C^1, C^2, \ldots, C^K\}$) or creating a new class for it. Therefore at the lower level, we have individual PFSA models for various quasi-stationary characteristics whereas at the higher level, we can learn how the system makes transition from one quasi-stationary characteristics to another. Hence, this learning scheme is termed as a hierarchical SDF framework. From inference computation, the likelihood function $Pr\left(\hat{S}_{\tau_j}|S^i\right)$ can be written as $Pr\left(\hat{S}_{\tau_j}|C^i\right)$ for class C^i. Then, the posterior probability for each class given the models learnt already is expressed as:

$$Pr\left(C^i|\hat{S}_{\tau_j}\right) \propto Pr\left(\hat{S}_{\tau_j}|S^i\right) \tag{8}$$

where first test epoch is initially assumed to be in first class, C^1.

Hierarchical SDF formulation begins with the possibility of a new class creation with the arrival of a new data epoch τ_j. This is based on the idea of the Chinese restaurant process (CRP), which is an induced distribution over partitions (Aldous, 1985). A weighted likelihood function is evaluated using the inference formulation described earlier for each class in C along with the probability of transitioning to a new class C^{K+1}. Weighting of data log-likelihoods derived from previous slow time epoch was proposed in Akintayo and Sarkar (2015). This process is summarized in Eqs. (9) and (10) with the CRP hyper-parameter γ upon selecting suitable parameter ε as:

$$\gamma = \frac{\varepsilon}{\left[\sum_{i=1}^{K} Pr\left(\hat{S}_{\tau_j}|S^i\right)\right] + \varepsilon} \tag{9}$$

γ is then used for making the class transition decision after normalizing as in Eq. (10):

$$Pr_\gamma\left(C^{K+1}|\hat{S}_{\tau_j}\right) = \gamma \ \text{ and } \ Pr_\gamma\left(C^i|\hat{S}_{\tau_j}\right) = \frac{1-\gamma}{\sum_{i=1}^{K} Pr\left(\hat{S}_{\tau_j}|S^i\right)} \times Pr\left(\hat{S}_{\tau_j}|S^i\right) \tag{10}$$

where $Pr_\gamma\left(C^i|\hat{S}_{\tau_j}\right)$ is the probability with which $\hat{S}_{\tau_j}$ is assigned to class C^i (including the new class).

With $K+1$ potentially existing classes, a stickiness factor is introduced to reduce rapidly generating many more new classes. Note that when a time-series segment at slow time epoch τ_{j-1} belongs to a certain class, $C^k \in C$, the probability that a new data epoch τ_j will belong to the last seen class, C^k, may be higher compared to probabilities for other classes, as real-life systems typically may not change operating point or parametric condition at every slow time epoch. This consideration is implemented by skewing the probability of the last seen class C^k as follows:

$$Pr_\gamma\left(C^k|\hat{S}_{\tau_j}\right) = \max\left\{\kappa, Pr_\gamma\left(C^k|\hat{S}_{\tau_j}\right)\right\} \tag{11}$$

where $0 < \kappa < 1$ is the stickiness factor. Finally, the class assignment distribution is renormalized again before using for decision-making.

The online algorithm for learning Hierarchical SDF is summarized below:

ALGORITHM: ONLINE LEARNING USING HIERARCHICAL SDF

Input Parameters: Stickiness parameter κ, and CRP parameter ε
Data Input: Symbol sequence, $\hat{S}_{\tau_1}$ for slow time epochs $\tau_1, \tau_2, \ldots$
Initialize: $C = \{C^1\}$
Initialize: All $N_{mn}^i = 0$ (m, n chosen based on $|Q|$ and $|\Sigma|$)
Compute N_{mn}^1 using $\hat{S}_{\tau_1}$
FORALL $\tau_2, \tau_3, \ldots$ DO
    Compute $\hat{N}_{mn}$ using $\hat{S}_{\tau_j}$
    Evaluate $Pr_\gamma\left(C^i|\hat{S}_{\tau_j}\right)$ using Eqs. (10) and (11)
    $\forall C^i \in C = \{C^1, C^2, \ldots, C^K\}$ and C^{K+1}
    Assign $\hat{S}_{\tau_j}$ to class C^i according to $Pr_\gamma\left(C^i|\hat{S}_{\tau_j}\right)$
    IF a new class is created
        Update C as $\{C^1, C^2, \ldots, C^K, C^{K+1}\}$
        Compute N_{mn}^{K+1} using $\hat{S}_{\tau_j}$
    ENDIF
ENDFOR

The algorithm described identifies different classes of quasi-stationary characteristics in an online fashion and online hierarchical SDF can represent those characteristics by different Π matrices. To eliminate the redundant classes created by the online algorithm, a periodic step (Akintayo and Sarkar, 2015) is included to merge different PFSAs that are close enough based on a metric (Mukherjee and Ray, 2014) defined below.

Definition {Distance Metric for PFSA} Let $P_1 = (Q_1, \Sigma, \delta_1, \Pi_1)$ and $P_2 = (Q_2, \Sigma, \delta_2, \Pi_2)$ be two PFSA with a common alphabet. Let $P_1(\Sigma^j)$ and $P_2(\Sigma^j)$ be the steady-probability vectors of generating words of length j from the PFSA P_1 and P_2, respectively, i.e., $P_1(\Sigma^j) \triangleq \Delta[P(w)]_{w \in \Sigma^j}$ for P_1 and $P_2(\Sigma^j) \triangleq \Delta[P(w)]_{w \in \Sigma^j}$ for P_2. Then, the metric for the distance between the PFSAs P_1 and P_2 is defined as:

$$\varphi(P_1, P_2) = \lim_{n \to \infty} \sum_{j=1}^{n} \frac{P_1(\Sigma^j) - P_2(\Sigma^j)_{l1}}{2^{j+1}} \tag{11}$$

where the norm $\| *_{l1} \|$ indicates the sum of absolute values of the elements in the vector $*$.

Thus, the revision step can merge two online hierarchical SDF matrices P_1 and P_2 when $(P_1, P_2) < C$, where $C > 0$.

3.2.3 Results

To demonstrate the algorithm efficacies, a 400,000 data points long time-series was randomly generated with two types of dynamics (one before bifurcation and one after, i.e., $K = 2$). The result of online hierarchical SDF (in red) is superimposed on the ground truth or actual data (in blue). Every slow time epoch is taken to be 1000 data points long (i.e., $N = 1000$) and there are 400 epochs for the entire data set. Results are shown in Fig. 6 with both noiseless and SNR $= 1$ scenarios. Similar accuracy can be obtained under both conditions after carefully choosing hyper-parameters, κ and γ. See Akintayo and Sarkar (2015) for further details on the hyper-parameters.

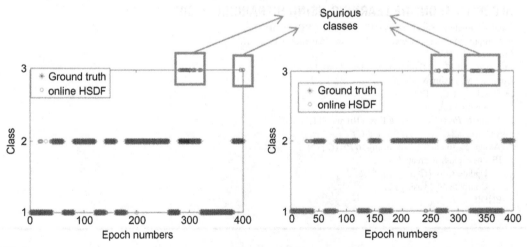

FIG. 6

Online hierarchical SDF learning results for noiseless case *(left)* and with SNR $= 1$ *(right)*.

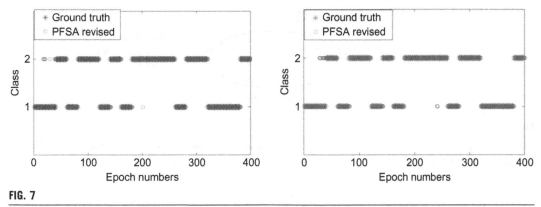

FIG. 7

Offline revision for hierarchical SDF learning scheme: noiseless case *(left)* and with SNR = 1 *(right)*.

Finally, with the experimental selection of $C = \dfrac{1}{2K}$ for K number of classes, an offline revision is performed on the online result for further reduction of the modeling error. The result obtained enhances the learning accuracy as shown in Fig. 7.

The results show significant promise in terms of automatically identifying and classifying various a priori unknown dynamics from time-series data. Furthermore, this technique is computationally significantly less intensive than competing sampling based methods, quantitatively more than three times faster compared to the competitive method described in Fox et al. (2011). Therefore investigations are currently being carried out to apply this method for real CPSs.

3.3 SPATIOTEMPORAL PREDICTIVE MODELING OF DISTRIBUTED CPSs

The two previous subsections have discussed spatial and temporal aspects of CPSs respectively. In this subsection, we extend the applications of PGM to model spatiotemporal interactions within a CPS, using a large wind farm (consisting of many wind turbines) as an example application. Wind turbines in a wind farm have complex spatiotemporal interconnections and have a significant impact on wind power prediction. This subsection presents the concept of spatiotemporal pattern network (STPN) (an extension of the SDF scheme) for capturing both temporal and spatial characteristics of wind turbines and their complex interactions.

Let us consider symbol sequences (1D for the sake of simplicity) from two wind turbines in order to quantify their spatiotemporal relations. According to the definition of a D-Markov machine (Sarkar et al., 2014) as described earlier, let Π^A and Π^B denote transition matrices for wind turbines A and B respectively. Similarly, let Π^{AB} and Π^{BA} be defined for xD-Markov machines (Sarkar et al., 2014) that capture the causal dependencies of turbine B on turbine A and of A on B, respectively. Note that such dependencies may not be symmetric, i.e., Π^{AB} and Π^{BA} are not necessarily the same. Features (e.g., state transition matrix) from D-Markov machines are termed as atomic patterns (AP) and those from xD-Markov machines are termed as relational patterns (RP) (Sarkar et al., 2014). The elements of the cross-state transition matrices Π^{AB} and Π^{BA} are shown as follows:

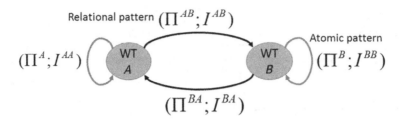

Relational pattern $(\Pi^{AB}; I^{AB})$

$(\Pi^{A}; I^{AA})$ WT A WT B Atomic pattern $(\Pi^{B}; I^{BB})$

$(\Pi^{BA}; I^{BA})$

FIG. 8

Extraction of atomic and relational patterns to characterize individual wind turbine behavior and interaction characteristics among different wind turbines.

$$\pi^{AB}_{k\ell} \triangleq P\left(q^B_{n+1}=\ell \mid q^A_n=k\right) \forall n$$
$$\pi^{BA}_{ij} \triangleq P\left(q^A_{n+1}=j \mid q^B_n=i\right) \forall n \tag{12}$$

where $j,k \in Q^A$ and $i,l \in Q^B$, Q^A, Q^B are respectively nonempty finite the sets of states. A cross-state transition matrix is learnt from symbol sequences generated from two wind turbines and every element in the matrix identifies the probability of transition from a state in one wind turbine to another state in the second wind turbine.

Furthermore, we quantify the information content of the atomic and relational features based on an information-theoretic measure. As shown in Fig. 8, the mutual information based importance metric for feature is denoted by I. For example, mutual information for the atomic pattern of wind turbine A is expressed as:

$$I^{AA}=I\left(q^A_{n+1};q^A_n\right)=H\left(q^A_{n+1}\right)-H\left(q^A_{n+1}\mid q^A_n\right) \tag{13}$$

where

$$H\left(q^A_{n+1}\right)=-\sum_{i=1}^{Q_A}P\left(q^A_{n+1}=i\right)\log_2 P\left(q^A_{n+1}=i\right)$$

$$H\left(q^A_{n+1}\mid q^A_n\right)=-\sum_{i=1}^{Q_A}P\left(q^A_n=i\right)H\left(q^A_{n+1}\mid q^A_n=i\right)$$

$$H\left(q^A_{n+1}\mid q^A_n=i\right)=-\sum_{i=1}^{Q_A}P\left(q^A_{n+1}=l\mid q^A_n=i\right)\log_2 P\left(q^A_{n+1}=l\mid q^A_n=i\right)$$

Note that the quantity I^{AA} signifies the temporal self-prediction capability (self-loop) of the wind turbine A. Similarly, mutual information for the relational pattern between wind turbines A and B is represented as:

$$I^{AB}=I\left(q^B_{n+1};q^A_n\right)=H\left(q^B_{n+1}\right)-H\left(q^B_{n+1}\mid q^A_n\right) \tag{14}$$

where:

$$H\left(q^B_{n+1}\mid q^A_n\right)=-\sum_{i=1}^{Q_A}P\left(q^A_n=i\right)H\left(q^B_{n+1}\mid q^A_n=i\right)$$

$$H\left(q_{n+1}^{B}\mid q_{n}^{A}=i\right)=-\sum_{i=1}^{Q_{B}}P\left(q_{n+1}^{B}=l\mid q_{n}^{A}=i\right)\log_{2}P\left(q_{n+1}^{B}=l\mid q_{n}^{A}=i\right)$$

Note that the quantity I^{AB} signifies wind turbines A's capability of predicting wind turbines B's outputs and vice versa for I^{BA}. Such a mutual information based metric can be used as the weights assigned for the patterns that can be used for rejecting patterns with low information content (for network pruning) as well as decision fusion. Interested readers can find more details in Sarkar et al. (2014).

3.3.1 Results

Validation results presented here are based on the Western Wind Integration data set available from the National Renewable Energy Laboratory (NREL) (NWTC, 2006). There are 12 wind turbines investigated in this example and their relative geographical location is shown in Fig. 9. Quantitative results for wind turbines 1, 5, 6, 7, and 10 are presented here.

First, the mutual information based causality metric (for the relational patterns) is calculated using the state transition matrices of the xD-Markov machines. In this context, the current symbol for one wind turbine is only dependent on the past one symbol of another wind turbine. With this setup, we investigate the effect of time lag and spatial distance on the causal relation between two turbines. As shown in Fig. 10, mutual information between any pair of wind turbines decreases along with increment of time lag from 1 to 10. This shows that the choice of memory of the Markov machines to be 1 maximizes the causality at time lag 1 for every case. The results in Fig. 11 show that with an increase in the spatial distance between the wind turbines, causality quantified by mutual information decreases. Fig. 12 shows the general decreasing trend of mutual information-based causality metric with the increase in the Euclidean distance between pairs of wind turbines. The above two observations demonstrate that the mutual information-based causality metric conforms to relative spatial locations of the wind turbines.

Fig. 13 show the comparisons between predicted and actual symbol sequences of wind turbine 5 using wind turbines 6 and 7, respectively. The results are obtained by using half year data of 2006 to train and then using the rest of the year data for testing. While the plots show strong prediction capability of xD-Markov machines, most of the errors come from the transient symbols as expected.

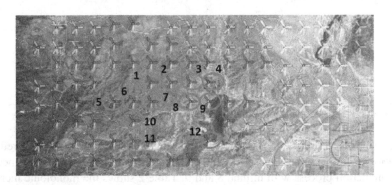

FIG. 9

Geographical information of 12 wind turbines.

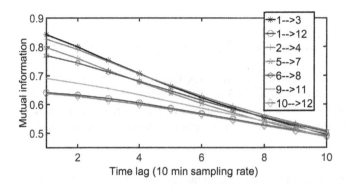

FIG. 10

Mutual information between multiple wind turbines decreases with increase in time lag. In the figure "→" represents the mutual information between two different wind turbines, while the direction indicates the causal dependence from the wind turbine on the left side to the wind turbine on the right side.

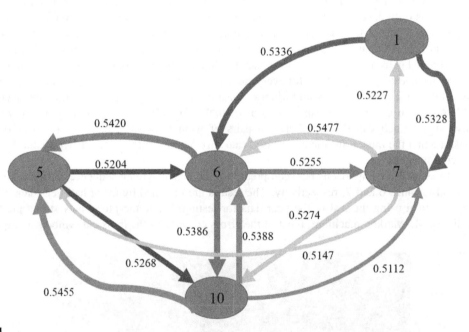

FIG. 11

Mutual information of relational patterns between pairs of wind turbines at different locations.

Furthermore, visual inspection can verify that turbine 7 performs slightly worse compared to predicting power generation by turbine 5 as suggested by the mutual information based causality metric. Continuing the same prediction process with turbine 8 and 9, Table 2 shows the monotonic increase in mean square errors (MSE) as spatial distance between turbine pairs increases (or mutual information decreases).

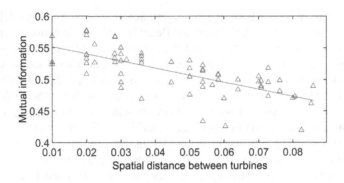

FIG. 12

Mutual information of relational patterns decreases with increase in the Euclidean distance between turbine pairs.

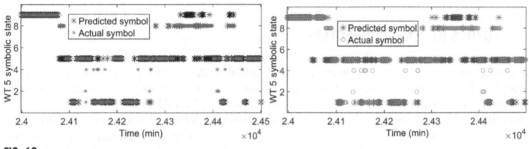

FIG. 13

Comparisons of symbol predictions of wind turbine 5 using wind turbine 6 *(left)* and wind turbine 7 *(right)*.

Table 2 Prediction Mean Square Errors Versus Spatial Distance Between Turbine Pairs	
Wind Turbine Behavior Predictions	**Means Square Error (MSE)**
Wt6 → Wt5	1.369
Wt7 → Wt5	1.885
Wt8 → Wt5	2.512
Wt9 → Wt5	2.724

4 CONCLUSIONS AND FUTURE WORK

4.1 CONCLUSIONS

This chapter discusses the applicability of PGMs such as Bayesian networks and SDF for modeling and analysis of complex CPSs. Three different use cases related to spatial modeling, multiscale temporal modeling and spatiotemporal modeling are presented. The use cases involve various applications such as detection of cyber-attacks and physical faults in complex building systems, hierarchical learning in dynamical modes of systems and wind turbine power prediction in large wind farms.

In all the problems presented here, it is demonstrated that data-driven models have particular advantages in terms of robustness, scalability and flexibility. Quick detection of anomalies in CPSs and distinguishing between physical faults and cyber-attacks is critical for identifying appropriate reactive and mitigating measures. Therefore Bayesian network modeling of spatially distributed CPSs can be extremely useful. Similarly, complex CPSs often show different dynamical behaviors and it is important to detect and identify various characteristics especially without labeled data for all possible situations. Hierarchical SDF-based modeling scheme provides a robust, unsupervised, online and computationally inexpensive method to enable such capabilities. Finally, characterizing complex spatiotemporal interactions among subsystems are critical in order to perform behavior prediction for a CPS. Although domain knowledge or physics based models can capture such interactions on a small scale, the task becomes extremely challenging for a large-scale system under various external disturbances and internal process uncertainties. Such a problem arises in farm-wide power prediction for a large wind farm and a novel type of PGM, namely STPN has been shown to be efficient in capturing spatiotemporal interactions among wind turbines for prediction purposes.

4.2 FUTURE RESEARCH DIRECTION

Most of the existing works on CPSs involving machine learning methods (PGMs in particular) perform modeling and analysis in order to perform health monitoring, diagnostics, prognostics and predictive analytics. The applications discussed here are also similar. The major ongoing and future research direction beyond these applications involves "closing the loop" on CPSs using PGMs. This means controlling CPSs based on the knowledge extracted from data generated from them for various purposes such as mitigating the impact of a fault, enabling resiliency, fault-tolerant control and life-extending control. The probabilistic nature of the algorithms and lack of provability are the main technical challenges in this context. However, PGMs offer rich capabilities of "what-if scenario" simulations via probabilistic programming that can be extremely useful. Finally, PGMs inherently capture the effects of uncertainties at various logical levels, from sensors to subsystems and, eventually, to the entire system level. Hence, PGMs present a powerful technological option in the context of complex CPSs for analysis, monitoring and control purposes.

ACKNOWLEDGMENTS

This work and chapter preparation have been partially supported by US NSF under Award No. 1464279. The authors would also like to thank the United Technologies Research Center for partially supporting this research.

REFERENCES

Aldous, D.J., 1985. Exchangeability and Related Topics in Lecture Notes in Mathematics, vol. 1117. Springer, Berlin Heidelberg.
Akintayo, A., Sarkar, S., 2015. A symbolic dynamic filtering approach to unsupervised hierarchical feature extraction from time-series data. In: Proceedings of American Control Conference, Chicago, IL, s.n, pp. 1–8.
ASHRAE, 2014. BACnet. [Online]. Available at: http://www.bacnet.org/ (accessed 18.08.15).

Beal, M.J., Ghahramani, Z., Rasmussen, C.E., 2002. The infinite hidden Markov model. Adv. Neural Inform. Process. Syst. 577–584.

Bradley, J., Barbier, J., Handler, D., 2013. Embracing the internet of everything to capture your share of $14.4 trillion. CISCO white paper.

Buhl, M., Kennel, M.B., 2005. Statistically relaxing on generating partitions for observed time-series data. Phys. Rev. E Stat. Nonlin. Soft Matter Phys. 71 (4), 1–14.

Cardie, C., Mooney, R., 2006. Markov Models and Hidden Markov Models. [Online]. Available at: http://www3.cs.stonybrook.edu/~ychoi/cse628/lecture/06-hmm.pdf (accessed 21.08.15.).

Chakraborthy, S., Sarkar, S., Ray, A., 2012. Symbolic identification for fault detection in aircraft gas turbine engines. Proc. I Mech. E Part G J. Aerospace Eng. 226 (4), 422–436.

Chakraborty, S., Sarkar, S., Gupta, S., Ray, A., 2008. Damage monitoring of refractory wall in a generic entrained-bed slagging gasification system. Proc. I Mech. E Part A J. Power Energy 222 (8), 791–807.

Chattopadhyay, I., 2014. Causality Network. Cornell University Library.

Chau, T., 2001. Marginal maximum nentropy partitioning yields asymptotically consistent probability density functions. IEEE Trans. Pattern Anal. Mach. Intell. 23 (4), 414–417.

Choi, A., Zheng, L., Darwiche, A., Mengshoel, O., 2011. A tutorial on Bayesian networks for system health management. Machine Learning and Knowledge Discovery for Engineering Systems Health Management 10 (1), 1–29. CRC Press.

Cox, M., Cox, T., 1994. Multidimensional Scaling. Chapman & Hall, London.

Daw, C.S., Finney, C., 2003. A review of symbolic analysis of experimental data. Rev. Sci. Instrum. 74 (2), 915–930.

Feng, N., Xie, J., 2012. A Bayesian network based security risk analysis model for information systems integrating the observed cases with expert experience. Sci. Res. Essays 7 (10), 1103–1112.

Ferguson, T.S., 1973. A Bayesian analysis of some nonparametric problems. Ann. Stat. 1 (2), 209–230.

Fitzgerald, J., et al., 2013. Model-based engineering of systems: the compass manifesto. COM-PASS technical report.

Fox, E., Sudderth, E., Jordan, M., Willsky, A.S., 2011. A sticky HDP-HMM with application to speaker diarization. Ann. Appl. Stat. 5 (2A), 1020–1056.

Freeman, P., Pandita, R., Srivastava, N., Balas, G., 2013. Model-based and data-driven fault detection performance for a small UAV. IEEE/ASME Trans. Mech. 18 (4), 1300–1309.

Frigault, M., Wang, L., 2008. Measuring network security using Bayesian network-based attack graphs. In: 2008 32nd Annual IEEE International Computer Software and Applications Conference, Turku, Finland.

Frigault, M., Wang, L., Singhal, A., Jajodia, S., 2008. Measuring network security using dynamic Bayesian network. In: Proceedings of the 4th ACM Workshop on Quality of Protection, 23–30, Alexandria, VA, USA.

Gershman, S.J., Blei, D.M., 2012. A tutorial on Bayesian nonparametric. J. Math. Psychol. 56 (1), 1–12.

Heller, K.A., Teh, Y., Gorur, D., 2009. Infinite Hierarchical Hidden Markov Models. Clearwater Beach, FL.

Houmb, S., Franqueira, V., Engum, E., 2010. Quantifying security risk level from CVSS estimates of frequency and impact. J. Syst. Softw. 83(9).

Jansson, M., 2004. Fault Isolation Using Bayesian Networks. KTH.

Jiang, Z., Sarkar, S., 2015. Understanding Wind Turbine Interactions Using Spatiotemporal Pattern Network. Columbus, OH.

Jin, X., et al., 2011. Anomaly detection in nuclear power plants via symbolic dynamic filtering. IEEE Trans. Nucl. Sci. 58 (1), 277–288.

Jollife, I.T., 1989. Principal Component Analysis. Springer-Verlag, New York.

Kesler, B., 2011. The vulnerability of nuclear facilities to cyber attack. Strategic Insights 10 (1), 15–25.

Kleissel, J., Agarwal, Y., 2010. Cyber-physical energy systems: focus on smart buildings. In: Proceedings of the 47th Design Automation Conference, pp. 749–754. Anaheim, CA.

Kohler, D., Friedman, N., 2009. Probabilistic Graphical Models: Principles and Techniques. MIT Press.

Kordy, B., Cambacedes, L., Schweitzer, P., 2014. DAT-based attack and defense modeling: don't miss the forest for the attack trees. Comput. Sci. Rev. 13, 1–38. Elsevier.

Kosut, O., Jia, L., Thomas, R.J., Tong, L., 2010, October. Malicious data attacks on smart grid state estimation: attack strategies and countermeasures. In: 2010 First IEEE International Conference on Smart Grid Communications (SmartGridComm). pp. 220–225.

Kramer, D., 2014. White House offers encouragement for cyberphysical systems. Phys. Today 67 (9), 20–22.

Krishnamurthy, S., Sarkar, S., Tewari, A., 2014. Scalable Anomaly Detection and Isolation in Cyber-Physical Systems Using Bayesian networks. San Antonio, TX.

Laboratory, D.S., 2013. GeNIe & SMILE. [Online]. Available at: https://dslpitt.org/genie/ (accessed 18.08.15.).

Liu, Y., Man, H., 2005. Network Vulnerability Assessment using Bayesian Netwoirks, pp. 61–71.

Liu, C., Ghosal, S., Jiang, Z., Sarkar, S., 2016. An unsupervised spatiotemporal graphical modeling approach to Anomaly detection in distributed CPS. In: Proceedings of the International Conference of Cyber-Physical Systems, Vienna, Austria.

Mukherjee, K., Ray, A., 2014. State splitting and merging in probabilistic finite starte automata for signal processing and analysis. Signal Process. 104, 105–119.

NIST, 2013. Strategic R&D Opportunities for 21st Century Cyber-Physical Systems.

NWTC, 2006. NREL. [Online]. Available at: https://nwtc.nrel.gov/ (accessed 18.08.15).

Pathria, R., 1996. Statistical Mechanics, second ed. Butterworth-Heinemann, Oxford, UK.

Rabiner, L., 1994. A tutorial on hidden Markov models and selected applications in speech recognition. Proc. IEEE 77, 257–286.

Rajagopalan, V., Ray, A., 2006. Symbolic time series analysis via wavelet-based partitioning. Signal Process. 86 (11), 3309–3320.

Rao, C., Ray, A., Sarkar, S., Yasar, M., 2009. Review and comparative evaluation of symbolic dynamic filtering for detection of anomalous patterns. Signal Image Video Process. 3 (2), 101–114.

Ray, A., 2004. Symbolic dynamic analysis of complex systems via wavelet-based partitioning. Signal Process. 84 (7), 1115–1130.

Roth, K.W., et al., 2005. Energy Impact of Commercial Building Controls and Performance Diagnostics: Market Characterization, Energy Impact of Building Faults and Energy Savings Potential. DOE.

Sarkar, S., 2011. Autonomous Perception and Decision Making in Cyber-Physical Systems. PhD thesis. The Pennsylvania State University Graduate School, State College, PA.

Sarkar, S., et al., 2008. Fault detection and isolation in aircraft gas turbine engines: part II—validation on a simulation test bed. Proc. I Mech E Part G J. Aerospace Eng. 222 (3), 319–330.

Sarkar, S., Mukherjee, K., Sarkar, S., Ray, A., 2009. Symbolic dynamic analysis of transient time series for fault detection in gas turbine engines. J. Dyn. Syst. Meas. Control.

Sarkar, S., Jin, X., Ray, A., 2011. Data-driven fault detection in aircraft engines with noisy sensor measurements. J. Eng. Gas Turb. Power. 133 (8).

Sarkar, S., Mukherjee, K., Jin, X., Ray, A., 2012. Optimization of symbolic feature extraction for pattern classification. Signal Process. 92 (3), 625–635.

Sarkar, S., Mukherjee, K., Sarkar, S., Ray, A., 2013a. Symbolic dynamic analysis of transient time series for fault detection in gas turbine engines. J. Dyn. Syst. Meas. Contr. 135 (1) p. 014506-1-6.

Sarkar, S., Srivastav, A., Shashanka, M., 2013, June. Maximally bijective discretization for data-driven modeling of complex systems. In: American Control Conference (ACC), 2013. IEEE, pp. 2674–2679.

Sarkar, S., Sarkar, S., Virani, N., Ray, A., Yasar, M., 2014. Sensor fusion for fault detection and classification in distributed physical processes. Front. Robot. AI 1, 16.1–16.9.

Saul, L., Rowels, S., 2000. An introduction to locally linear embedding. Unpublished, Toronto.

Sethuraman, J., 1994. A constructive definition of dirichlet priors. Stat. Sin. 4, 635–650.

Sridhar, S., Hahn, A., Govindarasu, M., 2012. Cyber–physical system security for the electric power grid. Proc. IEEE 100 (1), 210–224.

Sztipanovits, J., Ying, S., Cohen, I., Corman, D., Davis, J., Khurana, H., Mosterman, P.J., Prasad, V., Stormo, L., 2012. Strategic R&D Opportunities for 21st Century Cyber-Physical Systems. Technical Report for Steering Committee for Foundation in Innovation for Cyber-Physical Systems: Chicago, IL, USA, 13 March.

Velasco, J.M., 2012. A Bayesian network approach to diagnosing root cause of failure from trouble tickets. Artif. Intellig. Res. 1 (2).

Wakabayashi, K., Miura, T., 2012. Forward-backward activation algorithm for hierarchical hidden Markov models. Adv. Neural Inform. Process, 1–9.

Wang, L., et al., 2008. An Attack Graph-Based Probabilistic Security Metric. Springer-Verlag, pp. 283–296.

Wilks, S., 1963. Mathematical Statistics. John Wiley, New York.

Work, D.B., Bayen, A.M., 2008. National Workshop for Research on High-Confidence Transportation Cyber-Physical System: Automotive, Aviation and Rail. Impacts of the Mobile Internet on Transportation CyberPhysical Systems: Traffic Monitoring using Smart Phones, Washington, D.C.

Xie, P., Li, J.H., Ou, X., Liu, P., Levy, R., 2010, June. Using Bayesian networks for cyber security analysis. In: 2010 IEEE/IFIP International Conference on Dependable Systems and Networks (DSN). pp. 211–220.

MODEL-BASED TESTING OF CYBER-PHYSICAL SYSTEMS

19

A. Aerts*, M. Reniers*, M.R. Mousavi[†]

Eindhoven University of Technology, Eindhoven, The Netherlands* Halmstad University, Halmstad, Sweden[†]

1 MODEL-BASED TESTING
1.1 WHY, WHAT, AND WHEN

The design of a cyber-physical system (CPS) is typically characterized by the use of a system development life cycle (SDLC), such as the one depicted in Fig. 1 (the outer V depicted using solid black lines). This life cycle represents how an abstract artifact such as system requirements is gradually refined into a concrete implementation artifact (the bottom-most part of the V) and then integrated into a part of an operational system.

Ideally, each of these development activities is accompanied with testing activities, including validation and verification activities (the inner V depicted using solid gray lines in Fig. 1): validation makes sure that the artifacts developed in the activity are in-line with users' intentions and verification ensures that the developed artifacts are consistent with the artifacts of the other activities. Validation often involves users (or user models). Several challenges make validation and verification nontrivial: firstly, it is not always easy to translate user and consistency requirements into concrete test inputs; secondly, it is not straightforward to judge the outcome of tests. The latter problem is often called "the oracle problem" (Binder, 1999); constructing a complete and correct oracle is as difficult as building a complete system. These two problems are intensified in the context of CPSs due to their huge domain of test inputs (endless interactions with huge parameter space) and complex correctness criteria.

Model-based testing (MBT) provides solutions to these two problems. Models are exploited to generate test inputs automatically and models also define the correctness (conformance) criteria in order to produce a test verdict after (or while) executing test cases. MBT can be performed at various steps of the CPS development life cycle, as depicted in Fig. 1 in the inner V-path (gray), along the whole development cycle. Applying MBT at early development phases allows for early detection of faults, which often proves to be efficient and cost-effective (Vishal et al., 2012). In addition to the classical benefit of early testing, such as reduced debugging cost and effort (Boehm, 1981), MBT provides a structured and systematic method for testing at various stages of the CPS development life cycle.

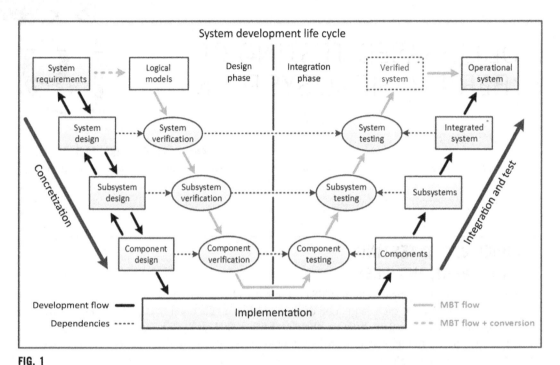

FIG. 1

System development life cycle of cyber-physical systems (V-model), inspired by Firesmith (2014).

1.2 HOW

MBT can be applied throughout the V-model, and is hence applicable on different types of artifacts. In the design phase (see Fig. 1), specifications are used in order to test the correctness of the design artifacts (see Section 2). A prerequisite of this verification task is a precise specification of the requirements. This is a highly nontrivial first challenge, i.e., incorrect specifications lead to incorrect designs and subsequently to invalid conformance testing verdicts throughout the development process. To overcome this challenge, techniques such as model-checking can be used (Baier and Katoen, 2008; Alur et al., 1995).

Conformance testing is the process of verifying the correctness of an artifact in the development cycle of a CPS against its model. Mathematically, a notion of conformance is defined as a relation between two models: an abstract one and a concrete one; in order to designate these two models, we refer to the abstract one as the specification and the concrete one as the implementation. (The implementation need not be the actual implementation, but can be any concretized artifact.) Hence, conformance testing relies on a fundamental assumption that the specification and the implementation are both representable by *some* model. Fig. 2 provides an overview of the conformance testing process and its mathematical underpinning (conformance relation, to be discussed in Section 3). The model of the implementation is often too large to be explicitly represented and is often assumed to be only partially known or even completely unknown. Hence, conformance testing, i.e., test-case generation (based on the specification model, see Section 4), test-case execution, and conformance analysis

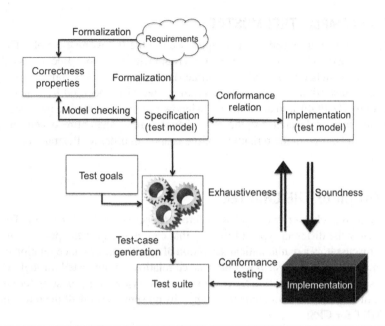

FIG. 2

Schematic view of conformance testing and conformance relation, inspired by Utting et al. (2012).

(evaluating the conformance relation, see Section 3) are used to establish the conformance relation without having direct access to the underlying model of the implementation. Typically the notions of conformance relation and conformance testing are proven to be equivalent through a *soundness* and *exhaustiveness* (also called completeness) theorem. In other words, it is proven that any conforming system passes all the test cases generated in the conformance testing process and any nonconforming system will fail at least one such test case.

In practice, specification models are often absent or out of date and, hence, model learning techniques (Aarts et al., 2015; Meinke et al., 2012) can be used to learn abstract models through interactions with the implementation under test. As stated before, one must ensure that the model does indeed represent the desired behavior described by the user or system requirements. To this end, models are formally checked (or tested) against more abstract properties or manually reviewed by the domain experts for confirmation or adjustments.

The application of MBT to support a V-model SDLC (see Fig. 1) can be interpreted as part of a systems engineering framework (Blanchard et al., 1990), which provides means for system development using a SDLC. Particularly, the domain of model-based systems engineering (MBSE), which includes MBT, is becoming increasingly popular. For an overview of MBSE methodologies, see Estefan et al. (2007). To support this discipline, the Systems Modeling Language (SysML) (Object Management Group, 2014) was created which, amongst others, provides verification and validation functionality for CPSs (Friedenthal et al., 2014). MBT of CPSs as considered in the remainder of this chapter can be seen as part of the verification activities. However, in order for SysML to utilize such techniques, transformations are required to exploit external (model) simulation environments such as Modelica (Modelica Association et al., 2005).

1.3 RUNNING EXAMPLE: THERMOSTAT

To illustrate the concepts introduced in this chapter, we use the following example. The example is a system that comprises a thermostat device located in a room with a window. Because the thermostat can be either fully ON or switched OFF, and no accurate (feedback) control is applied, the system is considered unregulated. Instead an acceptable temperature interval is specified by which the thermostat switches accordingly (ON/OFF). In addition, the window provides an input to the system by influencing the temperature increase/decrease in the room, hence making the entire system input dependent.

   This example will be used in the remainder of the chapter in order to illustrate the core concepts in the field of MBT.

1.4 ORGANIZATION OF THE CHAPTER

In the remainder of this chapter, we discuss different aspects of MBT for CPSs. To start with, in Section 2, we review the different types of models that can be used for the purpose of testing CPSs. In Section 3, we discuss how a mathematical definition of a CPS-conformance relation can be devised. Then, in Section 4, we study how such a conformance relation can be tested through test-case generation and test-case execution. In Section 5, we provide an overview of the existing tooling for this purpose and finally, in Section 6, we conclude the chapter by presenting some of the remaining challenges in the field of MBT for CPSs.

2 MODELS

In this section, we review the different types of models used for capturing the specification (and type of the purported model of the implementation) for different notions of conformance testing.

2.1 STATE-MACHINE-BASED MODELS

Finite state machines (FSMs) or finite automata have been used traditionally in hardware modeling and hardware testing (Lee, 1996). Extensions of such models with variables and data, e.g., extended FSMs and class FSMs, were also proposed for software testing (Hierons et al., 2009; Hong, 1995). In an FSM the history of interactions with the system (e.g., the current valuations of variables or signals) is represented by the system state and the transitions capture the system output as the result of providing an input to the system. Labeled transition systems (LTSs) (Uselton and Smolka, 1994) are variants of FSMs that remove the finiteness assumption and also relax the tight coupling between input and output (e.g., a sequence of several inputs may lead to a single output in an LTS while in FSMs inputs and outputs are paired). Input-Output Labeled Transition System (IOLTSs) (Fernandez et al., 1996) and I/O automata (Lynch and Tuttle, 1987) are useful extensions of LTSs and automata, respectively, that allow for explicit modeling of discrete input and output actions.

Example. In Fig. 3, the thermostat example introduced in Section 1.3 is modeled using an I/O automaton. Two states, namely "heater_on" and "heater_off," represent the switching behavior of the system based on a temperature reading from the room (rcv_tmp_low? and rcv_tmp_high?). Intuitively, such a model is a coarse abstraction of reality. Particularly, this model does not specify the

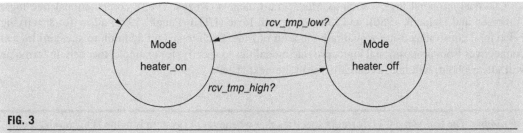

FIG. 3

Thermostat FSM.

system dynamics regarding temperature and the rules governing them. Hybrid (I/O) automata, introduced next, are often used to model CPSs and their system dynamics.

In order to model CPSs, it is often necessary to enrich (extended) FSMs with models of system dynamics, such as differential equations or piecewise continuous trajectories. Examples of such enriched models are hybrid automata (Alur et al., 1993) and hybrid LTSs (Cuijpers et al., 2002).

Example. In Fig. 4, a hybrid automaton model of the thermostat example is shown. The dynamics $\dot{x}(t)$ of each mode represent the temperature behavior in the room with the $x(t)$ variable modeling temperature and the x_1/x_2 variables being input-dependent temperature constants linked to the window position (input-dependent system). The temperature is regulated between x_{min} and x_{max} degrees leading to the corresponding guard conditions in Fig. 4. Hence, by opening and closing (to different extents) the window the dynamics of the system are affected.

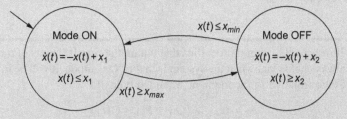

FIG. 4

Thermostat state-model with $x_{min} \leq x(t) \leq x_{max}$ as the configured desired room temperature, $\dot{x}(t)$ as the room (temperature) dynamics, x_1 and x_2 as temperature (input) variables that influence the dynamics and mode-dependent invariants on room temperature.

2.2 LOGICAL MODELS

Logical formulae are often used to specify an abstract constraint on the behavior of dynamical systems. Such formulae often refer to qualitative or quantitative time, e.g., the relative order of events or specific moments of time when certain events may occur or system dynamics may take specific values. Temporal logics are a class of logical formalisms that allow for such timed specifications (Manna and Pnueli, 1992).

Classical temporal logics, such as Modal mu-Calculus (Kozen, 1983), computational tree logic (Emerson and Halpern, 1986), and linear temporal logic (LTL) (Pnueli, 1977) allow for specifying behavioral constraints using a qualitative notion of time. Such logics, in addition to standard logical connectives from propositional logic, provide modalities to specify, for example, that certain formulae will always hold, will hold eventually, or in the next state.

Example. The following LTL formula specifies that whenever the system is in the "On" mode, it will eventually reach the "Off" mode. The fact that this formula should always hold is specified by the modality G; intuitively, $G\varphi$ means that formula φ holds in all reachable states of the system. Logical implication is denoted by $\Rightarrow$ and eventuality is denoted by modality F. Formula $F\varphi$ denotes that φ will eventually hold in the future:

$$\varphi_1 = G(\text{On} \Rightarrow F \text{ Off}).$$

The above-given formula specifies that in each state of the system, whenever On holds, eventually in the future Off will hold, i.e., Off is an inevitable consequence of On.

For CPSs, it is often useful to refer to quantitative time and also to the evolution of system dynamics. Examples of temporal logics that allow for specifications with reference to quantitative time and systems dynamics are metric temporal logic (MTL) (Koymans, 1990) and signal temporal logic (STL) (Maler and Nickovic, 2004), respectively. In particular, MTL extends LTL modalities with the possibility of specifying the time interval within which the ensuing property is satisfied. Additionally, using STL one can also refer to values.

Example. The following MTL formula specifies that within the first 100 units of time from the start of the system, it always holds that when the system is in the "On" mode, it will be in the "Off" mode within 5 units of time (relative to the moment when On holds):

$$\varphi_2 = G_{[0,\,100]}\left(\text{On} \Rightarrow F_{[0,\,5]}\ \text{Off}\right).$$

Similarly, the following STL formula specifies that the first 100 units of time, whenever the temperature falls below a certain threshold, it will again be above the threshold within 5 units of time:

$$\varphi_2 = G_{[0,\,100]}\left(x(t) < x_{\min} \Rightarrow F_{[0,\,5]} x(t) \geq x_{\min}\right).$$

2.3 STATE-BASED MODELS

State-based models typically specify a system in terms of a relation between valuation of model variables before and after different operations (state transformers). Examples of state-based modeling languages include Z (Spivey, 2008), B method and Event B (Abrial, 2010), and action systems (Back, 2003).

Example. The following action system specifies the discrete behavior of the thermostat system:

Thermostat $\triangleq$ |[

> **var** *mode**: {on, off} •
>> *mode* := off;
>
> **do**
>> *rcvTmpLow* = true $\rightarrow$ *mode* := on;   []
>>
>> *rcvTmpLow* = false $\rightarrow$ *mode* := off;
>
> **od**

]| : *rcvTmpLow*

The syntax is selfexplanatory: first a state variable (*mode*) is declared (the asterisk marks a variable that is exported, i.e., can be referenced in another action system). Subsequently, the state variable is initialized. Finally, it is specified how the state variable evolves by specifying a number of alternative guarded assignments. Finally, it is specified that *rcvTmpLow* is an imported variable.

State-based models have been extended to represent system dynamics, see, e.g., (Rönkkö et al., 2003; Back et al., 2001; Banach et al., 2015). Below we specify our thermostat example in the continuous action system style (Back et al., 2001).

Example. The following continuous action system specifies the discrete and dynamic behavior of the thermostat system:

ThermostatDynamics $\triangleq$

> |[

>> **var** *mode**: {*on, off*},
>>> *x*: *Real*$_+\rightarrow$*Real* •
>>
>> *mode* := *off*;
>>
>> *reset*(*x*);
>>
>> **do**
>>> *x. now* $\leq$ x$_{min}\rightarrow$*mode* := on;
>>>> $x = \lambda t \,.\, x.\ now + {}^{(t - now)^2}\big/_2 + x_1(t - now);$ []
>>>
>>> *x. now* $\geq$ x$_{max}\rightarrow$*mode* := off;
>>>> $x = \lambda t \,.\, x.\ now - {}^{(t - now)^2}\big/_2 + x_2(t - now);$
>>
>> **od**

>]|

The syntax of the above-specified action system is almost identical to the discrete specification. The main difference is that the system dynamics (variable *x*) is specified in terms of a trajectory (a real-valued function). The value of *x* is subsequently specified using lambda expressions (an expression $x = \lambda t \cdot f$ can be read as the equation $x(t) = f$).

3 CONFORMANCE RELATION

In MBT, models and implementations are either compared directly or indirectly with one another. These types of conformance are named conformance relation and conformance testing respectively.

A conformance relation is a well-defined mathematical relation between a specification and an implementation model. For this it is essential that both the specification and implementation can be represented by formal models. Of course, the formal model underlying the implementation may not be available. Nevertheless, it is assumed that the behavior of the implementation may be represented by some (possibly unknown) model. This assumption is called the *test assumption* (ISO/IEC JTC1/ SC21 WG7, ITU-T SG 10/Q.8, 1996). For discrete event systems, examples of conformance relations are defined in Lee (1996) and Tretmans (1996). Such definitions are extended to systems with data, e.g., in Frantzen et al. (2006) and with time, e.g., in Briones and Brinksma (2005). For surveys of such conformance relations, we refer to Hierons et al. (2009) and Broy et al. (2005).

Typically a conformance relation requires that the set of outputs produced by the implementation is a subset of those allowed by the specification. However, practically it is customary to utilize partial test models, which focus on certain aspects of the system behavior. In such a case, the conformance relation should be adapted so that it concerns only a subset of implementation's behavior, namely the subset that is included in the specification, e.g., see Tretmans (1996).

In order to establish the conformance relation, a *conformance testing* technique is used. Such a technique obtains a test suite from the specification and executes the test cases in that test suite on the implementation. Subsequently, conformance testing establishes whether or not the implementation passes the test cases and the test suite (as expected). Hence, it should be clearly defined what it means for an implementation to pass a test case (and hence a test suite). A conformance testing technique is *sound* if any conforming implementation passes all the obtained test cases (and hence, the test suite) and it is *complete* when every nonconforming implementation fails at least one test case. (We refer to Fig. 2 for a schematic presentation of this process.)

Example. Consider the I/O automaton of the thermostat example of Fig. 3 as a specification model. An implementation model for this system is provided in Fig. 5. This implementation model does not conform to the specification since initially (assuming that "heater_on" is the initial mode), it is in mode "heater_off" while according to the specification the mode "heater_on" should be active. Also, after the rcv_tmp_high? input, the specification only allows for the "heater_off" mode while the implementation is in mode "heater_on."

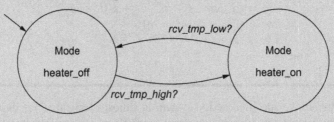

FIG. 5

Implementation model of a thermostat example in the form of an I/O automaton (deduced from the specification model in Fig. 3; however, note that the modes are switched).

In order to test CPSs, it is often essential to exploit models incorporating system dynamics and adapt the notions of conformance relation and conformance testing for such models. Involving system dynamics, one concern is how to compare the output behavior (trajectories) of specification and implementation. One may distinguish between approaches where behaviors of specification need to be mimicked precisely by the implementation or only approximately. The hybrid input-output conformance relation defined by Osch (2006) is an example of the former, whereas the approximate notion of conformance proposed in Abbas et al. (2014a,b) is an example of the latter. Concerning the latter relation, a notion of closeness between two output sequences is defined that allows for deviations in value and time between outputs up to the specified values of ε and τ, respectively. The conformance degree, given some allowed time deviation τ between two systems, is defined as the minimal value for ε such that the systems are (τ, ε)-conformant. Based on this notion of closeness the authors of this chapter developed a prototype tool for MBT of CPSs (Aerts et al., 2015). In the literature one may find other works on approximate equivalence of hybrid systems, such as Girard and Pappas (2007, 2011) and Julius and Pappas (2006). We refer to Khakpour and Mousavi (2015) and Mohaqeqi et al. (2014) for a comparison of the abovementioned notions.

Example. Given the hybrid automaton (specification) model of the thermostat example in Fig. 4, an exemplary steady-state behavior for such a system is shown in Fig. 6 (using the solid line). In addition, there is also a steady-state behavior for a unique thermostat implementation model shown (dashed line). The output or temperature behavior of the specification model (solid line) is surrounded by conformance bounds (dotted line). The conformance bounds allow for deviations (of the implementation) from the specification in such a way that the room temperature still remains approximately between 20 and 25 degrees. In the same figure, we have indicated state transitions of the underlying hybrid automaton by solid arrowheads (upwards = "On" → "Off", downwards = "Off" → "On", see Fig. 4).

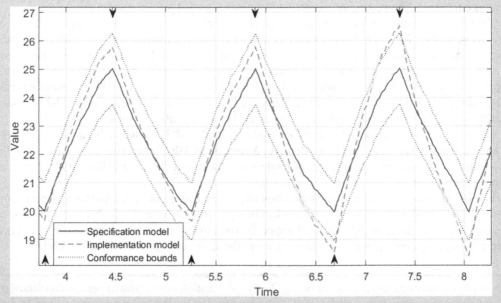

FIG. 6

Steady-state behavior of a thermostat example (see Fig. 4) of a specification and implementation model, visualization of a conformance relation.

Continued

CONT'D

As illustrated in Fig. 6, for approximately 6.4 time units, the implementation model remains within the conformance bounds of the specification model. However, beyond this point, nonconforming behavior is detected, i.e., the implementation repeatedly violates the conformance bounds as indicated in Fig. 6 by the highlighted areas.

Conformance relations for CPSs may be parameterized by a conformance (robustness) degree. Such a degree specifies a bound on the deviation of the implementation from the specification, and has been defined in Abbas et al. (2014a,b) for behavioral models (akin to sampling of hybrid automata). In addition, a robustness degree for temporal logic specifications has been introduced in Donzé and Maler (2010) to measure the "satisfaction degree" of temporal logic specifications (as specification model) on implementation models. The latter is particularly useful in test-case generation and is discussed in the next section.

4 CONFORMANCE TESTING
4.1 TEST-CASE GENERATION

In Section 2, an overview of the modeling languages used for MBT is presented. Assuming the presence of a test model in one such modeling language, the next step in a MBT workflow is to extract a number of test cases from the model. The test cases are subsequently applied to the implementation to assign a verdict that indicates whether the implementation conforms to the specification or not (i.e., whether the underlying model of the implementation is related to the specification model by the conformance relation). The process of extracting test cases is called *test-case generation*.

Two important attributes of a test-case generation algorithm are its soundness and exhaustiveness (Tretmans, 1996). *Soundness* means that a conforming implementation is never rejected. In other words, if the implementation conforms to the specification, then indeed the implementation passes all test cases that can be generated. *Exhaustiveness* of test-case generation signifies that all faulty implementations (possibly with respect to a given fault model) are rejected by some generated test case. Exhaustiveness is hard to obtain in practice, particularly without a well-defined fault model, because it usually requires an unbounded number of test cases, i.e., an infinite test suite.

Given the size of the systems to which one wishes to apply MBT, manual definition of test cases is an approach that is generally too labor-intensive. A strength of MBT is that sound test-case generation algorithms are devised for various types of models, e.g., Hierons et al. (2009), Broy et al. (2005), and Abbas et al. (2013).

For deterministic models, test cases are often expressed as sequences of inputs and expected outputs. For nondeterministic systems, test cases can be represented as trees. Each edge in the sequence or tree either provides an input (allowed by the specification) to the system under test and/or observes an output from the system in order to evaluate it using the allowed outputs by the specification.

Example. Consider the I/O automaton of Fig. 3 as a specification model, and the automaton model of Fig. 5 as an implementation model. Based on the specification model, a test tree can be generated as shown in Fig. 7. This test tree depicts the test cases for the implementation under test, and specifies conforming and nonconforming behavior.

In the case of the thermostat FSM example, when executing this test tree, the implementation will fail after the first test case. The implementation will be in mode "heater_on" as response to the test input rcv_tmp_high! and hence, the test case is terminated because of nonconforming behavior. Note that one can perform online and offline test-case generation, which mainly differs in the fact that online test-case generation will typically result in a sequence of test cases, i.e., the next step is determined depending on the behavior observed so far, while in offline test-case generation one may have to deal with all possible outcomes of every test input (which typically results in a test tree).

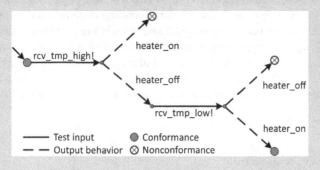

FIG. 7

Test tree for the thermostat FSM example (based on Fig. 3).

4.2 TEST-CASE SELECTION AND COVERAGE

CPS models typically have an infinite state space and hence, the sampling strategy in this infinite state space can have a major impact in the effectiveness of the generated test cases. In order to steer the sampling process, a notion of model coverage can be used. Traditional examples of such coverage metrics are branch coverage criteria in control-flow graphs and predicate and condition coverage in state-machine based models (Ammann and Offutt, 2008).

In Dang and Nahhal (2009), the authors present a test coverage measure that is based on the notion of star discrepancy (Beck and Chen, 1988). This coverage notion is used to quantify the extent to which a test suite covers the state space of the system. This coverage measure is then used to guide a method for test-case generation based on rapidly-exploring random trees (RRT) into areas of the state space where coverage is low.

For the domain of CPSs, a method to develop structural coverage based on symbolic computation of conditions on the inputs of the system (specified by means of differential dynamic logic) is to drive the specification in a certain structural region of its behavior (Zhang et al., 2013). This method uses counter-examples from model-checking algorithms to obtain test cases for specific structural regions.

In recent work by Dreossi et al. (2015) a novel approach for MBT of CPSs was proposed in which robustness analysis and coverage were combined. The authors propose an approach in which the system under test is modeled as a (nonautonomous, discrete-time) hybrid dynamical system and the requirements imposed on the system are formalized as formulas stated in STL (Donzé and Maler, 2010; Donzé et al., 2013). The algorithm used for hybrid test-case generation is an adapted version of the previously discussed RRT algorithm. In particular, decisions on what part of the state space to explore further, that are taken randomly in the original RRT algorithm, are instead based on a robustness analysis of the satisfaction of the requirement formulas.

Example. Given the hybrid automaton model of the thermostat example in Fig. 4, a test suite (in the form of a tree) can be generated as shown in Fig. 8. This test suite comprises a representation of a set of pass (solid) and fail (crossed) verdict nodes corresponding to correct and incorrect observed output trajectories. Note that the visualized output trajectories in Fig. 8 should not be taken literally, in fact each observed trajectory represents a subset of allowed or not allowed output trajectories. In practice testing based on these trajectories should allow for certain error margins both in timing and in the valuation of the variables in output trajectories. We refer to Abbas et al. (2013, 2014a) for examples of testing notions that accommodate such error bounds.

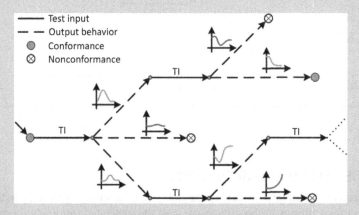

FIG. 8

Test tree for the thermostat hybrid automaton example (Fig. 4).

5 TOOLING

In order to apply MBT methodologies in practical cases, mechanized tools are required. In this section, we briefly review the existing tools that can be used to mechanize different stages of MBT. There is ample room for more mature tooling in this area and we hope to see more stable and robust tools that are based on published research results in this area.

One of the few test-case generation tools for hybrid systems is HTG (Dang, 2011). In this tool, hybrid automata models or SPICE netlists serve as input for coverage-guided test-case generation based on the star discrepancy notion from statistics and the RRTs algorithm from robotic motion planning. The generated test cases can be visualized with viewers such as Matlab or the GNU plot tool.

Another tool that includes the conformance testing functionality as explained in Section 3, is S-TaLiRo (Annpureddy et al., 2011). This tooling, designed in Matlab, utilizes arbitrary Matlab or Simulink models in order to perform conformance testing using stochastic techniques or Monte-Carlo methods for black-box test-case generation (no model knowledge is used). Hence, only coverage metrics defined over the input space are applicable. Moreover, the tooling is also compatible with logical models, in specific the falsification of MTL formulae on implementation models.

The Matlab toolbox RRT-REX (for RRT Robustness-guided EXplorer) (Dreossi et al., 2015) implements a "white-box testing" approach for logical models specified in terms of STL formulae (i.e., model knowledge is used for the test-case generation process). This method, as explained in Section 4, guarantees a global minimum of the robustness degree of its temporal logic formulae (Fainekos and Pappas, 2006; Fainekos et al., 2007). However, at the moment of writing, the current version of the tooling does not incorporate conformance analysis.

Regarding conformance testing of CPSs, only a few tools or toolboxes exist. One example worth mentioning is a tool prototype developed in Matlab (Aerts et al., 2015) to work with CPS models (and implementations) in Acumen (Taha et al., 2012). In this tool, the authors implemented a modified version of the conformance notion as presented in Abbas et al. (2014b), and proposed a white-box test-case generation algorithm based on hybrid automata models for CPS.

In order to provide a comparison of these tools, an overview can be found in Table 1, which lists all of the above tools and their respective features.

Regarding model-verification, several tools exists. Here, we briefly mention three examples of such tools. Firstly, the tool "SpaceEx" (Frehse et al., 2011) serves as a verification tool for continuous as well as hybrid dynamical systems. Secondly, "Strong" (Deng et al., 2013) is a trajectory-based verification toolbox for hybrid systems in Matlab, which utilizes a combination of simulation and formal verification algorithms. Lastly, "Breach" (Donzé, 2010) is a Matlab/C++ toolbox for simulation of hybrid systems, verification of temporal logic properties on hybrid systems, and reachability analysis of dynamical systems represented by ordinary differential equations or by external tools such as Matlab Simulink.

Table 1 Tool Overview

		Tools			
		HTG	**RRT-REX**	**S-TaLiRo**	**"Tool Prototype"**
1	Conformance testing	No	No	Yes	Yes
2	Temporal logic falsification	No	Yes	Yes	No
3	White-box test generation	Yes	Yes	No	Yes
4	Black-box test generation	No	No	Yes	No

6 CHALLENGES

Several challenges lie ahead regarding the applicability of MBT of CPSs. We categorize such challenges into theoretical challenges and practical applicability challenges; we refer to Khakpour and Mousavi (2015) for a more detailed description of these challenges.

Regarding theory, despite the existing literature, we still need more research concerning scalable conformance relations and conformance testing methods. Below we list a few challenging issues in designing such a conformance relation and conformance testing method:

- Component-wise and step-wise conformance: existing conformance relations are mostly noncompositional, i.e., composition of components that conform to their respective specifications may result in a system that does not conform to the composition of the specifications. Also many of the existing conformance relations are not transitive and hence, do not provide a natural means for step-wise refinement.
- Expressiveness of models: often conformance relations are defined in terms of very restrictive modeling assumptions. For example, models are assumed to be deterministic or full (i.e., at all states and all possible inputs are defined).
- Verified and robust discretization: for practical implementations inputs and outputs are discretized samples of their corresponding continuous trajectories in the model. There is very little work on providing verified discretization schemes that do not miss the relevant events and cover the model with respect to a given notion of coverage.

Applicability of sound scientific theories involves the integration of these scientific methods and techniques in stable and well-supported tools. If such tools are present, the proposed framework in Section 1, where MBT is positioned in a V-model SDLC, can be applied to adopt such an integrated MBT process. We envisage the following issues regarding practical applicability and tooling:

- User-friendliness: Tools should be intuitive to use for engineers of various backgrounds. Therefore, a comprehensible user interface and a simple input language should be present in which the tool configuration and operation can take place, including the specification of (formal) specifications. Up-to-date models are hardly present in practice and also often engineers do not feel comfortable with the abstractions in generic modeling formalisms. Hence, domain-specific languages can play an important role in developing models for such tools.
- Accuracy: There is a gap between mathematical theories and actual implementations of systems dynamic models. Having a reliable and verified basis for the calculations of numeric approximations of systems dynamics is an important challenge in tools for MBT.
- Real-time adapters: MBT tools often produce abstract test cases that need to be transformed into concrete test cases using the right data formats and protocols. Such adapters are often very complex and verifying their correctness is not a trivial task. Hence, providing reliable techniques for developing or extending adapters is an important issue in tooling. Accurate test execution and observations of system dynamics also requires real-time adapters with an accurate approximation of time, which is an extra challenge in tooling.
- Traceability: Many certification trajectories and standards (such as the ISO26262 for automotive and ARP-4761 and DO-178C/ED-12C for avionic systems) require traceability among

certification artifacts (e.g., from requirements to test cases). The traceability path passes through models in MBT and requires additional machinery in the tooling.

• White-box versus black-box testing: MBT is often used as a black-box testing method, i.e., the structural information of the model and the implementation are often not exploited in the process of test-case generation and execution. However, white-box (glass-box) testing is a viable alternative and can potentially improve the effectiveness of the test process in order to avoid gaps and redundancies. Further research is required in order to build in such techniques in the MBT trajectory of CPSs.

REFERENCES

Aarts, F., Jonsson, B., Uijen, J., Vaandrager, F., 2015. Generating models of infinite-state communication protocols using regular inference with abstraction. Form. Method. Syst. Des. 46 (1), 1–41.

Abbas, H., Fainekos, G., Sankaranarayanan, S., Ivancic, F., Gupta, A., 2013. Probabilistic temporal logic falsification of cyber-physical systems. ACM Trans. Embed. Comput. Syst. 12 (S2), 1–95. 30.

Abbas, H., Hoxha, B., Fainekos, G., Deshmukh, J., Kapinski, J., Ueda, K., 2014a. Conformance testing as falsification for cyber-physical systems. arXiv. preprint arXiv:1401.5200.

Abbas, H., Mittelmann, H., Fainekos, G., 2014b. Formal property verification in a conformance testing framework. In: IEEE International Conference on Formal Methods and Models for Codesign (MEMOCODE), 2014 Twelfth ACM. IEEE, pp. 155–164.

Abrial, J.R., 2010. Modeling in Event-B: System and Software Engineering. Cambridge University Press, Cambridge, UK.

Aerts, A., Mousavi, M., Reniers, M., 2015. A tool prototype for model-based testing of cyber-physical systems. In: Proceedings of the 12th International Colloquium on Theoretical Aspects of Computing (ICTAC 2015). Springer, Berlin.

Alur, R., Costas, C., Henzinger, T., Ho, P., 1993. Hybrid automata: an algorithmic approach to the specification and verification of hybrid systems. In: Hybrid Systems, vol. 736. Springer, Berlin, Heidelberg, pp. 209–229.

Alur, R., Courcoubetis, C., Halbwachs, N., Henzinger, T.A., Ho, P.-H., Nicollin, X., et al., 1995. The algorithmic analysis of hybrid systems. Theor. Comput. Sci. 138 (1), 3–34.

Ammann, P., Offutt, J., 2008. Introduction to Software Testing. Cambridge University Press, Cambridge.

Annpureddy, Y., Liu, C., Fainekos, G., Sankaranarayanan, S., 2011. S-TaLiRo: a tool for temporal logic falsification for hybrid systems. In: Aziz Abdulla, P., Rusten, K., Leino, M. (Eds.), Tools and Algorithms for the Construction and Analysis of Systems. Springer, Berlin, pp. 254–257.

Back, R.-J., Petre, L., Porres, I., 2001. Continuous action systems as a model for hybrid systems. Nord. J. Comput. 8 (1), 2–21.

Back, R.J.R., von Wright, J., 2003. Compositional action system refinement. Formal Aspects of Computing 15 (2-3), 103–117.

Baier, C., Katoen, J., 2008. Principles of Model Checking. MIT Press, Cambridge, MA.

Banach, R., Butler, M., Qin, S., Verma, N., Zhu, H., 2015. Core hybrid event-B I: single hybrid event-B machines. Sci. Comput. Program. 105, 92–123.

Beck, J., Chen, W., 1988. Irregularities of Distribution. Cambridge University Press, Cambridge.

Binder, R.V., 1999. Testing Object-Oriented Systems: Models, Patterns, and Tools. Addison-Wesley Professional, Boston, MA.

Blanchard, B.S., Fabrycky, W.J., Fabrycky, W.J., 1990. Systems Engineering and Analysis. vol. 4, Prentice Hall, Englewood Cliffs, NJ.

Boehm, B., 1981. Software Engineering Economics. Prentice Hall, Englewood Cliffs, NJ.

Briones, L., Brinksma, E., 2005. A test generation framework for quiescent real-time systems. In: Formal Approaches to Software Testing—Proceedings of the 4th International Workshop (FATES 2004). Springer, Berlin, Heidelberg.

Broy, M., Jonsson, B., Katoen, J., Leucker, M., Pretschner, A., 2005. Model-Based Testing of Reactive Systems: Advanced Lectures (vol. 3472 of Lecture Notes in Computer Science). Springer, Berlin.

Cuijpers, P., Reniers, M., Heemels, W., 2002. Hybrid Transition Systems. TU/e, Eindhoven.

Dang, T., 2011. Model-based testing of hybrid systems. In: Zander, J., Schieferdecker, I., Mosterman, P.J. (Eds.), Model-Based Testing for Embedded Systems. CRC Press, Boca Raton, FL, pp. 383–424.

Dang, T., Nahhal, T., 2009. Coverage-guided test generation for continuous. Form. Method. Syst. Des. 34 (2), 183–213.

Deng, Y., Rajhans, A., Julius, A., 2013. STRONG: a trajectory-based verification toolbox for hybrid systems. In: Joshi, K., Siegle, M., Stoelinga, M., D'Argenio, P. (Eds.), Proceedings of the 10th International Conference on Quantitative Evaluation of Systems (QEST 2013). Springer, Berlin, Heidelberg, pp. 165–168.

Donzé, A., 2010. Breach, a toolbox for verification and parameter synthesis of hybrid systems. In: Touili, T., Cook, B., Jackson, P. (Eds.), Proceedings of the 22nd International Conference on Computer Aided Verification (CAV 2010). Springer, Berlin, Heidelberg, pp. 167–170.

Donzé, A., Maler, O., 2010. Robust satisfaction of temporal logic over real-valued signals. In: Proceedings of the 8th International Conference on Formal Modeling and Analysis of Timed Systems (FORMATS 2010). Springer, Berlin, Heidelberg, pp. 92–106.

Donzé, A., Ferrère, T., Maler, O., 2013. Efficient robust monitoring for STL. In: Proceedings of the 25th International Conference on Computer Aided Verification (CAV 2013). Springer, Berlin, Heidelberg, pp. 264–279.

Dreossi, T., Dang, T., Donzé, A., Kapinski, J., Jin, X., Deshmukh, J.V., 2015. Efficient guiding strategies for testing of temporal properties of hybrid systems. In: Proceedings of the 7th International Symposium on NASA Formal Methods (NFM 2015). Springer International, pp. 127–142.

Emerson, A.E., Halpern, J.Y., 1986. "Sometimes" and "not never" revisited: on branching versus linear time temporal logic. J. ACM 33 (1), 151–178.

Estefan, J.A., et al., 2007. Survey of Model-Based Systems Engineering (MBSE) Methodologies, vol. 25. Incose MBSE Focus Group, San Diego, CA.

Fainekos, G.E., Pappas, G.J., 2006. Robustness of temporal logic specifications. In: Havelund, K., Núñez, M., Roşu, G., Wolff, B. (Eds.), Formal Approaches to Software Testing and Runtime Verification. Springer, Berlin, pp. 178–192.

Fainekos, G.E., Anand, M., Lee, I., Pappas, G.J., 2007. Robust test generation and coverage for hybrid systems. In: Proceedings of the 10th International Workshop on Hybrid Systems: Computation and Control (HSCC 2007). Springer, Berlin, Heidelberg, pp. 329–342.

Fernandez, J., Jard, C., Jéron, T., Viho, C., 1996. Using on-the-fly verification techniques for the generation of test suites. In: Proceedings of the 8th International Conference on Computer Aided Verification (CAV96). Springer, Berlin, pp. 348–359.

Firesmith, D., 2014. Common System and Software Testing Pitfalls: How to Prevent and Mitigate Them: Descriptions, Symptoms, Consequences, Causes, and Recommendations. Addison-Wesley Professional, Boston, MA.

Frantzen, L., Tretmans, J., Willemse, T.A., 2006. Symbolic framework for model-based testing. In: Proc. of FATES/RV 2006. Lecture Notes in Computer Science, vol. 4262. Springer, Berlin, pp. 40–54.

Frehse, G., Le Guernic, C., Donzé, A., Cotton, S., Ray, R., Lebeltel, O., et al., 2011. SpaceEx: scalable verification of hybrid systems. In: Gopalakrishnan, G., Qadeer, S. (Eds.), Proceedings of the 23rd International Conference on Computer Aided Verification (CAV 2011). Springer, Berlin, pp. 379–395.

Friedenthal, S., Moore, A., Steiner, R., 2014. A Practical Guide to SysML: The Systems Modeling Language. Morgan Kaufmann, San Francisco, CA.

Girard, A., Pappas, G.J., 2007. Approximation metrics for discrete and continuous systems. IEEE Trans. Autom. Control 52 (5), 782–798.

Girard, A., Pappas, G., 2011. Approximate bisimulation: a bridge between computer science and control theory. Eur. J. Control. 17 (5–6), 568–578.

Hierons, R., Bogdanov, K., Bowen, J., Cleaveland, R., Derrick, J., Dick, J., et al., 2009. Using formal specifications to support testing. ACM Comput. Surv. 41 (2), 1–76.

Hong, H.S., Kwon, Y.R., Cha, S.D., 1995. Testing of object-oriented programs based on finite state machines. sl, IEEE CS, pp. 234–241.

ISO/IEC JTC1/SC21 WG7, ITU-T SG 10/Q.8, 1996. Information retrieval, transfer and management for OSI; framework: formal methods in conformance testing. ITU-T.

Julius, A., Pappas, G., 2006. Approximate equivalence and approximate synchronization of metric transition systems. In: 2006 45th IEEE Conference on Decision and Control. IEEE, pp. 905–910.

Khakpour, N., Mousavi, M., 2015. Notions of conformance testing for cyber-physical systems: overview and road-map. In: Proceedings of the 26th International Conference on Concurrency Theory (CONCUR 2015), vol. 42. LIPIcs–Leibniz International Proceedings in Informatics, Madrid, Spain, pp. 18–40.

Koymans, R., 1990. Specifying real-time properties with metric temporal logic. Real-Time Syst. 2 (4), 255–299.

Kozen, D., 1983. Results on the propositional mu-calculus. Theor. Comput. Sci. 27 (3), 333–354.

Lee, D.a., 1996. Principles and methods of testing finite-state machines. Proc. IEEE 84 (8), 1089–1123.

Lynch, N.A., Tuttle, M.R., 1987. Hierarchical correctness proofs for distributed algorithms. In: Proceedings of the Sixth Annual ACM Symposium on Principles of Distributed Computing. ACM, New York, pp. 137–151.

Maler, O., Nickovic, D., 2004. Monitoring temporal properties of continuous signals. In: Proceedings of the Joint International Conferences on Formal Modeling and Analysis of Timed Systems (FORMATS 2004) and Formal Techniques in Real-Time and Fault-Tolerant Systems (FTRTFT 2004). Springer, Berlin, pp. 152–166.

Manna, Z., Pnueli, A., 1992. The Temporal Logic of Reactive and Concurrent Systems: Specification. Springer, New York.

Meinke, K., Niu, F., Sindhu, M.A., 2012. Learning-based software testing: a tutorial. In: International Workshops on Leveraging Applications of Formal Methods, Verification, and Validation, vol. 336. Springer, Berlin, pp. 200–219.

Modelica Association et al., 2005. Modelica—a unified object-oriented language for physical systems modeling. Language Specification, Version 2.

Mohaqeqi, M., Mousavi, M., Taha, W., 2014. Conformance testing of cyber-physical systems: a comparative study. In: Proceedings of the 14th International Workshop on Automated Verification of Critical Systems (AVOCS 2014), vol. 70.

Object Management Group, 2014. OMG Systems Modeling Language (OMG SysML™), V1.4. Opgehaald van http://www.omg.org/spec/SysML/1.4/PDF.

Osch, M.V., 2006. Hybrid input-output conformance and test generation. In: Formal Approaches to Software Testing and Runtime Verification. Springer, Berlin, pp. 70–84.

Pnueli, A., 1977. The temporal logic of programs. In: 18th Annual Symposium on Foundations of Computer Science. IEEE, pp. 46–57.

Rönkkö, M., Ravn, A., Sere, K., 2003. Hybrid action systems. Theor. Comput. Sci. 290 (1), 937–973.

Spivey, J.M., 2008. Understanding Z: a specification language and its formal semantics. Cambridge University Press, Cambridge, UK.

Taha, W., Brauner, P., Zeng, Y., Cartwright, R., Gaspes, V., Ames, A., et al., 2012. A core language for executable models of cyber physical systems (preliminary report). In: Proceedings of the 32nd International Conference on Distributed Computing Systems Workshops (ICDCSW 2012). IEEE CS, Macau, China, pp. 303–308.

Tretmans, J., 1996. Test generation with inputs, outputs and repetitive quiescence. Softw. Concepts Tools 17, 103–120.

Uselton, A.C., Smolka, S.A., 1994. A compositional semantics for statecharts using labeled transition systems. In: Proceedings of the 5th International Conference on Concurrency Theory (CONCUR '94). Springer, Berlin, Heidelberg, pp. 2–17.

Utting, M., Pretschner, A., Legeard, B., 2012. A taxonomy of model-based testing approaches. Softw. Test. Verif. Reliab. 22 (5), 297–312.

Vishal, V., Kovacioglu, M., Kherazi, R., Mousavi, M., 2012. Integrating model-based and constraint-based testing using SpecExplorer. In: Proceedings of the 4th Workshop on Model-based Testing in Practice (MoTiP 2012). IEEE CS, Dallas, pp. 219–224.

Zhang, L., He, J., Yu, W., 2013. Test case generation from formal models of cyber physical system. Int. J. Hybr. Inform. Technol. 6 (3), 15–24.

CYBER-PHYSICAL SYSTEMS FOR WIDE-AREA SITUATIONAL AWARENESS

20

C. Alcaraz, L. Cazorla, J. Lopez
University of Malaga, Malaga, Spain

1 INTRODUCTION

Cyber-physical systems (CPSs) are collaborative systems composed of autonomous and intelligent devices capable of managing data flows and operations, and monitoring physical entities integrated as part of critical infrastructures (CIs). The interaction between cyber-physical devices (CPDs) is done through large, heterogeneous and interconnected communication infrastructures. Through these infrastructures, control systems can process and manage measurements and evidence produced in remote locations, and distribute and visualize control transactions. The interfaces that lead these activities are generally control devices that ensure the intermediation tasks between the acquisition world and the control world, such as gateways, servers or remote terminal units (RTUs). An RTU, typically working at ~22–200 MHz with 256 bytes–64 MB RAM, 8 KB–32 MB flash memory and 16–256 KB EEPROM, serves as a data collector or a front-end to reach remote substations equipped with sensors or actuators responsible for executing a specific action in the field.

Unfortunately, remote substations do not always envisage a holistic protection, which makes prevention from anywhere, at any time and at anyhow, a crucial issue. The vast majority of CIs are exposed daily to continuous changes, mainly from unforeseen faults, malfunctions, or deliberate disturbances as published by the Industrial Control Systems Cyber Emergency Response Team (ICS-CERT, 2015). So the technological competences of many CPSs should consist of the minimal protection services that traditional situational awareness (SA) systems demand, such as perception of the observed environment, and understanding their meaning and their projection in the near future, which was initially introduced in Endsley (1995).

However, SA solutions for critical control applications deployed in large distributions can be insufficient, as local protection through "dynamic services" is also required to ensure one of the eight priority areas defined by the National Institute of Standards and Technology (NIST, 2014). This area, known as wide-area situational awareness (WASA), not only focuses on monitoring critical components and their performance at all times, but also on automatically *anticipating, detecting and responding to unplanned faults*, and if necessary *restoring states* before major disruptions can arise. This means that WASA strategies should be consolidated in a methodological process that helps the underlying system extract, interpret and respond to threatening situations, as proposed in Alcaraz and Lopez (2013). However, this paper neglects the relevance of studying preventive and corrective measures, taking into account the properties of the context, the features of the underlying system and its technological capacities.

Generally, CPDs can be categorized according to their functional capacities: *weak*, *heavy-duty* and *powerful-duty* (Roman et al., 2007). Within the weak class are extremely constrained devices but ones with sufficient competences to run simple operations, such as ~4 MHz, 1 KB RAM, and 4–16 KB ROM (e.g., home-appliances, sensors). Devices belonging to the heavy-duty category are relatively expensive (e.g., handled-devices, smartphones) and are able to execute any simple or complex critical action. Their microprocessors are quite potent, working at around 13–180 MHz, 256–512 KB RAM, and 4–32 MB ROM, and within this category, we highlight the role of the RTUs, smart meters (~8–50 MHz, 4–32 KB RAM, and 32–512 KB flash memory) or industrial wireless sensor networks. Industrial sensors usually have slightly greater capacities than conventional ones, equipped with a ~4–32 MHz, 8–128 KB RAM, 128–192 KB ROM, and can protect the observed infrastructure through their sensorial modules and manage data streams. Finally, the powerful-duty class contains all those devices with significant capacity to address any action or application, such as servers, proxies or gateways. Considering this taxonomy, the main contribution here is to offer the necessary guidelines to build effective WASA solutions, which should rely on automatic and lightweight protection services that do not jeopardize the functions of the primary systems.

The chapter is organized as follows: Section 2 introduces the current prevention, response, and restoration solutions, examining all those features that make them suitable for devices deployed in cyber-physical contexts, thus promoting their applicability for future WASA applications. The exploration of the capacities that the different models can demand, and their ability to encourage accurate awareness and response to faults [Byzantines, transient, or fail-stop faults (Treaster, 2005)], is presented in Section 3 together with the conclusions and future work.

2 WASA: AUTOMATED PREVENTION, RESPONSE, AND RESTORATION
2.1 PREVENTION AND DETECTION

Within the field of prevention, it is important to consider detection techniques as specified in Chandola et al. (2009), Gyanchandani et al. (2012), and Kotsiantis (2007). Five groups of detectors are stressed: *data mining-based*, where the techniques directly depend on a set of data to find behavior pattern sequences; *statistical-based*, composed of interference tests to verify whether a specific instance data belongs to a statistical model; *knowledge-based*, which progressively acquires knowledge about specific threats; *information and spectral theory-based*, which is focused on the data itself, its order and its meaning; and all those other well-known *machine learning-based* techniques, such as artificial neural networks (ANNs), Bayesian networks (BNs), support vector machines (SVMs), rule-based, nearest neighbor-based, fuzzy logic, and genetic algorithms.

For each of the aforementioned classes, there is a set of machine learning subclasses. Within the data-mining category are the *classification-based* techniques; a set of classifiers capable of assigning data instances to normal/anomalous collections such as *classification and regression trees*. These structures are composed of tree-like constructions of fast computation, which is achieved by comparing their data instances against a precomputed model. *Association rule learning* is another subclass of data mining, which identifies the relationships between categorical variables using rules and thresholds to

prune. The effectiveness of the approach depends on the parameters that configure the pruning operations and their algorithms. Likewise, the *clustering-based algorithms* group data instances in clusters
through an unsupervised/semisupervised method where computing distances between data points is
required, like the *k*-means algorithm.

The statistical-based group contains the *parametric and nonparametric-based* models, such as
Gaussian-based models or histogram-based techniques, which control interfering data points according
to the data observed. These models are generally accurate and tolerant to noise and missing values,
providing a better picture of the confidential interval associated with the anomaly, but the accuracy
relies heavily on the complexity and length of their datasets. This class also comprises the operational
models based on computing counters whose values are bounded to predefined thresholds. This
constrained feature restricts their use to those dynamic scenarios in which their contexts are subject
to continuous changes and their values are not predefined properly. *Time series* and *Markov-based*
models are two further subclasses of this group. The former predicts behaviors through successive
and uniform time series, such as smoothing techniques based on weighted data instances and their variant exponential smoothing models. Although they are generally accurate and tolerant to insignificant
changes and missing values, they tend to produce weak models for medium and long-range forecasting
with a heavy dependence on past evidence and on the smoothing factor to forecast. Markov models are,
to the contrary, a mathematical representation whose quantitative values and operations are closely
related to a state transition (probabilistic) matrix. This matrix contains all the activity transactions without having to have knowledge of the situation, where the operational difficulty varies according to the
complexity and dimensionality of the situation, and its precision depends on the variations in the
activity sequence.

The techniques in knowledge-based detection progressively acquire knowledge about specific
threats, guaranteeing high accuracy with a low false-positive rate (FPR), flexibility and scalability
for expanding the detection engine with new knowledge. Gyanchandani et al. (2012) identify three
types of approaches in this field: *state transactions* through state transaction diagrams; *Petri nets* using
directed bipartite graphs restricted to conditions and events; and rule-based *expert systems* capable of
reasoning about provided knowledge. But despite the potential to autonomously recognize anomalies,
their intelligent implementations depend on the degree of granularity and maintenance of their knowledge. Regarding the information and spectral theory-based class, their statistical approaches are
planned in accordance with the irregularities in the data. For example, through the entropy it is possible
to identify anomalies whose feasibility is subject to the size of the dataset; and through spectral
schemes it is also possible to obtain time series and the characteristics of the communication channel
whose effectiveness is related to the degree of handling high-dimensional data and the complexity of
their approaches.

The intrinsic features of all these techniques are summarized in Table 1, and can be incorporated as
detection engines within intrusion detection systems (IDSs) to constantly monitor and evaluate evidence. They can be classified into three categories: *anomaly-based* so as to detect unforeseen deviations from normal behaviors; *signature-based* to perceive abnormal behaviors by matching each
instance to an updated database containing the threat models; and *specification-based* to detect behaviors according to the legitimate specifications of the system. These detection features are also analyzed
in Jokar (2012), stating that anomaly-based IDSs usually provide high FPRs with the ability to predict
unknown threats, but require complex training and tuning time. In contrast, signature-based IDSs have
low FPR, but more difficulty ensuring the detection of unknown threats; whereas specification-based

Table 1 Classification and Characteristics of the Preventive Techniques

Category	Subcategory	Examples	Low Complexity	Speed of Classification	Speed of Learning	Handles Parameters	Compressibility	Accuracy	Learning From Observation	Control — Interdep. Data	Control — Missing Data	Control — Redundancy	Control — Noise	Control — Subtle Changes	Control — Drastic Changes	Incremental Learning
Data mining-based	Classification-based	Classification trees	✓	✓	✓	✓	✓		✓	✗	✓	✗				✓
		Regression trees	✓		✓	✓			✓	✗		✗	✗			✗
	Association rule-learning based	Apriori, FP-growth, etc.	✗			✗	✓	✗		✓		✓				
	Clustering-based	K-means, hierarchical clustering, etc.	✓	✗		✓			✓	✗	✓		✓			✓
Statistical-based	Parametric and nonparametric	In general	✓		✗	✗	✓	✓			✓	✓	✓			
		Operational models	✓			✗	✗	✗					✗	✗	✗	
		Smoothing	✗			✗	✗	✓					✗	✓	✗	
	Time series		✗		✗	✗	✗	✓	✓						✗	
	Markov chains	Markov models, HMMs, hierarchical Markov models, etc.				✗	✗					✓	✗			✓
Knowledge detection-based						✓	✗							✓		
Information and spectral theory-based						✗	✗					✓	✗			
Other machine learning-based	ANN	In general	✗	✓	✗	✗	✗	✓	✓	✓	✗	✗	✗			
	BN	Naïve Bayes networks	✗	✓	✓	✓	✓		✓		✓	✗	✓			
	SVM		✓	✓	✗	✓	✓	✗	✓	✗	✓	✗				
	Ruled-based		✓	✓		✓	✗	✓	✓	✓		✓	✗			
	Rule learners		✓	✓	✗	✓	✓	✓		✗	✗	✗	✗			✗
	Nearest neighbor	K-nearest	✓	✗	✓	✓	✓	✗	✓	✗	✗	✗	✗			✗
	Fuzzy algorithm						✗	✗	✓							✗
	Genetic algorithm		✗	✓	✓	✓	✓	✗			✓		✓			✓

IDSs also guarantee low FPRs with the ability to notice new threats/vulnerabilities within a given system. However, this detection mode has high computational costs to implement predefined threat models, which are closely linked to the functional capacities of their devices.

2.2 RESPONSE

Although IDSs help detect the existence of faults, it is also imperative to take evasive and/or corrective actions to prevent the propagation of secondary effects caused by these faults (Stakhanova et al., 2007). Intrusion response systems (IRSs) are systems that have all the IDS capabilities but with the necessary support to stop incidents (Scarfone and Mell, 2007). Since, CPSs are characterized by complex interconnected systems, the avoidance of faults and consequent cascading effects through these preventive systems is of paramount importance. Traditionally, the response to a threat was manual and required a high degree of expertise, but the increasing complexity and speed of cyber-attacks, and the ramifications of faults show the acute need for complex intelligent dynamic IRSs (Stakhanova et al., 2007).

Countermeasures applicable by an IRS can be divided into: *passive* and *active* responses (see Table 2). Passive reactions are usually included in the normal operation of some IDSs (Stakhanova et al., 2007), and can be implemented in almost all CPDs according to their capacities (i.e., powerful devices would be able to implement sophisticated mechanisms; whereas weak devices would implement simpler methods). Within passive reactions there are two main categories: *administrator notification* and *prevention measures*. The first logs the system's information and state and alerts the system's administrator to control the situation. Notifications to administrators are the most common operations implemented in deployed IDS/IRS solutions. Alert systems do not require high computational power, thus they can be implemented by any CPD.

The presence of prevention measures depends on the computational capabilities of the devices. We have distinguished six chief kinds of preventive solutions. *Cryptography* is an effective approach to prevent attackers from understanding captured data (Xing et al., 2010); it is usually used in normal to powerful-duty devices. *Security policies* are the security measures taken by the organizations and they should be implemented by all the devices in the CPS regardless of their computational power; IRSs can follow these guidelines to identify violations in the security policies of the surveilled system (Scarfone and Mell, 2007). *Monitoring* tools (e.g., IDSs) supervise the local (host or network) operations and state in search for intrusive behaviors; IDSs range from heavy-duty to lightweight solutions, thus monitoring could be present in all types of devices in the CPSs.

Protective/defensive infrastructure comprises devices or system configurations designed for protection tasks (e.g., firewalls, demilitarized zones, and proxies) (Byres et al., 2005); this infrastructure is compatible with CPDs of all ranges of computational power. *Low-level preventive mechanisms* are physical security measures implemented at the lower layers of the communication systems to prevent intrusions [e.g., directional antennas in wireless devices, or synchronized clocks (Xing et al., 2010)]; ideal for weak-duty devices. *Session/communication measures* are techniques that add security at the session or communications levels [e.g., packet leashes or cookies (Gollmann, 2008)]; these protective methods are useful for powerful devices that have to deal with remote connections and queries via the Internet.

Concerning the *active reaction mechanisms*, we can divide them into two groups (Stakhanova et al., 2007): *host-based* and *network-based* response actions. Host-based responses refer to those local operations, which modify parameters or processes within the affected CPDs, e.g., operations on files,

Table 2 Classification and Characteristics of the Response Mechanisms

			Low Complexity	Easy to Implement	Low Use of Resources	Requires Additional HW	Adaptable to Changes	Low Impact of Response	Low-Risk Automation
Passive	Administrator notification	Generate system logs	✓	✓		*	✓	✓	✓
		Generate alarm	✓	✓	✓		✓	✓	✓
		Generate report	✓	✓	✓		✓	✓	✓
	Prevention measures	Cryptography	*	✓	*			✓	✓
		Security policies	✓	✓	✓			✓	✓
		Monitoring	✓	✓		*	✓	✓	✓
		Protective/defensive infrastructure	✓	✓	✓	✓		✓	✓
		Low-level preventive mechanisms	✓	✓		*			
		Session/communication measures		✓	✓				✓
Active	Host-based	Operations on files		✓	*			✓	
		Operations on user accounts		✓	*				✓
		Operations on processes and services			*				
		Trust mechanisms			*		✓	✓	*
	Network-based	Disable/block operations	✓	✓	✓		✓		
		Isolation actions	✓	✓	✓		✓		
		Routing		✓	✓	✓	✓	✓	
		Deceiver devices					✓		*

operations on user accounts, and operations on processes and services. Here, trust-based mechanisms, such as reputation, credit-based trust or token-based trust, are effective information protection methods in communication networks that can be implemented at different levels in the CPSs (Meghdadi et al., 2011). Network-based responses, conversely, correspond to those activities performed in the communications network that affect its services and parameters, e.g., disabling or blocking network operations (Ingols et al., 2009), isolating segments of the network (Meghdadi et al., 2011), modifying routing parameters (Karlof and Wagner, 2003), or setting up deceiver devices (Specht and Lee, 2004). These responses can be implemented in the CPS regardless of the computational power of its devices, since they focus on the network's operation.

Table 2 overviews the main active and passive response measures that IRSs can implement. For each identified method, we have analyzed several characteristics that determine the environment in which it can be deployed in terms of required resources, adaptability and performance. Parameters such as complexity of the solution and the implementation, the consumption of the resources of the infrastructure, the adaptability and impact of the responses, and the automation capabilities are of paramount importance to determine the applicability of given countermeasures (hence, IRSs) to critical systems such as CPSs. In Table 2, we analyze these characteristics for each of the main sets of responses and indicate the strengths of each mechanism (note that * indicates the dependence on the implementation of the countermeasure).

2.3 **RESTORATION**

Recovery mechanisms comprise all those actions related to resilience and fault-tolerance that help the underlying system to return to its natural state and operating configuration (Treaster, 2005; Bansal et al., 2012). *Replication-based* techniques are some of the most popular fault-tolerance techniques, which replicate functionalities and add redundancy to the system. The type of data consistency (linearizability, sequential, and casual) and the replication mode, *active* or *passive* are important. Passive replication activates the backup systems only when needed, where primary devices are the only ones that can manage replicas. In contrast, active replication constantly replicates evidence and configurations of the primary entity to preserve assets and maintain the backup elements updated at all times.

Although the active mode individually manages evidence in multicast mode to favor the response, the redundancy management causes complexities. When replicas need to be compared to identify Byzantine faults, a voting process in distributed networks is normally required to manage consensus according to detected events. Process level redundancy (PLR) is another example of active replication. It detects transient faults, which are less severe than Byzantines but harder to diagnose. For detection, their algorithms demand software-centric approaches to detect transient faults, resources to reduce overhead and redundant processes to schedule the processes across all system assets.

Rollback is another well-known recovery approach. It includes a *checkpoint-based rollback* with dependency on storage points containing current information, and a *log-based rollback* (also known as a message logging protocol) comprising checkpoints and a record of nondeterministic events. Within the checkpoint class, there are coordinated and uncoordinated approaches. The former synchronizes checkpoints to restrict the rollback propagation, but hampers the recovery time and their functionality in critical contexts. Uncoordinated checkpoints, to the contrary, individually execute checkpoints to later combine them with a message logging protocol, thereby guaranteeing a complete picture of the process's execution. According to Treaster (2005), there are three main log-based techniques: pessimistic/synchronous, optimistic/asynchronous, and causal. Pessimistic logging techniques consist of registering each message received by an entity to be subsequently re-sent, but only if necessary, during the rollback stages; whereas optimistic protocols register events to a volatile memory to later (only periodically) store them in disk. Although this protocol can simplify storage complexities in disk, the recovery process can become much more complex when the logs have not been stored properly over time. Finally, causal protocols log nondeterministic events in a casual manner, but they add the problem of optimistic protocols in which temporarily registered events may be lost unexpectedly.

As the checkpoints can be costly, experts (Ruchika, 2013; Veronese et al., 2009) recommend applying heterogeneous replication-based checkpoints to enhance performance and guarantee tamper-resistance for faults. But their implementation can bring about complexities in the recovery processes due to redundancy, a characteristic that has also been considered in *fusion-based* techniques. These techniques address the problem by relying on fewer backup devices as fusion points, instead of actively configuring replication-based approaches in all the devices. Nonetheless, this characteristic also tends to increase implementation costs and complexities for recovery by maintaining the fusion points up to date.

Table 3 encompasses all the aforementioned properties, which are also sustained by Bansal et al. (2012). These authors stress that the performance of each approach (denoted with * in Table 3) depends on a set of factors. Replication-based schemes vary according to the number of replicas produced within the system (the performance decreases as the number of replicas increases); checkpoints and rollback depend on the frequency and size of the checkpoints; fusion-based on the rate of faults (low rate of faults improves the recovery); and PRL on the set of faults. If, in addition, the approaches are equipped to incorporate multiple fault detectors, the level of reliability, accuracy and adaptation can become quite noteworthy, thereby favoring their use for cyber-physical contexts.

3 GUIDELINES FOR WASA: ANALYSIS AND DISCUSSION

To assess the applicability of each of the aforementioned methods for CPSs, it is necessary to take into account their computational features and the sensitive nature of the application context. Control systems generally demand (Alcaraz and Lopez, 2012) operational performance, reliability and integrity, resilience and security, and safety-critical. However guaranteeing these requirements also implies considering the complexities and characteristics of the different approaches (see Tables 1–3) and their suitability for support by the different CPDs.

For prevention, we consider parameters related to the complexity, the accuracy and the general performance of the models, understanding that supervised techniques have a better general performance in real-life problems than unsupervised ones (Sadoddin and Ghorbani, 2007). Solutions such as BNs are difficult to implement successfully in a complex constrained environment, since their computational cost and complexity of implementation increase with the complexity of the system being modeled. Whenever it is necessary to evaluate the correctness of a model, or to add a certain degree of expert knowledge, ANNs and SVMs are more restrictive methods; e.g., SVMs have low complexity models but the learning process's low speed adds implementation overhead. Thus the methods that are better suited for IDSs in constrained scenarios are those based on logic, such as decision trees, optimized rule learners and fuzzy logic, and on simplified computation models such as operational

Table 3 Classification and Characteristics of the Restoration Techniques

			Low Complexity	Capacity for Storage	Performance	Consistency	Accuracy	Responsiveness	Adaptability	High rate of Redundancy	Configurability	Handles N – Faults	Handles Byzantine Faults	Handles Transient Faults
Replication-based	Active	In general	✗	✓	*	✓	✓	✓	✓	✓		✓	✓	✓
		Voting	✗	✓	*	✓	✓	✓	✓	✓		✓		
		PLR		✓	*	✓	✓	✗	✗	✓				
	Passive			✓	*	✓		✗	✓	✗	✗	✓		
Rollback-based	Checkpoints	Coordinated	✗	✓	*	✓		✗		✓		✓		
		Uncoordinated	✗	✓	*			✗		✓		✓		
	Message logging	Replication-based	✗	✓	*	✓		✗		✓		✓		
		Pessimistic	✗	✓	*		✓	✓		✓		✓		
		Optimistic	✗	✓	*			✓	✓			✓		
		Casual	✗	✓	*	✓	✗	✗	✗			✓		
Fusion-based			✗	✓	*	✓	✗	✗	✗	✗		✓		

or rule-based models. Decision trees have well-balanced characteristics for this specific context, while rule learners have several drawbacks (e.g., accuracy) that can be easily overcome (e.g., implementing boosting algorithms), but they provide several capabilities (performance, comprehensibility, and ease in introducing rules by experts), that are vital in critical contexts. An example of such a system is available in the framework proposed in D'Antonio et al. (2009) where the IDS component has a classifier engine fed from the knowledge taken from other artificial intelligence modules such as processors and pattern recognition algorithms.

Knowledge-based and rule-based systems, optimized SVN and statistical methods are suitable for integration into heavy-duty devices because of their accuracy. Depending on the regularity of the traffic, the application context and its capacity for change, optimized parametric or nonparametric solutions, with the exception of the operational models, can be effective approaches since they are moderately complex and have a high efficiency for detection. Similarly, powerful-duty devices can also adopt the knowledge-based schemes and statistical methods because they can autonomously detect slight or abrupt anomalies with a high accuracy. Hidden Markov models (HMMs) are potential tools for detecting hidden dynamics and extracting knowledge where there may be obfuscation in the traffic received. An example of the use of HMMs in an IRS can be found in Haslum et al. (2007), where the HMM is used to represent the interaction between the attacker and the system's network. Nonetheless, other methods could be equally applicable for powerful-duty environments, although there are methods (e.g., SVN, rule learners, ANNs, genetic algorithms, etc.) with costly training processes that are less appropriate for dynamic networks with irregular traffic. Although they are present in traditional networks, as in Fessi et al. (2014), where the authors present an IRS using genetic algorithms for response selection, the frequent occurrence of asynchronous disturbances in the CPS dynamic scenarios may make the IDSs trigger the learning mechanisms more frequently, increasing the overhead in the underlying systems.

These restrictions can be translated to the IRSs, since the applicability of given countermeasures heavily depends on the characteristics of the environment where the response is launched. Currently, IRSs, designed specifically for critical systems, implement some of the response mechanisms mentioned in Section 2.2, particularly passive methods. However, most systems still lack important passive prevention mechanisms because of the legacy equipment, the proprietary protocols and components traditionally present in these environments (Alcaraz and Lopez, 2012). Nevertheless (see Table 2) most passive and preventive mechanisms are easy to implement and introduce few overheads. They are effective and suitable for application in CPSs regardless of the computational power of the environment, but are especially indicated for constrained contexts.

Active reaction solutions, however, are rarely present in these scenarios, since active responses usually imply the introduction of sophisticated mechanisms, implying a higher use of computational resources and equipment. Methods that modify the behavior of the system or the network are only suitable for those sections of the CPSs containing powerful equipment capable of devoting sufficient computational power to the IRS. Additionally, other methods that require extra hardware or need to run powerful intelligent algorithms are only applicable to powerful-duty environments of the CPS, since they need to perform complex operations with high requirements on computational power.

Regarding restoration, it is possible to note from Table 3 that the vast majority of techniques have significant computational and spatial complexities since they require high rates of redundancy. Logging events in an optimistic or casual manner, specification of multicast protocols and configuration of hybrid networks, in which the handling of replicas and checkpoints could be concentrated in

Table 4 Adapting WASA Techniques to Cyber-Physical Environments

	Weak	Heavy-Duty	Powerful-Duty
Prevention	Decision tress Rule learners Operational models Knowledge-based Rule-based Fuzzy logic	Knowledge-based SVN Rule-based Statistical-based	Knowledge-based Statistical-based
Response	Administrator notification (logs, alarms, reports) Security policies Low-level prevention Cryptography	Monitoring Session/communication measures Modification operations (files, accounts, processes) Routing (isolation, blocking)	Monitoring Operations on processes and services Deceiver techniques Trust mechanisms
Restoration	Passive replication Message logging (optimistic, casual)	Active replication Checkpoint replication-based Message logging (pessimistic, optimistic)	

some heavy-duty or powerful-duty nodes (RTUs, gateways, servers), could resolve the implicit overheads of the models built in constrained environments such as the decentralized architecture for CPS contexts described in Pradhan et al. (2014). External storage systems could benefit the data storage and the restoration phases in those nodes classified as weak. The nodes could connect to the cloud and leave backup instances of critical evidence favoring accountability and audits (Alcaraz and Lopez, 2014).

In less restrictive, distributed scenarios, where the rate of redundancy can be higher, more complex reparation mechanisms can be adapted such as the save-point, trees-based rollback presented in Koldehofe et al. (2013) for distributed environments. Even so, the provision of lightweight and dynamic fault tolerance systems composed of adaptive models in this type of application context is still required. Responsiveness, adaptability, accuracy and performance of the models are fundamental criteria when developing methods, without forgetting the need to launch optimized strategies that provide satisfactory average-time and with linear approximations as stated in Alcaraz and Wolthusen (2014).

Table 4 summarizes the analysis done in this section and concludes the chapter. The analysis has determined that it is still necessary to continue exploring new strategies that help simplify the implicit overheads in awareness and response tasks, providing more dynamic lightweight solutions where the rate of redundancy reaches minimum values, and the degree of accuracy and responsiveness reach high values. These goals should be part of future work where experts in the field of CI protection should combine efforts to foster the concept of WASA in all those sections that include a CI, without degrading the existing cyber-physical interdependencies and guaranteeing a suitable tradeoff between operational performance and protection (Alcaraz and Lopez, 2012).

ACKNOWLEDGMENTS

C. Alcaraz is supported by the "Ramón y Cajal" (RYC-2014-1631) research program financed by the Ministerio de Economía y Competitividad, and L. Cazorla is supported by a FPI fellowship from the Junta de Andalucía through the project FISICCO (P11-TIC-07,223). Additionally, this work has been partially supported by the project PERSIST (TIN2013-41,739-R) financed by the Ministerio de Economía y Competitividad.

REFERENCES

Alcaraz, C., Lopez, J., 2012. Analysis of requirements for critical control systems. Int. J. Crit. Infrastruct. Prot. 2 (3–4), 137–145.

Alcaraz, C., Lopez, J., 2013. Wide-area situational awareness for critical infrastructure protection. IEEE Comput. 46 (4), 30–37.

Alcaraz, C., Lopez, J., 2014. WASAM: a dynamic wide-area situational awareness model for critical domains in smart grids. Futur. Gener. Comput. Syst. 30, 146–154.

Alcaraz, C., Wolthusen, S., 2014. Recovery of structural controllability for control systems. In: Eighth IFIP WG 11.10 International Conference on Critical Infrastructure Protection, vol. 441. Springer, Virginia, USA, pp. 47–63.

Bansal, S., Sharma, S., Trivedi, I., 2012. A detailed review of fault tolerance techniques in distributed systems. Int. J. Internet and Distrib. Comput. Syst. 1 (1), 33–39.

Byres, E., Karsch, J., Carter, J., 2005. NISCC Good Practice Guide on Firewall Deployment for SCADA and Process Control Networks. Centre for the Protection of National Infrastructure.

Chandola, V., Banerjee, A., Kumar, V., 2009. Anomaly detection: a survey. ACM Comput. Surv. 41 (3), 15–58.

D'Antonio, S., Oliviero, F., Setola, R., 2006. High-speed intrusion detection in support of critical infrastructure protection. In: Lopez, J. (Ed.), Critical Information Infrastructures Security: First International Workshop, CRITIS 2006, Samos, Greece, August 31 – September 1, 2006. Revised Papers. Springer, Berlin, Heidelberg. ISBN 978-3-540-69084-9, pp. 222–234. http://dx.doi.org/10.1007/11962977_18.

Endsley, R., 1995. Toward a theory of situation awareness in dynamic systems. Hum. Factors 37 (33), 32–64.

Fessi, B.A., Benabdallah, S., Boudriga, N., Hamdi, M., 2014. A multi-attribute decision model for intrusion response system. Inf. Sci. 270, 237–254.

Gollmann, D., 2008. Securing web applications. Inf. Secur. Tech. Rep. 13 (1), 1–9.

Gyanchandani, M., Rana, J., Yadav, R., 2012. Taxonomy of anomaly based intrusion detection system: a review. Neural Netw. 2 (43), 1–14.

Haslum, K., Abraham, A., Knapskog, S., 2007. In: Third International Symposium on Information Assurance and Security. DIPS: A framework for distributed intrusion prediction and prevention using hidden markov models and online fuzzy risk assessment, pp. 183–190. http://dx.doi.org/10.1109/IAS.2007.67.

ICS-CERT, 2015. Years in review 2009–2014. https://ics-cert.us-cert.gov/Other-Reports (accessed 24.06.15).

Ingols, K., Chu, M., Lippmann, R., Webster, S., Boyer, S., 2009. Modeling modern network attacks and counter-measures using attack graphs. In: Computer Security Applications Conference, 2009. ACSAC '09. Annual, pp. 117–126. http://dx.doi.org/10.1109/ACSAC.2009.21. ISSN 1063-9527.

Jokar, P., 2012. Model-Based Intrusion Detection for Home Area Networks in Smart Grids. University of Bristol, Bristol, pp. 1–19.

Karlof, C., Wagner, D., 2003. Secure routing in wireless sensor networks: attacks and countermeasures. Ad Hoc Netw. 1 (2), 293–315.

Koldehofe, B., Mayer, R., Ramachandran, U., Rothermel, K., Völz, M., 2013. Rollback-recovery without checkpoints in distributed event processing systems. In: Proceedings of the 7th ACM International Conference on Distributed Event-based Systems (DEBS '13). ACM, New York, NY. ISBN 978-1-4503-1758-0, pp. 27–38. http://dx.doi.org/10.1145/2488222.2488259.

Kotsiantis, S.B., 2007. Supervised machine learning: A review of classification techniques. In: Proceedings of the 2007 Conference on Emerging Artificial Intelligence Applications in Computer Engineering: Real Word AI Systems with Applications in eHealth, HCI, Information Retrieval and Pervasive Technologies. IOS Press, Amsterdam, The Netherlands, pp. 3–24. ISBN: 978-1-58603-780-2. http://dl.acm.org/citation.cfm?id=1566770. 1566773.

Meghdadi, M., Ozdemir, S., Güler, I., 2011. A survey of wormhole-based attacks and their countermeasures in wireless sensor networks. IETE Tech. Rev. 28, (2).

NIST, 2014. Guidelines for Smart Grid Cybersecurity—Volume 1—Smart Grid Cybersecurity Strategy, Architecture, and High-Level Requirements. The Smart Grid Interoperability Panel Cyber Security Working Group (SGIP), National Institute of Standards and Technology, NISTIR 7628 Rev 1.

Pradhan, S., Otte, W., Dubey, A., Szabo, C., Gokhale, A., Karsai, G., 2014. Towards a Self-adaptive Deployment and Configuration Infrastructure for Cyber-Physical Systems. Institute for Software Integrated Systems, Nashville. Technical Report, ISIS-14-102. http://www.isis.vanderbilt.edu/sites/default/files/Pradhan_SEAMS_TechReport.pdf.

Roman, R., Alcaraz, C., Lopez, J., 2007. A survey of cryptographic primitives and implementations for hardware-constrained sensor network nodes. Mobile Netw. Appl. 12 (4), 231–244.

Ruchika, M., 2013. Schemes for surviving advanced persistent threats. Diss. Faculty of the Graduate School of the University at Buffalo, State University of New York.

Sadoddin, R., Ghorbani, A.A., 2007. A comparative study of unsupervised machine learning and data mining techniques for intrusion detection. In: Perner, P. (Ed.), Machine Learning and Data Mining in Pattern Recognition: 5th International Conference, MLDM 2007, Leipzig, Germany, July 18–20, 2007. Proceedings. Springer, Berlin, Heidelberg. ISBN 978-3-540-73499-4, pp. 404–418. http://dx.doi.org/10.1007/978-3-540-73499-4_31.

Scarfone, K., Mell, P., 2007. Guide to Intrusion Detection and Prevention Systems. Institute of Standards and Technology, Computer Security Division. NIST Special Publication (800–94), Gaithersburg.

Specht, S.M., Lee, R.B., 2004. Distributed denial of service: Taxonomies of attacks, tools, and countermeasures. In: Proceedings of the 17th International Conference on Parallel and Distributed Computing Systems, 2004 International Workshop on Security in Parallel and Distributed Systems. pp. 543–550.

Stakhanova, N., Basu, S., Wong, J., 2007. A taxonomy of intrusion response systems. Int. J. Inf. Comput. Secur. 1, 169–184.

Treaster, M., 2005. A survey of fault-tolerance and fault-recovery techniques in parallel systems. ACM Comput Res. Repos. 501002, 1–11.

Veronese, G.S., Correia, M., Bessani, A.N., Lung, L.C., 2009. Highly-resilient services for critical infrastructures. In: Proceedings of the Embedded Systems and Communications Security Workshop. http://citeseerx.ist.psu.edu/viewdoc/citations;jsessionid=47C8AD961F92582ECAC5A48317DF5839?

Xing, K., Srinivasan, S.S.R., Rivera, M.J.M., Li, J., Cheng, X., 2010. Attacks and countermeasures in sensor networks: A survey. In: Huang, C.-H.S., MacCallum, D., Du, D.-Z. (Eds.), Network Security. Springer US, Boston, MA. ISBN 978-0-387-73821-5, pp. 251–272. http://dx.doi.org/10.1007/978-0-387-73821-5_11.

APPLICATIONS

THE NEED OF DYNAMIC AND ADAPTIVE DATA MODELS FOR CYBER-PHYSICAL PRODUCTION SYSTEMS

21

C. Brecher[†], C. Ecker[†], W. Herfs[†], M. Obdenbusch[†], S. Jeschke*, M. Hoffmann[†], T. Meisen[†]

Institute Cluster IMA/ZLW & IfU, RWTH Aachen University, Aachen, Germany*
RWTH Aachen University, Aachen, Germany[†]

1 INTRODUCTION

The entry of digitalization into modern production has increased dramatically over recent years, resulting in the emergence of so-called cyber-physical production systems (CPPSs). Nowadays, embedded and automation devices such as programmable logic controls (PLC), sensors, actors, as well as production machinery—like robots or machines—are the building blocks of modern production systems. As a result of development and global ramp-up strategies, shortening product life cycles, and increasing variants, the complexity of such distributed manufacturing systems—as well as the complexity of the underlying production and manufacturing processes—have increased substantially. In order to handle the arising complexity, one strategy consists of increasing production system virtualization equipped with partly autonomous decision-making entities that can be interconnected consistently by means of information technology. With regard to the CPPS approach, by introducing selfawareness, semantic interfaces, and environmental interpretation, embedded systems make use of dynamic and adaptive data models to match the "cyber" requirements.

According to the digitalization strategy, CPPS forms the fundamental basis of the German government's high-tech strategy known as the "Industrie 4.0" initiative. Following the well-known three previous industrial revolutions (industrialization—*steam engine*, rationalization—*collaborative mass production*, and automation—*PLC and robots*) the fourth industrial revolution has already proclaimed in advance as the combination of global production technology with information and communication technology (ICT). As a general approach, Industrie 4.0 covers not only the usage of intelligent embedded devices and their vertical (business processes) and horizontal (value networks) interconnectedness, but also industrial processes in production, engineering, supply chain, or the overall product life cycle management (PLM) (Fig. 1). Continuous engineering over the whole value chain is available and required at the same time (Kagermann et al., 2013).

Cyber-Physical Systems. http://dx.doi.org/10.1016/B978-0-12-803801-7.00021-3

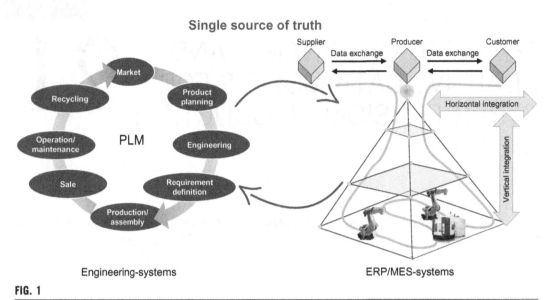

FIG. 1

Bi-directional information exchange between the product life cycle and the benefit chain.

Reference papers like (Kagermann et al., 2013) recommend eight indispensable fields of activity, which are, in particular: *standardization and reference architecture, control of complex systems, extensive broad band infrastructure for industry, safety, work organization and design, training, legal requirements, and resource efficiency.* The biggest challenge facing these areas is integrating different views on urgent aspects (e.g., individualization, energy efficiency, and information consistency), technologies (e.g., OPC UA), or paradigms (e.g., service-oriented architecture, viable systems model) into one consistent approach.

In 2014, the Laboratory for Machine Tools and Production Engineering (WZL) explored the topic Industrie 4.0—"The Aachen Approach"—at the Aachen Machine Tool Colloquium (Brecher et al., 2014). The complex situation was formed in cooperation with the WZL and an industrial advisory board to produce one overall picture that evolves from the state-of-the-art without calling for disruptive changes (Fig. 2). The four primary fields of action are single source of truth (SSoT), cooperation, automation, and IT globalization.

In this chapter, we focus on the fields of IT globalization and automation, in which specific challenges regarding the interoperability of different levels of currently hierarchically organized production systems exist. By gaining higher transparency in terms of machine diagnostics and real-time production states, a dynamic reconfiguration of the production is possible during execution. This opens up various perspectives in terms of product individualization and selfoptimizing CPPS—finally reaching the "Lot Size 1" goal for production systems. By enabling process transparency as well as consolidation of information in different levels of production, IT globalization gains increasing importance. The next step in propagating production information extends the scope from the Internet site of a company to a worldwide web approach, enabling the distribution of live process data from different production facilities. However, the current situation in production organization is still lacking profound interoperability in terms of vertical information exchange.

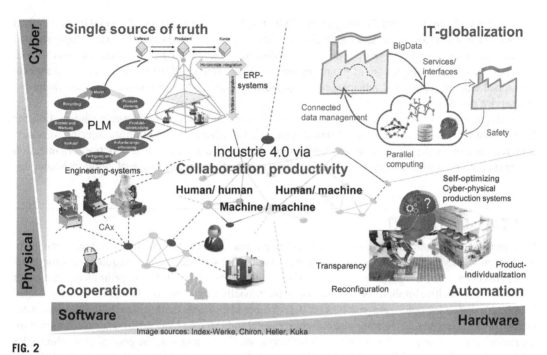

FIG. 2

"Industrie 4.0—The Aachen Approach" presented at the Aachen Machine Tool Colloquium.

To reach flexible, adaptive, and reconfigurable production in terms of Industrie 4.0, interoperability is needed between automation and control layer systems as well as those of management and resource planning. This can only be achieved by establishing both horizontal and vertical integration of information from various systems. The horizontal integration of data is mostly performed by proprietary systems and interface standards. Thus, information exchange in an encapsulated layer of the automation system is organized in terms of a fixed system with predetermined information. However, the vertical integration of information is not trivial in terms of a dynamic production environment. Traditionally, the well-known automation pyramid inspires information exchange between different layers of the information system. Such concepts serve as pseudo-standardized architectural system patterns in ICT for factory automation.

Hence, in order to use information that is provided by these sensors and actuators which act like CPPS along the entire production chain and in a holistic manner, the integration of information from the field level to production management systems at the top of the automation pyramid have to be carried out seamlessly—and without additional efforts. From a semantic point of view, the autonomous integration and use of these data for short and mid-term production planning and reconfiguration purposes requires an integrated understanding of the underlying information.

Based on the demand for "cyber"—or virtual—representations of different systems within CPPS in combination with the overall picture shown in Fig. 2, we also discuss the challenge of realizing a life cycle attending modeling and overarching model usage. Within the engineering phase aspects such as meta-modeling, conceiving of new features within simple geometry-based CAD-tools, data consistency along different disciplines, and the integration of plant and process models are of special interest. When dealing with new control paradigms, the challenge of matching reality and virtuality, back propagation of process information for model sharpening, product-centered manufacturing execution systems (MESs), and cloud-based technologies are of major importance to the impact of the entire process. All aspects mentioned lead to significant advantages in competition for producing companies, manufacturers, and service providers.

Exemplary applications covered within this article are model-based human-robot interaction for flexible assembly automation, a cloud-based approach for advanced condition monitoring, product-centered control in the WZL's Smart Automation Lab, and a virtualized production organization use case based on artificial intelligence and optimized operations by the usage of cloud technologies.

2 STATE-OF-THE-ART

In terms of the communication in CPPS, two major requirements have to be taken into account: first, technical requirements for communicating in sensor networks; and second, semantic requirements for information exchange. The technical requirements of the communication in CPPS can be implemented either implicitly or explicitly. Implicit communication is traditionally carried out by the use of tracking-related features like barcodes, quick response (QR) codes, or other identifiable marks that are attached onto a device or product. In general, tracking objects requires a central instance that consolidates the acquired information to some storage and propagates it into the system. Nevertheless, as the general idea behind CPS supposes a sort of autonomy for each entity in the network, tracking-related solutions are not suitable for the design of CPPS for adaptive, selfoptimizing environments. The communication in CPPS needs to be carried out in an explicit way using well-defined communication protocols.

An approach that is continuously gaining importance for machine-to-machine communication in industrial environments is the previously mentioned OPC Unified Architecture (OPC UA), which is the successor of the widely accepted OPC standard for the vertical integration of information in production environments. OPC UA has emerged from a de facto standard to a standardized guideline: the IEC 62541 (International Electrotechnical Commission, 2015). In traditional automation, low-level production facilities are integrated into the ICT in a manual—and rarely standardized—way. To connect field devices with higher systems of the automation pyramid, interfaces and drivers have to be designed specifically for each different device in the machine layer. Using OPC UA, the information modeling and interface capabilities are separated from each other. In this way, production components can be modeled using an information modeling approach, and accessed using services. By providing abstract services, OPC UA delivers basic functionalities to create a service-oriented architecture (SOA) of the automation system (Hensel, 2012; Leitner and Mahnke, 2006). OPC UA integrates different specifications of Classic OPC into a single service set—thus, components and automation networks based on OPC can be easily integrated (Hannelius et al., 2008).

Another advantage of OPC UA is the configurability of the new standard. The OPC UA specification only stipulates the message format of the information that is sent. Unlike to OPC Classic, OPC UA does not specify an API (Leitner and Mahnke, 2006). Hence, the user of the OPC UA server is able to use or implement an API of their personal preference, enriched by semantic information and domain-specific knowledge of a particular field—even though adapting the information model is still a complex task. All communication between OPC UA clients and servers is performed using the communication stack. There is a client-side and a server-side communication stack. Both stack APIs can be developed in individual programming languages as long as their concepts support the technology mapping given by the OPC UA specification (Leitner and Mahnke, 2006).

The described configurability of the OPC UA API provides the user with a freedom of choice—in terms of technologies or programming languages—in order to access information in OPC UA networks. This way, OPC UA delivers a flexible usability in different environments, and for different purposes. This availability is linked to several advantages:

- The application, the context, or the purpose of an information or automation system determines the properties of the OPC UA communication stack API in use—for instance, whether the application is complex for major server environments, or lightweight for usage in small, embedded devices.
- The communication with and in OPC UA networks can be integrated into existing automation and enterprise communication systems and through all levels of the enterprise network.
- Tool interoperability with other systems or components of the factory using OPC UA is guaranteed, as the programming/development language is free to choose. This facilitates an embedding of OPC UA systems into business processes, such as data integration chains or persistence layers.

These advantages of OPC UA are able to address challenges that are introduced by Industrie 4.0 scenarios. Thus, many use cases that are in the main scope of the fourth industrial revolution can be accomplished if automation systems are fully equipped with OPC UA capable devices and the corresponding infrastructures. The embedding of OPC UA into industrial environments is a first step towards the goals of Industrie 4.0. However, there are more challenges to meet in terms of syntactical heterogeneity in industrial production. Further steps consist of creating structural and semantic standards.

One major challenge is the embedding of automation and production data from the technical system to a corporate information system in such manner that flexible, dynamic, and selfoptimizing CPPSs are realized within the production environment. Thus, information modeling in CPPS is discussed next.

3 INFORMATION MODELING (META-MODELING) IN CPS/CPPS

The semantic information exchange for the communication in CPS is performed in terms of ideas that are inspired by the Internet of Things (IoT). A compound of open, dynamic, and autonomous entities represents the communication process of CPS. Unlike communication in networks similar to the Internet, it does not rely on a central server instance that manages or organizes the requests from clients. In terms of the communication in CPS, each entity is able to serve as a client and as a server, sending and receiving messages autonomously, as well as processing requests from incoming messages. However, since there is no fixed hierarchy or structure within these sorts of networks, the communication in CPS requires a higher degree of interoperability. The information that is exchanged in IoT-like (sensor) networks has to be structured in terms of a machine-readable and interpretable form. According to the interoperability definition of the IEEE organization, this implies that components or embedded devices as part of a CPS or an IoT network need to be capable of building ad-hoc networks and flexible compounds of communicating entities that have the ability to exchange information—and not just data—among each other, and to use this information directly and autonomously.

Such systems require some kind of semantic understanding of information in terms of context awareness and context sensitivity to provide the required extendibility and adaptive capabilities. Due to the high heterogeneity between the different associated systems (like sensors, application data, etc.), a simple data exchange is not sufficient. Nowadays, heterogeneity still represents one of the main challenges in information networks. One tempting solution to solve the problems of heterogeneity is standardization. Doan, Halevy, and Ives name different reasons why this approach often fails in practice (Doan, 2012): "It is often hard to agree on standards, since some organizations are already entrenched in particular schemas, and there may not be enough of an incentive to standardize. [...] In practice, standards work for very limited use cases where the number of attributes is relatively small, and there is strong incentive to agree on them (e.g., if exchanging data is critical for business processes). [...] Even in these cases, while the data may be shared in a standardized schema, each data source models the data internally with a different schema." When designing heterogeneous, extendable networks that convert and interoperate between different domains, one extendable and flexible standard (besides a communication and data exchange standard—which is one main reason for technical heterogeneity in production networks) that allows managing the information exchange is especially very hard—if not impossible—to define. Information modeling as a discipline deals with the formalization of a domain—or knowledge—and the data gathered and observed within it. Subsequently, we present and discuss our extended view of information modeling in CPPS.

Instead of using information models only to define a common language between engineers and business users—or to derive the logical data model—the information model is directly integrated as a central component into the technical, distributed system. Hereby, the information model is used to define concepts, relationships, axioms and constraints, as well as correlations and dependencies among service providers, consumers, and data providers. As already stated by Siau (1999), "Information modeling is the cornerstone of information systems analysis and design. Information models, the products of information modeling, [...] provide a formal basis for developing tools and techniques used in information system development." Hence, information modeling deals with the identification of the central concepts of a domain to reveal the underlying semantics of gathered data in that domain. Furthermore, following our approach, the concrete technical system implicitly uses the information model as basis for the underlying data and object models. Nevertheless, the main objective of information networks is not only to provide information and computation capacity for processing it, but intelligence with respect to intelligence assessment. This does not only include gathering, propagation, and storage of data, but also covers facilitating the system to use logical reasoning about the validity and reasonability of observations and data provisions. Therefore, the system has to provide needed inference capabilities to derive implicit information from data.

To make the information model useable to the underlying system, we need a formalized, machine-readable, computer-interpretable representation of the model. In recent years, ontologies have gained much attention regarding this task. In the context of communication in CPPS and OPC UA as underlying communication protocol, the next section describes how information models and OPC UA can provide an applicable and reliable information system for future industrial applications.

OPC UA consists of a basic information model, the so-called address space. In the address space, protocol features, information modeling rules, as well as basic object, variable, and data type definitions are provided. This basic meta-model is extensible by user-defined objects, variables, methods, and data types suitable to a certain production. In order to integrate this modeling approach into the communication infrastructure of an enterprise, OPC UA delivers the concept of aggregating servers (i.e., an information model that has been designed according to the information modeling

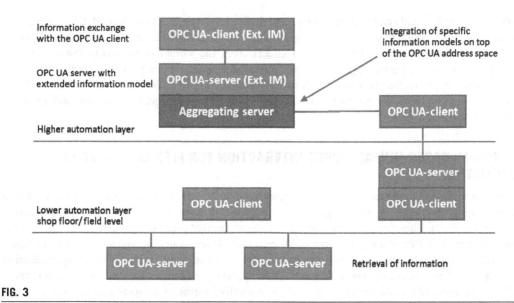

FIG. 3

OPC UA concept of the aggregating server for embedding of custom information models.

standards of OPC UA) that can be combined with the basic address space into an aggregated server. This aggregated OPC UA server is able to provide the information that can be represented based either on the basic address space or on custom information models at the same time. That way, OPC UA enables distinctive information model extensibility without losing any of the functionality provided by the base model. Fig. 3 depicts the concept of aggregating servers.

Using the aggregation of information models, custom and rich information models can be derived and easily integrated into the corporate OPC UA application. The corresponding information models can be designed for the mapping of complex infrastructures in manufacturing environments. For the OPC UA application, these user-defined information-mapping capabilities can be combined with the powerful basic meta-model that OPC UA already delivers. The combination of this powerful modeling functionality together with the strict separation from the communication application programming interface (API) makes OPC UA an adaptable, flexible, secure, robust, and extensible tool for realizing applications of Industrie 4.0. However, the modeling of domain-specific information into OPC UA meta-models is still lacking a uniform formalization of ontology-represented knowledge. Yet, there are no sufficient methodologies to integrate custom object types and variable types into the OPC UA meta-model in a consistent way. Interfaces for the docking of these domain-specific information models still have to be carried out properly. With the following use cases, we attempt to point out both the industrial relevance of the presented information modeling approaches and the problems of adapting present technologies to real-world demonstrators.

4 USE CASES AND APPLICATION

Based on the previous theoretical chapters, the following will depict three applications that all—in very different ways—can be defined as CPPS. First, a novel approach for human-robot interaction is presented. Extensive information or a data model from engineering enriches the conventional product or process information, and can be further completed during process execution. Second, the communication layer of a production machine (packaging machine in this context) is semantically and technologically extended for enabling cloud usage. Extracts from the virtual counterpart—the cloud—are presented to discuss the advantages of global interconnectedness. Finally, a new control paradigm as the answer to production processes with many variants and products with short product cycles is described. The very basis for product-centric control is built by intelligent, connected, and a semantically well-defined CPS.

4.1 MODEL-BASED HUMAN-ROBOT INTERACTION FOR FLEXIBLE ASSEMBLY AUTOMATION

Today's production processes (e.g., in the automotive industry) are already highly optimized, due to present expert knowledge iteratively gained over decades of research, development, and application (Davenport, 2010). Therefore, the involvement of these external process knowledge owners into the commissioning of CPPS is a key to success. Process knowledge owners—typically specialized small and medium-sized enterprises (SMEs)—try to leverage their experience when distributing production modules to different operating manufacturers. Thereby, new possibilities for distributing and contextually interpreting life cycle information lead to innovative forms of human-machine interaction—especially at the shop floor level.

A novel interaction concept for a flexible assembly process commissioning utilizes user knowledge concerning manual assembly tasks that are already present. It helps to increase productivity and process quality by giving assembly staff the ability to transfer tedious or nonergonomic assembly tasks to a flexible robotic system representing the CPPS. The system integrates multiple digital planning tools that are supporting the different development aspects of products as well as production resources interfacing at the particular production process. The integration is achieved by PLM, which—implemented as an IT system—combines different views upon virtually designed objects and manufacturing steps for different user groups (roles) (Stark, 2011). Thus, CPPS can make use of the linked planning information, and interpret the data in the specific context.

One critical step before ramp-up is the resource's commissioning—for example, for assembly tasks. Product variants call for flexibility and repeated process adaption usually performed by skilled labor. To increase productivity assembly, personnel is faced with innovative assistance systems that often include automated solutions like robotic process modules. A major disadvantage of today's process automation is the lack of possibilities for the shop floor user to change (adapt), for instance, a robot's behavior to match a certain product variant. On the one hand, this deficit is a deliberate consequence of standardization and deterministic line balancing to rule out any sources of manually induced uncertainty; on the other hand, digitally supported manufacturing can only increase flexibility if the user is able to access the collected information and make use of it during their decision-making process via dedicated interfaces (Wünsch et al., 2010).

A similar user interface is developed for a robotic assembly cell (see Fig. 4). It allows the assembly worker to define automated assembly tasks by direct (physical) interaction with assembly objects—such as within the user domain. The task definition is initiated from the user by demonstrating a manual assembly step. Thereby the manual processes are digitized by a sensor combination, which captures the user's motion, the assembly objects' locations, and the usage of specific assembly tools over time. Simultaneously, the recorded interaction information is digitally interpreted using model data of the

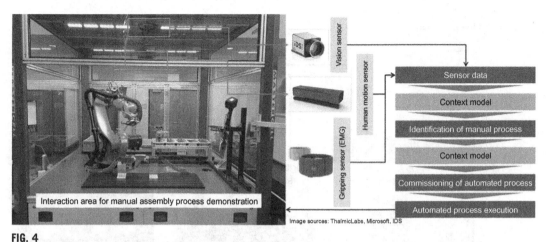

FIG. 4

Automated assembly cell for flexible process commissioning.

respective objects forming the interpretation context. To support the interpretation, assembly object models contain not only a 3D representation (CAD model), but also assembly-relevant aspects represented as assembly features that are geometrically linked to the object's shape during product development.

When the user manipulates two assembly objects, in reality their compatible assembly features virtually approach each other (e.g., a peg and a hole) so that a new assembly joint is formed and noted as an assembly step. Thus, the interaction outcome is a work plan for the underlying assembly task that consists of a sequence of several working steps. In order to then automatically plan and commission—according to the work plan—equivalent automated assembly processes, a conjunctive common data model is used (see Fig. 5). Thus, by physical interaction the user manipulates aspects of a virtual data model instance and creates the planning context for the automated application of virtual commissioning tools.

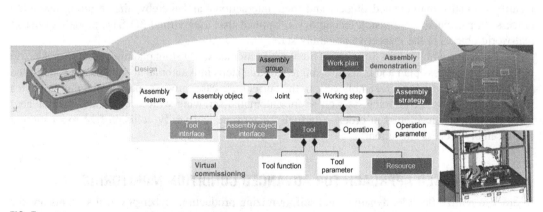

FIG. 5

Integrated context model to support the commissioning process.

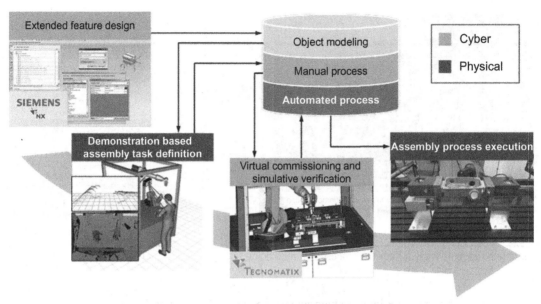

FIG. 6

Integrated tool chain for the creation and processing of assembly-relevant model data.

The commissioning workflow is represented by an integrated tool chain, which consists of virtual (cyber) planning and real-world (physical) interaction methods (see Fig. 6). To realize virtual planning, established software systems for product design (Siemens NX) and digital manufacturing (Siemens Tecnomatix) are extended by custom software modules to support the developed context model utilizing already available design functionality—for instance, for defining geometrical assembly feature specifications, and exporting enriched CAD models to a common database. The (physical) demonstration-based assembly task definition makes use of the specified context model in order to identify manually manipulated objects and their interaction—and thereby, the resulting assembly process. In particular, here vision technology is applied that uses virtual 3D STL models to detect real-world objects in the respective assembly state.

In interaction with Process Simulate of Siemens Tecnomatix, a custom planning module makes use of the manual process plan and translates single working steps into automated equivalent tasks for the previously modeled robotic assembly cell. In the current use case, process specific planning modules for robotic gripping, positioning, as well as for automated screwing with a robotically manipulated screwing tool are developed in order to virtually determine and verify collision-free robot motion and assembly process execution.

4.2 CLOUD-BASED APPROACH FOR ADVANCED CONDITION MONITORING

In order to realize flexible, dynamic, and selfoptimizing production, cyber-physical systems are the indispensable basis for building CPPSs (Reinhart et al., 2013; VDI/VDE-Gesellschaft Mess- und Automatisierungstechnik, 2014). At the same time, the adaption of existing technologies like cloud computing for enabling local CPPS with global intelligence is necessary (BITKOM et al., 2015). Furthermore, cloud computing is named as one of the key technologies for software-defined platforms and service platforms (Appelrath et al., 2014; Kagermann et al., 2015), and foreseen as a disruptive technology for present manufacturing. In this context, the guiding principles of Industrie 4.0—such as interconnection across life cycle phases, machines and companies, selfoptimizing CPPSs, and IT globalization—encourage conceptualizing new approaches. For example, Atmosudiro et al. (2014) deal with the challenge of realizing data consistency into cloud platforms and take aspects like protocols, machine connectivity or safety into consideration when launching the idea of a cloud gateway for the shop floor. Colombo et al. (2014) describe the adaption of a service-oriented architecture to industrial cloud-based CPS. With these approaches as a basis, first concepts for specific cloud-integrated applications like condition monitoring tasks develop (Bechhoefer and Morton, 2012; Eickmeyer et al., 2015; Liang et al., 2012).

Putting all the stated facets of Industrie 4.0—cloud computing, IoT, Internet of Services—together, the following use case presents a novel approach for cloud-integrated condition estimation for optimized machine operations. In applications with high cycle times like packaging, for example, a breakdown directly affects the overall equipment effectiveness (via productivity), and therefore should be avoided. In addition to wrong parametrization or general environmental influences, the breakage of components due to wear is very common. While the primarily used strategies still refer to the cyclic (i.e., time-based) change of components or firefighting strategies (change at breakdown), a condition-based replacement is preferable. To present an even more complex situation, the wear of certain components (belts, chains, drives, gears, etc.) already affects the optimal production process before the complete breakage occurs. For example, if a belt wears it causes slippage, since the adhesion between different elements is not defined any longer.

The reason for the sparse application of the condition-based strategy is mainly due to incomplete machine or component models: these have to be very detailed in order to predict the current condition correctly. In this context one of the most challenging tasks is to build these models—best during resource planning. This is possible with certain limits, since the data basis often is the mechanical model combined with some simulations. Real production data are available in the life cycle phase of production, and normally not fed back into conception (closed-loop engineering). In fact, some approaches with local learning algorithms exist—but each time a new local model is generated. Often the time necessary—or the archived accuracy—is not acceptable.

Another problem consists of a lack of models for an overarching view on a machine. In the packaging context, aspects like the packaging material (e.g., foil) or the product itself are usually neglected. Due to the interdependence between the named objects and the process, they all should be integrated for optimal operation. Until now there has been no automatism for receiving specific attributes or extensive models combined with methods for preventive condition monitoring.

With the introduction of cloud technology as an enabler of Industrie 4.0, new opportunities arise. The cloud can be seen as central data storage. Many different (similar) machines provide data about the production process or production environments. To date, real process information is very seldom fed back to the manufacturer or in the general engineering phase. On that basis, for instance, powerful algorithms can derive robust and universal component models. This promises great enhancements compared to a mere local data from one machine. With a new form of communication, not only homogenous machine data, but also, for example, expert knowledge about the current condition of a component can be provided and used for tagging specific measuring data. Companies selling

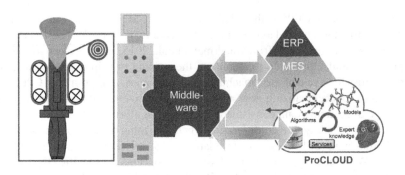

FIG. 7

Extract of a novel cloud-based reference architecture.

packaging materials can provide specific information about their products and data from the shop floor, since measured oscillation can be analyzed externally in cloud architecture by powerful algorithms for condition prediction. Finally, due to the application concept in clouds, the "out-of-the-box" functionalities of machines can be expanded significantly. During realization common advantages like performance services (software-as-a-service and platform-as-a-service), scalability or pay-per-use payment models can be established.

Fig. 7 shows an extract of the new cloud-based reference architecture for condition monitoring currently developed at the WZL. Conventional packaging machines can't be seen as CPS, since they have no virtual representation and don't provide the right—in terms of semantical description—communication interfaces. Therefore a middleware was developed that augments existing OPC DA interfaces with an extensive OPC UA information model. With this representation a new level of connectivity is reached. While state-of-the-art systems are embedded in a 1D vertical hierarchy, future CPS (here: packaging machine plus middleware) operate with complete new levels of connectivity. Through new interfaces, a horizontal access to the cloud for production (ProCLOUD) can be established via REST.

ProCLOUD is accessed via RESTful services and first collects machine and company overlapping data. Second, with powerful algorithms and external expert knowledge, models are built. Finally, these models are the basis for new services that analyze and evaluate sent data.

In times of growing—governmentally encouraged—requirements concerning energy efficiency, production machines should implement different process specific energy profiles ("stand-by"). Due to conditioning times of different auxiliary units, these profiles are not trivial to configure. ProCLOUD provides one service for thermal conditioning processes (CLOUDenergy, see Fig. 8). In a first step *energyprofile* collects data and parameterizes a multidimensional model (e.g., in modelica) describing the heating and cooling behavior. In a second step, this model can be used or accessed indirectly via the RESTful service *energysaving*. The following steps describe the complete information flow by means of a production break (see Fig. 8). The model for thermal conditioning was built before separately:

1. Determination of the production forecast by the MES.
2. Call of the OPC UA interface of the packaging machine with a certain predicted time.

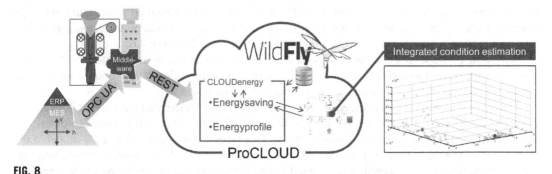

FIG. 8

Example of model usage for stand by and integrated condition estimation.

3. The machine itself cannot specify how long to shut down components like heaters, and when to switch them on again to be ready for operation at a specific point in time. Therefore, the machine uses the RESTful service CLOUDenergy with the application *energysaving*.
4. *Energysaving* starts the calculation with the data submitted and internal models.
5. The middleware polls the result continuously since RESTful services usually work asynchronously and derives local actions.

By the usage of OPC UA, different advantages arise. On the one hand, the transparency of available services on the shop floor layer increases. On the other, the information model dynamically changes with every new cloud service, which can be used by MES afterwards.

In the context of predictive maintenance or in general condition monitoring one essential challenge is to integrate wear-induced uncertainties into the analytic models. When moving components (drives, chains, belts, etc. are considered), the state usually influences the surrounding machine model. These aspects are not yet considered. Brecher et al. (2016) present an approach where acceleration data are sent to the cloud for analytical purposes. A cloud service consults implemented machine learning algorithms (k-means and maximum likelihood estimate) for determining the current condition of the film takeoff belts. The result can be passed to another service that uses the external state estimation to calculate optimal machine parameters, such as contact pressure.

In general, the cloud-based approach and the OPC UA middleware on shop floor allow implementing and providing different services for optimized machine operations. These are especially valuable, when required information, expert knowledge, profound data basis, or performance cannot be fulfilled locally.

4.3 SMART AUTOMATION LAB—INDIVIDUALIZED PRODUCTION

State-of-the-art production systems are generally achieved by hierarchically organized control architectures. In accordance with the automation pyramid, orders are managed and created in a rather static or manual way from the ERP and handed down to the process stations controlled by monolithic, embedded systems. The control and communication flow within such production systems is mostly organized through rigid communication layers or interface definitions. "How" processes or sequences have to be carried out is defined. The definition of the production process for new variants has to be determined manually. This leads to great efforts in adapting to changes in the general conditions or the configuration of the automation system. In addition to the layered automation architecture, the huge number of heterogeneous controllers and proprietary communication interfaces are technological barriers that need to be overcome in order to enable applicability of Industrie 4.0.

Motivated by these aspects, the overall challenge is to conceptualize new control strategies for products with many variants where an explicit description is not efficient. In contrast, an implicit model-based specification of "what" has to be produced is proposed.

The *Smart Automation Lab* at the WZL depicts a complete integrated and automated CPPS for the production of individualized products—didactically reduced duplo cubes in this case. Customers can influence different individualization features like surface information (pictures) or a 3D-model of the module via a tablet app (Fig. 9). The individualization features are deduced from the production system's capabilities, and can therefore be adapted to any expansion. Orders are translated to a product model and enriched with product data, which can be used within the ERP system for planning aspects. The data relevant for production contain the product model with CAD models of the different products to be manufactured, the related surface data, and information for the whole individualization process. The major contribution of the in-house developed MES is the control paradigm of a product-centric control. In conventional systems like the ones described above, the programming of different process cells is less flexible concerning production and assembly of customer-specific modules, which can hardly be adapted to changing demands like, for example, a new product. The Smart Automation Lab uses an OPC UA-based, service-oriented approach for keeping the overall process as flexible as possible. Services describe certain process cell functionalities that can be invoked by the overall MES as needed to fulfill the requirements defined in the product model. Through this approach the heterogeneous architecture is abstracted and standardized on a service level. At the same time, the technological basis for scalability and reconfiguration is established.

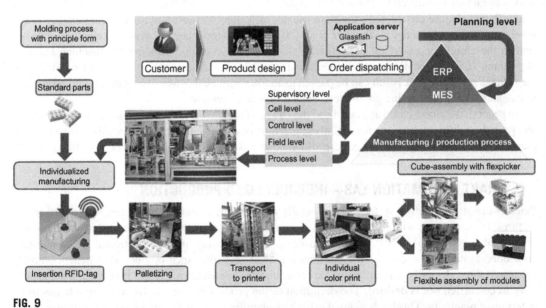

FIG. 9

Structure and process for the production of an individualized cube (BREC14).

Along different processes like forming, printing, assembly, or packaging, the customer-defined features are taken into consideration. Each station maps the product's requirements to its own features.

To allow adaptions to the production processes when there is complex material deployment, the product has to be identifiable at each station. Therefore the standard parts are equipped with specific radio frequency identification (RFID) tags (AutoID). The unique ID is read at every station, and connects the local brick to the central orders in the ERP. As a result, the connection to the specific CAD model or any other production information integrated can be resolved.

Fig. 10 shows an exemplary product-centered printing process within the CPPS. The deposition of new bricks (typically 18pc) within the printer grid causes an update (1) of the pallet model (PALM). This update triggers an event for the printer that subscribes any event of the PALM, for example. In the next step (2) the printer queries the whole pallet for receiving all necessary product model instances (PMI) in the next step (3). The PMIs contain all information relevant for the whole production process. Via self-perception—that is, knowledge of its own functionalities—the correct information can be extracted from the PMIs (4). In the given case, the station requests the surface information for the present brick's orientation and merges the single image parts to one printing image (5). Besides the image information from the product model, printer-specific color settings and optical position corrections are considered. After the printing process (6) the PMIs and with them the PALM is updated again. Since the handling robot for the material flow between several stations subscribes to the PALM as well, it receives the change notification.

The Smart Automation Lab integrates different capabilities of selfoptimization within different layers of the automation pyramid. For example, on the one hand, the central MES, considers the current

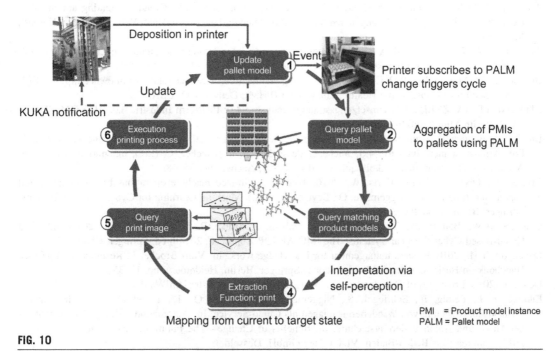

FIG. 10

Exemplary product-centric printing process.

order situation as well as the resources available, and executes new orders within the production system. On the other, local process cells like the assembly cell begin the parallel mounting of different orders for maximizing the probability of placing the next arriving brick on top of one cube to finish it, using the production capacity of the assembly robot in an optimal way. Since only the completion of one order notifies the MES, a deterministic planning of processes is not possible. Thus, new strategies for optimal resource usage have to be developed.

The Smart Automation Lab can be defined as a CPPS, mainly due to the combination of physical products/orders and virtual representations (product model) containing all information necessary for production. Additionally, the cells are CPS, since they integrate an extensive selfawareness, and provide their capabilities via flexible service interfaces.

ACKNOWLEDGMENT

The authors would like to thank the German Research Foundation (DFG) for its support of the research regarding *Individualized Production* and *Cloud-Based Approach for Advanced Condition Monitoring* within the Cluster of Excellence, "Integrative Production Technology for High-Wage Countries," as well as the German Federal Ministry for Economic Affairs and Energy (BMWi) for funding the project *MoDemo*.

REFERENCES

Appelrath, H.J., Kagermann, H., Krcmar, H., 2014. Future Business Clouds: Cloud Computing am Standort Deutschland zwischen Anforderungen, nationalen Aktivitäten und internationalem Wettbewerb. Utz, Herbert, München.

Atmosudiro, A., Faller, M., Verl, A., 2014. Durchgängige Datenintegration in die Cloud. Ein Konzept zur cloudbasierten Erfassung von Produktionsdaten. Werkstatttechnik Online 104 (3), 151–155.

Bechhoefer, E., Morton, B., 2012. Condition monitoring architecture: to reduce total cost of ownership. In: IEEE Conference on Prognostics and Health Management (PHM), Denver, CO, USA.

BITKOM, VDMA, ZVEI, 2015. Umsetzungsstrategie Industrie 4.0: Plattform Industrie 4.0. BITKOM/VDMA/ZVEI, Berlin-Mitte/Frankfurt am Main.

Brecher, C., Behnen, D., Brumm, M., Carl, C., Ecker, C., Herfs, W., et al., 2014. Virtualisierung und Vernetzung in Produktionssystemen. Industrie 4.0: Aachener Perspektiven. In: Brecher, C. (Ed.), Integrative Produktion: Aachener Werkzeugmaschinenkolloquium 2014. Shaker, Aachen, pp. 35–68.

Brecher, C., Obdenbusch, M., Herfs, W., 2016. Towards optimized machine operations by cloud integrated condition estimation. In: Niggemann, O., Beyerer, J. (Eds.), Machine Learning for Cyber Physical Systems. Springer, Berlin, Heidelberg, pp. 25–33.

Colombo, A.W., Bangemann, T., Karnouskos, S., Delsing, J., Stluka, P., Harrison, R., et al., 2014. Industrial Cloud-Based Cyber-Physical Systems: The IMC-AESOP Approach, 2014th ed. Springer, Cham.

Davenport, T.H., 2010. Process management for knowledge work. In: Vom Brocke, J., Rosemann, M. (Eds.), Handbook on Business Process Management 1. Springer, Berlin, Heidelberg, pp. 17–35.

Doan, A., 2012. Principles of Data Integration. Morgan Kaufmann, Amsterdam/Waltham.

Eickmeyer, J., Pethig, F., Schriegel, S., Niggemann, O., Givechi, O., Li, P., et al., 2015. Intelligente Zustandsüberwachung von Windenergieanlagen als Cloud-Service. In: Automation 2015. 16. Branchentreff der Mess- und Automatisierungstechnik. In: Benefits of Change—The Future of Automation; 11/12 June 2014, Kongresshaus Baden-Baden. VDI Verlag GmbH, Düsseldorf.

Hannelius, T., Salmenpera, M., Kuikka, S., 2008. Roadmap to adopting OPC UA. In: 2008 6th IEEE International Conference on Industrial Informatics (INDIN): Daejeon, South Korea.

Hensel, R., 2012. Industrie-4.0: Konzepte rütteln an der Automatisierungspyramide. Available from: http://www.ingenieur.de/Themen/Produktion/Industrie-4.0-revolutioniert-Produktion (accessed 04.10.13).

International Electrotechnical Commission, 2015. OPC Unified Architecture. International Electrotechnical Commission, Geneva.

Kagermann, H., Wahlster, W., Helbig, J., 2013. Umsetzungsempfehlungen für das Zukunftsprojekt Industrie 4.0: Abschlussbericht des Arbeitskreises Industrie 4.0. acatech - Deutsche Akademie der Technikwissenschaften e. V, München.

Kagermann, H., Riemensperger, F., Hoke, D., Helbig, J., Stocksmeier, D., Wahlster, W., Scheer, A.W., et al., 2015. Smart Service Welt. Umsetzungsempfehlungen für das Zukunftsprojekt Internetbasierte Dienste für die Wirtschaft. Acatech—Deutsche Akademie der Technikwissenschaften, Berlin/München.

Leitner, S.H., Mahnke, W., 2006. OPC UA: Service-Oriented Architecture for Industrial Applications. ABB Corporate Research Center, Ladenburg.

Liang, B., Hickinbotham, S., Mcavoy, J., Austin, J., 2012. Condition Monitoring Under the Cloud: Digital Research. Oxford Press, Oxford.

Reinhart, G., Engelhardt, P., Geiger, F., Philipp, T.R., Wahlster, W., Zühlke, D., et al., 2013. Cyber-Physische Produktionssysteme: Produktivitäts- und Flexibilitätssteigerung durch die Vernetzung intelligenter Systeme in der Fabrik. Werkstattstechnik Online 103 (2), 84–89.

Siau, K., 1999. Information modeling and method engineering. J. Database Manag. 10 (4), 44–50.

Stark, J., 2011. Product Lifecycle Management. Springer, London. pp, 1–16.

VDI/VDE-Gesellschaft Mess- und Automatisierungstechnik, 2014. CPS-basierte Automation. Forschungsbedarf anhand konkreter Fallbeispiele. VDI/VDE-Gesellschaft Mess- und Automatisierungstechnik.

Wünsch, D., Lüder, A., Heinze, M., 2010. Flexibility and re-configurability in manufacturing by means of distributed automation systems—an overview. In: Kühnle, H. (Ed.), Distributed Manufacturing: Paradigm, Concepts, Solutions and Examples. Springer, London, New York, pp. 51–70.

ABOUT THE AUTHORS

Prof. Dr.-Ing. Christian Brecher (1969) holds the Chair of Machine Tools at the Laboratory for Machine Tools and Production Engineering (WZL) at the RWTH Aachen University. After studying mechanical engineering and finishing his doctorate at the RWTH Aachen University he was in a leading position for machine development at DS Technologie Werkzeugmaschinenbau GmbH (2001–03). In 2004 he returned as professor to the RWTH Aachen University to take over the Chair of Machine Tools. Since 2013 Prof. Brecher has been the leading director of the WZL. Furthermore he is member of the Fraunhofer Institute for Production Technology's (IPT) directorate.

Dipl.-Ing. Christian Ecker (1986) is research assistant at the WZL's Department of Automation and Control. Until 2011 he studied mechanical engineering and production technology at the RWTH Aachen University as well as manufacturing at the University of Bath (UK). Since 2015 he has lead the group for automation.

Dr.-Ing. Werner Herfs MBA (1975) is Academic Senior Councilor and Executive Chief Engineer at the Chair of Machine Tools. After his studies in electrical engineering he graduated as a PhD student in the field of mechanical engineering. As Executive Chief Engineer he is coordinating the different departments of the Chair of Machine Tools.

Dipl.-Ing. Markus Obdenbusch (1986) is research assistant at the WZL's Department of Automation and Control. Before he received a degree in computer engineering at the RWTH Aachen University's Faculty of Electrical Engineering and Information Technology. From 2014 to 2015 he was leading the group for automation and since 2016 he is Chief Engineer of the Department of Automation and Control at the Chair of Machine Tools.

Prof. Dr. rer. nat. Sabina Jeschke (1968) is head of the Cybernetic-Cluster IMA/ZLW & IfU and Vice Dean of the Faculty of Mechanical Engineering at the RWTH Aachen University. She is also the Chairwoman of the Board of Management of the VDI Aachen and a member of the supervisory board of the Körber AG. Her doctorate in 2004 followed a junior professorship (both at the TU Berlin) with the construction and direction of its media center. Afterwards she became head of the Institute of Information Technology Services (IITS) for electrical engineering at the University of Stuttgart where she was also the director of the Central Information Technology Services (RUS) at the same time. Her main research areas are: Complex IT-systems, robotics and automation, traffic and mobility and virtual worlds for research alliances and education.

Dipl.-Ing. Max Hoffmann MBA (1985) has been a scientific researcher and doctoral student at the IMA since July 2012. Within his position, he is a member of the research group "Production Technology." In his current research, he focuses on ontologies, data exploration, and object-orientated software construction for usage in manufacturing environments. Max Hoffmann is member of the expert group "Multi-Agent Systems in Automation Technology" of the VDI/VDE GMA.

Prof. Dr.-Ing. Dipl.-Inform. Tobias Meisen (1981) is the Managing Director of the Institute for Information Management in Mechanical Engineering (IMA) and was appointed junior professor for "interoperability of simulations in mechanical engineering" at the RWTH Aachen University in October 2015. After studying computer science at the RWTH Aachen University, he worked in the Cluster of Excellence "Integrative Production Technology for High-Wage Countries" and finished his doctorate in 2012. Earlier, in 2011, he became leader of the research group "production technology" at the IMA. His research focuses on modern information management in cyber-physical systems and Industry 4.0—in particular, interoperability and artificial intelligence in heterogeneous system landscapes.

TRANSFORMATION OF MISSION-CRITICAL APPLICATIONS IN AVIATION TO CYBER-PHYSICAL SYSTEMS

22

L. Ren, H. Liao, M. Castillo-Effen, B. Beckmann, T. Citriniti

GE Global Research, Niskayuna, NY, United States

1 INTRODUCTION AND BACKGROUND

The aviation domain is characterized by complex, real-time and/or near-real-time, mission-and-safety-critical systems and applications with embedded sensors and controls that interact with the physical world, for example, aircraft, airspace, weather, and human operators. Traditionally, safety and performance guarantees have been achieved through physical system partition, redundancy, and point-to-point connections following strict, dedicated protocols. As traffic demand increases and the global nature becomes ever more important, under the rigid architecture, many systems and applications are approaching their capacity and performance limits. Developments in cyber-physical systems (CPSs) offer opportunities to smartly integrate computing, controls, sensing, and networking to transform the joint behavior of elements in aviation, and by such to achieve capacity, efficiency, and performance breakthrough beyond what incremental enhancement of individual airborne and ground elements can offer, all while ensuring safety.

1.1 EXISTING AVIATION SYSTEM ARCHITECTURE

Within the context of CPS, the current National Airspace System (NAS) can be considered as a decision triad of aircraft, aircraft operator's Flight Operations Control (FOC), and the Air Navigation Service Provider (ANSP), along with the environment and human operators. The system architecture can be illustrated by a dependency map of various applications in the NAS, as shown in Fig. 1. This map was developed based on the review of current operations in the NAS. Refer to Ren et al. (2013) for the list and description of applications shown in this figure. Here, dependency refers to requiring input from or relying on the functionality of another application. Dependencies between two applications may be unidirectional or bidirectional. In the figure, applications are first grouped by their owner entity in the decision triad of aircraft, FOC, and ANSP. ANSP applications are further grouped by major systems for operations in different airspace domains. Dependencies may represent connections via different communication channels, including air-ground voice, air-ground data link, commercial data

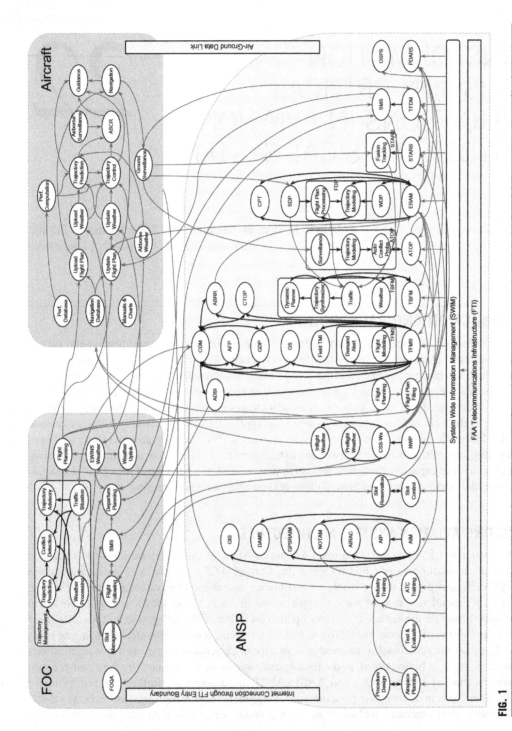

FIG. 1

Existing NAS architecture shown as a dependency map.

link, and ground networks. Connections may exist between applications within and between systems owned by the same entity. They may also exist between applications within and between systems owned by different entities, for example, between an application in an ANSP system and an application in an FOC system. It should be noted that for the sake of simplicity, sensors are not considered as separate applications; instead, integrated systems involving sensors are considered.

In Fig. 1, the grouping of applications and subsystems for the ANSP is done by aligning functions representing the same major system vertically, where possible. Dependencies between applications or subsystems within the same major system are denoted by *solid black arrows* while dependencies between systems or functions within different systems are denoted by *dark blue arrows* (*dark gray arrows* in print version).

In current operations, the supporting infrastructure consists of the Federal Aviation Administration (FAA) Telecommunications Infrastructure (FTI) for communications within the ANSP, as required by the FAA to satisfy its security and safety requirements. The FAA's System Wide Information Management (SWIM), currently being developed and tested, serves as the information technology (IT) enterprise infrastructure necessary for NAS systems to share and reuse information and increase interoperability. SWIM provides governance to NAS programs to ensure services are SWIM compliant and meet all FAA Service Oriented Architecture (SOA) standards. By providing this governance and a supporting common enterprise infrastructure, SWIM is envisioned by the FAA to reduce the cost and risk of rework for NextGen programs that develop and deploy services within the NAS. For this reason, the dependency map assumes that all the ANSP systems are interconnected through SWIM. This includes planned infrastructure development that is not yet fully operational. It is also assumed that some level of air-ground data link capability exists between ANSP and the aircraft in domestic airspace, as in current oceanic operations. Communications between FOC and the ANSP are Internet-based through controlled entry boundary to the FTI. Communications between airborne systems are assumed based on current data bus standards.

As seen in the dependency map, many connections exist. In addition to the ground network, these connections also include air-ground voice communications, ANSP-operated air-ground data link, and commercial data link. However, many of these connections are not yet integrated with automation. This inevitably creates barriers to fully utilizing the capabilities of current existing systems and causes operational inefficiencies. It is also clearly seen that disparity, gaps, and complexity in communication are not the only issues faced by current operations.

Ideally, all the ANSP applications, FOC applications, and airborne applications should rely on the same true and complete picture of the operational environment to make informed and robust decisions. Yet, each of these applications has been developed with its own processes, creating its own version of the picture with discrepancies among assumptions, input data, and capabilities. Often, these supporting processes cost much more to develop and operate than the core decision support capabilities, not to mention the many conflicts and inconsistencies in decision making associated with this issue. The dependency map provides a means to investigate this issue and, jointly with the analysis of cyber systems, provides insights into potential improvements that may be achieved by transforming into CPS.

1.2 AVIATION CPS AND PROBLEM DEFINITION

Sampigethaya and Poovendran (2013) provided a high-level review of the physical layer, cyber layer, cyber-physical interactions in aviation, at the aircraft level, the airport level, and the air traffic management (ATM) level, along with safety assurance and security for aviation CPS. The ability to interact with, and expand the capabilities and capacities of, aircraft and airspace facilities through increased computation and communication is critical in addressing issues faced by the NAS CPS, as described in the previous section.

Cloud computing, the most significant development in the cyber world in recent years, is intended for resource pooling and ubiquitous access. It allows for the client applications, processing, and data storage to be separated or loosely coupled. Such separation enables effective and consistent access to the same capability at multiple locations (or by multiple entities) in a synchronous or an asynchronous manner, so as to reduce duplication of development and data storage. It thus minimizes the need for processing, storage, and software to be resident at the location of use. The connectivity and sharing of information and computational resources provide scalable and virtually unlimited capacity for applications, and more optimal decision making and operations at individual aircraft, airspace facility, and system levels.

Recognizing the advantages of cloud computing and value to a full CPS transformation, a team of researchers (Simmon et al., 2013) envisioned a cyber-physical cloud computing (CPCC) concept as "a system environment that can rapidly build, modify and provision auto-scale cyber-physical systems composed of a set of cloud computing based sensor, processing, control, and data services." From the combination of CPS and cloud computing, CPCC provides unique benefits, such as modular composition and smart adaptation to environment at every scale. This makes the CPCC concept particularly relevant in the current discussion because of the dynamic, distributed, mobile, and global nature of physical assets in aviation CPS.

The aviation industry is moving towards CPS leveraging cloud computing for cost savings and increased system capacity and performance. The Airline IT Trends Survey (Airline Business and SITA, 2015) conducted in 2015 revealed that 38% of airlines are using Infrastructure as a Service, and 57% are using Software as a Service. These numbers are expected to increase to 74% and 88% in 2018, respectively. LH Systems' offers cloud-based flight operations systems and the planning solutions to airlines (Lufthansa Systems, 2013). In Apr., 2011, SITA launched its Air Transport Industry community Cloud (ATI Cloud), which hosts a number of aircraft operational applications including flight planning, flight tracking, and air-ground messaging (SITA, 2015). In addition to SWIM, the FAA (FAA, 2015) and the European Organisation for the Safety of Air Navigation (EUROCONTROL) (Brenner, 2015) are also moving forward with NAS level CPS initiatives.

However, a number of challenges and stakeholder concerns must be addressed before a full transformation of aviation mission-critical applications becomes a reality. This chapter provides an in-depth analysis of mission-critical applications in aviation that involve the decision triad of aircraft, FOC, and the ANSP. A characterization of operations and system requirements is carried out to provide bases for identifying CPS transformation opportunities. Transformation opportunities are examined from the perspectives of the aircraft, the FOC, and the ANSP, respectively. Challenges faced by this transformation and concerns from stakeholders about the envisioned transition are discussed, with potential path for addressing them identified.

2 CHARACTERIZATION OF MISSION-CRITICAL AVIATION APPLICATIONS IN THE CONTEXT OF CPS

The characterization of mission-critical aviation applications is carried out through an attribute-based analysis of the physical layer and the cyber layer. The analysis qualifies and quantifies the physical layer by its computational and network requirements on the one hand, and the cyber layer by its computational and network capabilities on the other.

2.1 ATTRIBUTE-BASED ANALYSIS APPROACH

Attribute-based allocation uses reasoning frameworks (either qualitatively or quantitatively) based on quality attribute-specific models (Klein et al., 1999). Attribute-based allocation is especially suitable for the characterization. This is because attribute-based allocation is also attribute specific. This provides reasoning to allocation decisions. SOA is commonly used to organize and standardize technical capabilities to allow for flexible accomplishment of constantly changing demands. Effectively, an SOA abstracts and encapsulates processes as shareable services, increasing system transparency while promoting process ownership and governance. The problem of allocating abstracted services to specific computing environments could thus be viewed as a CPS architectural synthesis/optimization problem (Hang et al., 2011).

Attribute characterization of the physical layer is performed by analyzing applications at ANSP, FOC, and aircraft. Those applications are viewed as abstracted services offering given technical capabilities. An application attribute schema is developed. Individual applications are analyzed against the schema, with a focus on those that are most significant to the transformation to the CPS environment. Quantitative attributes are used where possible.

Attribute characterization of the cyber layer is performed in a similar fashion. A cyber system attribute schema is developed. Corresponding to the attribute characterization of applications, a set of most significant cyber attributes are selected. A combination of qualitative analysis and experiments is employed to determine attribute values.

With the characterization of the physical layer and the cyber layer, an analysis is conducted to convey the relationship between application attributes and cyber system attributes. This is essentially a projection (or mapping) of cyber computing attributes into the application design space that are defined by the corresponding attribute characterization. With the attribute characterization, this analysis provides a formal reasoning for the identification of candidate applications for transformation. If a match exists, a candidate will be identified. This process starts with the most significant attributes first, and extends to more attributes as needed.

2.2 CHARACTERIZATION OF ATTRIBUTES FOR AVIATION SYSTEMS AND APPLICATIONS

The attribute schema for aviation applications consists of a subset of system attributes, shown as a tree structure in Fig. 2. The attributes are grouped in three major categories: organization, operations, and computing and network. Refer to Ren et al. (2013) for a formal definition of the attributes.

The attribute schema described above covers many aspects of aviation applications. Thorough and detailed characterization of the long list of applications using this schema is time-consuming and is only required when the transformation of specific applications is considered. To provide a system level view, a small set of the most significant attributes are established. The two most significant attributes are decision time horizon and criticality, as defined below.

The nominal required time interval between the execution of a specific function in an application and the realization of its impact on air traffic operations is called the *decision time horizon*. Possible attribute values span the planning, strategic traffic flow management (TFM), tactic air traffic control, monitoring, and analysis phases. Using time units, this attribute ranges from years, to months, weeks, days, hours, minutes, seconds (including subseconds), and post-operations, which have negative values.

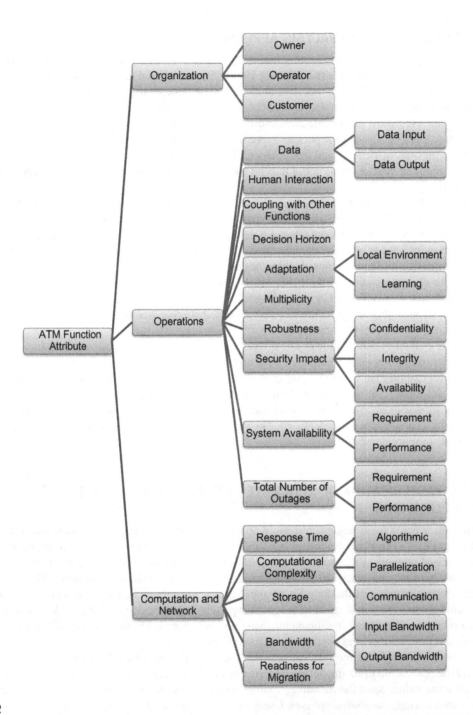

FIG. 2

Aviation application attributes.

A new attribute, *criticality*, which combines several aspects of air traffic operations, is introduced. Criticality is evaluated by inspecting the direct operational consequences of an application. It does not evaluate the impact of applications beyond the current day of operations. Possible values are defined as follows:

- *Class 1: Safety-critical.* Improper operation of the application has safety consequences.
- *Class 2: Wide impact.* Operation of the application itself may not have direct safety consequences but it is depended upon by many other applications, including potential safety-critical functions (although the impact may have been mitigated in those applications), thus resulting in wide impact on operations.
- *Class 3: Non-safety-critical.* Nonsafety-critical or with no immediate critical consequences, but aggregated effects over time may bring a system to a tipping point to seriously impact system-wide efficiency and performance.
- *Class 4: Non-operationally-critical.* Failure of these applications has no impact on the current day of operations.

The definition of criticality is formulated within the context of CPS. It is thus not the intention to directly correlate the criticality with assurance levels defined in RTCA DO-178C (SC-205, 2011a), nor with that defined in RTCA DO-278A (SC-205, 2011b).

The two most significant attributes defined above are related to readily identified system requirements. FAA's enterprise-level requirements for current ANSP applications are listed in the "National Airspace System Requirements Document" (NAS-RD-2012). The requirements are expressed in terms of functions that need to be performed and, where appropriate, with a quantitative qualifier characterizing the distribution of a relevant parameter, for example: "The NAS shall update the position of aircraft in en route airspace with a maximum time between updates of 12.1 seconds." Additionally, each requirement is assigned an inherent availability rating, as listed in Table 1.

Surveillance for separation assurance has been identified as the most demanding ANSP application in terms of latency. The update interval, latency, and service availability requirements for surveillance data service are shown in Table 2. In this table, aircraft position is selected as a representative functionality. These parameters are selected because they directly relate to decision time horizon and criticality attributes. The most demanding requirements come from terminal area surveillance, where NAS-RD-2012 establishes a 1.1-second maximum update interval, and a 2.2-second maximum latency. This update interval is comparable to the automatic dependent surveillance-broadcast (ADS-B) latency requirement specified in 14 CFR 91.227 (Jan. 1, 2014 edition), where it is mandated that the aircraft must transmit its position and velocity at least once per second while airborne or while moving on the airport surface. The ADS-B is allocated a 0.6-second maximum uncompensated latency

Table 1 Inherent Availability Ratings and Service Availability Requirements

Inherent Availability Rating	Minimum Service Availability
Safety-critical	0.99999
Efficiency-critical	0.9999
Essential	0.999
Routine	0.99

Table 2 Update Interval and Latency Requirements for Surveillance Data Services

Surveillance Functionality[a]	Update Interval (s)	Latency[b] (s)	Service Availability
Aircraft position on the airport surface area	≤ 1.1	≤ 2.2	≥ 0.99999
Aircraft position in flying closely spaced parallel approaches when runways are less than 3400 ft apart	≤ 1.125		
Aircraft position in precision approach airspace	≤ 2.4		
Aircraft position in terminal airspace	≤ 5.33		
Aircraft position in en route airspace	≤ 12.1	≤ 3.0	
Aircraft position in controlled airspace	≤ 13		
Aircraft position in remote areas		≤ 15	

[a]*The inherent availability rating for surveillance functionalities is safety-critical.*
[b]*Time between the display of surveillance data and their detection (or report from the aircraft in remote areas).*

from the time of detection to the time of transmission by the aircraft. This leaves 1.6 seconds for the ground-based surveillance data service to receive, process, and transmit an ADS-B report from an aircraft in the terminal area.

2.3 CHARACTERIZATION OF ATTRIBUTES FOR CYBER SYSTEMS

Cyber systems have many attributes that need to be considered when mapping applications and cyber system technologies. Similar to aviation systems and applications, attribute schemata are used to comprehensively classify each cyber system technology. Building on the attribute-based analysis, an application-specific technology capability recommendation can be programmatically generated for each of the architectural characterizations of interest. Three commonly used architectural characteristics are power, price, and performance. Fig. 3 illustrates high-level attributes related to performance. Power and price have similar hierarchical attributes, but specific details are omitted. This figure illustrates that a hierarchy of attributes contributes to the performance of a system. At the highest level, a system's stimuli, architectural parameters, and response all contribute to its performance. Moreover, each of the three performance subattributes has its own subattributes. Some of the subattributes can be quantitatively or qualitatively measured relative to the other technologies considered. Others will require further division.

According to the FAA, NAS systems, which are characterized by real-time and safety-critical operations, operate on a closed NAS FTI network (FAA, 2014a). The FTI landline networks provide Internet protocol (IP) communications with a high-availability design of optical backbone. The main requirements that characterize IP service quality are reliability, maintainability, and availability (RMA), diversity, and latency. RMA consists of multiple subparameters including availability, restoral time, mean time between outages, and maximum number of outages. Seven FTI RMA categories are defined (FAA, 2014b). Among these seven categories, RMA1 through RMA5 are provided 7 days per week, 24 hours per day (7×24). The relevant RMA parameters for the five 7×24 FTI RMA categories are shown in Table 3.

FTI latency is defined as the total time required to successfully transmit a unit of information across a connectivity from one service delivery point (SDP) to another. For IP services confined to the FTI Internet domain, latency is measured from the moment in time when the last bit in a packet departs the

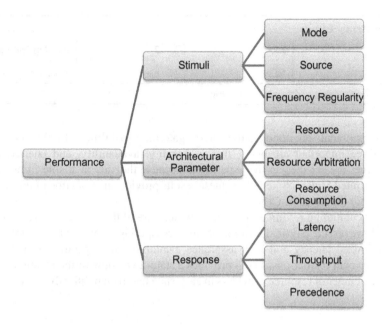

FIG. 3

Architectural characterization of performance.

Adapted from Klein, M.H., Kazman, R., Bass, L.J., Carrière, S.J., Barbacci, M., Lipson, H.F., 1999. Attribute-based architecture styles. Technical report CMU/SEI-99-TR-022, ESC-TR-99-022, Software Engineering Institute, Carnegie Mellon University, Pittsburgh, PA.

Table 3 FTI RMA Categories

Parameter	RMA1	RMA2[a]	RMA3	RMA4	RMA5
Minimum availability (over a 12-month period)	0.9999971	0.9999719	0.9998478	0.9979452	0.9972603
Maximum restoration time (min)	0.1	0.98	8.00	180	240
Maximum diversity/redundancy restoration time (min)	180	180	180	180	240
Maximum preventive maintenance service interruption time (h/year)	2	4	4	8	8
Minimum interval (h) between service-interrupting preventive maintenance	2190	2190	2190	2190	2190

[a]*FTI service configurations for RMA2 services were still under review and no IP RMA2 services had been implemented as of May 2010 (FAA, 2010).*

originating SDP to the moment in time when the last bit in the same packet arrives at the receiving SDP. Latency will also include the time required for retransmission of lost or corrupted frames or packets where the retransmission occurs wholly within the FTI network. Within FTI, seven latency limits (LLs) are identified (FAA, 2014b). LL-1 and LL-2 apply to FTI IP services (FAA, 2010). They are shown in Table 4.

Table 4 FTI Latency Limits

Parameter	LL-1	LL-2	FTI Optical Backbone
Latency limit (ms)	50	90	1 per 100 miles interconnect[a]

[a]A general rule that has been measured from FTI optical backbone.

Separately, experiments are conducted to investigate the capabilities of public cloud computing, with focus on three important attributes, that is, locality, availability, and latency. Public cloud computing resources are selected in the experiment because they are available from numerous providers within the United States. Nonetheless, these results provide an indication of the capabilities that may be available for the aviation CPS.

Users of cloud computing resources are generally unaware of the physical location of the resource they are using. However, the physical location of cloud computing resources is relevant to applications that require minimized latency and high assurance availability during failover, and are subject to regional data movement restrictions. Fig. 4 provides a state-level view of the location of cloud server resources and the number of resource providers in each state based on publically available information.

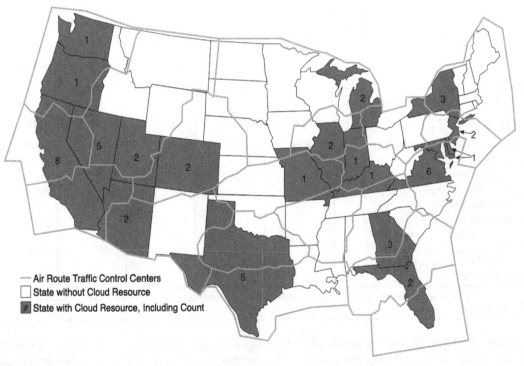

FIG. 4

Map of states with public cloud server resources overlaid with ARTCCC boundaries.

The figure also displays the boundaries of the 20 Air Route Traffic Control Centers (ARTCCs) in the contiguous United States. The map in Fig. 4 illustrates that cloud resources are not evenly distributed across the country. In fact, only 21 states contain all 51 locations where cloud server resources are available. Moreover, 47% of all of the cloud server resource facilities are located in four states: California, Texas, Utah, and Virginia. The regional concentration of public cloud resources might be a concern for latency-sensitive applications for which users are located throughout the country, should public cloud resources be used. However, increasing market size will encourage cloud providers to build new facilities in locations that are currently underserved, especially at locations near Internet backbone hubs. Of course nonpublic clouds could be installed and used as needed.

Cloud computing environments, although mature, are still subject to outages. These outages range from short-term "hiccups" where a cloud recovers quickly, to potential large-scale failures that cause complete, unscheduled, uncontrolled outages. Both cloud providers and third-party cloud monitoring services are collecting cloud service stability and performance statistics. These statistics can be used to programmatically control an application's elasticity and inform the cloud provider selection process. An example of aggregated cloud monitoring information, which measures uptime of 135 distinct public cloud services, is shown in Fig. 5. These services include content delivery networks (CDNs), infrastructure resources (servers, databases, and storage services), and platform applications. Measurements were collected using monitoring services provided by Panopta.com over a period of 90 days, ending on Mar. 7, 2013. The mean uptime of these services is 99.89% with a standard deviation of 0.8%. Fig. 5 illustrates that service uptime is not uniform. In fact, in this experiment there

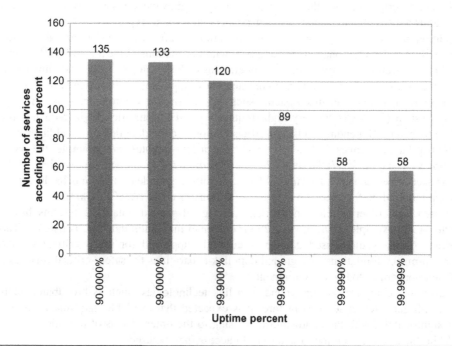

FIG. 5

Measured percent uptime of 135 cloud services over a period of 90 days.

were only 13 services whose recorded uptime was below the mean. Furthermore, 43% of the services tested had no measured downtime over the 90-day period. A historical perspective of aggregated stability data, like uptime, provides a quantitative measure by which competing public cloud services can be compared.

Network latency is measured by timing multiple HTTP GET requests for the same five-byte-long JavaScript file residing on each cloud resource. This measurement provides a round-trip latency that includes the outbound total network latency from the origin of the HTTP GET command to the cloud server, the cloud server response time, and the inbound network latency from the cloud server to the origin. In these experiments 22 CDNs, 54 cloud servers, 8 cloud storage services, and 7 platform applications were individually tested. All numerical measurements were based on the mean latency measured. The results are shown in Figs. 6 and 7. Fig. 6 illustrates the geographic separation of the experiment origin and target resources. The *large solid circle* denotes the experiment's origin location (Niskayuna, New York). *Smaller solid circles* denote resource locations. Resource locations are abstracted to the state level to simplify comparison. Solid arcs are used to delineate radial distance from the experiment origin. Each arc is labeled with its radial distance from the origin measured in miles. The pattern of each line connecting the origin to a resource location denotes the mean network latency of this connection, coded into one of four values. The four possible coding values are a latency: (1) less than 45 ms, (2) greater than or equal to 45 ms but less than 80 ms, (3) greater than or equal to 80 ms but less than 130 ms, or (4) greater than or equal to 130 ms. Fig. 6 is intended to be visually illustrative. More descriptive statistics are provided in Fig. 7. Inspection of Fig. 6 reveals that network latency is roughly proportional to the distance between the origin and the tested resource. This result is expected. Fig. 7 shows boxplots of resource latency partitioned by distance to the origin. Further, the same latency data sorted by cloud resource type is shown in Fig. 8. Fig. 8 suggests a difference in latency that is dependent on resource type.

In addition to ground-ground communications, communication between en route aircraft and cloud resources will face the additional latency and bandwidth limitations of the air-ground data link. Existing data link applications approved for flight operations include controller-pilot data link communication and automatic dependent surveillance-contract. These applications are currently implemented over character-oriented digital data link systems such as the Aircraft Communications Addressing and Reporting System (ACARS) via very high frequency (VHF) (domestic), high frequency (oceanic), or satellite (oceanic). Depending on the specific situation and the data link physical layer being used, the round-trip latency ranges from a few seconds to a few minutes. For example, the widely used subnetwork VHF data link (VDL) mode 2 is not standardized to support time-critical applications. Due to the access mechanism, VDL mode 2 latency exhibits a nondeterministic behavior and it cannot guarantee a required performance level in terms of transfer delay (AMCP, 2003).

Compared with other wireless networks, existing air-ground data link systems have limited bandwidth. As an example, VDL mode 2 25 kHz channel maximum data rate is listed in Table 5.

Satellite communication-based data links currently approved for safety-critical services have similar performance limitations. Air-ground data link capabilities for safety-critical services are thus a significant concern in today's environment.

In the future, as high-speed air-ground data link technologies (such as broadband satellite and cell-based inflight Internet service) mature, it is expected that available data link capabilities will increase significantly, with the bandwidth increasing to the order of tens of megabits per second or more. While the aircraft is on the ground, Wi-Fi access may be used.

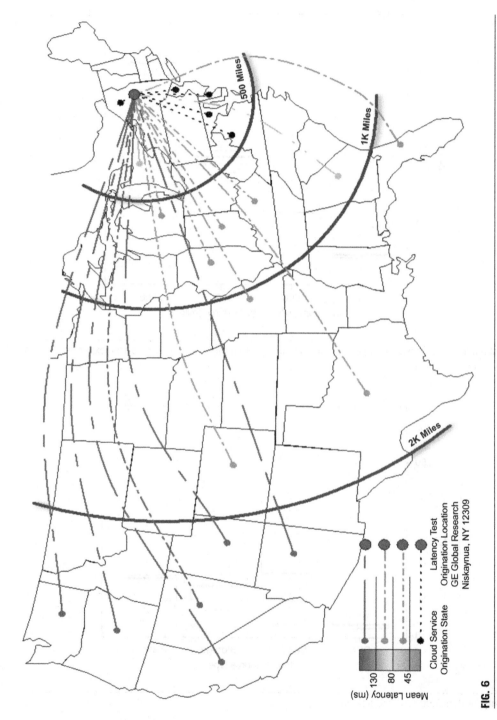

FIG. 6

Geographic representation of cloud server network latency results.

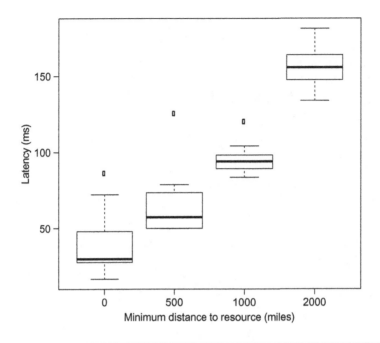

FIG. 7

Latency of publicly available cloud server resources binned by minimum radial distance from experiment origin to resource location.

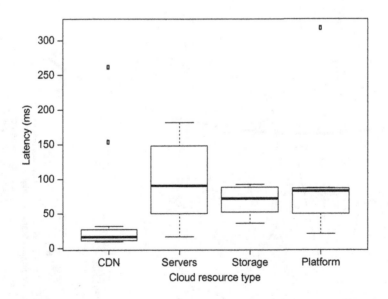

FIG. 8

Latency of cloud resource by resource type.

Table 5 VHF Data Link Mode 2 Bandwidth	
Data Link	**Maximum Bandwidth**
VHF data link mode 2, 25 kHz channel	32.5 Kbps

3 TRANSFORMATION OPPORTUNITIES OF AVIATION MISSION CRITICAL APPLICATIONS

It is expected that transformation opportunities can be identified by projecting the attributes of the cyber layer to the design space of the physical layer. An analysis is conducted to convey the relationship between application attributes and computing attributes. This is done by mapping (or projecting) cloud computing attributes into the application design space that is defined by the corresponding attribute characterization. A match identifies a transformation opportunity. This section presents the analysis results for ANSP, FOC, and aircraft applications respectively.

3.1 ANSP APPLICATION TRANSFORMATION OPPORTUNITIES

Fig. 9 is a projection of cloud computing capabilities into the application design space, with ANSP ATM systems and applications placed in the design space, and cloud computing capabilities represented as limitations. The application design space and cloud computing capabilities are defined by their most significant attributes, respectively. While the latency can be directly translated into limitations on the decision time horizon that can be supported by cloud computing, limitations on criticality are somewhat more complicated. The latter is correlated to the decision time horizon. As the decision time horizon extends, limitations will be relaxed as there will be more time available to catch up in case of a system outage. In this figure, major ANSP ATM systems are denoted by bold-line bubbles, while subsystem applications are denoted by thin-line bubbles.

From this projection, it can be seen that many of the ANSP ATM system can be supported by cloud computing. Surveillance applications in Advanced Technologies and Oceanic Procedures, En Route Automation Modernization, Standard Terminal Automation Replacement System, and Terminal Flight Data Manager are the most demanding ones in terms of latency and availability requirements. Surveillance definitely requires the highest availability and reliability. While the attribute is not represented in the design space shown in Fig. 9, weather systems are the most demanding ones in terms of bandwidth requirements, requiring multiple gigabytes of input each day.

Examples of ANSP transition opportunities includes:

- Common support services cloud—an FAA initiative:
  - Aeronautical Information Management Modernization
  - Common Support Services for Weather
  - Web Mapping Common Service
  - NextGen Weather Processor
- Cloud-Based Traffic Flow Management System
- Cloud-Based Time-Based Flow Management

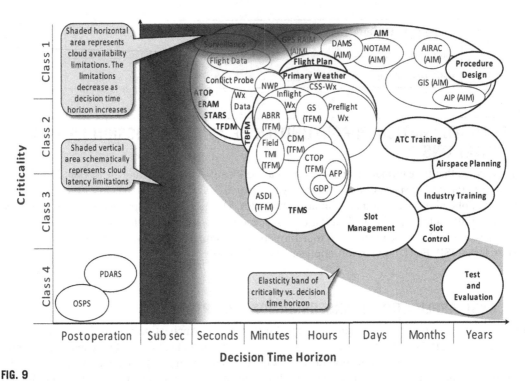

FIG. 9

Projection of cloud computing capabilities into ANSP ATM function design space.

3.2 FOC APPLICATION TRANSFORMATION OPPORTUNITIES

Fig. 10 is a projection of cloud computing capabilities into the application design space for FOC systems and applications. The limitations of cloud computing on the applications remain the same for FOC. From Fig. 10 it can be seen that unlike ANSP applications, FOC applications span a relatively narrow range of decision time horizon, from months to minutes, rather than from years to seconds or even subseconds. Even though there are Class 1 applications, they are normally not as critical as ANSP Class 1 applications. As such, most of the FOC applications can be supported by cloud computing. Another observation is that many of the FOC applications are already hosted by networked servers. For example, the flight following systems used by major airlines in the United States use the FAA's TFMData flight data feed as a major source of aircraft en route tracking information. Airlines may also include down linked ACARS reports in their flight following system. Some vendors are providing Internet-based flight planning services to commercial air carriers, where no flight planning software will be running on in-house computer systems.

Based on these observations, most of the FOC applications could be transitioned to cloud computing. However, some of these will benefit significantly from reductions in IT cost due to their high demanding computational needs, such as flight planning, while others will have relatively low gains due to their nominal requirements on computational capabilities, such as slot management. For those FOC

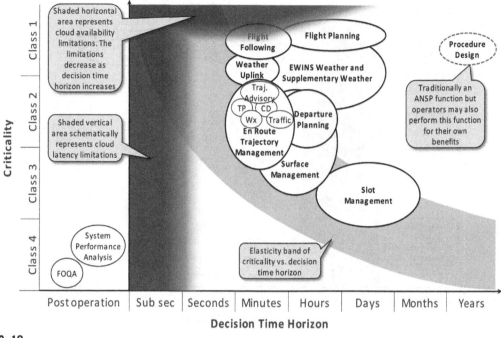

FIG. 10

Projection of cloud computing capabilities into FOC ATM function design space.

applications that do not require significant computational power, their transition to cloud computing may be part of a cloud-based data center initiative.

3.3 AIRCRAFT APPLICATION TRANSFORMATION OPPORTUNITIES

Fig. 11 is a projection of cloud computing capabilities into the application design space for aircraft applications. It should be noted that, although significant improvements of the air-ground data link are expected, aircraft application will continue to face more stringent latency and bandwidth (not reflected by the figure) limitations than those at the FOC and the ASNPS. For this reason, the *vertical shaded area* in Fig. 11 extends further to the right than that in Figs. 9 and 10. Furthermore, a number of aircraft applications' decision time horizon is down to subsecond, and all the functions identified are considered Class 1 functions.

Based on these observations, a smaller subset of aircraft applications is feasible for transition to cloud computing, as compared with ANSP and FOC ATM functions. There are also operational considerations other than computing and network connectivity. For example, the ground surveillance function is to provide information from the aircraft to the ground to support ground-based surveillance, thus it would not in any way be moved to a ground-based computing infrastructure. In any case, there are indeed aircraft application functions that could be transitioned to the cloud, such as trajectory control functions. The computation could be done in the cloud, with the new flight plan uplinked to the aircraft for execution. Similar to the case of ANPS applications or FOC applications, there could be significant cost savings by

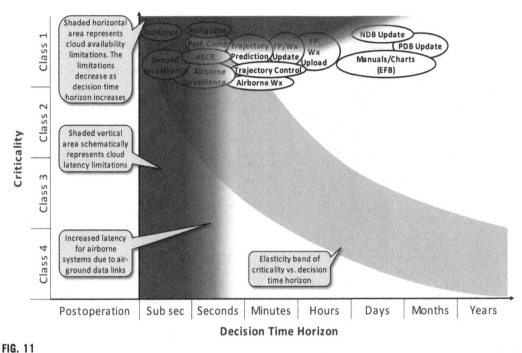

FIG. 11

Projection of cloud computing capabilities into aircraft ATM function design space.

transitioning aircraft applications to cloud computing. The savings could be from avoiding airborne equipment investment, which is more often dominated by the length and costly certification process of the equipment itself and maintenance and spares. The transition could potentially be supported by certified common communication and interfacing functionalities in airborne systems.

4 BENEFITS, CHALLENGES, AND POTENTIAL DIRECTIONS

The discussion of aviation CPS in this chapter focuses on integrated interacting networks of physical and cyber components at the NAS and global level. The global nature of these systems becomes ever important. Long range commercial jets can reach any major city within the world by a single hop of a flight, traversing airspace with drastically different environmental conditions and communication, navigation, surveillance and air traffic management (CNS/ATM) infrastructures. The operator's FOC needs to plan, follow, and support its aircraft as the flight progresses. Conversely, an ANSP facility may have to accommodate flights from around the globe, executed by aircraft with drastically different capabilities. As traffic demand increases, many procedural approaches have been adapted to balance the demand and system capacity at business airspace facilities, and to enable safe operations where CNS/ATM infrastructure lacks, such as in oceanic and remote airspace. On the other hand, air transportation is expected by the public to provide the highest level of safety, and consequently, aviation systems are subject to the most stringent certification requirements than any other mode of transportation. As such, the transformation of aviation mission critical systems to CPS, while providing much anticipated benefits, will face tough challenges.

This section will discuss the benefits first, and then inspect the challenges and potential directions in addressing those challenges to achieve the transformation of mission-critical applications.

4.1 BENEFITS OF TRANSFORMATION INTO CPS

The global nature of aviation requires highly integrated smart interacting networks of physical and cyber components both vertically from sensors, avionics, aircraft, fleet, to the NAS, and laterally across different operational domains from ground, terminal, to en route, and across different geographical regions and political boundaries. Coupled with new operational concepts and technologies, such a CPS transformation is anticipated to provide significant benefits, including:

- Reduced and predictable system operating cost
- Increased, elastic system capacity and flexibility
- Improved decision support and flight efficiency
- New safety regime
- Stakeholder social-cultural ecosystem
- Improved life cycle development efficiency and accelerated pace of technology transition

1. *Reduced and predictable system operating cost.* One immediate benefit of aviation CPS is reduced and predictable operating cost. This is largely achieved by leveraging cloud-based system integration and deployment. The cloud deployment model enables previous discrete on primes installations to be integrated into a streamlined system that is transparent to the user. Duplications in current systems can be reduced. High levels of resource pooling, including computing, data, deployment, and maintenance and support can be achieved with similar and often better performance and at much lower cost. An analysis of a notional arrival sequencing and scheduling system shows 31–89% 3-year total cost of ownership (TCO) savings without considering software system adaptation cost, and 26–58% 3-year TCO savings considering software system adaptation cost (Ren et al., 2014). Such savings are enabled by the availability of an NAS wide ADS-B integration network (FAA, 2011) along with other aeronautical services whose costs are separate from the cost of the application system in question both before and after the transition.
2. *Increased, elastic system capacity and flexibility.* Real-time integration of sensor information, system state, and decision support capabilities/autonomy enables airspace constraints to be dynamically determined instead of using rigid, predefined procedural parameters. For instance, the integration of weather sensor information and aircraft state information may be used to determine dynamic wake vertex separation. System down time associated with the evaluation of current condition (e.g., runway condition during snow storm) or that associated with airspace configuration switch may be reduced. Utilization of system resources can be maximized thus increases system capacity. CPS transformation also enables real-time or near real-time performance-based resource assignment, and thus provides certain levels of elastic capacity and flexibility based on real-time demand.

3. *Improved decision support and flight efficiency.* CPS transformation enables increased level of system-wide situation awareness by making integrated, comprehensive system state information available to all entities involved, including aircraft, FOC, and ANSP. Flight operation decisions can be made at a longer time horizon and at a more strategic level. Networked cyber components enable increased remote access of decision support capabilities by different users. Informed coordination and negotiation can be made to ensure high levels of success of such decisions. Flight efficiency is thus increased due to reduced waste caused by shorter time horizon, tactical, reactive decisions.

4. *New safety regime.* With the integration of best available physical and computational components, CPS provides potential backup in a timely manner when adverse conditions occur in the primary system, and thus provides a new safety regime. Two recent commercial aircraft accidents exemplify this concept. In both, Air France 447 (BEA, 2012) and Malaysia 370 (Malaysian Team, 2015), adverse conditions occurred in the primary systems. The former lost the airspeed sensor, and the latter lost communications. Investigation indicates alternative capabilities, either airborne or on ground, were available but the existing physical and cyber components associated with those alternative capabilities were separated. Should CPS integration be in place, the accidents could either be avoided, or the search and rescue be expedited.

5. *Stakeholder social-cultural ecosystem.* Social-cultural aspects have been an important factor in aviation. The advancement in aviation has significantly improved the quality of life for countless people, contributed to the economy, and revolutionized global cultural and economic interaction. However, aviation is still a highly specialized profession. Interactions between stakeholders frequently run into issues due to disconnections and misunderstanding. Stakeholder models, especially of consumers and the general public, may be integrated with the remainder of the aviation system to form an aviation CPS ecosystem to further improve aviation system efficiency, the quality of life and the world economy. This is made possible by big data and ubiquitous mobile technology, and it is already happening in many other industries.

6. *Improved life cycle development efficiency and accelerated pace of technology transition.* Architecture and component abstraction and standardization enable seamless and flexible integration of physical and cyber components. This maximizes component reuse. The design and development of an application system can thus focus on its core algorithm capabilities instead of spending precious resources on developing mission specific supporting services and interfaces as in conventional discrete system development regime. Individual components can also be developed and replaced without having to wait for all the components supporting the application to be designed and developed all at once. Improved life cycle development efficiency and accelerated pace of technology transition can thus be achieved.

4.2 CHALLENGES AND POTENTIAL DIRECTIONS

The increased network connection and the increased architecture and component abstraction also presents unique challenges to aviation CPS. The discussion of aviation CPS would not be complete without considering these challenges and potential directions in addressing them, including:

- New failure modes and system validation and verification
- Heterogeneous networks and air-ground system architecture

- Situation awareness of complex abnormal conditions design abstraction—requires human (operators/controllers) in the loop (physical), and human factor study of operational procedures with heterogeneous information channels and human machine interactions
- Degraded system operations
- Security and privacy
- Current and planned NextGen modernization investment cycle
- Stakeholder interests and sustainable business models

1. *New failure modes and system validation and verification.* The increased use of reusable networked, distributed physical and cyber components introduces new failure modes and thus present new challenges to system validation and verification. Unlike in conventional discrete application design, although integration interfaces may be standardized, specific physical and cyber components in an aviation CPS application system may be developed using different approaches or different model frameworks. The boundary between system design and software design is also become less distinctive in CPS application systems due to increased level of software and hardware integration and more complex "firmware" design. Interactions between increasingly networked components, either within the same physical or cyber layer or across different layers, may lead to conditions not experienced in more isolated environment. A failure or non-normal behavior in a component or the network is likely to have wider impact than in conventional discrete applications. Research into these problems will provide directions system and software design, verification and validation guidance, and updates to DO-178C and DO-278A certification objectives.

2. *Situation awareness of complex abnormal conditions.* While CPS improves situation awareness for normal operations by increased and integrated access of sensory and system state information, it will be a significant challenge under complex abnormal conditions when certain information becomes unavailable or erroneous information is present. Extensive use of abstracted information in human-machine interface might make it more difficult for the human to regain situation awareness and to engage in degraded operation should such information becomes unavailable or unreliable. Human factor study is needed to provide system design and training guidelines.

3. *Degraded system operations.* When component or system level anomaly occurs, a CPS application system must still provide basic operational and safety functionalities in degraded modes. It is critical to clearly indicate and manage what information is still reliable, what capabilities are still available, what lost information or capabilities can be compensated by available functionalities, what functionalities need to be shutdown to contain the impact of the abnormal conditions, and what procedures need to be established. While generic frameworks for addressing this challenge is useful, analysis of specific system anomaly scenarios for a particular aviation CPS application is important to developing degraded modes in the system and establishing degraded operational procedures.

4. *Heterogeneous networks and air-ground system architecture.* Unlike ground-based systems, air-ground data link is a unique component of aviation CPS. Most available air-ground data link technologies are still based on specific protocols and are subject to significant bandwidth and latency limitations. Heterogeneous networks and protocols have to be used to complete a single end-to-end air-ground message transmission. The physical layer of the air-ground network may have to switch from one protocol to another as a flight traverses from one region to another or from one operation domain to another. Increasingly available inflight Internet services are still not yet certified for flight deck applications. The interim solution is recommended logical layer air-

ground system architectures and protocols to serve specific applications, as exemplified by those in RTCA DO-359 (SC-206, 2014) and DO-350 (SC-214, 2014). Additional work is needed to finalize those recommendations as criteria for airworthiness approval and operational authorizations.

5. *Security and privacy.* One of the unique aspects of aviation CPS is that some of the currently used common air-ground data links are not yet provisioned with the same level of security measures developed for ground-based networks. For example, the 1090 MHz Mode S data link used by ADS-B and the VDL are not encrypted. Current measures include installation of ground stations at secured locations, monitoring network activities, and tracking and detecting data inconsistencies (Gandolfi and Mora-Castro, 2014). Research and developments are underway to provide long-term solutions and establish standards. Privacy is another important aspect. A balanced approach needs to be taken in information and data sharing to achieve system level efficiency objectives while protecting operator proprietary information. This balance requires a combination of technical approaches and policy.

6. *Long system life cycle and investment decisions.* Because of the long system life cycle typical to aviation mission-critical systems, many of the current system modernization programs are still based on conventional system set-up and infrastructure. Even if the benefits of CPS design can be justified, the transformation has to be aligned with investment decisions that are already in place and development or deployment that are already underway. For new development or modernization programs, CPS design is normally incorporated when the technical, schedule, and financial risks are within the manageable level. In this case, the approach is often to start with system architecture transformation first to ensure long-term goals. For programs that are already under development or deployment, CPS design is normally considered in incremental enhancement packages or technical refresh programs. In this case, it becomes more of an evolution process to ensure short-term benefits where possible.

7. *Stakeholder interests and sustainable business models.* A revolutionary CPS transformation is likely to have significant impact on stakeholder interests and the business landscape. The long development cycle and relatively high capital investment requires stakeholder interests to be carefully considered to ensure an effective transformation, a healthy ecosystem, and sustainable business models in the long run. In addition to the privacy concern mentioned above, intellectual property and software licensing in CPS environment also needs to be considered. The long life cycle of aviation systems also means that mixed fleet and mixed equipage operations will continue and need to be accommodated. If the cost/benefit can be justified, fleet and/or equipage modernization can be accelerated to alleviate this issue. For example, the general aviation segment has a relatively old fleet but it has seen accelerated avionics retrofit in recent years due to the benefits offered by GPS based navigation and flight deck automation.

5 CONCLUSION

This chapter has provided an in-depth analysis of mission-critical applications in aviation that involve the decision triad of aircraft, FOC, and the ANSP. CPS transformation opportunities exist for many of these mission-critical applications to achieve great benefits for aircraft operators, the ANSP, and the general public. However, several challenges have to be dealt with during the transformation process. If technology readiness, program schedule, and investment are aligned, certain applications may achieve an early transformation, probably at the architectural level first with full integration of physical and cyber components over time. Examples include both ground-based automation and airborne systems. For other systems and the industry as a whole, the transformation would likely be gradual, especially when the transformation time scale is compared with consumer facing industry sectors. The aviation community has been collectively working on overcoming some of the key challenges, and has started to seriously look into others. Leveraging progress in CPS and relevant enabling technologies, benefits are being realized, although initially on a smaller scale, ultimately an industry-wide transformation will be achieved.

REFERENCES

Airline Business & SITA, 2015. The Airline IT Trends Survey 2012—Executive Summary. Airline Business & SITA, Geneva, Switzerland.

AMCP, 2003. Analysis of VHF data links for point to point communications with a focus on VDL mode 4. In: Aeronautical Mobile Communications Panel (AMCP) 8th Meeting. ICAO, Montreal, Canada.

BEA, 2012. Final report on the accident on 1st June 2009 to the Airbus A330-203 registered F-GZCP operated by Air France Flight AF 447 Rio de Janeiro—Paris. BEA, Aéroport du Bourget, Le Bourget, France.

Brenner, F., 2015. Looking back, moving forward. Skyway Magazine (Autumn/Winter), 7–9.

FAA, 2010. FAA telecommunications infrastructure (FTI) operational network IP users' guide. Revision 2D, Redacted for Public Release, FAA, Washington, DC.

FAA, 2011. Surveillance and broadcast services description document. SRT-047, Revision 01, FAA, Washington, DC.

FAA, 2014a. FAA cloud services program objectives. Attachment L-6 to FAA Cloud Services (FCS) acquisition SIR No. DTFACT-13-R-00013, FAA, Washington, DC.

FAA, 2014b. FAA telecommunications services description (FTSD). Attachment J-2b to FAA Cloud Services (FCS) acquisition SIR No. DTFACT-13-R-00013, FAA, Washington, DC.

FAA, 2015. FAA Cloud Services (FCS) acquisition SIR No. DTFACT-13-R-00013. as amended by Amendment 1-15, FAA, Washington, DC.

Gandolfi, C., Mora-Castro, M., 2014. Technical Issues in the Implementation of Regulation (EC) No 29/2009 (Data Link), Version 1.1. European Aviation Safety Agency, Köln, Germany.

Hang, C., Manolios, P., Papavasileiou, V., 2011. Synthesizing cyber-physical architectural models with real-time constraints. In: Computer Aided Verification. Lecture Notes in Computer Science, vol. 6806/2011. Springer-Verlag, Berlin, Germany, pp. 441–456.

Klein, M.H., Kazman, R., Bass, L.J., Carrière, S.J., Barbacci, M., Lipson, H.F., 1999. Attribute-based architecture styles. Technical report CMU/SEI-99-TR-022, ESC-TR-99-022, Software Engineering Institute, Carnegie Mellon University, Pittsburgh, PA.

Lufthansa Systems, 2013. IT That Makes Your Life Easier, Key Figures 2012. Lufthansa Systems, Kelsterbach, Germany.

Malaysian Team, 2015. Factual information safety investigation for MH370—Malaysia Airlines MH370 Boeing B777-200ER (9M-MRO) 08 March 2014. Malaysian ICAO Annex 13 Safety Investigation Team for MH370, Ministry of Transport, Malaysia.

Ren, L., Beckmann, B., Citriniti, T., Castillo-Effen, M., 2013. Cloud computing for air traffic management—framework analysis. In: Proceedings of the 32nd Digital Avionics Systems Conference. pp. 5D6-1–5D6-31.

Ren, L., Beckmann, B., Citriniti, T., Castillo-Effen, M., 2014. Cloud computing for air traffic management—framework and benefit analysis. Final report, NASA contract NNA12AB81C, GE Global Research, Niskayuna, NY.

Sampigethaya, K., Poovendran, R., 2013. Aviation cyber-physical systems: foundations for future aircraft and air transport. Proc. IEEE 101 (8), 1834–1855.

SC-205, 2011a. Software considerations in airborne systems and equipment certification. DO-178C, RTCA, Washington, DC.

SC-205, 2011b. Software integrity assurance considerations for communication, navigation, surveillance and air traffic management (CNS/ATM) systems. DO-278A, RTCA, Washington, DC.

SC-206, 2014. Architecture recommendations for aeronautical information (AI) and meteorological (MET) data link services. DO-349, RTCA, Washington, DC.

SC-214, 2014. Safety and performance standard for baseline 2 ATS data communications (baseline 2 SPR standard). DO-350, RTCA, Washington, DC.

Simmon, E., Kim, K.-S., Subrahmanian, E., Lee, R., de Vaulx, F., Murakami, Y., Zettsu, K., Sriram, R.D., 2013. A vision of cyber-physical cloud computing for smart networked systems. NISTIR 7951, NIST, Washington, DC.

SITA, 2015. Aircraft operational applications. SITA Product Brochure.

ENSURING DEPENDABILITY AND PERFORMANCE FOR CPS DESIGN: APPLICATION TO A SIGNALING SYSTEM

23

E. Soubiran*, F. Guenab*, D. Cancila†, A. Koudri‡, L. Wouters‡

Alstom Transport, Saint-Ouen, France* CEA, LIST, Gif-sur-Yvette, France† IRT SystemX, Palaiseau, France‡

1 INTRODUCTION

The railway urban sector currently undergoes a great transformation. The Observatory of Automated Metros (UITP) states that "in 2013 there are 674 km of automated metro in operation." This number is expected to triple in the next 10 years to reach 1800 km in 2025 (Observatory of Automated Metros, 2013). Still, outside the mass urban transportation segment, the sector is heavily challenged in terms of competitiveness and attractiveness by other transportation industries. This fact is mostly visible through the travelers' demands for further improvements regarding speed and comfort, but also, punctuality, availability, and reliability. From a business perspective, this means a heavier and heavier pressure from the market for a leap forward in terms of performance, infrastructure and, of course, safety. To this end, the railway industry invests almost 1000 Million € each year in research and development (http://www.unife.org/research/overview.html). There are also strong expectations from the stakeholders for the delivery of cost-effective services for intermediate and final users: more intelligent, integrated and autonomous systems, safe access to vehicles, reduction of the CO_2 footprint, etc.

From an engineering perspective, the pressure has definitely been felt for a decade. This is visible through the progressive change from VAL (*Véhicule Automatique Léger*) systems to the more recent CBTC (Communication Based Train Control) systems. The realization and the maintenance of a VAL system heavily emphasizes the role of the wayside infrastructure. The new generation of CBTC urban transportation systems lessens this role by integrating more functions within the trains themselves and by promoting train-to-train communications. The rationale for this decision is the cost benefits coming from a reduction in the burden that had been put on the infrastructure. However, this leads to more and more complex train systems with the development of a distributed and autonomous intelligence for their control. Assuring the predictability of the global behavior of an autonomous train system is now a major challenge for the industry.

From a technical perspective, a signaling system has to guarantee the safety of trains' movements. Its role, however, is not solely limited to the safety of the system and its passengers. It also provides

Cyber-Physical Systems. http://dx.doi.org/10.1016/B978-0-12-803801-7.00023-7

operational and supervision functionality. In particular, the system's performance requires a great deal of attention because it provides value to the system's operator, i.e., the industry's client, as well as its passengers. For instance, a minimal interval between two trains is a performance requirement that provides value in terms of fluidity and flexibility during the system's operation.

The evolution of the train systems described above from the engineering and technical perspectives becomes obvious when focusing on the railway signaling system. The new generation of signaling systems is primarily implemented by a set of physical and software components ranging from train doors and bogies to control-command sub-system. As a result, signaling systems are a capital example of a cyber-physical system (CPS). In the context of the train industry, the design, implementation and maintenance of a signaling system are also constrained by safety norms and standards.

This chapter presents a case study of the design of a CPS in the train industry. It addresses the design methodology on a simplified functionality which encompasses both software, hardware and physical components. The chapter also presents challenges encountered by engineering teams when they integrate data coming from different disciplines and how they can be addressed by a federation technique. To this end, Section 2 gives a bird's eye view of the railway engineering artifacts methodology. Then, Section 3 develops the railway use case. Finally, Section 4 concludes with feedbacks, as well as perspectives for the future.

2 METHODOLOGY FOR ENGINEERING ARTIFACTS

This section presents a methodology for the design of a train system. It also highlights the key points regarding the collaborative aspects of the engineering and how they are addressed.

Fig. 1 illustrates the overall methodology and process that supports the realization of the case study in Section 3. This process is compliant with the first six phases of the CENELEC safety standards (CENELEC, 2012, 2011, 2003). It is divided into three major stages:

1. System definition: Capture and structure the *user needs* (i.e., the system's operators and its passengers) into a set of *services* that the system shall offer under safety and performance constraints. Perform preliminary hazard analysis.
2. System requirement and architecture: Define the *functional and system architecture* that fulfills the previously defined *services*. Perform system hazard analysis.
3. System design and implementation: Define the *physical architecture* with a breakdown into hardware and software components that will implement the *functions* previously defined. Perform testing and data assessment.

An important aspect of the process is its inherently iterative nature. As shown in Fig. 1, each of the three major stages have micro-iteration cycles between their respective engineering activities. The process also supports macro-iteration cycles across the major stages. This iterative nature reflects the interactions that happen in the real world between the engineering activities. It also acts as an indicator of two underlying key points:

1. The collaborative nature of the process across engineering activities.
2. The continuous verification of the engineering choices.

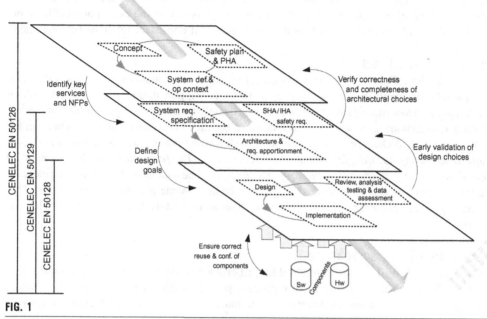

FIG. 1

Combining collaborative engineering and the CENELEC process.

These two points are not directly captured by the process and yet they are instrumental to the success of designing a CPS such as the one presented in our case study.

Collaborative engineering

At each stage, the involved teams share a common objective. They produce engineering artifacts that hold different information about the system under study. For example, the overall objective of the first stage is to define services that the system shall offer to end-users (operator and passengers). These services are constrained by nonfunctional properties, including, besides others, safety and performance ones. Still, the produced artifacts are not entities that exist in a vacuum independently from the rest of the world. All of them have relations with the others. For example, in the first phase, there must be traceability relations between the preliminary hazard analysis and the identified services. With more and more complex systems, it is critical to capture and manage those relations.

Continuous verification

In addition, the stages and activities presented in the process above can have some concurrent overlaps and therefore have continuous feedback loops. The activities in a stage can shed new light on the decisions taken in its upper stage, leading to the early detection of errors in the design or in the implementation. For example, a performance requirement that must be satisfied by an operational scenario is derived at the end into constraints on the performance of some actuators or software functions. Particular implementation choices can lead to the infeasibility of earlier design decisions; these kinds of inconsistencies will be detected and reported. In an iterative process supported by the appropriate tools, the continuous verification of the engineering choices precisely consists of the examination of the input artifacts at all stages and activities, as well as the maintenance of consistent relations between them.

Relating engineering artifacts

The critical element for efficient collaborative engineering and its continuous verification is the management of the relations between the engineering artifacts. To support it, the proposed methodology adopts a technique able to semantically relate engineering data that are exposed by the tools and used by the engineers across technological spaces. This technique is called "federation" and relies here on the Semantic Web and Linked Data. *In the use case, we exploit a web interface as a means to access data from requirements, safety hazards, functional architecture, etc... The federation enables a controlled acquisition of the data by preserving the full property (of data).* The data shared by all the engineering teams and their colleagues can be leveraged to perform business-wide analysis and trade-offs. This capability reinforces the engineering and business decisions, ultimately leading to more value for the end users.

Comparison with existing approaches

Today, railway (critical) artifacts are submitted to a rigorous process to ensure the expected safety integrity levels. The high cost related to this process prompted engineers to take strict control over the introduction of mechatronics innovations. In many cases, then, two artifacts present small differences.

At the same time, the high competition between constructors demands innovative functionality and a whole cost-reduction of the artifact. A top-down approach, as described in the CENELEC norms, is not competitive anymore. Often, constructors combine top-down with bottom-up approaches, for example in the reuse of software, system design, electronic or mechanical devices. Interfaces between components acquires a relevant role to fix the objectives of performance, innovation, cost and safety. A change in the interface could impact other components and *de facto* avoids a correct reuse, thus increasing the overall cost. The current practice synchronizes information between teams at precise times in order to avoid this.

Our methodology does not aim to change the adopted tools or the knowhow of each company, but only to suggest a possible solution to combining collaborative engineering with the CENELEC rigorous process by reducing the go-to-market timing. The proposed approach can be extended to other application domains. First preliminary results, showed on the railway domain, received a positive and encouraging feedback by an avionic company. Currently a feasibility study is under evaluation.

3 DESIGNING CPS

This section applies the design process described in Section 2 to the system part that allows the exchange of passengers at a railway station. Broadly speaking, *passenger exchange* addresses getting on and off a train. It is, then, an important operational scenario to be supported by an urban autonomous train and one of the most obvious ones from the perspective of the public. As a result, substantial engineering efforts have to be spent on this precise scenario. The interaction between passengers and the

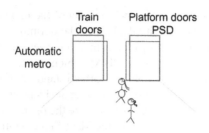

FIG. 2

Technically System's Boundary.

train system is full of potential safety hazards that have to be identified, analyzed, traced, and finally fixed by technical means. This scenario involves hardware, software, and physical components, such as doors, doors' actuating motors, human presence detectors, and software for train control and for train line's supervision.

Fig. 2 contextualizes the following nominal scenario. An autonomous train arrives at a given station. The station is equipped with **Platform-Screen Doors (PSD)** preventing the passengers on the platform from inadvertently falling onto the tracks. The train proceeds along the platform until its **Rolling-Stock Doors (RSD)** are aligned with the PSD. When the train stops with its doors aligned, it opens the RSD and the PSD simultaneously. The passengers can then get off and on the train. After a fixed duration, audio and visual systems warn passengers that the doors are going to close. For safety reasons, PSD are closed with a small delay before that RSD are fully closed. The safety argumentation is based on the probability that the required behavior for passengers ("go off" and then "go on") is respected. Independently for the three scenarios (full synchronization, PSD before RSD, RSD before PSD), sensors are deployed on the doors, platform and train to detect if a passenger is stuck between the two doors. Finally, the train leaves the station if and only if there is not a passenger stuck and PSD and RSD are verified closed.

3.1 SYSTEM DEFINITION

The main objective of the first stage is to capture and to structure user needs into a set of services that the system shall offer under safety and performance constraints. User needs captured and structured into requirement modules are called, *System User Need Requirements*. These later represent a common input for each activity in the first stage. *For example, system engineers elicit requirement Req_Timing-DoorsDetection from the following user need*

> **User need:** Whenever a train in operation safely stops at a station to carry out a passenger exchange, it shall open train and platform doors if and only if they are aligned with an error margin of $x\%$. The opening sequence should take less than 4 s and shall never take more than 4.5 s. An audio and visual signal shall warn doors of movement.
>
> **Req_TimingDoorsDetection:** During the passenger exchange phase in a station equipped with PSD, the maximum duration between the detection of train safe stop and the effective door opening shall be: at best <4 s in 95% of cases and at worst <4.5 s in all cases (100% of cases).

The *Concept* activity in Fig. 1 addresses the specification of the main engineering plans and the system's boundaries (Fig. 2). The next activity deals with the definition of operational contexts, which allow safety and design engineering teams to focus their analyses on common hypotheses. Safety engineers perform a preliminary hazard analysis (PHA), which identifies risks, accident scenarios, and mitigation barriers, in compliance to the CENELEC 50126 standard. A mitigation barrier is a physical, software, or procedural mean to reduce the severity of an accident. It is always associated to a safety requirement. In the meantime, system engineers formalize the operational behavior of the system, i.e., they specify possible scenarios that can occur in the identified operational contexts and perform operational analyses. The main goal of the scenarios is to clearly identify the interactions between the system and its environment, they are modeled thanks to message sequence charts where the system is considered as a blackbox entity (Ferrogalini and Le Bastard, 2012). It allows system designers to outline a specification of the functional services that the system will provide. Safety and system engineering activities are concurrent. However, the results of the safety analysis could impact the choices in the system space. Indeed, safety results tag existing user needs in the design space as "safety-related" and also could introduce new safety-related requirements—thus forcing a change in the design space.

For example, the *passenger exchange* scenario could occur in the following *operational context*:

- **Mode**: The train is in an unattended train operation (UTO) mode.
- **Phase**: The train movement is stopped.
- **Zone**: The train is at a station, aligned with a platform.

Starting from the operational context and applying PHA methodology, safety engineers identify the hazards and hence the accidents. For example:

- **Hazard**: the train doors could still be open while the platform doors are closed.
- **Accident**: passenger is trapped between the train and the platform.

Intuitively, the hazard means that the passengers could try to get off the train and position themselves behind the train doors, but not on the platform because the platform's doors are closed.

To avoid the accident, safety engineers define safety barriers that can introduce new safety-related requirements, such as:

> **PHA-ClosedTrainDoorsCondition:** train doors corresponding to deactivated PSD will remain closed and locked.
> **PHA-BlockageDetectionIntegration:** train doors and PSD will integrate doors blockage detection mechanism. The blockage will be reported to signaling system.

All this information is traced as shown in Fig. 3. Note that the accident scenario is a conjunction of a context, a tolerable hazard rate (THR) and the barrier that addresses it.

System designers must take into account the new safety-related requirements, potentially even impacting the previous operational scenario. For example, requirement *PHA-BlockageDetectionIntegration* forces the system engineers to modify the operational scenario by introducing a new element in the door closing sequence:

"A passenger trying the get off the train in a hurry is blocked by the closing platform doors (while the train doors are still open). When the train doors start to close, the sensors detect the passenger and abort the closing command. They try to close every 5 s until no obstacle is detected."

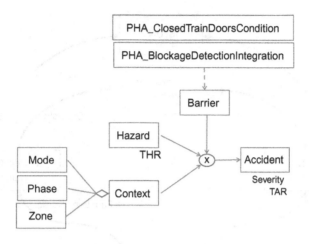

FIG. 3

Accident scenario in PHA analysis.

This modification of the operational scenario impacts the set of services that will be fulfilled by the system. Now the train and platform doors must detect obstacles and have a complex closing sequence that can retry multiple times.

System engineers and safety engineers work in an iterative cycle in order to converge toward a stable specification. The two teams use different tools for the production of different artifacts. Yet, these artifacts refer to each other in some way, as shown in Fig. 4. The system engineers produce user needs in a requirement management tool. The safety engineers use a custom in-house tool. They rely on the federation mechanism described in Section 2 for sharing data with each other. In this way, they can trace the relations between the user needs, the identified system services and the safety analyses. The information of requirements *PHA-ClosedTrainDoorsCondition* and *PHA-BlockageDetectionIntegration* are physically stored in the requirement management tool and yet they can be traced to the accident scenario in the safety tool because they are shared through the federation space. The ability to directly trace relations between different artifacts eliminates the risk of having discrepancies between them. The traceability relations become an integral part of the engineering artifacts and are therefore managed in versions and baselines. Ultimately, this makes the engineering teams more efficient.

At the end of the first stage, the engineers produce the set of services that must be fulfilled by the system under development.

3.2 SYSTEM REQUIREMENTS AND ARCHITECTURE

The goal of the second stage is to produce the functional and system architecture that will fulfill the services identified in the first stage.

The functional architecture includes the system functions and the data-flows between them. For example, Fig. 5 shows the functional architecture that fulfills the passenger exchange service. The architecture is rooted on **F1** that corresponds to the global control-command function. A sub-function of

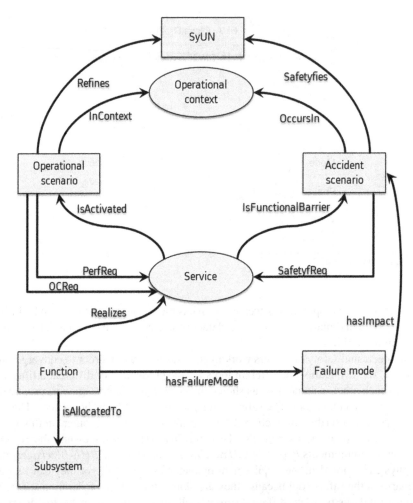

FIG. 4

Traceability between the engineering artifacts.

interest is **F1.1** which operates the train and platform doors (i.e., opening, closing, warning, etc.). Engineers also identify the inputs and outputs of **F1.1**. They are respectively:

F1.1.Input: state of the platform and train doors (are they open or closed)
F1.1.Output: command for the actuation of the platform and train doors (open or close).

From the functional architecture, the safety engineers perform a system hazard analysis (SHA) that identifies the possible failures of each function. It is conducted as a failure mode and effect analysis (FMEA).

With the assumption of a stable functional architecture, the system engineers specify the product breakdown structure (PBS) (Fig. 6). Functions are allocated to identified sub-systems. The sub-systems'

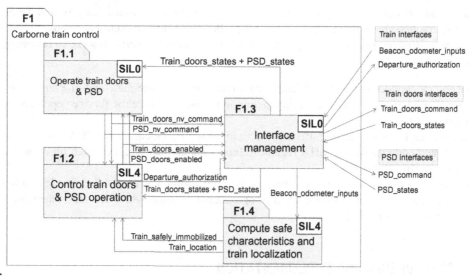

FIG. 5

Functional architecture.

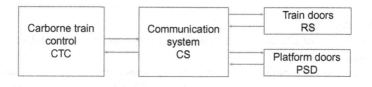

FIG. 6

Product breakdown structure.

interfaces are specified and for each one the safety team performs an interface hazard analysis (IHA). In this case, the PBS identifies four sub-systems: the carborne train control (CTC), the communication system, the PSD, and the rolling stock doors (RS).

Knowing that functions inherit service's performance and safety constraints, the main challenge here is to refine the initial constraint into an end-to-end one applicable to the integration of the four sub-systems. This new constraint is then split into coherent sub-system performance requirements:

PlatformScreenDoorsSystem: in the nominal operational context (any perturbation from passengers), the delay between incoming closing request on Interface XX and the emission on the Interface YY of the PSD closed status will be of 3 s in 95% and 3.5 s in 100%.
ComunicationSystem: the quality service will allow a message transmission with maximum latency 100 ms and a maximum of one message can be missing.
CarborneTrainControlSystem: function **F1.1** (operation) and function **F1.2** (control) shall be synchronized in a common execution rate of 100 ms.

The functional architecture, the system hazard analysis, the interface hazard analysis, and the product breakdown structure are all shared through the same federation space, as mentioned above. The semantic traceability between the engineering artifacts allows engineers to perform some overarching verification and reasoning:

- Consistency between services and functional architecture, i.e., the matching of every input and output of an interface function to an interaction in a scenario where the service depending on that function is activated.
- Completeness of SHA regarding PHA by verifying that all accidents in the PHA have a root cause in SHA, and completeness of SHA regarding functional architecture, i.e., each function has a complete dysfunctional model.
- Consistency of hazard analyses with the system specification, i.e., each failure mode of a function shall trace potential accidents that occur in operational contexts where services depending on that function are activated.

3.3 SYSTEM DESIGN AND IMPLEMENTATION

In this third stage, the sub-systems identified in Fig. 6 are dispatched to sub-system engineering teams. The following sub-sections complete the story and show how system requirements are transformed into low-level requirements for both the physical and cyber sides.

3.3.1 The physical side

The physical components of interest are doors with their actuators, sensors and door control unit (e.g., DCU), and I/O modules (Fig. 7).

The I/O module allows communication between the doors control unit (DCU) and carborne train control (CTC) through the communication system (CS). DCU system contains basic speed controllers used to command DC motor speed.

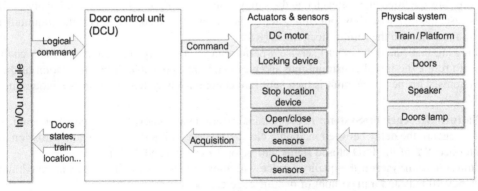

FIG. 7

Physical implantation.

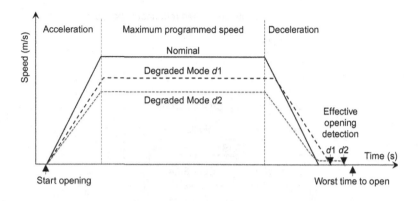

FIG. 8

Speed trapezoidal profile: d1 detected degraded mode, d2 undetected degraded mode.

Three phases are considered in a door opening sequence: (i) the acceleration, (ii) the constant speed movement, and (iii) the deceleration (Fig. 8). With modern hardware, features considered in each phase/speed profile are adjustable, i.e., the maximum door speed and acceleration/deceleration curves may be customized. This allows physical system to meet the allocated time to opening and closing operations. It also ensures that passengers are not injured when doors are closing. The system engineering team has allocated **4** s for the opening sequence in nominal condition and **3.5** s at worst to handle exceptional condition. Considering the IO module and DCU response time, the applied speed profile shall allow the door to traverse the distance (its width) in **2.8** s. This requirement will export some physical constraints to the motor, allowing its dimensioning (especially the settling time).

In the nominal case, the speed meets the timing parameter within an acceptable error interval. In faulty cases, we consider a loss of efficiency a fault of the actuator. The real speed of DC motor is less than the expected one, because it doesn't reach the maximum speed (Fig. 8).

In the case where a diagnostic mechanism is implemented allowing the detection of such faults, a fault tolerance strategy is implemented at the control stage to compensate for such faults without changing the speed profile. It also achieves the maximum speed within the expected time. This type of fault tolerance strategy is used in particular when loss of efficiency faults are minor. If the diagnostic mechanism fails to detect the minor loss then the opening operation continues at low speed until the opening is confirmed by the sensor (d2 in Fig. 8). The opening must be confirmed before the predefined worst case time (i.e., 3.5 s). In the case of a major loss of efficiency fault that cannot be accommodated, a new degraded profile may be applied. However, the time of opening will be a little more extended (d1 in Fig. 8) or even unreachable. In all cases, in order to ensure that the opening is operating within an expected time, a corresponding predefined tolerable loss of efficiency of **4%** is taken into account in the design.

3.3.2 The cyber side

The engineers use a component-based design approach relying on a layered model. This approach is supported by an execution product, depicted in Fig. 9 providing basic services such as: real-time scheduling, memory management, communication protocol stacks, spatial isolation and also, railway-specific services (e.g., redundancy and safety management). Regarding the tooling, engineers rely on a tool for the selection and assembly of software components, as well as another for the deployment

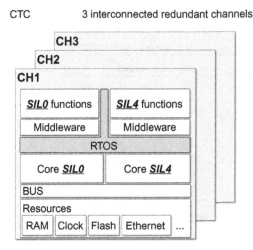

FIG. 9

2oo3 architecture.

(assistance). A description of the software environment is not within the scope of this paper. Rather, we are interested in the connection between this environment and the system specification environment. In particular, we are interested in the engineering continuity between business rules and their implementation in the cyber side, with a strong focus on the verification of nonfunctional properties. For each subsystem function, the engineers build an assembly of implementation components selected from an inhouse library. In this case, such an assembly does not exist (new functionality or nonmatching constraints), new implementation components shall be developed according to the CENELEC 50128 rules.

The seamless and early verification is based on a double approach, which exploits the standard GSN (global structuring notation) (GSN Community Standard, 2011) and CBD (contract-based design) (Sangiovanni-Vincentelli et al., 2012; Derler et al., 2013; Cancila et al., 2010; Cimatti et al., 2016; Cimatti and Tonetta, 2015). CBD is a means to deal with a correct integration of (heterogeneous) components at functional level. Moreover, CBD allows a correct integration between the software and hardware components. Engineers specify a GSN-tree to support safety-related requirements, then they use a CBD to structure the link between a goal and the related evidence. For example, a contract matches the assumption at software component level (independence between SIL0 and SIL4) and the guarantee at the execution platform (spatial and temporal isolation).

4 CONCLUSION

This chapter presented a view of the development process of a CPS in the railway industry at Alstom. For this purpose, it started from the simple scenario of passengers getting off and on an autonomous train at a station. By applying the process described in Section 2, the design, implementation and safety analyses of the corresponding CPS was explained in Section 3. Throughout this application the chapter emphasized two key points of the considered process: (1) the collaborative nature of the engineering activities, and (2) the continuous verification of the engineering choices. As demonstrated in the case study, these two points are instrumental in the realization of safety-critical systems at Alstom.

Consequently, this case study highlighted two important factors for the success of the design of a CPS. First, the strong support for an iterative methodology that actually fosters the collaborative nature of the engineers' work. A CPS is the quintessence of a multi-disciplinary system. Problems encountered during the design of a complex system are exacerbated in this case. The ability of the engineers to efficiently share their data then becomes even more critical. Second, the technical support for this collaboration. One cannot underestimate the time consuming, budget wasting and demoralizing effect of rework due to poor communication and inefficient methods of data sharing. As shown throughout Section 3, this case study relied in the "federation" technique. It has proved itself useful in its ability to empower the engineers to share meaningful data, ultimately enabling business-wide analysis that supports the continuous verification of the engineering choices.

Finally, it has to be noted that the "federation" technique is still an active research subject. Efforts in this direction are exemplified by Technological Research Institute SystemX. In the same way, the full application of iterative and agile processes are currently being investigated in an industrial context at Alstom. Similar investigations in other industries would provide great insight for the better design of CPSs.

ACKNOWLEDGMENTS

This work is supported by the Technological Research Institute (IRT) SystemX (http://irt-systemx.fr/). It is partially funded by the French public program *"Investissement d' Avenir."*

REFERENCES

Cancila, D., Passerone, R., Vardanega, T., Panunzio, M., 2010. Toward correctness in the specification and handling of non-functional attributes of high-integrity real-time embedded systems. IEEE Trans. Ind. Inf. 6 (2), 181–194.

CENELEC, 2003. Railway applications—communications, signaling and processing systems Safety related electronic systems for signaling. European Standards.

CENELEC, 2011. Railway applications—communications, signalling and processing systems—software for railway control and protection systems. European Standards.

CENELEC, 2012. Railway applications—the specification and demonstration of reliability, availability, maintainability and safety (RAMS)—Part 2: Systems approach to safety. European Standards.

Cimatti, A., Dorigatti, M., Tonetta, S., 2016. Othello Contracts Refinements Analysis (OCRA). http://www.goalstructuringnotation.info/.

Cimatti, A., Tonetta, S., 2015. Contracts-refinement proof system for component-based embedded systems. (Part 3) Sci. Computer Program 97, 333–348.

Derler, P., Lee, E.A., Torngren, M., Tripakis, S., 2013. Cyber-physical system design contracts. In: International Conference on Cyber-Physical Systems (ICCPS 2013), Philadelphia, USA.

Ferrogalini, M., Le Bastard, J., 2012. Return of experience on the implementation of the system engineering approach at ALSTOM. In: Complex System and Design Management 2012 International Conference.

GSN Community Standard, 2011. Goal Structuring Notation. Version 1. Available to http://www.goalstructuring-notation.info/documents/GSN_Standard.pdf.

Observatory of Automated Metros, 2013. Annual World Report. UITP. http://metroautomation.org/about-the-report/.

Sangiovanni-Vincentelli, A., Damm, W., Passerone, R., 2012. Taming Dr. Frankenstein: contract-based design for cyber-physical systems. Eur. J. Control 3, 217–238.

INTEGRATING RENEWABLE ENERGY RESOURCES IN SMART GRID TOWARD ENERGY-BASED CYBER-PHYSICAL SYSTEMS

24

P. Moulema*, S. Mallapuram*, W. Yu*, D. Griffith[†], N. Golmie[†]

Towson University, Towson, MD, United States[*] National Institute of Standards and Technology, Gaithersburg, MD, United States[†]

1 INTRODUCTION

Generally speaking, cyber-physical systems (CPSs) (Lee and Seshia, 2011) are built upon the integration of information and communication technologies and physical components. Smart grid is often described as an energy-based CPS as it integrates a physical power grid with the cyber process of computing and communication components. Via the integration of information communication techniques, the two-way interaction between consumers and utility providers can be enabled and the efficient and reliable operation of power grid can be supported. Also, the integration of renewable energy resources in the power grid is one the key endeavor of the smart grid, with the goal of reducing the dependency on the traditional energy resources and making the power grid cost effective and environment friendly (U.S. Department of Energy, Office of Energy Delivery and Energy Reliability, 2016).

Integrating renewable energy resources into the power grid has a potential to progressively replace the depleting fossil fuel resources, increase the overall energy supply, and, consequently, lead to a cost-efficient and environmentally friendly power grid (National Institute of Standards and Technology (NIST), 2016; U.S. Department of Energy, Office of Energy Delivery and Energy Reliability, 2016). Nonetheless, the integration of renewable energy resources is also subject to serious challenges. Particularly, the intermittency of renewable energy resources can seriously impact the stability and reliability of the power grid, as the generated energy from renewable resources depend upon factors (e.g., weather, etc.) (Gul and Stenzel, 2005; Mills, 2010b). Therefore, it is critical not only to quantify the impact of the massive integration of distributed renewable energy resources, but also to develop monitoring and control mechanisms to mitigate the impact of the variability of renewable energy resources and maintain the overall stability of the power grid.

A number of research efforts have been devoted to integrating renewable energy resources and energy storage devices into the power grid. For example, techniques (e.g., technology diversity, geographic diversity, etc.) were considered to smooth the variability of power generation from solar and wind power plants (Eichman et al., 2013; Sims et al., 2011; Hoste et al., 2009). Because solar and wind plants largely depend on weather and climate changes, they can only be deployed in specific geographic locations in order to achieve sustainable power output. Consequently, the geographic diversity of deployment will require additional transmission infrastructures to bring the power supply near to the demand, incurring transmission costs and transmission line power losses. Furthermore, geographic diversity and technology diversity focus on mitigating the variability of the power generation from renewable energy resources and do not provide an effective way to address failures or performance degradation of solar or wind plants. Therefore, there is the need for developing efficient control mechanisms to monitor and control distributed energy resources.

In this chapter, we first review renewable energy resources and existing efforts toward their integration into the power grid. We then design simulation scenarios with the consideration of different types of renewable energy resources and deployment techniques. We also propose a control mechanism that is capable of mitigating their inherent variability. Using GridLAB-D and GridMat simulation tools, we implement these scenarios and conduct a performance evaluation of mixing different types of renewable energy resources and deployment techniques, as well as our proposed control mechanism. Our simulation results show that mixing different types of renewable energy resources and energy storage can effectively reduce the power generation cost and power distribution loss. Our simulation results also show that the proposed control mechanism is effective in responding to failures and contingencies by monitoring and controlling distributed energy resources.

The remainder of the chapter is organized as follows: We review the related work in Section 2. We describe our approach, including design rationale, simulation model, scenarios, and control mechanism in Section 3. We present the performance evaluation of the integration of renewable energy resources and the proposed control mechanism in Section 4. Finally, we conclude the paper in Section 5.

2 RELATED WORK

In this section, we conduct a brief literature review of existing efforts on the integration of renewable energy resources in the power grid.

There have been a number of research efforts devoted to understanding the role of renewable energy resources in the power grid and developing techniques for integrating them into the power grid (Hossain et al., 2016; Phuangpornpitak and Tia, 2013; Mwasilu et al., 2014; Lund et al., 2014; Chen et al., 2016; Henninger et al., 2015; Mathiesen et al., 2015; Sharma et al., 2015; Eltigani and Masri, 2015). For example, Phuangpornpitak and Tia (2013) studied the role of renewable energy and distributed generation in the smart grid system and explored the challenges of effectively integrating renewable energy resources into the smart grid system. Chen et al. (2016) proposed technologies for scheduling, monitoring and control in integrating different types of renewable energy resources. Henninger et al. (2015) investigated the smart power plant based on micro-grids that consist of renewable energy resources and energy storage devices to make the energy management process efficient.

Prior to the most recent research work mentioned above, extensive research efforts have been made on studying the factors, requirements, and effects of integrating renewable energy resources and energy storage into the power grid (Connolly et al., 2012; Denholm et al., 2012; Hoste et al., 2009; Halamay et al., 2011; Delille et al., 2012; Levron and Shmilovitz, 2012; Evans et al., 2012; Hill et al., 2012; Oren et al., 2012; Sims et al., 2011; Miller et al., 2011). For example, Connolly et al. (2012) studied the capability of large-scale energy storage to compensate the variability of renewable energy resources. Their research efforts showed that by using pumped heat electrical storage (PHES), wind energy could supply 100% of the load in a high cost. Denholm et al. (2012) investigated the role and impact of energy storage devices in the grid when the number of renewable energy resources integrated to the power grid increases. Hoste et al. (2009) conducted a case study on the California electricity grid and their results showed that a combination of independent wind and solar power plants with hydraulics could be effective in complementing their individual intermittencies and increasing the effectiveness of renewable energy resources. Halamay et al. (2011) analyzed the interaction between the variability of the utility load, wind power generation, solar power generation, and ocean wave power generation. Their results showed that the variability of renewable energy resources could be significantly reduced by combining different types of renewable resources. Delille et al. (2012) leveraged the characteristics of energy storage devices and developed a dynamic frequency control mechanism to reduce the negative impact of solar and wind power generation.

Also, Levron and Shmilovitz (2012) developed an approach to minimize the peak hour power generation through an optimal management of energy storage devices. Evans et al. (2012) reviewed energy storage technologies and compared the performance of different integration schemes. Hill et al. (2012) investigated the effectiveness of energy storage devices to deal with variability, frequency, and voltage control issues related to renewable energy resources. They also proposed the integration of energy storage devices in the energy market with a goal of increasing the economic benefits of renewable energy resources. Oren et al. (2012) investigated the efficiency, reliability, economic, and environmental impacts of the massive integration of wind power plants into the grid, as well as different approaches to mitigate their negative impact on the grid. Sims et al. (2011) reviewed future energy systems and discussed benefits and challenging issues of massive integration of renewable energy resources. Taneja et al. (2012) investigated a real-time blend of power supplies in the California electricity grid and assessed the impact of demand shaping, storage, and agility of the hybrid power system. Their study showed that energy storage and load shifting could significantly increase the efficiency and reliability of the power grid when a number of renewable energy resources are integrated into the power grid. Miller et al. (2011) investigated the impacts and benefits of wind plant controls, which are capable of mitigating frequency variability issues in the grid.

3 OUR APPROACH

In this section, we first give an overview of our approach. We then describe the simulation model. Finally, we present scenarios used for our simulation study and the proposed control mechanism.

3.1 OVERVIEW

We now briefly review renewable energy resources, including solar power and wind power. One typical technique for solar power is photovoltaic electricity, which relies on photovoltaic cells and the solar-thermal electricity. The key factors that impact on solar photovoltaic performance include solar radiation, panel efficiency, panel size, and panel orientation. Wind energy generation utilizes a turbine, which converts the kinetic energy from the wind into electrical power. The effectiveness of wind plants depends on factors including the speed and the force of wind, the altitude of site, the size of rotor, etc.

It is worth noting that the integration of renewable energy resources into the grid can provide the following benefits to the power grid: (1) *Reduction of generation and transmission costs*—the deployment of small and medium wind and solar plants near to the demand can significantly reduce the amount of bulk power generation and can counterpoise the need of extra power generation at peak hours, thereby leading to the reduction of generation and transmission costs. (2) *Reduction of transmission and distribution power losses*—according to the US Energy Information Administration (EIA), transmission and distribution losses represent an amount of 6% of the total electricity transmitted and distributed in the United States in a year (U.S. Energy Information Administration (EIA), 2016). Therefore, the proximity of distributed renewable energy resources to the demand can not only reduce the losses on distribution lines, but also offset the need for additional distribution and transmission infrastructures. (3) *Sustainable and clean power grid*—the increase of the shared load of renewable energy resources in the power grid can reduce the dependency on fossil fuel resources, leading to the reduction of greenhouse gas emissions (Sims et al., 2003).

Nonetheless, challenges related to the integration of renewable energy resources are raised by their intrinsic characteristics. Renewable energy resources (e.g., wind turbines, solar plants, etc.) are intermittent by nature as their generated power depends heavily on factors such as the radiance of the sun, the speed and the force of the wind, etc. (Gul and Stenzel, 2005; Mills, 2010b; Sims et al., 2011). To deal with this issue, we design a power grid simulation model and investigate deployment techniques for renewable energy resources. We also propose a control mechanism that is capable of monitoring the performance of renewable energy resources regularly, and taking action when the performance degradation of renewable energy resources occurs. We use GridLAB-D (a power modeling and simulation tool) (Pacific Northwest National Laboratory (PNNL), 2016) and GridMat (a co-simulation tool that integrates Matlab and GridLAB-D) (University of California, Irvine. Electrical Engineering and Computer Science Department, 2016) to validate the deployment techniques of renewable energy resources and the effectiveness of the proposed control mechanism.

3.2 SIMULATION MODEL

In our simulations, we use two simulation tools: GridLAB-D and GridMat. Generally speaking, GridLAB-D is a power distribution system simulation and analysis tool, developed by the Pacific Northwest Laboratory (PNNL) (Pacific Northwest National Laboratory (PNNL), 2016). GridMat is a co-simulation tool (University of California, Irvine. Electrical Engineering and Computer Science Department, 2016), developed by the Department of Electrical Engineering and Computer Science of the University of California, Irvine. GridMat leverages the capabilities of GridLAB-D to model and simulate the physical power grid, and the capabilities of Matlab as a control engine to the physical power system. We used GridMat toolbox to implement the proposed control mechanism.

Fig. 1 shows the simulation model, which consists of the following six major components:

(1) *Substation:* It consists of a swing bus with a nominal voltage of 33,000 V and a power rating of 20 MVA. A transformer installed between the generation and the transmission and distribution components will step the voltage down to 2401.7 V.

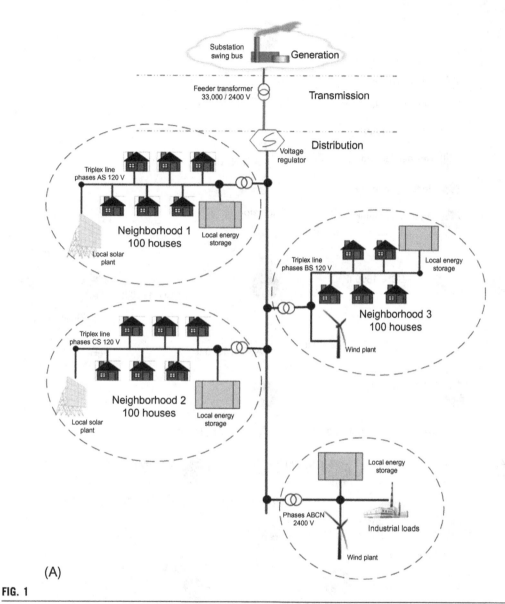

(A)

FIG. 1

Simulation topology. (A) High-level view. (B) Electric network.

(Continued)

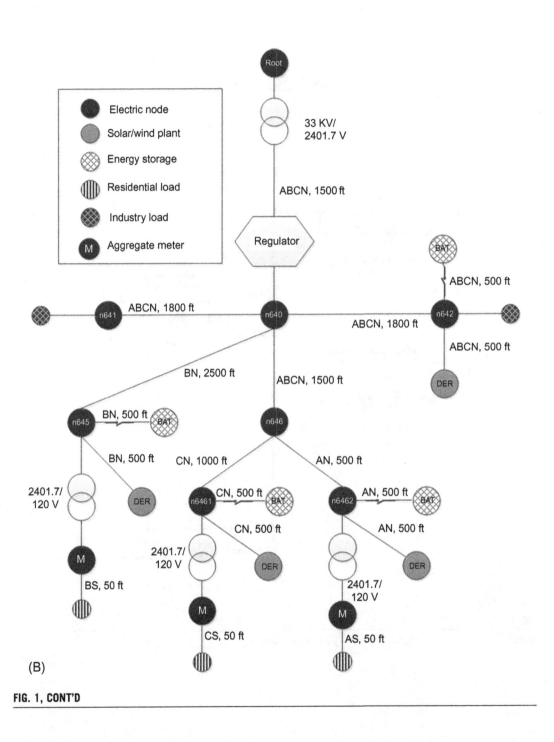

(B)

FIG. 1, CONT'D

(2) *Transmission and distribution:* It consists of overhead lines with different electric characteristics to deliver power flows.

(3) *Residential load:* It consists of 300 houses, which are divided equally into three neighborhoods. In each neighborhood, there are 100 houses, a local solar or wind power plant, and a local energy storage. A meter is installed in each neighborhood to collect the aggregated load consumed in the area and a neighborhood transformer steps the voltage down from 2401.7 V to 120 V.

(4) *Industrial load:* It consists of three phases industry load of 45 kW in average.

(5) *Local renewable energy plant:* It can be either a solar or wind plant, where the solar plant is made of a single crystal silicon panel with 3000 square feet that delivers approximately 25% amount of the neighborhood total load in average and the wind plant is based on a Vestas V82 turbine model that delivers 1.7 MW maximum power.

(6) *Local storage:* It consists of a Lithium-Ion battery that can be automatically charged or discharged based on the schedule.

3.3 SCENARIOS

In the following, we describe several scenarios in our simulation study.

3.3.1 Scenario 1: Integration of distributed energy resources

The goal of this scenario is to evaluate the effectiveness of integrating distributed renewable energy resources in the smart grid. To this end, we consider three cases. The first case considers a traditional power grid without renewable energy resources and storage devices, the second case considers a power grid with only renewable energy resources, and the third case uses a power grid with both renewable energy resources and energy storage devices. To observe the impact of seasons on the performance of the grid, simulations will be run over a year's time period from January 2011 to December 2011. We consider the climate and weather parameters of Baltimore, Maryland, USA. For a realistic simulation, we define a variety of residences with different characteristics, including floor area that ranges from 700 to 3500 square feet. Residences are also equipped with all necessary house appliances (e.g., refrigerator, freezer, washer, dryer, microwave, heating, ventilation, and air conditioning (HVAC) system, etc.).

We consider the following key metrics:

(1) *Quantity of bulk power generation:* It is defined as the amount of power that the substation feeder supplies in order to meet the energy demand.

(2) *Quantity of power loss in transmission and distribution lines:* It is defined as the aggregated value of resistive losses and corona losses during the transmission and distribution process. Notice that the resistive loss is raised by the resistance of power lines and is proportional to the length of power lines, while the corona loss is the power discharge raised by the electric field that surrounds conductors.

(3) *Quantity of transformer power loss:* It represents the combination of heat losses and eddy currents in the primary and secondary conductors of the transformer. It is measured as the difference between the input power and the output power in the generation and distribution step-up transformers.

3.3.2 Scenario 2a: Integration of distributed energy resources with geographic diversity

In this scenario, we consider the deployment of renewable energy resources through geographic diversity, meaning that multiple solar plants are dispersed in different sites within the same region in order to reduce the variability of aggregated power generation (Sims et al., 2011; Mills, 2010a). When solar plants are dispersed in different geographic areas, their aggregated power generation can be improved as it is unlikely that clouds will shade all plants simultaneously or it is unlikely that the strength of wind will decline in all sites simultaneously. In this scenario, we simulate the power grid with solar plants with various climate and weather patterns. This scenario can help to understand how much the geographic diversity can help mitigate the uncertainty of renewable energy resources by reducing the variability of aggregated power generation from distributed energy resources. Evaluation metrics include the quantity of bulk generation and aggregated power losses.

3.3.3 Scenario 2b: Integration of distributed energy resources with technology diversity

Technology diversity aims to reduce the variability of aggregated power generation from multiple distributed energy resources via combining different deployment technologies in the same region (Sims et al., 2011; Hoste et al., 2009; Taneja et al., 2012). For example, wind plants and solar plants can be deployed in the same neighborhood such that when the power output of a solar plant decreases to zero at night, the wind plant can still provide power as winds may be strong during the night. Therefore, renewable energy resources can provide additional power to the grid over long-term scales reduction.

3.3.4 Scenario 3: Control mechanism

Fig. 2 shows the workflow of control mechanism. Notice that this scenario aims to demonstrate the effectiveness of the developed control mechanism. Particularly, in GridMat toolbox, we create a controller for each renewable energy plant, which embeds the control mechanism implemented by C++. The meter at the plant station record the total amount of power generated by the renewable energy plant. In our experiment, the shared load of renewable energy resources is set to 25% of the total load in the neighborhood. Nonetheless, the grid will tolerate variation within the range 20–25%, where 20% is set as threshold. The meter reading for the renewable energy generator, the backup generator, and the switch status of the renewable energy plan in GridLAB-D are collected and sent to the controller in GridMat every hour. Notice that intervals for meter readings can be set to different time scales according to the level of monitoring needs. Once these values are received, the controller will compute the status of renewable energy plants and control backup generators based on the decision of the control mechanism. The detailed description of the developed control mechanism can be found in Section 3.4.

In the designed control mechanism, the controller will check whether the output of renewable energy generation is less than the threshold (e.g., 20% of the total load). The controller will then issue the appropriate command to either keep the renewable energy plant to be connected in the grid or disconnect the renewable energy plant from the grid. If the decision is to disconnect the renewable energy plant temporally from the grid, the controller will issue a concurrent command to turn ON a dispatching backup generator, which is used to substitute the renewable energy plant. At the same time, the power generated by the renewable energy resources, if there is any, will be distributed to the local storage

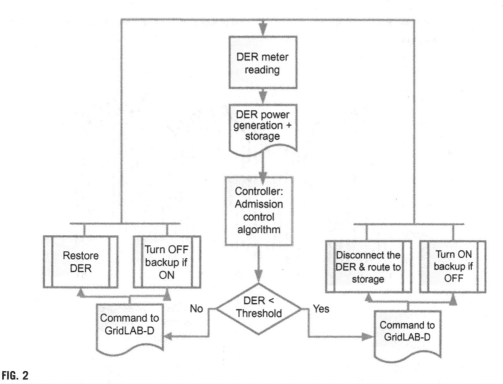

FIG. 2

Workflow of control mechanism.

device. When the renewable energy plant is restored to the grid, the controller will issue a turn OFF command to the dispatching backup generator (Figs. 3–6).

3.4 CONTROL MECHANISM

We now present the control mechanism, which can be used to help the effective integration of renewable energy resources in the power grid. The control mechanism checks the reliability and availability of the renewable energy plant, and make a decision on whether to maintain or restore it, or to disconnect it from the power grid. By doing so, we can enable a reliable and stable operation of the power grid, although distributed renewable energy resources have numerous uncertainties. We assume that renewable energy plants and the energy distribution management at the substation are connected through communication networks. As such, renewable energy plants measure the output of generated power using smart meters and send the information to the controller regularly. When the controller receives the information, it can make a decision about whether to keep the connection of distributed energy resources or disconnect it from the power grid temporarily. We assume that each renewable energy plant is attached to a local storage device that stores the power generated when it is disconnected from the grid.

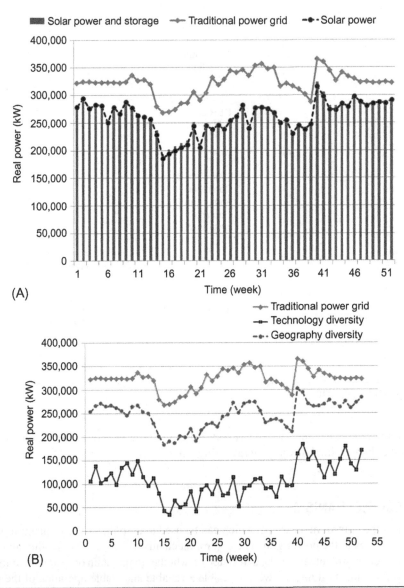

FIG. 3

Bulk power generation. (A) Solar plant and energy storage. (B) Technology and geographic diversity.

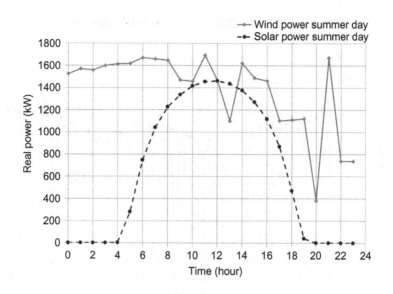

FIG. 4

Solar and wind power generation (summer day).

ALGORITHM 1 CONTROL ALGORITHM

1: BP_s: Status of the backup generator
2: RE_s: Status of the solar or wind plant
3: P_l: Power penetration level (%)
4: T_h: Threshold value (say, 20% of total load)
5: RE_g: Solar or wind plant power generation
6: *Time*: Current time
7: *Load*: Current energy demand
8: ————————————
9: $BP_S \leftarrow 0$
10: $RE_S \leftarrow 1$
11: $T_h \leftarrow P_l * Load$
12: $Time \leftarrow currenttime$
13: $BACKUP_S \leftarrow 0$
14: $Load \leftarrow current_load$
15: **while** $(Load > 0 \&\& Time in[5AM - 6PM])$ **do**
16:    **if** $(RE_G < T_h)$ **then**
17:      $RE_S \leftarrow 0$
18:      RE power output routed to local storage
19:      **if** $(BP_S == 0)$ **then**
20:        $BP_S \leftarrow 1$
21:      **end if**
22:    **elseIf** $(BP_S == 1)$
23:        $BP_S \leftarrow 0$
24:    **end if**
25:    Wait for next time interval
26:    $Load \leftarrow$ update current load
27: **end while**

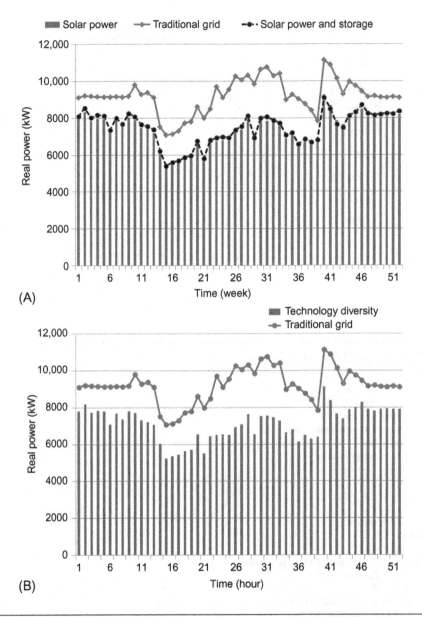

FIG. 5

Total power losses. (A) Solar plant and energy storage. (B) Technology diversity.

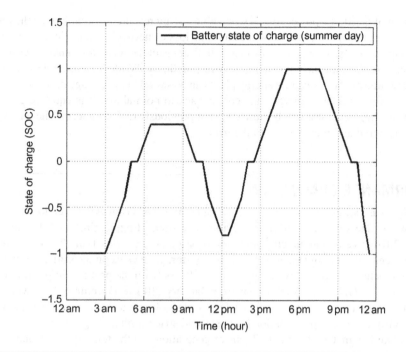

FIG. 6

Battery state of charge (summer day).

We now describe the inputs and outputs of the control mechanism in detail. As the input of control mechanism, we have the following parameters:

(1) *Total load:* It represents the aggregated amount of the actual energy demand.
(2) *RE power output (RE_G):* It represents the measured amount of actual power output of the renewable energy plant.
(3) *Power penetration level (P_l):* It represents the amount of energy generated from solar or wind power as the ratio of the total energy used (the higher the power penetration, the higher the risk of grid instability if a failure of the renewable energy plant occurs).
(4) *Control time frame (Time):* It represents the duration when renewable energy plants are controlled and depends on the geography and weather condition of site, as well as the deployment technology and type of renewable energy resources. For example, it is irrelevant to check the power output of a solar plant at midnight as it is predictable that the power output from the solar plant at that time is nearly zero.
(5) *Reliability threshold (T_h):* It is used to check the reliability of renewable energy resources. Notice that even when the solar plant is operating normally, the power output is not constant.

For the purpose of realistic control and monitoring, we define a variation range, in which the renewable energy plant will be considered operable and reliable. Additional parameters include the following: (1) *backup status:* it indicates the current status of the backup generator; and (2) *renewable energy plant status:* it indicates whether the renewable energy plant is in the network or disconnected from the power grid.

If the output of generated power from a renewable energy resource is smaller than the threshold due to factors (e.g., contingencies, mechanical failure, or loss of network connection, etc.), its status will be marked as unavailable. Then, the controller will disconnect the renewable energy resource from the power grid and turn ON the backup generator correspondingly. While it is disconnected from the grid, the power generated by the renewable energy plant can be stored in the storage device deployed locally. If the output of the renewable energy plan comes back to normal and it is equal or greater than the threshold, the backup generator will be turned OFF if previously ON. As a result, the renewable energy plant will be restored and reconnected to the grid.

4 PERFORMANCE EVALUATION

Our simulation was carried out on a personal computer and used GridLAB-D version 3.1 and the Grid-Mat toolbox (University of California, Irvine. Electrical Engineering and Computer Science Department, 2016). We first conducted simulations based on scenarios 1 and 2 to evaluate the impact of integrating renewable energy resources and energy storage devices on the grid, as well as the impact of deployment technology diversity and geographic diversity on the power grid performance. We ran the simulation using GridLAB-D for 1 year from January 2011 to December 2011 with real climate data. To collect output data, we use GridLAB-D recorders objects to collect and write data in different files on an hourly basis. For presentation purposes and space limits, we group the data in weeks and show results. Our key metrics are the bulk power generation and the total losses, which include transformer losses, and transmission and distribution losses.

We then conducted a simulation study to evaluate the performance of our proposed control mechanism. To this end, we used two simulation tools—GirdLAB-D and Matlab—which are integrated into the Grid-Mat framework and controlled through the GridMat interface. The power grid topology is implemented in a GridLAB-D script file and the control mechanism is implemented in a Matlab script, which serves as the controller. Prior to running the simulation, we select GridLAB-D objects (e.g., electric nodes, smart meters, transformers, switches, etc.). The properties and values of these objects are used as inputs and outputs variables in the Matlab script file. GridLAB-D objects are defined as unidirectional or bidirectional. A unidirectional GridLAB-D object either sends values to the controller or receives new controlling signals from the controller, whereas bidirectional objects are capable of both. For example, smart meters that report power output values are unidirectional while control switches are bidirectional. The simulation will pause every hour. Input data are collected in GridLAB-D and sent to the GridMat core. The Matlab script is run to determine new controlling signals or the status of control switches. Once these values are computed, they are written into GridLAB-D and the simulation resumes until the next time step.

An additional code in the Matlab script enables the logging of control switches status in log files. Specifically, we first simulated a failure of solar plant without the control mechanism in place for 24 hours. In this setting, we purposely disconnected the solar plant from 8 am to 10 am, and then from 1 pm to 3 pm using a control switch. We then simulated a power grid with the control mechanism enabled for 72 hours from August 24 to August 26, 2011. The key metrics include the quantity of bulk power generation, the current status of the renewable energy resources over time, and voltage output. The status of the solar plant and backup generator are recorded by GridLAB-D collector object in a CSV file and are recorded by Matlab in a log file, respectively. In the following, we show evaluation results reflected in different scenarios.

4.1 SCENARIO 1: INTEGRATION OF RENEWABLE ENERGY RESOURCES AND ENERGY STORAGE

4.1.1 Bulk power generation

Fig. 3A shows the variation of the bulk power generation over a year. With a traditional grid, the total power generated is 16,733.72 MW. When a solar power is in place, for the same load, the bulk power generation is reduced to 13,530.82 MW, which is a significant decline of 381.02 MW or 19.14%. By adding energy storage, the bulk power generation is 13,756.22 MW, which is 17.79% decrease. Our data confirms that integrating renewable energy resources and energy storage to the grid can offset the need for energy generation from fossil fuel plants, which can further reduce the cost for power generation.

4.1.2 Impact of energy storage

Fig. 7 illustrates the impact of an energy storage device on the power grid during a summer day. The operation of the battery is defined by a specific schedule shown in Fig. 6. As we can see from the figure, from 0 am to 6 am, the battery is in the discharge mode. From 6 am to 10 am, as the energy demand declines, the battery is charged. From 11 am to 1 pm, which is the time of high energy use, the stored energy can be discharged to the power grid to support the power generation and then go back to the charge mode from 4 pm to 10 pm. As shown in Fig. 7, we observe that the storage devices effectively discharge to the power grid between 0 am and 6 am when the solar power output is near zero. As the sun's radiance increases, the solar generation increases as well. Nonetheless, during this period (from 6 am to 10 am), the demand is very low so that the storage device is in charge mode to store the excess power produced by the solar plant. From 11 am to 1 pm, the demand is high, but the power from sunlight starts to decline. Therefore, the storage device switches to the discharge mode and

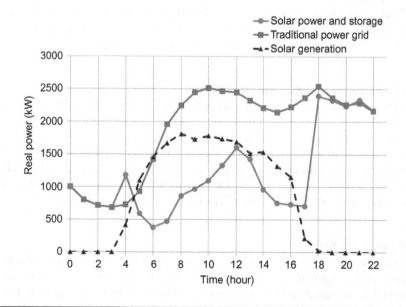

FIG. 7

Impact of energy storage.

compensates the low performance of the solar plant. Indeed, the energy storage device can help smooth the variability and the mismatch between the solar power generation and the energy demand.

4.1.3 Total power losses

Fig. 5A illustrates the variation of total power losses when renewable energy resources and energy storage devices are integrated into the power grid. By deploying renewable energy resources close to demanded nodes, the total losses can be significantly reduced. In the case of renewable energy resources (e.g., solar plant), the power losses were reduced by 20.30% (97 kW) as compared to the traditional power grid whereas adding the storage device yields 18.90% reduction.

4.2 SCENARIO 2: TECHNOLOGY/GEOGRAPHIC DIVERSITY

4.2.1 Bulk power generation

Fig. 3B illustrates the variation of bulk power generation over the whole year in terms of deployment technology diversity (combining distributed energy resources and storage, in addition to bulk generator) and geographic diversity (deploying distributed energy resources in various geographical areas), respectively. As we can see from the figure, the deployment technology diversity leads to the most significant impact on the grid with the amount of bulk generation of 5600.15 MW, as a 66.53% decrease shown in Fig. 3C. The amount of bulk power generation with the geographic diversity is 12,815.9 MW with a decrease of 23% in comparison with the traditional power grid. The wind and solar power can complement each other so that when the solar power output decreases during the night, the wind power increases as winds are stronger at night time. Our results confirm that the deployment technology diversity can be a viable choice over the geographic diversity not only because the magnitude of its impact on the performance of power grid, but also because geographic diversity can introduce additional transmission and distribution costs and power losses.

4.2.2 Complementarity of solar and wind power

Fig. 4 illustrates the effectiveness of deployment technology diversity on a summer day. As shown in the figure, solar and wind power generations complement each other well. Early in the morning from 0 am to 4 am, the solar generation is zero, while the wind power could deliver about 1600 kW. Between 12 pm and 2 pm, the variation of wind power generation is compensated by the solar plant, which is at the highest end. At the end of the afternoon, as the sunlight declines, the decrease of solar power output can be covered by the wind power generation. Our data show that the deployment technology diversity is an efficient solution to mitigate the variability of renewable energy resources.

4.2.3 Total power losses

Fig. 5B represents the variation of power total losses over the whole year where the deployment technology diversity and geographic diversity are considered. Because these scenarios can reduce the amount of the bulk power generation needed, the power losses can be reduced as well. For the deployment technology diversity, we observe that the losses are 370.84 kW, which is 22.43% decline.

4.3 SCENARIO 3: CONTROL MECHANISM

4.3.1 Solar failure without control mechanism enabled

Fig. 8 represents the outputs for a solar failure when the control mechanism is not enabled. In Fig. 8A, we observe that there is a sudden drop of power at the time when the solar plant is disconnected from the grid, between 8 am and 10 am and between 1 pm and 3 pm, as the disconnection of the solar power leads to a shortage of available energy in the grid. The impact of this shortage will be proportional to the

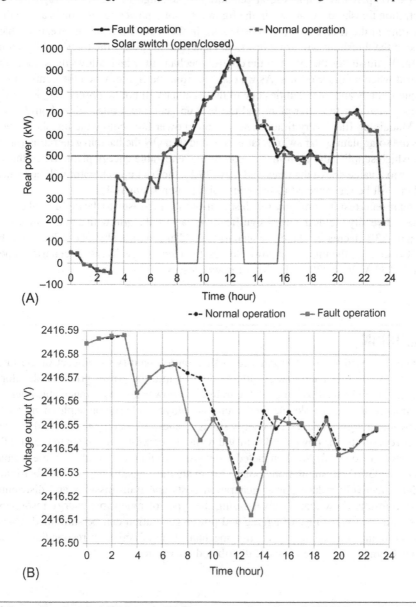

FIG. 8

Solar plant failure without control (24 hours). (A) Power generation phase A. (B) Voltage output phase A.

shared load assigned to the renewable energy plant. As shown in Fig. 8B, the disconnection of the solar plant is demonstrated by a drop of voltage output when failures occur.

4.3.2 Solar failure with control mechanism enabled

Fig. 9 illustrates the behavior of the power grid over 72 hours when the control mechanism is in place. In Fig. 9A, we compare the 72 hours solar generation from August 24, 2011 to August 26, 2011 to the power penetration threshold. As shown in the figure, the solar power generation is above the threshold most of the time in the first 24 hours. Nonetheless, in the second day, as the weather changes, solar power is above the threshold only for a few hours and remains below the threshold for almost the last 24 hours. Fig. 9B illustrates the effect of the control mechanism, which allows the smooth operation of the power grid when a failure occurs. As we can see from the figure, when the control mechanism is used, multiple failures and uncertainties due to the weather will not be observable. Fig. 9C shows the complementary status of the solar plant and the backup generator collected by both GridLAB-D recorder and Matlab log file. Every time when the solar power falls below the threshold, the controller disconnects the solar plant (solar switch status is 0) and turn ON the backup generation (backup switch status is 1), which supplies the 25% shared load of the renewable energy plant. When the solar power output returns to a value above the threshold (say, 20%), the backup generator will be turned OFF and the solar plant will be reconnected to the grid (solar switch status is 1).

Table 1 represents results summary for all scenarios in comparison to the traditional grid. As shown, the technology diversity yields the most significant impact on the grid with a bulk power generation of 5600 MW, a 66.53% decrease. The other scenarios yield approximately 25% reduction of bulk power generation. Concerning the impact on power losses, the technology diversity scenario outperforms the others with 370.84 MW or 22% reduction of power losses.

5 CONCLUSION

In this chapter, we conducted a performance evaluation to quantify the impact of the integration of distributed energy resources and energy storage devices into the power grid. We developed several scenarios (e.g., the traditional power grid, the traditional power grid with renewable energy resources and energy storage, the renewable energy resources deployment based on deployment technology diversity, renewable energy resources and energy storage, renewable energy resources deployed based on geographic diversity, and the power grid with and without a control mechanism to monitor the status of renewable energy resources). We conducted extensive simulations based on these scenarios using GridLAB-D and GridMat simulation tools. Our evaluation results show that distributed energy resources can effectively reduce power generation costs and transmissions and distribution losses. Our evaluation results show that by diversifying the types of renewable energy resources deployed in the power grid, the variability of aggregated power generation can be smoothed. Our evaluation results also show that the developed control mechanism is effective in responding to failures and contingencies by monitoring the status of distributed energy resources.

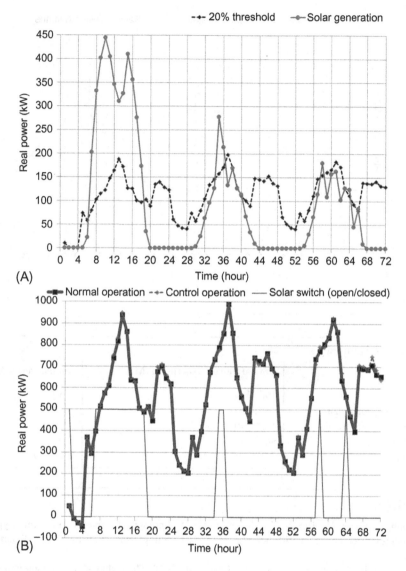

FIG. 9

Solar plant with control (72 hours). (A) Solar generation with 20% threshold. (B) Power generation phase A. (C) Solar and backup status.

(Continued)

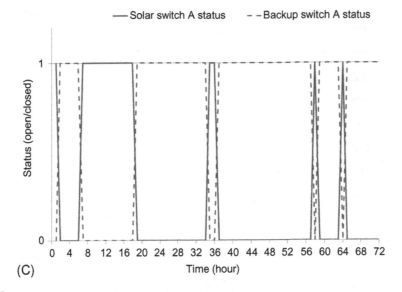

FIG. 9, CONT'D

Table 1 Results Summary

Scenarios	Annual Power Generation (MW)	Annual Power Losses (kW)
Traditional grid	16,733.72	478.11
Renewable energy (solar plant)	13,530.82	381.02
Renewable energy with storage	13,756.01	387.70
Technology diversity	5600.15	370.84
Geographic diversity	12,815.90	NA

REFERENCES

Chen, H., Xuan, P., Wang, Y., Tan, K., Jin, X., 2016. Key technologies for integration of multitype renewable energy sources research on multi-timeframe robust scheduling/dispatch. IEEE Trans. Smart Grid 7 (1), 471–480.

Connolly, D., Lund, H., Mathiesen, B., Pican, E., Leahy, M., 2012. The technical and economic implications of integrating fluctuating renewable energy using energy storage. Int. J. Renew. Energy 43, 47–60.

Delille, U., Francois, B., Malarange, G., 2012. Dynamic frequency control support by energy storage to reduce the impact of wind and solar generation on isolated power system's inertia. IEEE Trans. Sustainable Energy 3 (4), 931–939.

Denholm, P., Ela, E., Kirby, B., Milligan, M., 2012. The role of energy storage with renewable electricity generation: technical report. In Proceedings of ACM/IEEE 3rd International Conference on Cyber-Physical Systems (ICCPS).

Eichman, J.D., Mueller, F., Tarroja, B., Schell, L.S., Samuelsen, S., 2013. Exploration of the integration of renewable resources into California's electric system using the Holistic Grid Resource Integration and Deployment (HIGRID) tool. Energy 50, 353–363.

Eltigani, D., Masri, S., 2015. Challenges of integrating renewable energy sources to smart grids: a review. Renew. Sustain. Energy Rev. 52, 770–780.

Evans, A., Strezov, V., Evans, T.J., 2012. Assessment of utility energy storage options for increased renewable energy penetration. Renew. Sustain. Energy Rev. 16 (6), 4141–4147.

Gul, T., Stenzel, T., 2005. Variability of wind power and other renewables: management options and strategies. In: International Energy Agency.

Halamay, D., Brekken, T.K., Simmons, A., McArthur, S., et al., 2011. Reserve requirement impacts of large-scale integration of wind, solar, and ocean wave power generation. IEEE Trans. Sustainable Energy 2 (3), 321–328.

Henninger, S., Rubenbauer, H., Jaeger, J., 2015. An advantageous grid integration method and control strategy for renewable energy sources and energy storage systems. In: Proceedings of ETG-Fachbericht-International ETG Congress 2015.

Hill, C., Such, M.C., Chen, D., Gonzalez, J., Grady, W.M., et al., 2012. Battery energy storage for enabling integration of distributed solar power generation. IEEE Trans. Smart Grid 3 (2), 850–857.

Hossain, M., Madlool, N., Rahim, N., Selvaraj, J., Pandey, A., Khan, A.F., 2016. Role of smart grid in renewable energy: an overview. Renew. Sustain. Energy Rev. 60, 1168–1184.

Hoste, G., Dvorak, M., Jacobso, M., 2009. Matching Hourly and Peak Demand by Combining Different Renewable Energy Sources: A Case Study for California in 2020.

Lee, E.A., Seshia, S.A., 2011. Introduction to Embedded Systems: A Cyber-Physical Systems Approach. LeeSeshia.org.

Levron, R., Shmilovitz, D., 2012. Power systems optimal peak-shaving applying secondary storage. Electr. Power Syst. Res. 89, 80–84.

Lund, H., Werner, S., Wiltshire, R., Svendsen, S., Thorsen, J.E., Hvelplund, F., Mathiesen, B.V., 2014. 4th Generation District Heating (4GDH): integrating smart thermal grids into future sustainable energy systems. Energy 68, 1–11.

Mathiesen, B.V., Lund, H., Connolly, D., Wenzel, H., Østergaard, P.A., Möller, B., Nielsen, S., Ridjan, I., Karnøe, P., Sperling, K., et al., 2015. Smart energy systems for coherent 100% renewable energy and transport solutions. Appl. Energy 145, 139–154.

Miller, N., Clark, K., Shao, M., 2011. Frequency responsive wind plant controls: impacts on grid performance. In: Proceedings of 2011 IEEE Power and Energy Society General Meeting.

Mills, A., 2010. Implications of Wide-Area Geographic Diversity for Short-Term Variability of Solar Power. Lawrence Berkeley National Laboratory.

Mills, A., 2010. Understanding Variability and Uncertainty of Photovoltaics for Integration With the Electric Power System. Lawrence Berkeley National Laboratory.

Mwasilu, F., Justo, J.J., Kim, E.K., Do, T.D., Jung, J.W., 2014. Electric vehicles and smart grid interaction: a review on vehicle to grid and renewable energy sources integration. Renew. Sustain. Energy Rev. 34, 501–516.

National Institute of Standards and Technology(NIST), 2016. Smart Grid. http://www.nist.gov/smartgrid/nistandsmartgrid.cfm (accessed March 2016).

Oren, S., Callaway, D., Mount, T., Zhang, M., Thomas, R., Gross, G., Dominguez-Garcia, A., 2012. Renewable Energy Integration and the Impact of Carbon Regulation on the Electric Grid. Power Systems Engineering Research Center (PSERC).

Pacific Northwest National Laboratory (PNNL), 2016. GridLABD, http://www.gridlabd.org (accessed March 2016).

Phuangpornpitak, N., Tia, S., 2013. Opportunities and challenges of integrating renewable energy in smart grid system. Energy Procedia 34, 282–290.

Sharma, S., Dua, A., Prakash, S., Kumar, N., Singh, M., 2015. A novel central energy management system for smart grid integrated with renewable energy and electric vehicles. In: Proceedings of 2015 IEEE International Conference on Transportation Electrification Conference (ITEC).

Sims, R., Mercado, P., Krewitt, W., Bhuyan, G., Flynn, D., Holttinen, H., Jannuzzi, G., Khennas, S., Liu, Y., Nilsson, L.J., et al., 2011. Integration of renewable energy into present and future energy systems. In: IPCC Special Report on Renewable Energy Sources and Climate Change Mitigation.

Sims, R.E., Rogner, H.H., Gregory, K., 2003. Carbon emission and mitigation cost comparisons between fossil fuel, nuclear and renewable energy resources for electricity generation. Energy Policy 31 (13), 1315–1326.

Taneja, J., Katz, R., Culler, D., 2012. Defining CPS challenges in a sustainable electricity grid. In: Proceeding of the 3rd the ACM/IEEE Third International Conference on Cyber-Physical Systems (ICCPS).

University of California, Irvine. Electrical Engineering and Computer Science Department, 2016. GridMat. http://www.sourceforge.net/projects/gridmat (accessed March 2016).

U.S. Department of Energy, Office of Energy Delivery and Energy Reliability, 2016. Smart Grid. http://energy.gov/oe/services/technology-development/smart-grid/ (accessed March 2016).

U.S. Energy Information Administration (EIA), 2016. How Much Electricity is Lost in Transmission and Distribution in the United States. http://www.eia.gov/tools/faqs/ (accessed March 2016).

AGRICULTURE CYBER-PHYSICAL SYSTEMS

25

W. An*, D. Wu†, S. Ci‡, H. Luo§, V. Adamchuk¶, Z. Xu*

Chinese Academy of Sciences, Beijing, China*
University of Tennessee at Chattanooga, Chattanooga, TN, United States†
University of Nebraska-Lincoln, Lincoln, NE, United States‡
Yahoo, San Jose, CA, United States§
McGill University, Montreal, QC, Canada¶

1 PRECISION AGRICULTURE AND CYBER-PHYSICAL SYSTEMS

1.1 PRECISION AGRICULTURE

Precision agriculture is a management strategy that employs detailed, site-specific information to precisely manage production inputs. Specifically, based on the soil and crop characteristics unique to each part of the field, precision agriculture aims to optimize the production inputs within small portions of the field, such as reduced use of water, fertilizers, herbicides and pesticides besides farm equipment. Precision agriculture techniques can improve the economic and environmental sustainability of crop production. Another term frequently used to refer to these techniques is precision farming, which is defined as the farm management strategy utilizing precise information and information gathering technology to increase profit and reduce environmental impact.

Precision agriculture distinguishes itself from traditional agriculture by its level of management (Grisso et al., 2002). An agricultural production system is an outcome of a complex interaction of seed, soil, water and agro-chemicals (including fertilizers) (Patil et al., 2012). Judicious management of all the inputs is essential for the sustainability of such a complex system. Based on information technologies, instead of managing whole fields as a single unit, management is customized for small areas within fields. This increased level of management emphasizes the need for sound agronomic practices. Precision agriculture often has been defined by the technologies that enable it and is often referred to as GPS (global positioning system) agriculture or variable-rate farming. As important as the devices are, information is the key ingredient for precise agriculture/farming. Farmers who effectively use information earn higher returns than those who don't. Modern tools and technologies are becoming available to bring information technology and agricultural science together for improved economic and environmentally sustainable crop production (Davis et al., 1998; Rains and Thomas, 2000).

1.2 ARCHITECTURE OF AGRICULTURE CYBER-PHYSICAL SYSTEMS

Cyber-physical systems (CPSs) can create more modern and precise agriculture. CPSs are engineered systems that are built from, and depend upon, the seamless integration of computational algorithms and

Cyber-Physical Systems. http://dx.doi.org/10.1016/B978-0-12-803801-7.00025-0

physical components. Advances in CPSs will enable capability, adaptability, scalability, resiliency, safety, security, and usability that will far exceed those of simple embedded systems of today. Agriculture cyber-physical systems (ACPSs), the CPSs designed and applied in agriculture, can collect fundamental and timely information about the climate, the soil, and the crops with high granularity, in order to realize more accurate systems of agricultural management. ACPSs can also constantly monitor different resources, such as watering, humidity, plant health and others, through sensors and, thus, maintain the ideal environmental values through actuators and facilities.

Precision agriculture provides an effective scenario for applying ACPSs and a strong impetus for the development of ACPSs. ACPSs are expected to address effective and precise agricultural management concerning both system level issues (i.e., adaptability, autonomy, efficiency, functionality, reliability, and safety) and final user needs (i.e., communication reliability and robustness, user friendliness, versatile and powerful graphical user interfaces).

Fig. 1 is a typical architecture of ACPSs, which includes four parts: sensors and sink nodes, network, control center and farming facility:

- *Sensor and sink nodes*: Sensor nodes are spatially distributed over the field of interest to monitor physical or environmental conditions, such as temperature, sound, pressure, etc. Sink node aggregates the collected data from sensor nodes and relays them to the users via the network.
- *Network*: Communication devices in the network relay the data from the sink node to the control center.
- *Control center*: Remote control center processes the collected data from sensor nodes and makes decisions on what kind of commands should be sent to the destination farming facility.

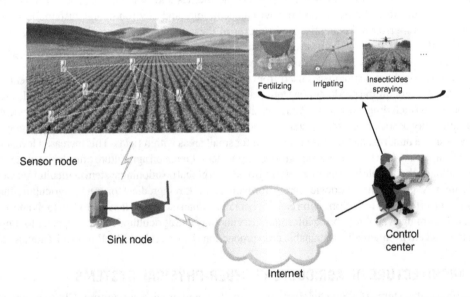

FIG. 1

The architecture of a typical agriculture cyber-physical system.

- *Farming facilities*: Agricultural facilities are usually equipped with a control system, which receives the command from the remote control center and implements the command with behaviors such as fertilizing, irrigating, insecticides spraying, etc.

2 APPLICATIONS OF AGRICULTURE CYBER-PHYSICAL SYSTEMS

A variety of ACPSs have been developed for different applications. Those applications could be soil moisture monitoring for irrigation scheduling, soil mineral content monitoring for fertilization scheduling, weather monitoring for frost prevention, crop growth monitoring for disease prevention and harvest management, etc. Next, some typical ACPSs and their applications will be reviewed.

2.1 ENVIRONMENTAL INFORMATION GATHERING

Some ACPSs were developed for environmental information gathering. Wark et al. (2007) applied sensor network technology to farming by developing a large-scale, outdoor, pervasive computing system based on the Fleck hardware platform. Comprising static and animal-borne mobile nodes, the system measures the state of a complex, dynamic system comprising climate, soil, pasture, and animals. These data support prediction of the land's future state and improved management outcomes through closed-loop control. Wireless sensor networks were applied to monitor farmland environment for irrigation and fertigation scheduling (Patil et al., 2011; Roy and Bandyopadhyay, 2008). Srbinovska et al. (2015) proposed a wireless sensor network architecture for vegetable greenhouse in order to achieve scientific cultivation and lower management costs from the aspect of environmental monitoring, in which a practical and low-cost greenhouse monitoring system is designed based on wireless sensor network technology in order to monitor key environmental parameters such as the temperature, humidity and illumination.

Architectures and protocols for WSNs in precision agriculture have also been investigated. Dedicated energy-efficient MAC and network layer protocols have been designed for a WSN for automated collection of soil data from a farm-field (Sahota et al., 2010, 2011). Besides aboveground WSNs, underground WSNs have also been considered (Dong et al., 2013; Silva and Vuran, 2010), which usually consist of wirelessly connected underground sensor nodes communicating through soil and monitoring soil conditions such as water and mineral content.

ACPSs were also used in fields where farms need to monitor the climate condition to acquire real-time data or information and make decisions. The Discovery Channel (Discovery Channel, 2003) reported that about 65 sensor motes were deployed in 1-acre of land in BC, Canada. The sensor motes were installed to remotely monitor and transmit the information about the temperature, sun light and moisture to a central PC periodically. This made it easy for the owner to monitor in real time each area in the field to prevent frost, have an idea when to apply fertilizers and set up a harvest schedule and also to manage irrigation.

Soil water monitoring and crop modeling often facilitate improved irrigation scheduling. Various methods have been used to focus on the temporal and spatial variability of the quantity of irrigation water. Adamchuk et al. (2010) deployed a few sets of soil water sensors with wireless communication capability in the locations of interest. Since water storage capacity depends on the properties of the soil profile and the potential for surface water runoff, high-density measurement of apparent soil electrical conductivity (ECa) and field elevation can be used to define field locations with different levels of water available to the crop during the growing season. To obtain high-resolution maps of apparent electrical conductivity and elevation, a Veris 3150 unit (Mobile Sensor Platform, Veris Technologies, Inc., Salina, Kansas) equipped with an RTK-level AgGPS 442 GNSS receiver (Trimble Navigation Limited, Sunnyvale, California) was used to map a 37-ha field located at the University of Nebraska-Lincoln Agricultural Research and Development Center near Mead, Nebraska, USA.

2.2 PLANT INFORMATION GATHERING

As mentioned previously, most of the prior works only focus on gathering environmental information, which can be measured and collected by off-the-shelf physical and chemical sensors easily. But plant information, which is even more important than environmental information, such as disease and growth, cannot be sensed directly due to its inherent characteristics. In the following, two typical plant information gathering systems are introduced.

Walsh et al. (2013) simulated a scenario where wheat producers would use soil moisture data obtained from a set of sensors installed in their field, in addition to crop reflectance measurements, to obtain an estimate of yield potential and to generate nitrogen fertilizer rates. The sensor based nitrogen rate calculator utilizes the Normalized Difference Vegetation Index (NDVI) and the in-season estimated yield (INSEY) as the estimate of biomass to assess yield potential and to generate nitrogen recommendations based on estimated crop need.

Improving fruit farm profitability through integrated pest management programs is always an important issue to modern agriculture systems. Jiang et al. (2013) developed an automatic infield monitoring system to efficiently capture long-term and up-to-the-minute environmental fluctuations in a fruit farm with the objective to enhance integrated pest management programs against *Bactrocera dorsalis*. A remote agro-ecological monitoring system built upon WSNs was exploited to provide precision agriculture services with large-scale, long-distance, long-term, scalable, and real-time infield data collection capabilities. Pest population forecast results are also provided so that farmers and government officials would be able to accurately respond to infield variations.

Wireless Sensor and Actuator Networks (WSAN) are emerging as the new generation of sensor networks and providing a boost towards the development of autonomous systems. Yoo et al. (2007) deployed an automated agriculture system in greenhouses with melon and cabbage in Dongbu Handong Seed Research Center. The system was employed to monitor the growing process of them and control the environment of the greenhouses.

2.3 UAV-BASED ACPSs

Unmanned aerial vehicles (UAVs) are uninhabited and reusable motorized aerial vehicles, which are remotely controlled, semiautonomous, autonomous, or have a combination of these capabilities, and that can carry various types of payloads, making them capable of performing specific tasks within the earth's atmosphere, or beyond, for a duration, which is related to their missions. UAVs for agriculture have gained rapid development in recent years due to its tremendous potential uses in precision agriculture. Xiang and Tian (2011) developed a remote sensing system based on an easily transportable helicopter platform that is capable of acquiring multispectral images at desired locations and times. Based on the navigation data, and the waypoints generated by the ground station, the UAV could be automatically navigated to the desired waypoints and hover around each waypoint to collect field image data. Makynen et al. (2012) developed a UAV imaging system. The Fabry-Perot Interferometer based hyperspectral imager was used in a UAV imaging campaign for forest and agriculture tests during the summer 2011. The system that includes both the high spatial resolution color-infrared camera and a light-weight hyperspectral imager can provide all necessary data with just one UAV flight over the target area. Tokekar et al. (2013) studied the problem of coordinating a UAV and an unmanned ground vehicle (UGV) for a precision agriculture application, in which the ground and aerial measurements are used for estimating nitrogen levels on-demand across a farm and guide fertilizer application with these estimates in turn. They presented a method to identify points whose probability of being misclassified is above a threshold, and then maximized the number of such points visited by an UAV subject to its energy budget.

2.4 GIS-BASED ACPSs

As a software application for managing and analyzing spatial data, geographic information systems (GISs) is mainly used to store, retrieve and transform spatial information relating to productivity and agronomy (Blackmer and White, 1998). This information can be derived from a number of data sources including digital maps, digitized maps and photographs, soil and crop surveys, sensor data with positioning information, point analytical data and/or yield maps. As for the sugar industry, using GPS and GIS to chase the benefits of selective harvesting is potentially worthwhile. Sorting out the spatial and temporal interactions between yield and commercial cane sugar values will be a critical research issue. Palaniswami et al. (2011) adopted these technologies in a precision conservation philosophy to assist in managing the interactions between cane farming and environmental protection in the sensitive coastal floodplain ecosystems. Based on remote sensing and GIS, Castaño et al. (2010) presented a method of quantifying groundwater abstractions for irrigation based on the analysis of multitemporal and multispectral satellite images. The process begins with a highly detailed classification of irrigated crops; these data are entered in a GIS, overlain with a correct estimate of the irrigation requirements of the crop, and corrected in accordance with the agricultural practices of the area.

GIS is also adopted recently to estimate average soil loss that would generally result from splash, sheet, and rill erosion from agricultural plots. Erdogan et al. (2007) effectively integrated the GIS-based procedures into the Universal Soil Loss Equation (USLE) to predict soil losses and to plan control practices in agricultural watersheds in the Kazan Watershed located in the central Anatolia, Turkey. The USLE/GIS methodology can predict soil erosion risks for planning conservation measures in the site. Most of the wetlands in Illinois were lost during the conversion of the landscape to agriculture and urban use. McCauley and Jenkins (2005) used GIS to estimate the spatial extent, density, pattern, and sizes of former and extant depressional wetlands in Champaign County. A GIS combined with a pollutant generation and transport model, can be used to identify and rank critical pollutant source areas on a regional basis. Hamlett et al. (1992) used a GIS-based, statewide screening model to rank the agricultural pollution potential of 104 watersheds in Pennsylvania.

3 SENSOR DEPLOYMENT FOR FIELD INFORMATION COVERAGE IN ACPSs

As mentioned earlier, CPSs have been widely used in various applications, such as monitoring, surveillance, tracking, event detection, and process management. Numerous research works have been carried out on sensor deployment with different objectives, such as extending the overall network lifetime by increasing energy efficiency (Xu and Sahni, 2007; Xu et al., 2008; Yang and Qiao, 2010) and improving field physical coverage (Chakrabarty et al., 2002; Li et al., 2005; Ma and Yang, 2007; Shu and Liang, 2005; Wang et al., 2006; Xu and Sahni, 2007; Yang and Qiao, 2010; Zhang and Wang, 2009). However, among these work little effort has been made to address the related issues of the coverage of field information, e.g., the measurement of apparent soil electrical conductivity (Adamchuk et al., 2010), which usually describes the significant characteristics residing in the field in precision agriculture. In other words, how to deploy sensor nodes for providing the required coverage of field information is very important in precision agriculture.

In the rest of this chapter, we will study the issue of sensor deployment for field information coverage in precision agriculture by taking into account an ACPS-enabled automatic irrigation system. In a piece of given agricultural land, the efficiency of a center-pivot irrigation system depends on its ability to meet the water demands of the crops that are growing (Sadler et al., 2005). While limited water supplies can reduce crop yield, excessive irrigation leads to resource waste and even decrease of agricultural output. ACPS provides a useful tool for monitoring the soil spatial variability in order to assist decision making for field irrigation.

Effective sensor deployment has gained increasing research interests in many aspects, such as improving field area coverage (An et al., 2009; Li et al., 2005; Shu and Liang, 2005; Wu et al., 2006), extending the overall network lifetime (Xu and Sahni, 2007; Xu et al., 2008; Yang and Qiao, 2010). Sensor deployment affects not only the overall network performance but also the cost of the whole network. In the literature, sensor deployment strategy is usually based on the concept of *physical coverage* (Wang et al., 2007) where the sensing range of a sensor is assumed to be fixed and the coverage range of the sensor is the area centered at itself with a radius equal to the sensing range. Any location is said to be covered if and only if it is within the sensing range of a sensor. However, different physical locations within a small piece of farming land may possess almost the same *field information*, which can be utilized to determine the occurrence frequencies of events, e.g., water irrigation. Here, field information can be diverse, including the apparent soil electrical conductivity, elevation, soil profile, soil salinity and sodicity, and water storage capacity, etc. (Adamchuk et al., 2010; Corwin and Lesch, 2005; Farahani et al., 2004; Ganjegunte and Braun, 2010). Consequently, to monitor only a portion of locations that correspond to the full range of field information is sufficient to monitor the whole field. Therefore monitoring the events associated with the field information provides a new perspective on sensor deployment.

In this research project on the ACPS-enabled irrigation system, field information is the water storage capacity of the locations, which is closely associated with the apparent soil electrical conductivity (ECa). Since the field information changes spatially, it is necessary to deploy some sensor nodes such that the water content in the soil can be monitored and the optimum quantity of irrigation water can be derived. The soil temperature and moisture sensor nodes are deployed in the selected locations to collect the field information. The concept of *information coverage* will be discussed in detail in Section 3.1. Because multiple locations share the same value of water storage capacity, sensor deployment based on information coverage can remarkably decrease the number of required sensor nodes. Consider a simple example shown in Fig. 2 where areas A and D share the same field information. This implies that areas A and D consume the water at the same speed. Assume that the initial water content of two areas be identical. Then, one single sensor deployed in A or D is sufficient to collect the water consumption information for both areas.

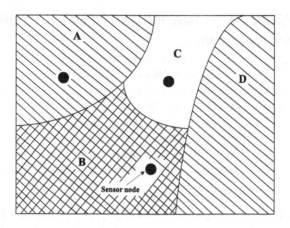

FIG. 2

Illustration of information coverage, where different physical locations could share the same field information.

Similarly, deploying one sensor in each area of *B* and *C* can also sufficiently monitor the water scarcity of these two areas.

3.1 INFORMATION COVERAGE

In a given field, the field information value $\widetilde{y}$ can be generally represented as:

$$\widetilde{y} = f(x_1, x_2, x_3), \tag{1}$$

where x_1 and x_2 are the horizontal and vertical coordinates, respectively, and x_3 is the elevation of the location. For instance, as discussed in our previous work (Adamchuk et al., 2010), the field information (i.e., water storage capacity) is closely associated with the ECa of the locations in the field. Thus, $\widetilde{y}$ may represent the ECa value of location (x_1, x_2). Given a crop field, the field usually contains considerable different information values. For these information values, we give the definition as follows.

Definition 1 (Information Coverage) A field information value is said to be covered if at least one location associated with this value is monitored by one sensor node. The field information is completely covered, namely *complete information coverage*, if all the information values are monitored by the deployed sensor nodes. The field is called *information q-coverage*, if each information value is monitored by at least *q* sensor nodes at a certain location in the field.

There are significant differences between the proposed information coverage and conventional physical coverage. First, complete field information coverage usually needs far fewer sensor nodes than complete physical coverage, e.g., areas *A* and *D* can be completely covered by one sensor node as illustrated in Fig. 2, because multiple physical locations usually share almost the same field information values. Thus, information coverage is preferred from the cost-effectiveness perspective. Second, for information *q*-coverage, an information value is covered by *q* sensor nodes each of which may be deployed in any location of the field, while *q*-physical coverage requires all *q* sensor nodes deployed in the vicinity of the covered location.

3.2 FIELD INFORMATION DISCRETIZATION AND FIELD PARTITION

3.2.1 Field information discretization

A discretization strategy is adopted for tackling numerous field information values. Without losing generality, assume $[\pi^-, \pi^+]$ is the range of values for field information y. Considering the fact that the information values are usually continuous and uneven, we process them by discretizing the range $[\pi^-, \pi^+]$ at step size τ, i.e., as shown in Fig. 3, $[\pi^-, \pi^+]$ can be uniformly discretized as $y_0 = \pi^- < y_1 < y_2 < \cdots < y_m = \pi^+$, where $y_i = y_0 + i \cdot \tau, i = 1, 2, \dots, m$. For the sake of description, y_i is called *information point*. As for field information, the given field can be uniformly sampled to achieve a great number of locations with the related data, represented by $\tilde{y}_1, \tilde{y}_2, \dots, \tilde{y}_r$ where r is the number of information points that appears in the field. For each information point y_i, it appears in the field if, and only if, there exists one information value $\tilde{y}_{i_0}$ satisfying $\left| y_i - \tilde{y}_{i_0} \right| < \frac{1}{2}\tau$.

Suppose the field is divided into n parcels, denoted as $\left\{ \mathcal{L}_j \right\}_{j=1}^n$. Based on the discretization, a column vector ϕ_j for each parcel $\mathcal{L}_j$ of the field can be formed as $\phi_j^T = (\phi_{j1}, \phi_{j2}, \dots, \phi_{jm})$, where ϕ_j^T denotes the transpose of ϕ_j, and $\phi_{ji} = 1$ if information point y_i is in $\mathcal{L}_j$ and otherwise $\phi_{ji} = 0$. Then, the vectors $\phi_1, \phi_2, \dots, \phi_n$ of all the parcels form a matrix $\Phi = (\phi_1, \phi_2, \dots, \phi_n)$:

3.2.2 Field partition

For the sake of convenience of irrigation management, the given field is partitioned into a number of same-size square parcels with side length a, as shown in Fig. 4. Within each parcel, we deploy an array of soil temperature and moisture sensor nodes for collecting the water storage information, and one

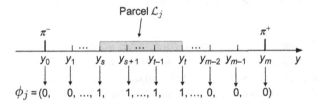

FIG. 3

Discretizing the range $[\pi^-, \pi^+]$ of field information values to derive the corresponding vector ϕ_j for parcel $\mathcal{L}_j$.

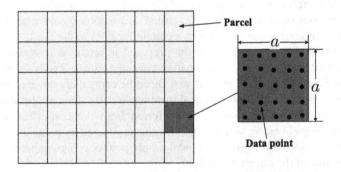

FIG. 4

Illustration of the field partition, where the given area is divided into many square parcels of the same size.

relay node for relaying the collected data to the sink node. For the sake of description, we call all the nodes deployed in one parcel as a whole *parcel node*. It is worth noting that the number of sampled field information data (ECa) within the square parcels increases with the increase of a, which results in the decrease of the number of parcel nodes required for field information monitoring. However, this also leads to the decrease of the sensing accuracy and degrades the overall performance of the field monitoring because each parcel node covers too many information points in each parcel, which are usually fused as one proper information value for representing the field water storage information and sending to the sink node. On the other hand, when a gets small, the collected data will be more accurate, but the number of parcel nodes deployed for field monitoring will increase. This unavoidably results in the increase of the deployment cost. Therefore how to determine the proper size of the parcels is a critical issue to achieve a good tradeoff between monitoring quality and deployment cost.

Let λ_j be the standard deviation of all the field information points $\{y_i\}_{i=1}^p$ in parcel $\mathcal{L}_j$, ie:

$$\lambda_j = \sqrt{\sum_{y_i \in \mathcal{L}_j} [y_i - \bar{y}]^2}, \tag{2}$$

where $\bar{y} = \dfrac{1}{p} \sum_{y_i \in \mathcal{L}_j} y_i$. Since, as discussed above, the λ_j is significantly affected by a, λ_j can be controlled by adjusting a. Now, to guarantee the monitoring accuracy of the field, a small real number $\delta > 0$ is given as the threshold of the λ_j. Then, the proper value a_j^* can be derived by:

$$a_j^* = \arg\max\left\{a_j : a_j \text{ satisfies } \lambda_j < \delta\right\}. \tag{3}$$

For all the parcels and given δ, the proper a^* can be calculated as:

$$a^* = \min_{1 \leq j \leq n}\left\{a_j^*\right\}, \tag{4}$$

where n is the total number of parcels.

3.3 COMMUNICATION CONSIDERATIONS

In general, the prerequisite for enabling communication between any two nodes is that the Euclidean distance between them is smaller than the communication radius R. Suppose A and B are two nodes with coordinates (x_1^A, x_2^A, x_3^A) and (x_1^B, x_2^B, x_3^B), respectively, the distance between them can be calculated as:

$$d(A, B) = \sqrt{\sum_{i=1}^3 \left(x_i^A - x_i^B\right)^2} \leq R. \tag{5}$$

Aside from distance, however, the communication between nodes A and B is also significantly affected by the blockage from the field surface and high-density crops. For the blockage from the field surface, we make a reasonable assumption that any two nodes can communicate with each other only when their line-of-sight does not intersect with the field surface. Then, the blockage from the field surface can be known by judging whether the segment between two relay nodes intersects with the field surface. As for the blockage from crops, extensive onsite tests are conducted in the field to derive the real communication radius of the relay nodes, which is usually much smaller than advertised. The detail of the onsite communication radius tests can be found in Section 4.

3.4 PROBLEM FORMULATION

For a given field, based on the aforementioned discretization, the appearance times of information point $y_1, y_2, \ldots, y_m$ in the whole field can be calculated as:

$$\psi_0 = \sum_{j=1}^{n} \phi_j = \left(\sum_{j=1}^{n} \phi_{j1}, \sum_{j=1}^{n} \phi_{j2}, \ldots, \sum_{j=1}^{n} \phi_{ji}, \ldots, \sum_{j=1}^{n} \phi_{jm} \right)^T, \tag{6}$$

where the appearance times of y_i in the field is represented by the ith component $\psi_0^{(i)}$ of ψ_0 with $\psi_0^{(i)} = \sum_{j=1}^{n} \phi_{ji}$. For simplicity, in this work we assume the field is divided into n same-size parcels.

A set of 2-tuples $\left\{ (\mathcal{L}_j, \mathcal{N}_j) \right\}_{j=1}^{\tau}$ is adopted to represent the sensor deployment scheme $\mathcal{S}$, where 2-tuple $(\mathcal{L}_j, \mathcal{N}_j)$ implies that parcel node $\mathcal{N}_j$ is deployed in parcel $\mathcal{L}_j$. We define vector $\mathbf{c}_{\mathcal{S}}$ for the scheme $\mathcal{S}$ such that $c_j = 1$ if $\mathcal{N}_j$ is deployed in parcel $\mathcal{L}_j$; otherwise $c_j = 0$. Further, the appearance times of $y_1, y_2, \ldots, y_m$ in the scheme $\mathcal{S}$ can be calculated as:

$$\psi_{\mathcal{S}} = \Phi \cdot \mathbf{c}_{\mathcal{S}} = \left(\sum_{j=1}^{\tau} \phi_{j1}, \sum_{j=1}^{\tau} \phi_{j2}, \ldots, \sum_{j=1}^{\tau} \phi_{jm} \right)^T. \tag{7}$$

Based on Eqs. (6) and (7), we define the monitoring efficiency as follows.

Definition 2 (Monitoring Efficiency) The monitoring efficiency η for information q-coverage is defined as $\eta(\mathbf{c}_{\mathcal{S}}, q) = \sum_{i=1}^{m} h_q \left(\psi_{\mathcal{S}}^{(i)} \right) / \sum_{i=1}^{m} h_q \left(\psi_0^{(i)} \right)$ where $\psi^{(i)}$ denotes the ith component of $h_q(\psi^{(i)})$ is a function with ψ. $h_q \left(\psi^{(i)} \right) = 1$ if $\psi^{(i)} \geq q$, otherwise $h_q \left(\psi^{(i)} \right) = 0$.

With this definition, monitoring efficiency η is the ratio between the number of information points covered q times by $\mathcal{S}$ and the total number of information points in the field with $0 \leq \eta(\mathbf{c}_{\mathcal{S}}, q) \leq 1$. As for connectivity, it is required to guarantee that any parcel node $(0 \leq \eta(\mathbf{c}_{\mathcal{S}}, q) \leq 1)$ in scheme $\mathcal{S}$ can transmit the collected data to sink node $\mathcal{N}_{\mathrm{SN}}$, i.e., any node is connected to the sink node, denoted by $\mathcal{N}_j \leftrightarrow \mathcal{N}_{\mathrm{SN}}$. The optimization problem is to derive the sensor deployment scheme $\mathcal{S}$ with the minimum number of parcels such that it reaches the given monitoring efficiency and maintains the connectivity between parcel nodes and sink node $\mathcal{N}_{\mathrm{SN}}$, ie:

$$\min_{c_{\mathcal{S}} \in \{0, 1\}^n} \| \mathbf{c}_{\mathcal{S}} \|_1$$
$$\text{Subject to}: \begin{cases} \eta(\mathbf{c}_{\mathcal{S}}, q) \geq 1 - \in \\ \mathcal{N}_j \leftrightarrow \mathcal{N}_{\mathrm{SN}} \text{ for } \forall c_j \neq 0 \end{cases} \tag{8}$$

where $\| \mathbf{c}_{\mathcal{S}} \|_1 = \sum_{i=1}^{n} |c_i|$, and $\in > 0$ be a small real number. This problem is an NP-hard integer programming problem, one of Karp's 21 NP-complete problems (Karp et al., 1975).

3.5 FINDING APPROXIMATE INFORMATION-COVERAGE-BASED SENSOR DEPLOYMENT SCHEME

In this research project, sensor nodes include two types of nodes: relay nodes and parcel nodes. Parcel nodes are much more expensive than relay nodes because they include an array of temperature and moisture sensor nodes. Considering cost and complexity jointly, we first achieve the approximate sensor deployment scheme based on the *Set Covering Theory* (Karp et al., 1975; Vazirani, 2001) without consideration of communication, and then develop an optimal algorithm for deploying additional relay nodes for the connectivity of the network.

3.5.1 Information 1-coverage

To derive the information 1-coverage sensor deployment scheme of the formulated problem Eq. (8), we give the following definition.

Definition 3 (Minus Operation) The minus operation is defined for vectors ϕ_j and ϕ_k as

$$\phi_j \ominus \phi_k = \begin{pmatrix} \phi_{j1} \ominus \phi_{k1} \\ \phi_{j2} \ominus \phi_{k2} \\ \vdots \\ \phi_{jm} \ominus \phi_{km} \end{pmatrix}$$

where, for $1 \leq l \leq m$, $\phi_{jl} \ominus \phi_{kl} = \begin{cases} \phi_{jl} - \phi_{kl} & \text{if } \phi_{jl} > \phi_{kl} \\ 0 & \text{otherwise} \end{cases}$.

Let **b** be an m-dimensional vector such that $b_i = 1$ if y_i is included in the given field, otherwise $b_i = 0$. Considering the NP-hardness of problem Eq. (8), a greedy strategy is exploited to select the parcel containing the largest number of uncovered information points of $\{y_i\}_{i=0}^m$ at each step. Then, we develop Algorithm 1 to derive the information 1-coverage sensor deployment scheme.

ALGORITHM 1 DERIVING THE INFORMATION 1-COVERAGE SCHEME

1: $\in > 0$ and $\mathcal{S} = \varnothing$
2: $I = \{j\}_{j=1}^n$ and $\Lambda = \varnothing$
3: $\mathbf{b}^T = (1, 1, ..., 1)$;
4: WHILE $\eta(\mathbf{c}_S, 1) < 1 - \in$
5: (a) Select parcel $\mathcal{L}_j$ from $\{\mathcal{L}_k : k \in I \backslash \Lambda\}$, whose vector ϕ_i minimizes $\left\{ \|\mathbf{b} \ominus \phi_j\|_1 : j \in I \backslash \Lambda \right\}$
6: (b) Let $\mathcal{S} = \mathcal{S} \cup \{(\mathcal{L}_j, \mathcal{N}_j)\}$ and $\Lambda = \Lambda \cup \{j\}$;
7: (c) $\mathbf{b} = \mathbf{b} \ominus \phi_j$;
8: ENDWHILE
9: RETURN $\mathcal{S}$

In Algorithm 1, selecting parcel $\mathcal{L}_j$ from $\{\mathcal{L}_k : k \in I \backslash \Lambda\}$ takes $O(n)$ time, where n is the number of the parcels and where Λ is the set of indices of the selected parcels. We select the parcel $\mathcal{L}_j$ at most n times. Therefore, the computational complexity of Algorithm 1 is $O(n^2)$.

As for the theoretical performance, motivated by Trevisan's work (Trevisan, 2011), the following theorem is given.

Proposition 1 *The sensor deployment scheme determined by Algorithm 1 is a* $\ln \dfrac{m}{|OPT|}$ *approximation to OPT, where OPT and |OPT| are the optimal sensor deployment scheme and the number of nodes in OPT, respectively.*

3.5.2 Information q-coverage

For efficient field monitoring, a strategy for information q-coverage is adopted to cover each information points with at least q parcel nodes. For effective information q-coverage, one way is to deploy q parcel nodes in different parcels rather than covering one information point q times in one parcel. To this end, we achieve the information q-coverage with Algorithm 2.

ALGORITHM 2 DERIVING THE INFORMATION Q-COVERAGE SCHEME

1: $\in > 0$ and $\mathcal{S} = \varnothing$
2: $I = \{j\}_{j=1}^{n}$ and $\Lambda = \varnothing$
3: $\mathbf{b}^{T} = (q, q, ..., q);$
4: WHILE $\eta(\mathbf{c}_{S}, 1) < 1 - \in$
5: (a) Select parcel $\mathcal{L}_{j}$ from $\{\mathcal{L}_{j} : j \in I \setminus \Lambda\}$, whose vector ϕ_{j} minimizes $\left\{ \|\mathbf{b} \ominus \phi_{j}\|_{1} : j \in I \setminus \Lambda \right\}$
6: (b) Let $\mathcal{S} = \mathcal{S} \cup \{(\mathcal{L}_{j}, \mathcal{N}_{j})\}$ and $\Lambda = \Lambda \cup \{j\};$
7: (c) $\mathbf{b} = \mathbf{b} \ominus \phi_{j};$
8: ENDWHILE
9: RETURN $\mathcal{S}$

The computational complexity of Algorithm 2 is $O(n^2)$, where n is the number of parcels. Since Algorithm 1 is repeated q times, each information point is covered by at least q parcels. Thus, one necessary condition for the existence of sensor deployment scheme S is that for any information point, we can find q parcels that cover the information point. Due to the limited number of parcels, the number q must have an upper bound. Some work (Abrams et al., 2004; Slijepcevic and Potkonjak, 2001) investigated the Set q-Cover for the physical coverage in the field monitoring. Provided that one parcel node is deployed in each parcel, the algorithms proposed by Abrams et al. (2004) can also be utilized to derive the upper bound of q.

3.6 OPTIMAL SOLUTION FOR DEPLOYING RELAY NODES

As for communication between nodes, if the resulting scheme S derived with Algorithms 1 and 2 is connected, i.e., for any $\mathcal{N}_{j} \in \mathcal{S}$, we have $\mathcal{N}_{j} \leftrightarrow \mathcal{N}_{\mathrm{SN}}$, then we do not need to deploy additional relay nodes for relaying the collected data to the sink node $\mathcal{N}_{\mathrm{SN}}$. Usually, scheme S is composed of v connected components, as $\{\mathcal{C}_{1}, \mathcal{C}_{2}, ..., \mathcal{C}_{v}\}$. Here, the connected component is a subnetwork in which any two nodes are connected with each other. Especially, a single parcel node is also viewed as a connected component. To guarantee the data transmission from any parcel node in S to the sink node, it is required to make sure that $\{\mathcal{C}_{1}, \mathcal{C}_{2}, ..., \mathcal{C}_{v}\}$ and $\mathcal{N}_{\mathrm{SN}}$ are included in one connected component by deploying relay nodes.

The distance between two connected components $\mathcal{C}_j$ and $\mathcal{C}_k$ is defined as:

$$d(\mathcal{C}_j, \mathcal{C}_k) = \min_{\forall \mathcal{N}_j \in \mathcal{C}_j, \mathcal{N}_k \in \mathcal{C}_k} \sqrt{\sum_{i=1}^{3} \left(x_i^{\mathcal{N}_j} - x_i^{\mathcal{N}_k} \right)^2}, \qquad (9)$$

where $\left(x_1^{\mathcal{N}_j}, x_2^{\mathcal{N}_j}, x_3^{\mathcal{N}_j} \right)$ and $\left(x_1^{\mathcal{N}_k}, x_2^{\mathcal{N}_k}, x_3^{\mathcal{N}_k} \right)$ are the deployed locations of parcel node $\mathcal{N}_j$ and $\mathcal{N}_k$, respectively. In this project, the sink node $\mathcal{N}_{SN}$ is assumed to be deployed at the specific place, denoted as connected component $\mathcal{C}_0$. Let $\mathcal{S} = \{\mathcal{C}_j\}_{j=0}^{v}$. Now Algorithm 3 is developed for deploying additional relay nodes such that each parcel node can transmit the collected data to the sink node $\mathcal{N}_{SN}$.

ALGORITHM 3 DEPLOYING ADDITIONAL RELAY NODES

1: WHILE there exists a component $\hat{\mathcal{C}}$ disconnected to $\mathcal{N}_{SN}$
2: Calculate the d among all the connected components in $\mathcal{S}$ and select $\mathcal{C}_j$ and $\mathcal{C}_k$ with the smallest d, denoted as d_{min}. Let d_{min} is achieved at $\mathcal{N}_{j0} \in \mathcal{C}_j$ and $\mathcal{N}_{k0} \in \mathcal{C}_k$
3: Let $\mathcal{S} = \mathcal{S} \backslash \{\mathcal{C}_j, \mathcal{C}_k\}$
4: Deploy $\lceil d_{min}/\mathcal{R} \rceil$ relay nodes to connect $\mathcal{N}_{j0}$ and $\mathcal{N}_{k0}$, where $\mathcal{R}$ is the communication radius. Then, $\mathcal{C}_j, \mathcal{C}_k$ and the added $\lceil d_{min}/\mathcal{R} \rceil$ relay nodes constitute a new connected component $\mathcal{C}'_j$
5: $\mathcal{S} = \mathcal{S} \cup \{\mathcal{C}'_j\}$
6: ENDWHILE
7: RETURN $\mathcal{S}$

In Algorithm 3, the complexity of calculating the d among all the connected components and selecting $\mathcal{C}_j$ and $\mathcal{C}_k$ with the smallest d_{min} is $O(v^2)$, where v is the number of the connected components. We repeat above procedure at most $v - 1$ times. Therefore the computational complexity of Algorithm 3 is $O(v^3)$.

Noting that in scheme $\mathcal{S}$ derived from Algorithm 3, all sensor nodes in each connected component can deliver the collected data to the sink node $\mathcal{N}_{SN}$. Here, this algorithm is motivated by the idea of *Prim's minimum spanning tree algorithm* (Papadimitriou and Steiglitz, 1998). As a result, Algorithm 3 requires the least number of relay nodes deployed for connecting the connected components.

---

4 FIELD EXPERIMENT AND PERFORMANCE ANALYSIS

The proposed method was applied in an ACPS-enabled automatic irrigation system and its performance was evaluated.

To obtain high resolution maps of ECa, a Veris 3150 unit (Mobile Sensor Platform, Veris Technologies, Inc., Salina, Kansas) equipped with an RTK-level AgGPS 442 GNSS receiver (Trimble Navigation Limited, Sunnyvale, California) is used to map a 37 ha field located at the University of Nebraska-Lincoln Agricultural Research and Development Centre near Mead, Nebraska, USA (Adamchuk et al., 2010). The shallow measurements of ECa, i.e., 0–30 cm, are exploited in the research project. Both ECa and elevation data are collected with 1Hz mapping frequency while moving at an approximate speed of 1.5 m/s with a 13.7-m swath width, which resulted in about 30,000 data points in the given land.

FIG. 5

The communication node (i.e., relay node) and the array of soil water potential and temperature sensor nodes.

As shown in Fig. 5, the eKo eN 2100 sensor nodes are used as the relay nodes to relay the data over the wireless sensor network and the eKo eS 1101 Watermark sensor nodes are used to monitor soil moisture and temperature (eKo, 2016). Considering the interference and the blockage from field surface and crops, we test the real communication radii of the eN2100 sensor nodes in the field. Since the sprinkler system for field irrigation is 1.5 ft. above the canopy, as shown in Fig. 6, we select five

FIG. 6

Three paths and five heights for testing the wireless communication radii in the field.

Table 1 The Communication Distances are Tested in the Corn Field, Where the Unit of Communication Distance is Meter

Height	#1	#2	#3	#4	#5
Path 1	0–47.55	0–85.95	0–91.44	0–182.88	0–273.10
Path 2	0–27.43	0–58.52	0–128.93	0–187.45	0–283.46
Path 3	0–26.52	0–77.72	0–99.67	0–111.56	0–365.76

relative heights compared with the canopy: $-3., -2., -1., 0.$ and 1 ft., where "$-$" denotes that the height is lower than the canopy. Also, as to the influence from the density of crops, we select three paths to test the communication distance among relay nodes. Table 1 shows the results of the tested communication radii.

Considering the field information of water storage capacity and irrigation requirements of the field, the deviation δ of information points of all the parcels are restricted by 0.5. The field is partitioned with $a = 10$ m. To distinguish the parcels, based on the extensive experiments, the range [0.23, 14.56] of the ECa values of this field is discretized into 150 information points, i.e., $y_0 = 0.23 < y_1 < y_2 < \cdots < y_{150} = 14.56$ with step size $\tau = 0.096$ as this step size can sufficiently distinguish the most of parcels. $R = 80\ m$ is selected for data transmission among the deployed nodes. Algorithm 1 is applied to derive the approximate sensor deployment scheme, shown as the red stars in Fig. 7, where the sink node is deployed at a given position.

Further, we evaluate the performance of the proposed algorithms by comparing them with our previous method (Adamchuk et al., 2010). Here, we carry out two sets of experiments. First, considering that there are 9 nodes in our previous sensor deployment scheme, we also derive a 9-node sensor deployment scheme by using the proposed method. The proposed 9-node sensor deployment scheme achieves the monitoring efficiency (η) of 81.66% with 6.76% performance gain over our previous sensor deployment scheme. Second, since monitoring efficiency η of our previous scheme is 74.90%, we investigate the minimum number of nodes that can reach the same or even better performance in regard to the same monitoring efficiency 74.90%. By utilizing the proposed method, the proposed 7-node sensor deployment scheme reaches 70.88% in the monitoring efficiency, while the proposed 8-node sensor deployment scheme reaches 75.43%. Therefore, the proposed method achieves better performance in monitoring efficiency with fewer nodes than our previous method.

As shown in Fig. 8, under communication radius $R = 80$ m, our previous 9-node sensor deployment scheme requires deploying additional 12 relay nodes to relay the collected data to the sink node. Then, the total number of nodes is 21. The results of monitoring efficiency with the proposed method under different number of nodes from 10 to 21 are shown in Table 2. In this table, the proposed method can meet different requirements on monitoring efficiency. Therefore the proposed method achieves significant improvement over the previous method and thereby is more flexible to various applications.

5 CONCLUSIONS AND FUTURE RESEARCH

In this chapter, a typical architecture of ACPSs for precision agriculture is introduced. To optimize sensor deployment in ACPSs, a novel notion of information coverage is defined. Based on the definition, the sensor deployment problem is specified as partitioning a field into multiple parcels and

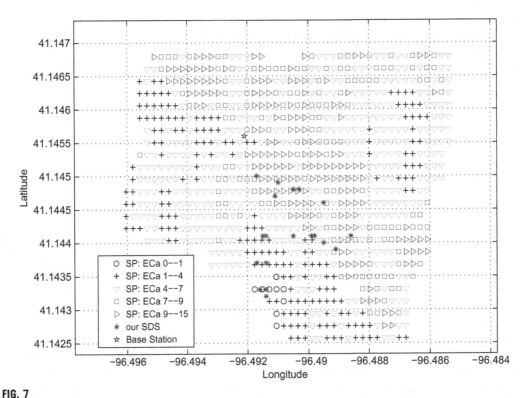

FIG. 7

The distribution of the proposed sensor deployment scheme in the field, where the ECa values are simply represented by five intervals with different markers, and each marker zone has several parcel nodes deployed in it.

selecting some parcels for sensor node deployment such that the covered field information meets the requirement. Without considering communication, two effective polynomial-time algorithms have been developed to determine the deployed locations of parcel nodes for the information 1-coverage and q-coverage, respectively. Then, by taking into account communication requirements, a polynomial-time algorithm has been developed for determining the deployed locations of relay nodes. The algorithm requires a minimized number of relay nodes. Extensive experimental results show the performance improvement achieved by the proposed method.

The research of ACPSs is promising, interesting, and profitable as ACPSs are believed as an effective way for labor saving, food crisis solving, and production efficiency improvement. Future research on ACPSs can be focused on the following aspects. (1) Precision agriculture requires proper hardware and software design of ACPSs for adapting to field environments. (2) The accuracy is always an essential part of precision farming. Although many works have been done on designing ACPSs and optimizing their use, improving the accuracy of sensing and operation is still challenging but necessary work from a cost-effectiveness perspective. (3) In precision agriculture, one issue in ACPSs is that sensing nodes are usually deployed within a large field. Effective data and signal processing approaches, such as compressive sensing, are needed to achieve the best tradeoff between cost of sensor nodes and accuracy of field sensing. (4) Energy is critical for running ACPSs. Powering an ACPS using

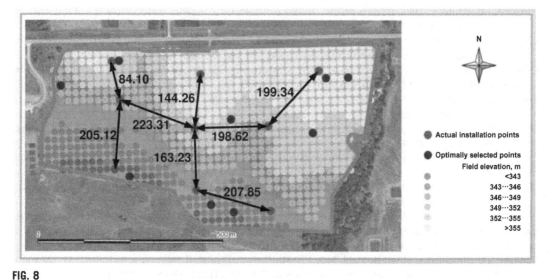

FIG. 8

The distances among parcel nodes in our previous sensor deployment scheme.

Table 2 The Monitoring Efficiency η and the Gains of the Proposed Sensor Deployment Schemes With Different Node Numbers

Node Number	10	11	13	15	17	19	21
Monitoring efficiency (η)	0.8384	0.8541	0.8773	0.9279	0.9405	0.9881	0.9997
Gain (%)	8.94	10.51	12.83	17.89	19.15	23.91	25.07

renewable energy comes with significant design and optimization challenges by exploiting the characteristics of energy available from renewable sources in agricultural environment. (5) ACPSs need to be autonomous, robust, and resilient. To achieve these objectives, the sensing and control loop of ACPSs needs to be further investigated by taking into account both uncertainties of environment and characteristics of agricultural facilities and equipment.

REFERENCES

Abrams, Z., Goel, A., Plotkin, S., 2004. Set k-cover algorithms for energy efficient monitoring in wireless sensor networks. In: IEEE/ACM International Symposium on Information Processing in Sensor Networks, Berkeley, CA USA.

Adamchuk, V.I., Pan, L., Marx, D.B., Martin, D.L., 2010. Locating soil monitoring sites using spatial analysis of multilayer data. In: World Congress on Social Simulation, Brisbane, Australia, pp. 1–6.

An, W., Shao, F.M., Meng, H., 2009. The coverage-control optimization in sensor network subject to sensing area. Comput. Math. Appl. 57, 529–539.

Blackmer, A., White, S., 1998. Using precision farming technologies to improve management of soil and fertiliser nitrogen. Aust. J. Agric. Res. 49, 555–564.

Castaño, S., Sanz, D., Gómez-Alday, J., 2010. Methodology for quantifying ground water abstractions for agriculture via remote sensing and GIS. Water Resour. Manag. 24, 795–814.

Chakrabarty, K., Iyengar, S.S., Qi, H., Cho, E., 2002. Grid coverage for surveillance and target location in distributed sensor networks. IEEE Trans. Comput. 51, 1448–1453.

Discovery Channel, 2003. Pro seris-eko node for environment monitoring. http://www.exn.ca/video/? video=exn20030925-wine.asx.

Corwin, D.L., Lesch, S.M., 2005. Apparent soil electrical conductivity measurements in agriculture. Comput. Electron. Agric. 46, 11–43.

Davis, G., Casady, W., Massey, R., 1998. Precision Agriculture: An Introduction. University of Missouri Extension.

Dong, X., Vuran, M.C., Irmak, S., 2013. Autonomous precision agriculture through integration of wireless underground sensor networks with center pivot irrigation systems. Ad Hoc Netw. 11, 1975–1987.

eKo, 2016. Pro seris-eko node for environment monitoring. http://www.memsic.com/wireless-sensor-networks/.

Erdogan, E.H., Erpul, G., Bayramin, I., 2007. Use of USLE/GIS methodology for predicting soil loss in a semiarid agricultural watershed. Environ. Monit. Assess. 131, 153–161.

Farahani, H.J., Buchleiter, G.W., Brodahl, M.K., 2004. Characterization of apparent soil electrical conductivity variability in irrigated sandy and non-saline fields in Colorado. Trans. ASAE 48, 155–168.

Ganjegunte, G., Braun, R., 2010. Application of electromagnetic induction technique for soil salinity and sodicity appraisal. In: International Conference on Environmental Engineering and Applications, Singapore.

Grisso, R.D., Jasa, P.J., Schroeder, M.A., Wilcox, J.C., 2002. Yield monitor accuracy successful farming magazine case study. Appl. Eng. Agric. 18, 147–151.

Hamlett, J., Miller, D., Day, R., Peterson, G., Baumer, G., Russo, J., 1992. Statewide GIS-based ranking of watersheds for agricultural pollution prevention. J. Soil Water Conserv. 47, 5399–5404.

Jiang, J.A., et al., 2013. Application of a web based remote agro-ecological monitoring system for observing spatial distribution and dynamics of *Bactrocera dorsalis* in fruit orchards. Precis. Agric. 14, 323–342.

Karp, R.M., Miller, R.E., Thatcher, J.W., 1975. Reducibility among combinatorial problems. J. Symb. Log. 40, 618–619.

Li, S., Xu, C., Pan, W., Pan, Y., 2005. Sensor deployment optimization for detecting maneuvering targets. In: 8th International Conference on Information Fusion, Philadelphia, USA.

Ma, M., Yang, Y., 2007. Adaptive triangular deployment algorithm for unattended mobile sensor networks. IEEE Trans. Comput. 56, 946–958.

Makynen, J., et al., 2012. Multi- and hyperspectral UAV imaging system for forest and agriculture applications. In: Proc. SPIE 8374, Next-Generation Spectroscopic Technologies.

McCauley, L.A., Jenkins, D.G., 2005. GIS-based estimates of former and current depressional wetlands in an agricultural landscape. Ecol. Appl. 15 (4), 1199–1208.

Palaniswami, C., Gopalasundaram, P., Bhaskaran, A., 2011. Application of GPS and GIS in sugarcane agriculture. Sugar Tech 13, 360–365.

Papadimitriou, C.H., Steiglitz, K., 1998. Combinatorial Optimization: Algorithms and Complexity. Dover Publications Inc.

Patil, P., et al., 2011. Wireless sensor network for precision agriculture. In: The International Conference on Computational Intelligence and Communication Networks.

Patil, M., Patil, V., Khosla, R., 2012. Precision nutrient management in cotton: a case study from Karnataka. In: The Third National Conference on Agro-Informatics and Precision Agriculture (AIPA 2012), Hyderabad, India.

Rains, G., Thomas, D., 2000. Precision Farming—An Introduction. The University of Georgia Extension.

Roy, S., Bandyopadhyay, S., 2008. Agro-sense: precision agriculture using sensor-based wireless meshnetworks. In: The Conference on Innovations in NGN: Future Network and Services, Geneva, Switzerland.

Sadler, E.J., Evans, R.G., Stone, K.C., Camp, C.R., 2005. Opportunities for conservation with precision irrigation. J. Soil Water Conserv. 60, 371–379.

Sahota, H., et al., 2010. An energy-efficient wireless sensor network for precision agriculture. In: IEEE Symposium on Computers and Communications, Riccione, Italy.

Sahota, H., et al., 2011. A wireless sensor network for precision agriculture and its performance. Wirel. Commun. Mob. Comput. 11, 1628–1645.

Shu, H., Liang, Q., 2005. Fuzzy optimization for distributed sensor deployment. In: IEEE Wireless Communications and Networking Conference, New Orleans, USA, pp. 1903–1908.

Silva, A.R., Vuran, M.C., 2010. Communication with aboveground devices in wireless underground sensor networks: an empirical study. In: IEEE International Conference on Communications, Cape Town, South Africa.

Slijepcevic, S., Potkonjak, M., 2001. Power efficient organization of wireless sensor networks. In: IEEE International Conference on Communications, Helsinki, Finland.

Srbinovska, M., et al., 2015. Environmental parameters monitoring in precision agriculture using wireless sensor networks. J. Clean. Prod. 88, 297–307.

Tokekar, P., Hook, J.V., Mulla, D., Isler, V., 2013. Sensor planning for a symbiotic UAV and UGV system for precision agriculture. In: IEEE/RSJ International Conference on Intelligent Robots and Systems (IROS), Tokyo, Japan, pp. 5321–5326.

Trevisan, L., 2011. Combinatorial Optimization: Exact and Approximate Algorithms. Stanford University.

Vazirani, V.V., 2001. Approximation Algorithms. Springer-Verlag.

Walsh, O.S., Klatt, A.R., Solie, J.B., Godsey, C.B., Raun, W.R., 2013. Use of soil moisture data for refined greenseeker sensor based nitrogen recommendations in winter wheat. Precis. Agric. 14, 343–356.

Wang, G., Cao, G., Porta, T.L., 2006. Movement-assisted sensor deployment. IEEE Trans. Mob. Comput. 5, 640–652.

Wang, B., Chua, K.C., Srinivasan, V., Wang, W., 2007. Information coverage in randomly deployed wireless sensor networks. IEEE Trans. Wirel. Commun. 6, 2994–3004.

Wark, T., et al., 2007. Transforming agriculture through pervasive wireless sensor networks. IEEE Pervasive Comput. 6, 50–57.

Wu, C.H., Lee, K.C., Chung, Y.C., 2006. A Delaunay triangulation based method for wireless sensor network deployment. In: 12th International Conference on Parallel and Distributed Systems, Minneapolis, USA.

Xiang, H., Tian, L., 2011. Development of a low-cost agricultural remote sensing system based on an autonomous unmanned aerial vehicle (UAV). Biosyst. Eng. 108, 174–190.

Xu, X., Sahni, S., 2007. Approximation algorithms for sensor deployment. IEEE Trans. Comput. 56, 1681–1695.

Xu, X., Sahni, S., Rao, N., 2008. Minimum cost sensor coverage of planar regions. In: 11th International Conference on Information Fusion, Cologne, Germany, pp. 1–8.

Yang, G., Qiao, D., 2010. Multi-round sensor deployment for guaranteed barrier coverage. In: IEEE International Conference on Computer Communications, San Diego, USA, pp. 1–9.

Yoo, S., Kim, J., Kim, T., Ahn, S., Sung, J., Kim, D., 2007. A2s: automated agriculture system based on WSN. In: IEEE International Symposium on Consumer Electronics, Irving, TX, USA, pp. 1–6.

Zhang, Y., Wang, L., 2009. A sensor deployment algorithm for mobile wireless sensor networks. In: Chinese Control and Decision Conference, Guilin, China, pp. 4606–4611.

AN EMERGING DECISION AUTHORITY: ADAPTIVE CYBER-PHYSICAL SYSTEM DESIGN FOR FAIR HUMAN-MACHINE INTERACTION AND DECISION PROCESSES

26

T. Töniges*, S.K. Ötting*, B. Wrede, G.W. Maier, G. Sagerer

Bielefeld University, Bielefeld, Germany

1 INTRODUCTION

New and emerging technologies will allow cyber-physical systems (CPSs) to take more information from the physical environment of production lines into account than has previously been available. This development enables CPSs to be adaptive and to selfoptimize with regard to not only performance criteria, but also safety and user acceptance.

One important factor in the physical production environment is the human operator. To optimize the system's performance and safety, the system has to adapt to the respective user in order to be able to respond to his or her special needs, habits and abilities. A user should, for example, be limited in the amount of physically or mentally highly demanding work processes that he or she carries out. In this example, it is thus necessary to monitor the user's current physical and mental state in order to prevent dangerous situations, and in order to be able to react to the user's behavior. From a technical perspective this requires the extension of current state-of-the-art systems: by incorporating a user model that is able to represent mental states, and an extended process model that takes this information into account.

While adapting to the user, the system will be making decisions concerning the user (e.g., the assignment of tasks, shift schedules, or even industrial safety). These system decisions (the same as human decisions) will cause reactions, most likely by constraining the user's autonomy and thereby affecting him or her (especially if these decisions are based on an individual behavior analysis). The users not only can be either satisfied or dissatisfied with the system's decisions, they can also

---

*Shares first authorship.

Cyber-Physical Systems. http://dx.doi.org/10.1016/B978-0-12-803801-7.00026-2

perceive these decisions as either fair or unfair. These perceptions of fairness, or the corresponding principles of organizational justice, have consistently been shown to affect satisfaction, well-being, and performance of employees (e.g., Greenberg, 2011). Hence, fairly designed and fairly communicated decisional processes in technical systems may have a positive impact not only on the system's acceptance, but also on the system's performance.

Current developments only rarely include the system's user in a comparable way as envisioned above. Until now this has not been a problem for the fully autonomous systems without any, or hardly any, user interface. The increasing interaction between users and CPSs will need to be taken into account as a future major factor for system decision-making processes. Therefore current human-machine interaction strategies and the development processes of CPSs have to be extended by principles derived from interdisciplinary research with a human-centered perspective.

In this chapter we will show examples of how users' basic needs can be integrated in CPSs to optimize the system's performance, safety, and acceptance. At first, we will show how the user's states (especially the mental states) can be measured and modeled to predict the user's behavior. To do so, we introduce a representation of users in computational systems to be able to anticipate their behavior. Based on the observed user states, we subsequently describe user-specific adaptation strategies, with a focus on adaptation design and decision-making processes. Importantly, we include considerations of human reactions by introducing criteria of just decision-making and the perception of machine fairness. Based on this analysis we derive implications for the system's design and close with an outlook on further research.

2 MODELING THE USER

As a basis for adaptation strategies (with regard to the special needs and abilities of each individual user), the system needs to model, analyze and represent the user. In other words, adaptations need to be based on predictions of the user's behavior. Knowledge of the user's mental state facilitates and improves these predictions. However, information about the user's mental states cannot be accessed directly. The only observable information is the user's behavior. Specific time-dependent behavior patterns can be used to infer information about the hidden mental states.

There are two main pattern recognition methods towards inferring hidden (not directly observable) states: the unsupervised learning method and the supervised learning method. On the one hand, if no predefined states are available (with relevancy to system safety, acceptance and performance), unsupervised statistical approaches can be used to find patterns in user behavior. Once these patterns have been determined, they can be recognized in new behavior data from potential different users. These patterns, however, cannot be explicitly named or defined. On the other hand, research in psychology (e.g., Dinges et al., 1998) already identified and defined mental states (both cognitive and affective), along with identifiable, distinct behavior patterns. An example of these kinds of patterns is microsleep (roughly identified by closing of the eyes and dropping of the head) as an indicator of fatigue. If training data are available (i.e., behavior patterns that have been annotated with respect to these categories), these categories can explicitly be learned by the system and can be used later on for recognition.

As unsupervised approaches provide results that are difficult to deal with by the process model (i.e., what adaptation action has to be performed given a specific observed state?), we focus on supervised approaches. In the following, we will present several methods through which mental states can be inferred from user behavior.

2.1 TASK-RELATED FEATURE ANALYSIS

The most obvious modality to be used in behavior interpretation is a direct analysis related to the current task or work step. This analysis depends strongly on the scenario and the used hardware. Examples are the pace at which the user completes a work step, the amount of user errors and their temporal structure, and the usage patterns of computational input devices. For example, Sun et al. (2014) show, that muscle stiffness changes, caused by stress, influence mouse movements, and that stress can be detected by analyzing these movements. Salmeron-Majadas et al. (2014) correlate mouse and keyboard interaction patterns with the categories "pleasure" and "arousal."

As far as we know, only a few related research publications have described the usage of task-oriented features to gain information about human states by detecting changes in the user's behavior and thereby extracting information about his or her current state. Obviously, a task-related feature analysis is not feasible if the user simply monitors the system and no direct input is necessary.

2.2 VOICE ANALYSIS

The human voice is an important resource to analyze the user's current state. Some spectral characteristics of the human voice are highly correlated with physical activation, while others are more closely related to deliberate control and thus higher-level human states. Voice features have, for example, been used to detect certainty (Liscombe et al., 2005), depression severity (Yang et al., 2013) and stress (review see Giddens et al., 2013).

Yet, while the human voice is a good indicator of human states, its analysis is of no use in industrial settings, as there are often many noise sources around CPSs. In addition, the user's voice can obviously only be analyzed when he or she is speaking, thus leaving large sequences without sensory input from this channel.

2.3 VITAL PARAMETER ANALYSIS

Vital parameters can be seen as direct correlates of human states and are consequently of high interest. They are mainly obtained by wearable biosensor systems (for an overview see Pantelopoulos and Bourbakis, 2010, or Schirner et al., 2013). There are also a few applications that are able to extract vital parameters in a remote way, e.g., by measuring the cardiac pulse by facial color changes (Poh et al., 2010).

The analysis of vital parameters is a reasonable choice in CPSs as it provides a reliable source of information about the current user (e.g., physical strain). A potential field of application would be the integration of these sensors in work clothes to monitor the user and to minimize health risks.

2.4 HEAD ANALYSIS

The main directions of research in the area of head analysis are head pose (Murphy-Chutorian and Trivedi, 2009), gaze direction (Hansen and Ji, 2010), and facial displays and facial expression analysis (De la Torre and Cohn, 2011). Only in rare situations (like the wearing of safety clothing that covers the face), can head analysis methods not be applied. Recently, methods detecting physical pain (Ashraf et al., 2009; Kaltwang et al., 2012; Prkachin and Solomon, 2008) and depression (Scherer et al., 2013) from facial images were presented. In the field of marketing, users' reactions to new products, advertisements, or movie trailers are forecasted based on an analysis of the head. Researchers were able to: estimate viewer's attentiveness while watching TV (Takahashi et al., 2013), extract movie ratings by observing the movie's audience (Navarathna et al., 2014), and detect whether a participant likes the medium and desires to view it again (McDuff et al., 2013, 2014). In online tutoring scenarios, head analysis methods are used to recognize frustration, boredom, or confusion (Whitehill et al., 2011). The detection of fatigue is especially important in the car industry (Dong et al., 2011) and can be detected based on head analysis (Ji et al., 2004; Saeed et al., 2007; Vural et al., 2007). For example, accidents in a driving simulator can be predicted earlier if the face is analyzed (Jabon et al., 2011).

Thus there are already many applications related to head analysis whose methods and algorithms can be integrated into the CPSs. Currently available systems are able to work in specially designed environments. There is a lack of general systems that are able to handle complex situations, such as an interaction between users and CPSs over a long period of time. Nevertheless, the analysis of the human state based on an analysis of the user's head is a promising direction of research.

2.5 MULTIMODAL ANALYSIS

Instead of using single features, multimodal approaches that use a variety of modalities promise more robustness, which is especially important in a realistic environment. Banda and Robinson (2011) and D'Mello et al. (2007) use a multimodal approach to detect the mental state of a participant (absorbed, interested, stressed, bored, confused, or frustrated). Malta et al. (2011) integrate different features to detect car drivers' level of frustration. Kim et al. (2013) use deep learning methods for the task of emotion classification.

It can be seen that multimodal approaches outperform single feature methods. Therefore the usage of different input sources in CPSs would be the desirable approach. Nevertheless, it is not yet known if the right approach is to use as many modalities as possible or if certain modalities share the same variance, which would make simultaneous use recommended and necessary. Furthermore, it is not clear whether there is convergence between the different modalities or if the different modalities are equally appropriate for the measuring of mental states. A detailed analysis still needs to be carried out.

3 HUMAN-CENTERED ADAPTATION OPTIONS

If a CPS is able to model human states, it is able to use this model to adapt its behavior to the user. After describing in the previous section the different possibilities of how data can be obtained to model the user, we now want to show some examples of possible representation forms of human states in CPS and possible adaptations of the system.

For a robust interpretation of these data and suitable reactions towards the user's mental states, it is crucial to consider different temporal scales.

The user's behavior can be analyzed in a *short-time* manner (up to the duration of a single work-process step) to directly connect: the current user, the current step in his or her workflow, the current task, and the current state of the system. The main purpose of a short-time analysis is to obtain information on the current work-step. It can be used to analyze whether the system provides either too few or too many instructions and information. The system can then either provide additional support or remove unnecessary information respectively. A short-time analysis based on user behavior can also be used to check if important warnings are noticed. A manual confirmation by the user (e.g., pushing a button) would become unnecessary; the user could, for example, simply nod. This leads to a more natural interaction between the user and the CPS. The short-time user state analysis can, furthermore, be used to learn the user's normal reactions to single work-steps, and thereby to forecast his or her behavior in order to react to abnormal patterns.

A further approach is a *medium time-period* analysis (ranging from more than one work-step to a whole workday). It can be used to adapt the workload assignment by suggesting a task change or a break if the user is, for example, getting frustrated, annoyed, bored, or tired. In addition, the user can be encouraged to stay focused and monotony can be prevented. This medium time period analysis information can also be used to support workplace and selfdirected learning. The ability to analyze the user, for example enables the CPS to automatically offer training or more complex work-steps if the user is not sufficiently challenged.

It will also be possible to take *long time periods* (ranging from months to years) into account. This will lead to a more general perspective of user behavior in working environments. A system will be able to detect specific temporal patterns (e.g., work patterns during the week) and will be able to optimize the task processes with respect to efficiency, safety, health and motivation, as well as maintenance and growth of competence.

We described different time scales above with respect to a single user of a CPS. A major capability of a system that is equipped with the described abilities is the capability to combine all single user analyses to a comprehensive interpretation of the work abilities of a production site as a whole. This information can be used to improve not only the efficiency of a single user, but also the efficiency of all workers in a production facility, and even the efficiency of all available production sites. For example, whole production processes can be shifted to another site if the system detects general periods of behavioral changes.

4 PERCEPTION OF MACHINE-FAIRNESS

As we can see from the previous section, it is very clear that CPSs will be able to adapt and respond to the user's behavior and even the user's mental states and therefore emerge as a new decision authority in organizations. Yet, it is not clear how these adaptations and decisions affect the users they are concern with and how these users will react to them. But, as individuals in organizations are subject to multiple decisions a day, it will be important to specify how system decisions should be made.

One important human reaction towards decisions is the perception of fairness or unfairness. Fairness in organizations is a well-researched topic in work and organizational psychology as it is an important factor in efficiency and work satisfaction. However, as far as we know, fairness has not yet been investigated in the context of artificial systems, in contrast that of supervisor decisions, or as a principle in system design.

In the following section we introduce the concept of organizational justice and its importance for employee behavior and attitudes before applying it to the context of system design in the next section of the chapter.

4.1 CONCEPT OF ORGANIZATIONAL JUSTICE

A rich body of research shows that the perception of fairness is contingent on the principles of organizational justice (for current overviews: Colquitt and Zipay, 2015; Greenberg, 2011). These principles are composed of four dimensions, namely: distributive, procedural, interpersonal, and informational justice. *Distributive justice* focuses on the justice of decision outcomes. It is based on "Adams' equity theory" (Adams, 1965), which states that a decision outcome is perceived as fair if the ratio of one's contributions to one's outcomes is equivalent to that of a comparison other. *Procedural justice* describes the justice of the decision-making procedures. A decision process is perceived as fair, if the recipients have voice during the process or influence over the outcome, and if justice criteria (such as consistency, lack of bias, accuracy, correctability, and ethicality) are given in the process. The last two dimensions concern the justice of the decision communication: *interpersonal justice* describes the justice of the decision authority's behavior towards the recipient—the communication of a decision is perceived as interpersonally fair, if it is respectful, polite and dignified; and *informational justice*, which refers to the adequacy of the information used to explain how the decision process is formed—it is fostered where information is truthful, well-reasoned, specific and timely.

These dimensions of organizational justice have been shown to be related to a number of affective (e.g., supervisor satisfaction, outcome satisfaction, job satisfaction, organizational commitment, trust) and behavioral (e.g., turnover intentions, withdrawal, counterproductive work behavior, work performance, organizational citizenship behaviors) outcomes (Cohen-Charash and Spector, 2001; Colquitt et al., 2001, 2013). Organizational justice is also an important factor to the successful mastery of organizational change, relating, for example, to acceptance, readiness and commitment to change (Korsgaard et al., 2002; Paterson and Cary, 2002), resistance to change and organizational cynicism (Bernerth et al., 2007; Shapiro and Kirkman, 1999), motivation to learn (Liao and Tai, 2006), and creative behavior (Streicher et al, 2012). In summary, research shows that organizational justice is important for the employees' job satisfaction and their positive attitude towards their organization. Furthermore, it is positively related to work performance and organizational citizenship behaviors and therefore fosters efficiency. Organizational justice can also constrain negative behaviors and support change and learning.

These relationships between fair treatment by the supervisor or the organization and outcomes (such as satisfaction with the supervisor, commitment to the organization or acceptance of change) may also be valid to the fair treatment by technical systems. Following the group engagement model of Tyler and Blader (2003), it could be expected that users treated fairly by a system should identify with it. Therefore the users accept the new technology, are more willing to comply with the system's decisions or are more satisfied with the system as new decision authority.

5 IMPLICATIONS FOR SYSTEM DEVELOPMENT

As we have seen from the previous section, taking a human-centered perspective on and integrating human-oriented principles into CPSs is essential, not only for the system's performance, but also for safety and acceptance. In order to consider the described human-centered perspective, existing system development principles have to be extended.

Fair system development is a matter not only of the design of communication strategies but also, and maybe more importantly, of the design of the system itself. This especially includes the design of the decision processes. Considering the principles of organizational justice introduced previously, several requirements should be met in the design of automatically adapting systems. This section describes these requirements and their implications for system design and research.

The first requirement is to design the phrasing of the communication in a polite and respectful tone. This mostly concerns the system's interface and how it communicates with the user. The user reaction to this way of communication with CPSs is yet to receive further scholarly attention. Furthermore, it is not clear if a polite and respectful communication strategy coming from an artificial system may even lead to a slower working speed and consequently a loss in efficiency.

The second requirement is to enable the system to communicate the decision, and the reasons behind it, in a timely, specific, well-reasoned and understandable manner. This means that it will not suffice to simply provide the sheer output, action or decision; each system component needs to be able to provide a reason for its output, action or decision. The path that the decision-making process takes has to be transparent. This requirement makes the explicit modeling of the user's mental states mandatory. As described previously, there are basically two different general approaches to automatic learning of the mental states. If pure statistical methods are used, the path of decision processes cannot be inferred. If a rationale of a decision has to be named by the system, the hidden human state categories have to be learned explicitly. Furthermore, this requirement leads to the demand for an aggregator. In a modern CPS, many decisions are made in seconds. Not all of these processes need to be passed on to the user. The system needs to be able to aggregate the different reasons, and, in a second step, to extract the currently most valuable information that needs to be passed on to the user (if deemed necessary). In general, we implicate that it is important for users to be aware of the system changes concerning them. This assumption builds on research in system transparency, stating that especially context-aware systems need to be transparent to enable users to trust the behavior of the system and develop necessary efficient skills (Cheverst et al., 2005; Frese, 1987). Research shows that users generally prefer transparent system recommendations (Herlocker et al., 2000). Yet, it remains unanswered, from what point onwards users perceive a system adaptation as a decision that is concerning them, and further, which decisions should be communicated and which system changes can be made without the knowledge of the user.

The third requirement is that the system has to be able to present a means for the user to raise objections to the decision (or the decision process), to make suggestions for improvements, or to make corrections. Especially adaptive systems could ensure voice by providing natural, conversation-like dialogs. Therefore backchannels are needed to handle the input of the user and to change the behavior of the system according to the user's input. This leads to an important change in terms of system design. The system has to cope with suggestions of the user, even if these suggestions are not optimal in terms of efficiency or objectivity. A balance has to be found between an efficient output of the system and the well-being of the user. It has to be discussed, if efficiency is truly the most important factor in work environments. A highly efficient environment, in which nobody wants to work for a longer period of time, can be just as bad as a nonefficient environment in which the users love to work. Furthermore, the way in which the system should react to conflicting goals like safety (e.g., detecting fatigue and suggesting a break), user acceptance (e.g., suggesting a different task according employee preferences), and productivity (e.g., produce as much as possible in a short time span) remains an open question. It is a trade-off between the well-being of the user and profitability.

The fourth requirement is that the user has to be informed, satisfied, and convinced that the procedures are: applied consistently across people and time, free of bias, based on accurately collected information, and conforming to prevailing standards of morality. On the one hand, this provides a chance for CPS development, because the automatic decisions may (more so than human-made decisions) be perceived as bias-free and accurate due to the usage of objectively collected data. On the other hand, additional requirements, such as the capability to adapt to the current user and to provide a higher amount of voice, could also restrict, for example, the consistency across people and time. Therefore, the different justice criteria cannot be considered independently, but are influencing each other. Previous research showed, for instance, that the control over the decision process can outweigh control over the decision outcome: individuals were willing to give up control over the decision outcome as long as they had control over the decision process (Thibaut and Walker, 1975). Research also points to a conflict between voice and consistency, uncovering that the possibility of voice might constrain the consistency of the decision process between individuals (Douthitt and Aiello, 2001). Yet research about other conflicting justice criteria is still mostly missing. In the system's design, this leads to a need for bounding values for the adaptation possibilities to a single user. For example, if a system is able to identify monotony and suggest breaks for one user, the system has to cope with the possibility that another user might not need the same amount of breaks, but might still react dissatisfied because of noticing the colleague's comparatively more regular breaks.

The fifth requirement is that the system not only needs to consider its current user, but also his or her relationship and comparison with colleagues. Because people compare the ratio of their contributions to their outcomes with the respective ratio of their comparable colleagues, the system should be able to monitor these ratios and maintain equality. Until now it is totally unknown, how such a general analysis of the user and his or her workplace will influence the work itself. This remains an open research question and has to be investigated in the future.

6 CONCLUSION AND OUTLOOK

In this chapter we presented and suggested principles meant to strongly improve not only system performance, but also industrial safety and the user's experience while working with modern human-oriented (human-centered) CPSs. Therefore we described approaches of modeling the user by analyzing his or her behavior and thereby inferring hidden mental states. Subsequently, we introduced adaptation possibilities based on these approaches, followed by the introduction of the concept of machine fairness. These principles gain particular importance in the case of automatic system decisions regarding user-specific behavior. Considering justice criteria in system design should lead to perceptions of fairness and therefore reduce the experience of autonomy constraints. This can result not only in higher user satisfaction, but also in better system performance. Therefore by combining the research results of psychology and computer science, we imposed requirements for the design of CPSs that assist, rather than patronize, the user.

The modeling and analyzing therein of user behavior and mental states raises further and more general questions, which are highly interdisciplinary.

One important question relates to the optimization criteria of such a described system. In practice, fair and adaptive systems will most likely positively affect efficiency, safety, work motivation, and user satisfaction. However, empirical research has to prove these assumptions in detail and investigate possible conflicting goals and principles. Moreover, an individual balance between these goals has to be found in each system through practical experience.

In addition, there are ranges of open questions related to ethical and legal concerns: What happens to the collected data? Is an automatic analysis of the user desired? Is it legally and ethically correct to infer decisions based on a permanent automatic analysis of the user behavior? How, and in which way, are safety and health regulations affected by automatic decisions? For example, if users are able to individualize their working habits, regulations of working hours per week may be affected.

There are further questions relating to possible changes in organizational structures caused by the application of these human-centered systems: If systems are able to plan workflow and process management, are supervisors no longer needed to make these decisions? Which decisions remain for the supervisors to make?

The described systems offer the promising opportunity to actively handle challenges, such as the inclusion of persons with physical and mental disabilities, or demographic changes. These systems will be able to offer advice and support with physically or mentally demanding tasks and help to retain employee health so that they will be able to work for longer. The workers will experience a feeling of still being needed, and, in the end, this may even lead to an improvement in quality of their life. Here, research needs to clarify how exactly adaptive and fair CPSs can be applied to address the special needs of these populations. Furthermore, due to the postulated backchannels of the described systems, the users are able to train these systems and the systems will strongly benefit from the users' experiences. To know that the systems need the users' support will lead to a much better working atmosphere than if the users are being treated as the minions of machines while working on assigned tasks that the machines are not yet capable of completing independently.

If users and machines are to cooperate as companions in future scenarios of the world of work, the described principles need to be applied. Otherwise, the users will sooner or later refuse to accept and use CPSs.

ACKNOWLEDGMENTS

The authors are part of the research program "Design of Flexible Work Environments—Human-Centric Use of Cyber-Physical Systems in Industry 4.0", by the North Rhine-Westphalian funding scheme "Fortschrittskolleg", affiliated to the Institute for Cognition and Robotics (CoR-Lab), Bielefeld University, and supported by the Cluster of Excellence Cognitive Interaction Technology "CITEC" (EXC 277) at Bielefeld University, funded by the German Research Foundation (DFG).

REFERENCES

Adams, J.S., 1965. Inequity in social exchange. In: Berkowitz, L. (Ed.), Advances in Experimental Social Psychology, vol. 2. Academic Press, New York, pp. 267–299.

Ashraf, A.B., Lucey, S., Cohn, J.F., Chen, T., Ambadar, Z., Prkachin, K.M., Solomon, P.E., 2009. The painful face: pain expression recognition using active appearance models. Image Vis. Comput. 27 (12), 1788–1796. http://dx.doi.org/10.1016/j.imavis.2009.05.007.

Banda, N., Robinson, P., 2011. Multimodal affect recognition in intelligent tutoring systems. In: D'Mello, S., Graesser, A., Schuller, B., Martin, J.-C. (Eds.), In: Affective Computing and Intelligent Interaction, vol. 6975. Springer, Berlin, Heidelberg, pp. 200–207.

Bernerth, J.B., Armenakis, A.A., Feild, H.S., Walker, H.J., 2007. Justice, cynicism, and commitment: a study of important organizational change variables. J. Appl. Behav. Sci. 43 (3), 303–326. http://dx.doi.org/10.1177/0021886306296602.

Cheverst, K., Byun, H.E., Fitton, D., Sas, C., Kray, C., Villar, N., 2005. Exploring issues of user model transparency and proactive behaviour in an office environment control system. User Model. User-Adap. Inter. 15 (3–4), 235–273. http://dx.doi.org/10.1007/s11257-005-1269-8.

Cohen-Charash, Y., Spector, P.E., 2001. The role of justice in organizations: a meta-analysis. Organ. Behav. Hum. Decis. Process. 86 (2), 278–321. http://dx.doi.org/10.1006/obhd.2001.2958.

Colquitt, J.A., Zipay, K.P., 2015. Justice, fairness, and employee reactions. Annu. Rev. Organ. Psychol. Organ. Behav. 2 (1), 75–99. http://dx.doi.org/10.1146/annurev-orgpsych-032414-111457.

Colquitt, J.A., Conlon, D.E., Wesson, M.J., Porter, C.O.L.H., Ng, K.Y., 2001. Justice at the millennium: a meta-analytic review of 25 years of organizational justice research. J. Appl. Psychol. 86 (3), 425–445. http://dx.doi.org/10.1037//0021-9010.86.3.425.

Colquitt, J.A., Scott, B.A., Rodell, J.B., Long, D.M., Zapata, C.P., Conlon, D.E., Wesson, M.J., 2013. Justice at the millennium, a decade later: a meta-analytic test of social exchange and affect-based perspectives. J. Appl. Psychol. 98 (2), 199–236. http://dx.doi.org/10.1037/a0031757.

D'Mello, S., Picard, R.W., Graesser, A., 2007. Toward an affect-sensitive AutoTutor. Intell. Syst. 4, 53–61. http://dx.doi.org/10.1109/MIS.2007.79.

De la Torre, F., Cohn, J.F., 2011. Facial expression analysis. In: Moeslund, T.B., Hilton, A., Krüger, V., Sigal, L. (Eds.), Visual Analysis of Humans: Looking at People. Springer, London, pp. 377–409.

Dinges, D.F., Mallis, M.M., Maislin, G., Powell, I.V., 1998. Evaluation of techniques for ocular measurement as an index of fatigue and the basis for alertness management (No. HS-808 762).

Dong, Y., Hu, Z., Uchimura, K., Murayama, N., 2011. Driver inattention monitoring system for intelligent vehicles: a review. IEEE Trans. Intell. Transp. Syst. 12 (2), 596–614. http://dx.doi.org/10.1109/TITS.2010.2092770.

Douthitt, E.A., Aiello, J.R., 2001. The role of participation and control in the effects of computer monitoring on fairness perceptions, task satisfaction, and performance. J. Appl. Psychol. 86 (5), 867–874. http://dx.doi.org/10.1037/0021-9010.86.5.867.

Frese, M., 1987. A theory of control and complexity: implications for software design and integration of computer systems into the work place. In: Frese, M., Ulich, E., Dzida, W. (Eds.), Psychological Issues of Human-Computer Interaction in the Work Place. North-Holland and Sole Distributors for the U.S.A. and Canada, Elsevier Science, Amsterdam and New York, pp. 313–337.

Giddens, C.L., Barron, K.W., Byrd-Craven, J., Clark, K.F., Winter, A.S., 2013. Vocal indices of stress: a review. J. Voice 27 (3), 390.e21–390.e29.

Greenberg, J., 2011. Organizational justice: the dynamics of fairness in the workplace. In: Zedeck, S. (Ed.), Maintaining, Expanding, and Contracting the Organization, vol. 3. American Psychological Association, Washington, DC, pp. 271–327.

Hansen, D.W., Ji, Q., 2010. In the eye of the beholder: a survey of models for eyes and gaze. IEEE Trans. Pattern Anal. Mach. Intell. 32 (3), 478–500. http://dx.doi.org/10.1109/TPAMI.2009.30.

Herlocker, J.L., Konstan, J.A., Riedl, J., 2000. Explaining collaborative filtering recommendations. In: The 2000 ACM Conference, pp. 241–250.

Jabon, M.E., Bailenson, J.N., Pontikakis, E., Takayama, L., Nass, C., 2011. Facial expression analysis for predicting unsafe driving behavior. Pervasive Comput. 10 (4), 84–95. http://dx.doi.org/10.1109/MPRV.2010.46.

Ji, Q., Zhu, Z., Lan, P., 2004. Real-time nonintrusive monitoring and prediction of driver fatigue. IEEE Trans. Veh. Technol. 53 (4), 1052–1068. http://dx.doi.org/10.1109/TVT.2004.830974.

Kaltwang, S., Rudovic, O., Pantic, M., 2012. Continuous pain intensity estimation from facial expressions. In: Bebis, G., Boyle, R., Parvin, B., Koracin, D., Fowlkes, C., Wang, S., Choi, M.-H., Mantler, S., Schulze, J., Acevedo, D., Mueller, K., Papka, M. (Eds.), Advances in Visual Computing, vol. 7432. Springer, Berlin, pp. 368–377.

Kim, Y., Lee, H., Provost, E.M., 2013. Deep learning for robust feature generation in audiovisual emotion recognition. In: IEEE International Conference on Acoustics Speech and Signal Processing, pp. 3687–3691.

Korsgaard, M.A., Sapienza, H.J., Schweiger, D.M., 2002. Beaten before begun: the role of procedural justice in planning change. J. Manag. 28 (4), 497–516. http://dx.doi.org/10.1177/014920630202800402.

Liao, W.-C., Tai, W.-T., 2006. Organizational justice, motivation to learn, and training outcomes. Soc. Behav. Personal. 34 (5), 545–556. http://dx.doi.org/10.2224/sbp.2006.34.5.545.

Liscombe, J., Hirschberg, J.B., Venditti, J.J., 2005. Detecting certainness in spoken tutorial dialogues. Proceedings of INTERSPEECH, pp. 2837–2840.

Malta, L., Miyajima, C., Kitaoka, N., Takeda, K., 2011. Analysis of real-world driver's frustration. IEEE Trans. Intell. Transp. Syst. 12 (1), 109–118. http://dx.doi.org/10.1109/TITS.2010.2070839.

McDuff, D., el Kaliouby, R., Demirdjian, D., Picard, R., 2013. Predicting online media effectiveness based on smile responses gathered over the Internet. In: International Conference on Automatic Face & Gesture Recognition, Shanghai, China, pp. 1–7.

McDuff, D., el Kaliouby, R., Senechal, T., Demirdjian, D., Picard, R., 2014. Automatic measurement of ad preferences from facial responses gathered over the Internet. Image Vis. Comput. 32 (10), 630–640. http://dx.doi.org/10.1016/j.imavis.2014.01.004.

Murphy-Chutorian, E., Trivedi, M.M., 2009. Head pose estimation in computer vision: a survey. IEEE Trans. Pattern Anal. Mach. Intell. 31 (4), 607–626. http://dx.doi.org/10.1109/TPAMI.2008.106.

Navarathna, R., Lucey, P., Carr, P., Carter, E., Sridharan, S., Matthews, I., 2014. Predicting movie ratings from audience behaviors. In: Winter Conference on Applications of Computer Vision, Steamboat Springs, CO, USA, pp. 1058–1065.

Pantelopoulos, A., Bourbakis, N.G., 2010. A survey on wearable sensor-based systems for health monitoring and prognosis. IEEE Trans. Syst., Man, Cybern. C, Appl. Rev. 40 (1), 1–12. http://dx.doi.org/10.1109/TSMCC.2009.2032660.

Paterson, J.M., Cary, J., 2002. Organizational justice, change anxiety, and acceptance of downsizing: preliminary tests of an AET-based model. Motiv. Emot. 26 (1), 83–103. http://dx.doi.org/10.1023/A:1015146225215.

Poh, M.-Z., McDuff, D.J., Picard, R.W., 2010. Non-contact, automated cardiac pulse measurements using video imaging and blind source separation. Opt. Express 18 (10), 10762–10774. http://dx.doi.org/10.1364/OE.18.010762.

Prkachin, K.M., Solomon, P.E., 2008. The structure, reliability and validity of pain expression: evidence from patients with shoulder pain. Pain 139 (2), 267–274. http://dx.doi.org/10.1016/j.pain.2008.04.010.

Saeed, I., Wang, A., Senaratne, R., Halgamuge, S., 2007. Using the active appearance model to detect driver fatigue. In: International Conference on Information and Automation for Sustainability, Melbourne, Australia, pp. 124–128.

Salmeron-Majadas, S., Santos, O.C., Boticario, J.G., 2014. An evaluation of mouse and keyboard interaction indicators towards non-intrusive and low cost affective modeling in an educational context. Procedia Comput. Sci. 35, 691–700.

Scherer, S., Stratou, G., Mahmoud, M., Boberg, J., Gratch, J., Rizzo, A., Morency, L.-P., 2013. Automatic behavior descriptors for psychological disorder analysis. In: International Conference on Automatic Face & Gesture Recognition, Shanghai, China, pp. 1–8.

Schirner, G., Erdogmus, D., Chowdhury, K., Padir, T., 2013. The future of human-in-the-loop cyber-physical systems. Computer 46 (1), 36–45. http://dx.doi.org/10.1109/MC.2013.31.

Shapiro, D.L., Kirkman, B.L., 1999. Employees' reaction to the change to work teams: the influence of "anticipatory" injustice. J. Organ. Chang. Manag. 12, 51–67. http://dx.doi.org/10.1108/09534819910255315.

Streicher, B., Jonas, E., Maier, G.W., Frey, D., Spießberger, A., 2012. Procedural fairness and creativity: does voice maintain people's creative vein over time? Creat. Res. J. 24, 358–363. http://dx.doi.org/10.1080/10400419.2012.730334.

Sun, D., Paredes, P., Canny, J., 2014. MouStress: detecting stress from mouse motion. In: Proceedings of the SIGCHI Conference on Human Factors in Computing Systems, pp. 61–70.

Takahashi, M., Naemura, M., Fujii, M., Satoh, S., 2013. Estimation of attentiveness of people watching TV based on their emotional behaviors. In: Humaine association conference on affective computing and intelligent interaction, Switzerland, Geneva, pp. 809–814.

Thibaut, J.W., Walker, L., 1975. Procedural Justice: A Psychological Analysis. L. Erlbaum Associates and Distributed by the Halsted Press Division of Wiley, Hillsdale and New York.

Tyler, T.R., Blader, S.L., 2003. The group engagement model: procedural justice, social identity, and cooperative behavior. Personal. Soc. Psychol. Rev. 7, 349–361. http://dx.doi.org/10.1207/S15327957PSPR0704/textunderscore.

Vural, E., Cetin, M., Ercil, A., Littlewort, G., Bartlett, M., Movellan, J., 2007. Drowsy driver detection through facial movement analysis. In: Lew, M., Sebe, N., Huang, T.S., Bakker, E.M. (Eds.), Human-Computer Interaction, vol. 4796. Springer, Berlin, pp. 6–18.

Whitehill, J., Serpell, Z., Foster, A., Lin, Y.-C., Pearson, B., Bartlett, M., Movellan, J., 2011. Towards an optimal affect-sensitive instructional system of cognitive skills. In: Computer Society Conference on Computer Vision and Pattern Recognition Workshops, Colorado Springs, CO, USA, pp. 20–25.

Yang, Y., Fairbairn, C., Cohn, J.F., 2013. Detecting depression severity from vocal prosody. IEEE Trans. Affect. Comput. 4 (2), 142–150.

SCHEDULING FEASIBILITY OF ENERGY MANAGEMENT IN MICRO-GRIDS BASED ON SIGNIFICANT MOMENT ANALYSIS

27

Z. Shi*, N. Yao[†], F. Zhang[†]

C3 IoT, CA, United States* Georgia Institute of Technology, Atlanta, GA, United States[†]

1 INTRODUCTION

Real-time scheduling is playing an important role in cyber-physical systems (CPSs). Examples of CPSs range from small systems, such as medical equipment and automobiles, to large systems like national power grid. CPSs require real-time management of both the physical and computing components including energy, computers, and networks. Such close integration requires all components from both the physical world and cyber world to operate efficiently in real-time (Rajkumar et al., 2010). The correctness of operations in CPSs depends not only on the logical results of operations, but also on the time when these results are computed and transmitted to the physical plants. Therefore real-time scheduling in CPSs is crucial for system stability and performance (Liu and Layland, 1973; Abdelzaher et al., 2004; Buttazzo et al., 2002; Lehoczky, 1990; Mok, 1983).

Cyber-physical energy system (CPES) studies the CPS in energy domain (Morris et al., 2009; Macana et al., 2011). An important problem in CPES is the scheduling of the electric devices in real-time under power consumption/supply constraints. A properly scheduled CPES should balance the power usage well to avoid unfavorable conditions, such as peak energy consumption. Real-time scheduling of energy flow for CPES is a new topic that has emerged in recent years (Mohsenian-Rad et al., 2010; Palensky and Dietrich, 2011; Conejo et al., 2010; Mathieu et al., 2013). The primary goal of real-time energy management is to schedule the operations of electric loads in electrical grids according to energy supply and user requirements (Subramanian et al., 2012). However, real-time energy management has many new challenges that are different from scheduling in the traditional real-time operating system (RTOS). First, the available energy supply in CPES varies with respect to time, while the available computation resource in RTOS is fixed over the entire time period. Second, multiple electric loads may be scheduled at the same time in CPES, while RTOS devotes all resources to a single operation at any time instant. Finally, electric loads can be either preemptive or nonpreemptive depending on their functionalities.

Due to the above challenges, the scheduling analysis for RTOS cannot be directly applied to CPES. A number of pieces of work have been carried out on the scheduling analysis of CPES. Facchinetti et al.

Cyber-Physical Systems. http://dx.doi.org/10.1016/B978-0-12-803801-7.00027-4

analyzed a type of electric load that controls the physical process (Facchinetti and Della Vedova, 2011). The feasibility analysis in this work checks whether electric loads can be scheduled by real-time energy management so that specific constraints on the physical process are satisfied. Nghiem et al. studied the feasibility of scheduling electric loads under a given constrained peak power (Nghiem et al., 2011). The results were further extended in Li et al. (2011), in which it was proved that the feasibility relies on the initial condition of electric loads.

In our work, we focus on the real-time energy management for a small and selfsustained CPES called smart micro-grid (Lasseter and Paigi, 2004). The micro-grid is a newly emerging CPES that uses advanced IT technology for the intelligent management of distributed power and energy-storage devices (Huang et al., 2008). One feature of the micro-grid is that it utilizes renewable energy such as sunlight, wind, tides, and waves (Katiraei et al., 2008), which are more sustainable and environmentally friendly. With the integration of renewable energy, micro-grids are able to completely isolate themselves from the national electric grid, and function as a stand-alone grid to improve energy utilization, efficiency, and reliability. However, using renewable energy involves the challenge of providing a consistent power supply, because the renewable energy relies on the environmental power resources. Since the power supply in micro-grids is highly variable, it is even more challenging to schedule the electric loads in micro-grids. Plenty of research work has been done to study the energy management in micro-grids (Morais et al., 2010; Lee et al., 2011; Liu et al., 2015; Arboleya et al., 2015; Barklund et al., 2008; Qin et al., 2012).

The feasibility analysis for micro-grids is primary in order to guarantee independent operation. Operators of micro-grids need to check whether the electric loads can be scheduled under fluctuating renewable energy supply. In this chapter, we extend our previous work on the task of scheduling for CPSs (Zhang et al., 2008, 2009, 2013; Shi, 2014; Shi and Zhang, 2013; Wang et al., 2015; Zhang and Shi, 2009) and introduce a novel real-time scheduling analysis method, named significant moments analysis (SMA). SMA serves as a centralized scheduling algorithm for coordinating discharge and charge of devices. Our contribution is summarized as followings: (1) SMA analyzes scheduling behaviors of electric loads and gives out a sufficient and necessary condition for scheduling feasibility in micro-grids; (2) with the real-time sensor data and electric load demand, SMA determines the activation of each electric load based on their priorities; and (3) SMA also provides accurate online predictions regarding how much power is insufficient for the independent operation of the micro-grid at any particular time instant. Such information allows the operators of micro-grids to take necessary and preventive measures in advance. To the best of our knowledge, these contributions have not been documented in the literature.

This chapter is organized as follows. In Section 2 we introduce the concept of micro-grid. In Section 3 we present the mathematical definitions and expression for different components in the micro-grids. Based on the mathematical models, SMA is introduced in Section 4. We use the state vector of SMA to describe the dynamic scheduling behavior in real-time energy management systems. In Section 5 the sufficient and necessary condition for scheduling feasibility is proposed based on SMA. Simulation results using SMA are presented in Section 6. Section 7 is the conclusion and future work.

2 SMART MICRO-GRIDS

Micro-grids are modern, local, small-scale versions of CPES. With the integration of real-time data and control, micro-grids can operate independently of the national electric grid, which provides more security against terrorism and natural disasters (Kroposki et al., 2008). Moreover, the use of renewable energy in micro-grids is environmentally friendly. These benefits have greatly stimulated the adoption of micro-grids in various applications. For example, the military bases are actively deploying micro-grids in order to ensure a reliable power supply without relying on the national power system (Hayden, 2013). Many educational institutions also have extensively built micro-grids on their campuses. In the University of California, San Diego, 90% of the annual electricity generation comes from micro-grids. As discussed in de Souza Ribeiro et al. (2011), micro-grids have become the best solution to isolated, stand-alone areas that may never be connected to the national electric grid due to their remoteness.

2.1 INFRASTRUCTURE OF MICRO-GRIDS

In Fig. 1 we consider a typical micro-grid, which consists of electric loads, on-site generations, and energy-storage.

The left hand side of Fig. 1 shows the on-site power supply, which comes from both the renewable energy (Huang et al., 2008) and fossil fuel generators. The power from on-site generations is clean and low cost, but highly variable. The right hand side of Fig. 1 shows electric loads and battery banks. Batteries are energy-storage devices, which can store energy whenever the supply exceeds the load demand and then provide energy whenever on-site generation is insufficient. The batteries provide a bridge to balance the supply and demand in micro-grids. The solid arrows represent the energy flow of micro-grids.

In micro-girds, all physical components are controlled by a centralized computer. The sensor data are transmitted to the computer. Based on the collected data, the computer calculates the control commands for battery bank and electric loads, and then sends the commands back to the loads. The dashed arrows represent the data flow in the cyber world of micro-grids.

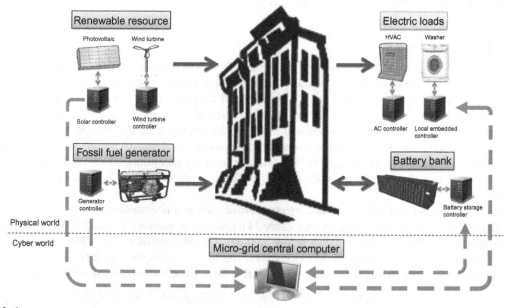

FIG. 1

An example of a micro-grid.

2.2 INDEPENDENT OPERATION

A fully evolved micro-grid must have the ability to operate independently from the main electric grid for an extended period of time. A successful independent operation requires that the on-site generated power and the battery storage should meet the demand of the electric loads. This requirement can be easily satisfied if all electric loads are deferrable.

However, in real applications, some electric loads in the micro-grid may become nondeferrable during the process of operation. We say an electrical load is nondeferrable if (1) the electrical load is nonpreemptive and it is currently in the middle of execution; and (2) the electrical load cannot complete its execution before the deadline if not executed immediately. In this case, the total energy supply from the on-site generation and battery storage should be at least more than the total demand of all nondeferrable electric loads. However, identifying nondeferrable electric loads is not easy because an electrical load can switch between a deferrable and nondeferrable state during its operation.

In the following sections, we will establish a mathematical model that describes the status evolution of electric loads in the micro-grids. Based on this model, we can identify all nondeferrable electric loads at any time t, which will facilitate the feasibility analysis of real-time energy management.

3 MODELS OF MICRO-GRIDS

In this section, we develop a set of models capable of describing different physical components in the micro-grids. Different from the previous real-time modeling of a micro-grid (Subramanian et al., 2012; Facchinetti and Della Vedova, 2011), the models presented in this section capture the complicated functionalities of deferrable and nondeferrable electric loads. Based on these models we can analyze the dynamic behaviors of micro-grids in the next section.

3.1 ELECTRIC LOADS

Without loss of generality, we assume that the micro-grid contains a set of electric loads $\Gamma = \{\tau_1, ..., \tau_N\}$. Each electrical load τ_n in Γ consists of a sequence of instances, and each instance corresponds to one operation request of τ_n. We use $\tau_n[k]$ to denote the kth instance of τ_n. $\tau_n[k]$ can be characterized by the requested time $\alpha_n[k]$, interrequest time interval $T_n[k]$ between $\alpha_n[k]$ and $\alpha_n[k+1]$, relative deadline $D_n[k]$, operational power $E_n[k]$, operation time $C_n[k]$, preemption $F_n[k]$, and priority $P_n[k]$. Note $F_n[k]=0$ denotes that t_n is nonpreemptive during operation and $F_n[k]=1$ denotes that τ_n is preemptive. The smaller value of $P_n[k]$ denotes a higher priority. In the following part of this section, we will study the representation of different types of electric loads using the above notations.

First, we give an example of a simple electrical load such as a rice cooker.

Example 1 A rice cooker operates once every 24 h and each operation will take 1 h. The operation is requested at 09:00 am and must finish before 7:00 pm. The operational power is 310 W. The operation is nonpreemptive once started. The rice cooker has the second highest priority.

Based on the above requirement, we can characterize the rice cooker with the following parameters:

$$\alpha_n[k] = 24k+9, \ T_n[k] = 24, \ D_n[k] = 10, \ E_n[k] = 310, \ C_n[k] = 1, \ F_n[k] = 0, \ P_n[k] = 2 \tag{1}$$

Table 1 Dishwasher Specification						
Phase	**Prewash**	**Wash**	**1st Rinse**	**Drain**	**2nd Rinse**	**Dry**
Operation power (W)	64.20	1517.8	103.8	8.2	1872.3	1.9
Operation time (h)	0.25	0.54	0.17	0.07	0.31	0.86
Preemption	Yes	Yes	No	Yes	No	Yes

where $F_n[k] = 0$ denotes that the electrical load is nonpreemptive during its operation.

Second, we introduce electric loads with multiple internal operation phases. For example, the operation of dish washers go through five phases as: prewash, wash, first rinse, drain, second rinse, and dry. The five operation phases must follow a strict sequential order. Each operation phase may have different execution time, power, and preemption property. Table 1 shows a detailed specification of a dishwasher. According to the specification, we can express the dishwasher as:

$$E_n[k] = [64.20, 1517.8, 103.8, \quad 8.2 \quad, 1872.3, 1.9]$$

$$C_n[k] = [\, 0.25, 0.54 \quad, \, 0.17, \quad 0.07, \quad 0.31, 0.86]$$

$$F_n[k] = [\quad 1 \,, \quad 1 \quad, \quad 0 \,, \quad 1 \,, \quad 0 \,, \quad 1]$$

Third, we introduce electric loads subject to the precedence constraints with other electric loads. For instance, a dryer machine cannot start its operation until a washing machine has completed its operation. To model this, we can view any group of precedence constrained electrical loads as a whole comprehensive electrical load, whose parameters $\{E_n[k], C_n[k], F_n[k], P_n[k]\}$ contain the characteristics of each electrical load. Therefore the mathematical expression of comprehensive electric loads is similar to electric loads with internal operation phases (type 2). Note that $P_n[k]$ is a vector for comprehensive electric loads as each individual load has a different priority.

Finally, we introduce electric loads with dynamics changing according to the physical environment. Consider air conditioners (AC) as an example. We use x to denote the house temperature inside the house, and TP_{out} denotes the outside temperature. According to the dynamic model in Meliopoulos et al. (2013), we can represent the operation of AC as:

$$\dot{x} = -\frac{G_{out}}{C_h}(x - TP_{out}) + \frac{1}{C_h}n_{ac}P_{ac}u \tag{2}$$

where G_{out} is the thermal conductance between the house and outside environment, C_h the thermal capacitance of the house, n_{ac} the coefficient of performance of AC, P_{ac} the power of AC, and u is the duty cycle of AC. Note that AC will cycle on and off periodically. Therefore the duty cycle of AC is the percentage of one period in which AC is on. By controlling the duty cycle u, AC will guarantee that the house temperature will stay within a bounded range such that $TP_{min} \leq x \leq TP_{max}$. Suppose that x is currently at a stable point x_{stable} such that $TP_{min} \leq x_{stable} \leq TP_{max}$. To guarantee that x always stays at this point, we must have that $\dot{x} = 0$, ie:

$$0 = -\frac{G_{out}}{C_h}(x - TP_{out}) + \frac{1}{C_h}n_{ac}P_{ac}u \tag{3}$$

which implies that:

$$u = \frac{G_{out}}{n_{ac}P_{ac}}(x_{stable} - TP_{out}) \tag{4}$$

Given the duty cycle u, we have the execution time of AC as:

$$C_n[k] = T_n[k]u = T_n[k]\frac{G_{out}}{n_{ac}P_{ac}}(x_{stable} - TP_{out}) \tag{5}$$

where $T_n[k]$ is the period of one on and off cycle. As it shows, the execution time of AC will dynamically change according to the outside temperature TP_{out}.

According to the above discussions, we can represent different types of electric loads as the tuple $\{C_n[k], E_n[k], D_n[k], T_n[k], F_n[k], P_n[k]\}$. Since we want to model the dynamics of the real-time energy management at any time t, we define the characteristics of electric loads in a continuous time domain as follows:

Definition 1 At any time t, an instance of t_n is effective if and only if it is the nearest instance released before or at time t, i.e., $t_n[k]$ is effective at time t if and only if:

$$\alpha_n[k] \leq t < \alpha_n[k+1] \tag{6}$$

Definition 2 At any time t, $C_n(t)$, $E_n(t)$, $D_n(t)$, $T_n(t)$, $F_n(t)$, and $P_n(t)$ are respectively defined as the operation time, power, relative deadline, interrequest time, preemption, and the priority of the effective instance of τ_n, ie:

$$\begin{aligned}&\text{if } \alpha_n[k] \leq t < \alpha_n[k+1]\\ C_n(t) = C_n[k],\ &E_n(t) = E_n[k],\ D_n(t) = D_n[k],\ T_n(t) = T_n[k],\ F_n(t) = F_n[k],\ P_n(t) = P_n[k]\end{aligned} \tag{7}$$

Therefore electric loads in the micro-grid can be represented in a continuous time domain as $\{C_n(t), E_n(t), D_n(t), T_n(t), F_n(t), P_n(t)\}_{n=1}^{N}$.

3.2 ON-SITE GENERATION AND BATTERY BANK

In micro-grids, a noticeable portion of electricity is generated on-site from different sources of energy. We formally define on-site generation of electricity as follows:

Definition 3 At any time t, $EG(t)$ is defined as the on-site electricity generation in a micro-grid.

$EG(t)$ includes the electricity from both the fossil fuel generator and the renewable energy. Fig. 2 shows the power generation of one wind turbine from the Alberta Electric System Operator on Jul. 12, 2011. The maximum wind power generation at 10:00 am is twice as much as the minimum wind power generation at 6:00 am. As it shows, the electricity generated from renewable energy is highly variable, so the total on-site electricity generation $EG(t)$ changes with respect to time.

Battery bank is used in micro-grid applications where the generation and load demand cannot be exactly matched. It increases the stability and reliability of the micro-grids. We formally define the battery bank as follows:

Definition 4 At any time t, $SOC(t)$ is defined as the state of charge of the battery bank in the micro-grid. B_{power} is defined as maximum charge/discharge rate of the battery bank. $B_{capacity}$ is defined as the capacity of the battery bank.

$SOC(t)$ indicates the percentage of energy remaining in the battery bank. In real applications, batteries should not be discharged below 20% of its SOC. Otherwise, the battery life will be significantly shortened. Therefore we put the following constraints on the battery operation such that:

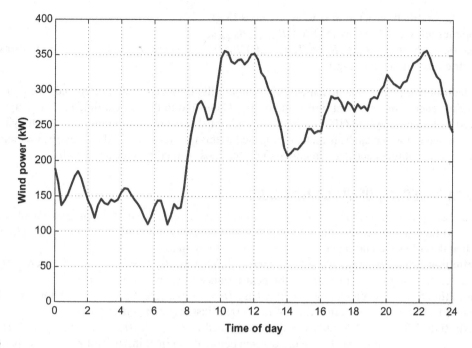

FIG. 2

Wind power generation reported by Alberta Electric System Operator.

$$20\% \leq \text{SOC}(t) \leq 100\% \tag{8}$$

According to the above constraint, we know that the maximum power output of the battery bank at any time t depends on its SOC. When the state of charge is larger than 20%, we have the maximum output power of the battery bank as B_{power}, ie:

$$\text{when SOC}(t) > 20\%, \text{ maximum power output of the battery bank is } B_{\text{power}} \tag{9}$$

On the other hand, when the state of charge drops to 20%, the batteries are not allowed to output any power, ie:

$$\text{when, SOC}(t) \leq 20\% \text{ maximum power output of the battery bank is } 0 \tag{10}$$

According to Eqs. (9) and (10), the maximum power output of the battery bank at any time t is expressed as:

$$\text{sgn}(\text{SOC}(t) - 20\%)B_{\text{power}} \tag{11}$$

4 SIGNIFICANT MOMENT ANALYSIS

In the previous sections, we have shown that the micro-grid consists of three major components represented as follows:

1. On-site electricity generation is represented as $EG(t)$
2. Batteries are represented as $\{SOC(t), B_{power}, B_{capacity}\}$
3. Electric loads $\{C_n(t), E_n(t), D_n(t), T_n(t), F_n(t), P_n(t)\}_{n=1}^{N}$, where N denotes the total number of electric loads in the micro-grid.

The establishment of these mathematical expressions will allow us to analytically study the dynamics of real-time energy management in the micro-grid. Compared to the critical time instant analysis introduced by Liu and Layland (1973), which is a traditional real-time scheduling analysis studying the worst-case scenarios, the analysis method proposed in this chapter studies all the significant moments whenever a scheduling decision is required. This analysis method is referred as SMA.

4.1 STATE VECTOR OF ELECTRIC LOADS

We consider a micro-grid with a set of electricity loads $\{\tau_1, ..., \tau_N\}$. To study the progression of electric loads under the real-time energy management, we introduce a state vector $Z(t) = [S(t), R(t), O(t)]^T$ for SMA that describes the current status of $\{\tau_1, ..., \tau_N\}$ at any time t.

Definition 5 The dynamic interrequest time is defined as $S(t) = [s_1(t), ..., s_N(t)]$, where $s_n(t)$, for $n = 1, 2, ..., N$, denotes how long after t the next instance of τ_n will be requested.

Definition 6 The remaining time is defined as $R(t) = [r_1(t), ..., r_N(t)]$, where $r_n(t)$, for $n = 1, 2, ..., N$, denotes the remaining operation time of the current instance of τ_n *after* time t.

Definition 7 The dynamic response time is defined as $O(t) = [o_1(t), ..., o_N(t)]$, where $o_n(t)$, for $n = 1, 2, ..., N$, denotes how much time has elapsed before the current instance of τ_n finishes operation.

Based on the above definitions, the progression of electric loads under the real-time energy management can be represented through the evolution of the SMA state vector $Z(t)$. In the following part, we will study the evolution of $Z(t) = [S(t), R(t), O(t)]^T$ under real-time energy management.

4.2 NONDEFERRABLE ELECTRIC LOADS

According to Section 2.2, two types of electric loads are considered as nondeferrable. In this subsection, we can easily identify all the nondeferrable loads using the state vector $Z(t)$ of SMA.

The first type of nondeferrable electric load could not complete execution before deadline if further delayed. This type of electrical load is represented as:

$$\{i | o_i(t) + r_i(t) = D_i(t)\} \tag{12}$$

where $o_i(t)$ denotes the delay of t_i since its time of request, $r_i(t)$ denotes the remaining operation of τ_i, and $o_i(t) + r_i(t) = D_i(t)$ indicates that τ_i can complete before its deadline if and only if it continuously executes from now on.

The second type of nondeferrable electric load is nonpreemptive and currently in the middle of operation. This type of electric load is represented as:

$$\{i | F_i(t) = 0, \text{ and } 0 < r_i(t) < C_i(t)\} \tag{13}$$

where $F_i(t) = 0$ denotes that τ_i is nonpreemptive, and $0 < r_i(t) < C_i(t)$ denotes that τ_i is in the middle of operation.

Definition 8 We use NonDefer(t) to denote a set of nondeferrable electric loads that must be executed immediately at time t. NonDefer(t) can be represented as:

$$\text{NonDefer}(t) = \{i | o_i(t) + r_i(t) = D_i(t)\} \cup \{i | F_i(t) = 0, \text{ and } 0 < r_i(t) < C_i(t)\} \tag{14}$$

4.3 REAL-TIME ENERGY MANAGEMENT

In this section, we rigorously define real-time energy management using mathematical expression. Let $\{1, ..., N\}$ denote the set of indices of total electric loads in the micro-grid and OP(t) denote the set of indices of electric loads executing at time t.

Definition 9 A *real-time energy management* is a set-valued map between R^+ and the collection of all subsets of $\{1, ..., N\}$. It is parameterized as OP(t): $R^+ \rightarrow 2^{\{1, ..., N\}}$, where $t \in R^+$.

At any time t, OP(t) should at least contain nondeferrable electric loads. If the energy supply is larger than the demand of all nondeferrable electric loads, the real-time energy management will activate the execution of other deferrable electric loads in terms of their priorities.

At any time t, we can construct the real-time energy management OP(t) through three steps as follows.

Step 1: Initialization. Initialize OP(t) as a set of indices of nondeferrable electric loads at time t, ie:

$$\text{OP}(t) = \text{NonDefer}(t) \tag{15}$$

Moreover, we initialize an scheduling pool SCH(t) as a set of deferrable electric loads that have remaining operation time, ie:

$$\text{SCH}(t) = \{i | r_i(t) > 0\} - \text{NonDefer}(t) \tag{16}$$

where $\{i | r_i(t) > 0\}$ denotes a set of electric loads with a remaining operation time and "$-$" denotes the subtraction between two sets.

Step 2: Scheduling. In this step, we decide whether the highest priority electrical load in SCH(t) can be scheduled to operate at time t. The highest priority electrical load in SCH(t) can be denoted as τ_n such that $n = \min\limits_{i \in \text{SCH}(t)} P_i(t)$. If τ_n satisfies the following condition:

$$E_n(t) + \sum_{i \in \text{OP}(t)} E_i(t) \leq \text{EG}(t) + \text{sgn}(\text{SOC}(t) - 20\%)B_{\text{power}} \tag{17}$$

where $\sum\limits_{i \in \text{OP}(t)} E_i(t)$ represents the demand of electric loads that have been scheduled to execute at time t, $E_n(t)$ represents the demand of τ_n, and the right hand side represents the total supply of on-site generation and the battery storage. Eq. (17) indicates that the total supply at current time t is enough to satisfy the demand of both electric loads in OP(t) and the electrical load τ_n. In this case, τ_n will be scheduled to execute at time t, ie:

$$\text{OP}(t) = \text{OP}(t) + \{n\} \tag{18}$$

On the other hand, if the electrical load τ_n satisfies the following condition:

$$E_n(t) + \sum_{i \in \text{OP}(t)} E_i(t) > \text{EG}(t) + \text{sgn}(\text{SOC}(t) - 20\%)B_{\text{power}} \tag{19}$$

which indicates that the total energy at current time t is NOT enough to satisfy the demand of both electric loads in OP(t) and the electrical load τ_n. In this case, τ_n will NOT be scheduled to execute at time t, ie:

$$\text{OP}(t) = \text{OP}(t) \tag{20}$$

Step 3: Update and check. In this step, we update scheduling pool SCH(t) and check the number of electric loads in SCH(t). Since the execution of electrical load t_n has been decided in the second step, we need to remove t_n from the scheduling pool, ie:

$$\text{SCH}(t) = \text{SCH}(t) - \{n\} \tag{21}$$

If the scheduling pool is NOT empty, i.e., $\text{SCH}(t) \neq \emptyset$, real-time energy management needs to decide the execution of other electric loads in SCH(t). In this case, go back to Step 2. On the other hand, if the scheduling pool is empty, i.e., $\text{SCH}(t) \neq \emptyset$, the execution of all deferrable electric loads have been decided and the construction of OP(t) finishes.

Algorithm 1 gives a detailed implementation of real-time energy management. The input of the algorithm are the current SMA state vector $Z(t)$, on-site generation, and battery SOC. The output of the algorithm is a set of electric loads that are scheduled to execute at time t. In the next section, we will show that at any time t, an electrical load t_n has different evolution dynamics in SMA, depending on whether t_n is scheduled to execute (i.e., $n \in \text{OP}(t)$) or not (i.e., $n \notin \text{OP}(t)$).

Algorithm 1: Real-time Energy Management

Data: $Z(t)$, SOC(t), EG(t), $\mathscr{B}_{\text{power}}$, $\{C_n(t), E_n(t), D_n(t), T_n(t), F_n(t), P_n(t)\}_{n=1}^{N}$
Result: OP(t)
1   NonDefer(t) = $\{i | o_i(t) + r_i(t) = D_i(t)\} \cup \{i | F_i(t) = 0, \text{and } 0 < r_i(t) < C_i(t)\}$;
    /*1st Step: Initialization*/
2   OP(t) = NonDefer(t);
3   SCH(t) = $\{i | r_i(t) > 0\}$ − NonDefer(t);
4   **while** SCH(t) $\neq \emptyset$ **do**
       /*2nd Step: Scheduling*/
5       $n = \min_{i \in \text{SCH}(t)} P_i(t)$;
6   **if** $\sum_{i \in \text{OP}(t)} E_i(t) + E_n(t) \leq \text{EG}(t) + \text{sgn}(\text{SOC}(t) - 20\%)\, \mathscr{B}_{\text{power}}$ **then**
7   OP(t) = OP(t)+$\{n\}$;
8   **else**
9   OP(t) = OP(t);
       /*3nd Step: Update and Check*/
10   SCH(t)=SCH(t)−$\{n\}$;
11 **return** OP(t);

4.4 EVOLUTION OF ELECTRIC LOADS IN SMA

The SMA state vector $Z(t)$ contains the status for a set of N electric loads. Therefore the evolution of $Z(t)$ can be obtained through the evolution of each electrical load. For each electrical load τ_n, its state vector $[s_n(t), \ r_n(t), \ o_n(t)]^{\text{T}}$ can evolve in both continuous and discrete ways.

First, we discuss the discrete evolution of $[s_n(t), \ r_n(t), \ o_n(t)]^T$. If a new instance of τ_n arrives at time t, i.e., $s_n(t) = 0$, the state vector will update according to the characteristics of the new instance. Thus, we have that:

$$\text{if } s_n(t) = 0: \begin{bmatrix} s_n(t) \\ r_n(t) \\ o_n(t) \end{bmatrix} = \begin{bmatrix} T_n(t) \\ C_n(t) \\ 0 \end{bmatrix} \tag{22}$$

Next, we discuss the continuous evolution of $[s_n(t), \ r_n(t), \ o_n(t)]^T$. According to Section 4.3, the execution of τ_n will continue at time t when $n \in \text{OP}(t)$ and stop at time t when $n \notin \text{OP}(t)$. Therefore the continuous evolution of $[s_n(t), \ r_n(t), \ o_n(t)]^T$ depends on the execution status of τ_n.

Case 1 τ_n will execute at time t, i.e., $n \in \text{OP}(t)$. Then the continuous evolution of $[s_n(t), \ r_n(t), \ o_n(t)]^T$ at time t is:

$$\text{if } s_n(t) \neq 0 \text{ and } n \in \text{OP}(t): \begin{bmatrix} \dot{s}_n(t) \\ \dot{r}_n(t) \\ \dot{o}_n(t) \end{bmatrix} = \begin{bmatrix} -1 \\ -1 \\ 1 \end{bmatrix} \tag{23}$$

Case 2 τ_n will NOT execute at time t, i.e., $n \notin \text{OP}(t)$. In this case, $s_n(t)$ will still continuously decrease. And since the execution of τ_n stops at time t, the remaining time $r_n(t)$ will NOT decrease in this case because, i.e., $r_n(t + \Delta t) = r_n(t)$. Finally, the evolution of $o_n(t)$ from $t + \Delta t$ is a little involved. It depends whether the current instance of τ_n has completed its execution before time t. If the current instance of τ_n has completed execution before time t, i.e., $r_n(t) = 0$, the dynamic response time $o_n(t)$ keeps constant. On the other hand, if the current instance of τ_n has NOT completed execution before time t, i.e., $r_n(t) > 0$, the dynamic response time $o_n(t)$ increases, i.e., $r_n(t + \Delta t) = r_n(t)$. In all, the continuous evolution of $[s_n(t), \ r_n(t), \ o_n(t)]^T$ at time t can be expressed as:

$$\text{if } s_n(t) \neq 0 \text{ and } n \notin \text{OP}(t): \begin{bmatrix} \dot{s}_n(t) \\ \dot{r}_n(t) \\ \dot{o}_n(t) \end{bmatrix} = \begin{bmatrix} -1 \\ 0 \\ \text{sgn}(r_n(t)) \end{bmatrix} \tag{24}$$

4.5 BATTERY STATE OF CHARGE EVOLUTION

As discussed earlier, the batteries in the micro-grid have two modes of operation. In the first mode, the on-site electricity generation exceeds the load demand and the batteries store extra energy. This happens when the following condition is satisfied $\sum_{i \in \text{OP}(t)} E_i(t) < \text{EG}(t)$. The extra power stored in batteries at time t is $\text{EG}(t) - \sum_{i \in \text{OP}(t)} E_i(t)$. In the second mode, the on-site energy is insufficient to supply load demands and the batteries will provide power to electric loads. It happens when the following condition is satisfied: $\sum_{i \in \text{OP}(t)} E_i(t) > \text{EG}(t)$. The power provided by batteries is $\sum_{i \in \text{OP}(t)} E_i(t) - \text{EG}(t)$.

We use Δt to denote an arbitrarily small time. Then, for both two modes, we have that:

$$(\text{SOC}(t+\Delta t) - \text{SOC}(t))B_{\text{capacity}} = \left(\text{EG}(t) - \sum_{i \in \text{OP}(t)} E_i(t) \right) \Delta t \tag{25}$$

which implies that:

$$\dot{\text{SOC}} = \lim_{\Delta t \to 0} \frac{\text{SOC}(t+\Delta t) - \text{SOC}(t)}{\Delta t} = \frac{\text{EG}(t) - \sum_{i \in \text{OP}(t)} E_i(t)}{B_{\text{capacity}}} \tag{26}$$

5 FEASIBILITY ANALYSIS

In this section, we will propose a necessary and sufficient feasibility analysis enabled by SMA that checks whether the requirement for the independent operation of the micro-grid can be satisfied. Based on this feasibility analysis, we can also accurately predict when and how much power is insufficient for the independent operation.

5.1 NECESSARY AND SUFFICIENT FEASIBILITY ANALYSIS

Our feasibility analysis of real-time energy management checks whether the micro-grids can operate independently from the main electric grid. Here, we propose a necessary and sufficient condition for feasibility analysis as follows:

Claim *A micro-grid can operate independently within $[t_a, t_b]$ if and only if it satisfies the following conditions for all $t \in [t_a, t_b]$:*

$$\sum_{i \in \text{NonDefer}(t)} E_i(t) \leq \text{EG}(t) + \text{sgn}(\text{SOC}(t) - 20\%)B_{\text{power}} \tag{27}$$

Proof According to Definition 8 and Eq. (14), NonDefer(t) denotes a set of nondeferrable electric loads that must be executed at time t, which can be identified by SMA. Since $E_i(t)$ denotes the power demand of one nondeferrable electrical load τ_i, the left hand side of Eq. (27) denotes the demand of all nondeferrable electric loads at current time t.

On the other hand, EG(t) denotes the on-site generation according to Definition 3, and $\text{sgn}(\text{SOC}(t) - 20\%)B_{\text{power}}$ denotes the maximum power output of the battery storage. Hence, the right hand side of Eq. (27) denotes the total supply from both the on-site generation and battery storage at time t.

Therefore the micro-grid can run independently at time t if and only if the left hand side (power demand) is smaller or equal to the right hand side (power supply). The micro-grid can run independently within $[t_a, t_b]$ if and only if the inequality condition is satisfied for any time $t \in [t_a, t_b]$. $\square$

Note. The *Claim* seems to be quite straightforward. But checking whether the condition (Eq. 27) is satisfied for all time t is a very difficult task. The main difficulties are: first, the set of NonDefer(t) cannot be determined unless we know the SMA states and, second, the batter state SOC(t) needs to be predicted. The SMA Eqs. (22)–(24), together with the SOC Eq. (26) and Algorithm 1 are used to determine NonDefer(t) and SOC(t) so that condition (Eq. 27) can be checked. This is the main contribution of our work.

5.2 DEFICIENCY IN POWER SUPPLY

A successful feasibility analysis in Claim in Section 5.1 checks whether the micro-grid can run independently from the main electric grid. Based on this analysis, we can predict when and how much power is insufficient for the independent operation.

Case 1 The power demand of all nondeferrable electric loads is smaller or equal to the power supply from the on-site generation and the battery storage, ie:

$$\sum_{i \in \text{NonDefer}(t)} E_t(t) \leq \text{EG}(t) + \text{sgn}(\text{SOC}(t) - 20\%)B_{\text{power}} \tag{28}$$

In this case, the micro-grid can run independently from the main electric grid.

Case 2 The power demand of all nondeferrable electric loads is larger than the power supply from the on-site generation and the battery storage, ie:

$$\sum_{i \in \text{NonDefer}(t)} E_t(t) > \text{EG}(t) + \text{sgn}(\text{SOC}(t) - 20\%)B_{\text{power}} \tag{29}$$

In this case, the micro-grid cannot run independently and the amount of insufficient power supply can be expressed as:

$$\sum_{i \in \text{NonDefer}(t)} E_t(t) - \text{EG}(t) - \text{sgn}(\text{SOC}(t) - 20\%)B_{\text{power}} \tag{30}$$

Based on the above analysis, we have the following claim about the insufficient power supply of the micro-grid.

Claim *At any time t, the amount of insufficient power for the independent operation of micro-grid can be expressed as:*

$$\max\left\{0, \sum_{i \in \text{NonDefer}(t)} E_i(t) - \text{EG}(t) - \text{sgn}(\text{SOC}(t) - 20\%)B_{\text{power}}\right\} \tag{31}$$

6 NUMERIC SIMULATION

In this section, we will use numeric simulations to verify our feasibility analysis of real-time energy management in the micro-grid. All simulation results are implemented in MATLAB.

6.1 SIMULATION SETUP

First, we consider a micro-grid with the on-site renewable energy and fossil fuel generation. The renewable energy has the generation as shown in Fig. 2 and the fossil fuel can output the constant power of 100 kW.

Then, we consider four types of electric loads in the micro-grid. The first type of electric load has simple characteristics as follows:

$$C_1(t) = 0.5 E_1(t) = 80 \quad D_1(t) = 2T_1(t) = 2F_1(t) = 0 P_1(t) = 1$$

$$C_2(t) = 0.5 E_2(t) = 120 \quad D_2(t) = 2T_2(t) = 3F_2(t) = 1 P_2(t) = 2 \tag{32}$$

$$C_3(t) = 1.0 E_2(t) = 160 \quad D_2(t) = 4T_2(t) = 4F_2(t) = 1 P_2(t) = 3$$

The second type of electric load has multiple internal operation phases and can be represented as:

$$C_4(t) = [0.5, \ 1] \qquad E_4(t) = [150, \ 160] \qquad D_4(t) = 4T_4(t) = 4F_4(t) = [0, \ 0] \ P_4(t) = 4$$

$$C_5(t) = [1, \ 0.5, \ 1] \qquad E_5(t) = [120, \ 80, \ 180] \qquad D_5(t) = 4T_5(t) = 5F_5(t) = [0, \ 0, \ 0] \ P_5(t) = 5 \tag{33}$$

The third type of electric load has precedence constraints and they formulate a comprehensive electric load that can be represented as:

$$C_6(t) = [0.5, \ 1, \ 0.5, \ 1.5, \ 0.5] \ E_6(t) = [50, \ 120, \ 30, \ 140, \ 80]$$

$$D_6(t) = 6T_6(t) = 6F_5(t) = [0, \ 1, \ 0, \ 1, \ 0] \ P_6(t) = [6, \ 7, \ 8, \ 9, \ 10] \tag{34}$$

The last type of electrical load is an AC operating dynamically according to the outside temperature TP_{out}. According to Eq. (5), we have this electrical load represented as:

$$C_7(t) = T_7(t)u \ E_7(t) = 120 \ D_7(t) = 2T_7(t) = 2F_7(t) = 0 \ P_7(t) = 11 \tag{35}$$

where u is the duty cycle as

$$u = \frac{1}{400}(70 - TP_{out})$$

Finally, we assume the battery bank has the nominal voltage of 400 V and the capacity of 450 Ah. The C-rate of the battery bank is 0.5. The value of $B_{capacity}$ and B_{power} are as following:

$$B_{capacity} = 180 kWh \ B_{power} = 90 kWh \tag{36}$$

Fig. 3 shows the power demand of the above electric loads in the micro-grid under the real-time energy management. The solid blue line represents the total on-site energy generation including the wind power and fossil fuel. The area with cross line represents the power demand of all electric loads within 24 h. As it shows, the power demand exceeds the on-site generation at some time points. At these time points, if the battery can provide enough energy to cover the extra power demand, the micro-grid will still be able to run independently. Otherwise, the micro-grid cannot run independently.

6.2 FEASIBILITY VERIFICATION

In this section, we will show that our feasibility analysis can accurately predict when and how much power is insufficient for the independent operation of the micro-grids.

As shown in Fig. 3, the power demand exceeds the on-site generation at some time points. To guarantee the independent operation of the micro-grid, the battery storage will provide extra energy to electric loads. Fig. 4 shows the battery dynamics within 24 h. The solid line represents the SOC of the battery. The evolution of the battery SOC is derived as part of the dynamic timing model in Section 3.2. The battery SOC increases when the battery is charging and decreases when the battery is discharging. For example, we can see from Fig. 3 that the battery is charging within the time interval [14.4, 16] because the on-site

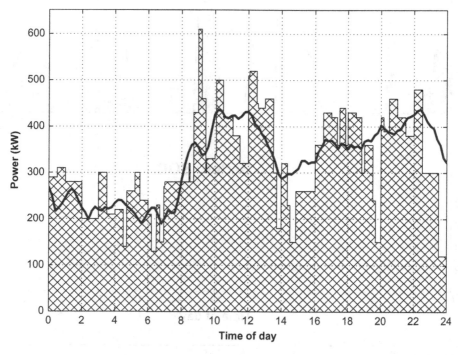

FIG. 3

Power demand under real-time energy management.

generation is larger than the total power demand. This observation exactly matches the increase of the battery SOC within [14.4, 16] as shown in Fig. 4. Moreover, the solid red area represents the discharging power of the battery. The faster the battery SOC decreases, the larger the discharging power is.

However, even with the supply from the battery storage, the micro-grid may still not be able to run independently. Fig. 5 shows when and how much power is insufficient for the independent operation of micro-grids. It is derived from our feasibility analysis studied in Section 5.

To verify the result of the feasibility analysis, we mark the results of Figs. 4 and 5 using different colors, as illustrated in Fig. 6. The black color above the solid line denotes the discharging power of the battery, which is same as that in Fig. 4. The light gray color denotes the amount of insufficient power for the independent operation of the micro-grids, which is same as that in Fig. 5. As we can see, the combination of the black area and gray area exactly cover the extra power demand exceeding the on-site energy generation (solid line). Therefore our feasibility analysis can accurately predict when and how much power is insufficient for the independent operation of the micro-grid.

7 CONCLUSIONS AND FUTURE WORK

The main contribution of this work is the novel SMA for real-time energy management systems. We introduce SMA state vectors to describe dynamic behaviors of multiple operations scheduled in smart micro-grids. Based on SMA, we can easily find out all the nondeferrable electric loads and schedule the

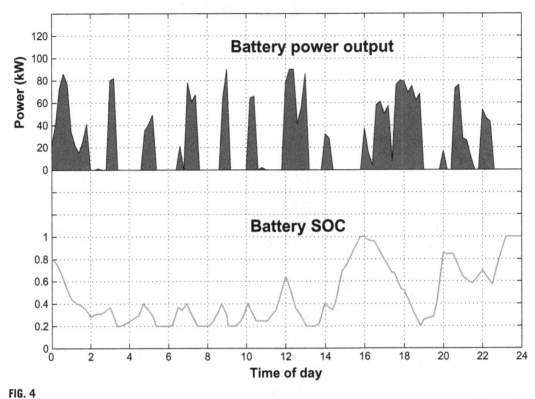

FIG. 4

Battery output and SOC.

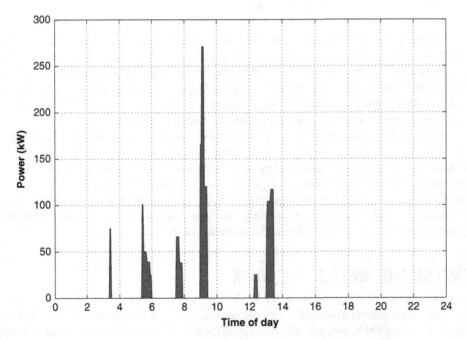

FIG. 5

Power deficiency.

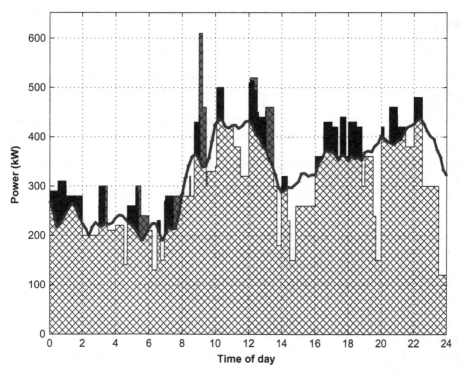

FIG. 6

Feasibility analysis result from SMA.

deferrable loads according to the current power supply at any moment. The necessary and sufficient feasibility analysis based on SMA is straightforward and easy to implement. SMA analysis provides a general model that is not limited to the micro-grid. It also could be applied to large-scale CPES energy management or other types of dynamic resource allocation problems between the consumer and the supplier, such as the pricing and resource scheduling in the cloud computing (Tsai and Qi, 2012; Tsai et al., 2010).

Feasibility analysis is a fundamental step that guarantees the normal operation of the micro-grid under appropriate energy management. One of the most important motivations behind the CPES is the economic benefit. The future research direction of energy management can focus on optimizing the operation or implementation cost of CPES, under the feasibility constraint.

ACKNOWLEDGMENTS

This work is partially supported by ONR grants N00014-10-10712 (YIP) and N00014-14-1-0635; and NSF grants OCE-1032285, IIS-1319874, and CMMI-1436284. The authors want to thank for the support.

REFERENCES

Abdelzaher, T., Sharma, V., Lu, C., 2004. A utilization bound for aperiodic tasks and priority driven scheduling. IEEE Trans. Comput. 53 (3), 334–350.

Arboleya, P., Gonzalez-Moran, C., Coto, M., Falvo, M., Martirano, L., Sbordone, D., Bertini, I., Pietra, B., 2015. Efficient energy management in smart micro-grids: zero grid impact buildings. IEEE Trans. Smart Grid 6 (2), 1055–1063.

Barklund, E., Pogaku, N., Prodanovic, M., Hernandez-Aramburo, C., Green, T., 2008. Energy management in autonomous microgrid using stability-constrained droop control of inverters. IEEE Trans. Power Electron. 23 (5), 2346–2352.

Buttazzo, G., Lipari, G., Caccamo, M., Abeni, L., 2002. Elastic scheduling for flexible workload management. IEEE Trans. Comput. 51 (3), 289–302.

Conejo, A., Morales, J., Baringo, L., 2010. Real-time demand response model. IEEE Trans. Smart Grid 1 (3), 236–242.

de Souza Ribeiro, L., Saavedra, O., de Lima, S., de Matos, J., 2011. Isolated micro-grids with renewable hybrid generation: the case of Lencois Island. IEEE Trans. Sustainable Energy 2 (1), 1–11.

Facchinetti, T., Della Vedova, M., 2011. Real-time modeling for direct load control in cyber-physical power systems. IEEE Trans. Ind. Inf. 7 (4), 689–698.

Hayden, E., 2013. Introduction to microgrids. Securicon Report, pp. 1–13.

Huang, J., Jiang, C., Xu, R., 2008. A review on distributed energy resources and microgrid. Renew. Sust. Energ. Rev. 12 (9), 2472–2483.

Katiraei, F., Iravani, R., Hatziargyriou, N., Dimeas, A., 2008. Microgrids management. IEEE Power Energy Mag. 6 (3), 54–65.

Kroposki, B., Lasseter, R., Ise, T., Morozumi, S., Papathanassiou, S., Hatziargyriou, N., 2008. Making microgrids work. IEEE Power Energy Mag. 6 (3), 40–53.

Lasseter, R., Paigi, P., 2004. Microgrid: a conceptual solution. In: 2004 IEEE 35th Annual Power Electronics Specialists Conference. (IEEE Cat. No. 04CH37551).

Lee, P., Lai, L., Chan, S., 2011. A practical approach of energy efficiency management reporting systems in microgrid. In: 2011 IEEE Power and Energy Society General Meeting.

Lehoczky, J., 1990. Fixed priority scheduling of periodic task sets with arbitrary deadlines. In: IEEE 11th Real-Time Systems Symposium, pp. 201–209.

Li, Z., Huang, P., Mok, A., Nghiem, T., Behl, M., Pappas, G., Mangharam, R., 2011. On the feasibility of linear discrete-time systems of the green scheduling problem. In: 2011 IEEE 32nd Real-Time Systems Symposium.

Liu, C., Layland, J., 1973. Scheduling algorithms for multiprogramming in a hard-real-time environment. J. ACM 20 (1), 46–61.

Liu, B., Zhuo, F., Zhu, Y., Yi, H., 2015. System operation and energy management of a renewable energy-based DC micro-grid for high penetration depth application. IEEE Trans. Smart Grid 6 (3), 1147–1155.

Macana, C., Quijano, N., Mojica-Nava, E., 2011. A survey on cyber physical energy systems and their applications on smart grids. In: 2011 IEEE PES Conference on Innovative Smart Grid Technologies (ISGT Latin America).

Mathieu, J., Koch, S., Callaway, D., 2013. State estimation and control of electric loads to manage real-time energy imbalance. IEEE Trans. Power Syst. 28 (1), 430–440.

Meliopoulos, A., Polymeneas, E., Tan, Z., Huang, R., Zhao, D., 2013. Advanced distribution management system. IEEE Trans. Smart Grid 4 (4), 2109–2117.

Mohsenian-Rad, A., Wong, V., Jatskevich, J., Schober, R., Leon-Garcia, A., 2010. Autonomous demand-side management based on game-theoretic energy consumption scheduling for the future smart grid. IEEE Trans. Smart Grid 1 (3), 320–331.

Mok, A., 1983. Fundamental design problems of distributed systems for the hard-real-time environment. Massachusetts Institute of Technology, Cambridge, MA.

Morais, H., Kada, P., Faria, P., Vale, Z.A., Khodr, H.M., 2010. Optimal scheduling of a renewable micro-grid in an isolated load area using mixed-integer linear programming. Renew. Energy 35 (1), 151–156.

Morris, T., Srivastava, A., Reaves, B., Pavurapu, K., Abdelwahed, S., Vaughn, R., McGrew, W., Dandass, Y., 2009. Engineering future cyber-physical energy systems: challenges, research needs, and roadmap. In: 41st North American Power Symposium.

Nghiem, T., Behl, M., Mangharam, R., Pappas, G., 2011. Green scheduling of control systems for peak demand reduction. In: IEEE Conference on Decision and Control and European Control Conference.

Palensky, P., Dietrich, D., 2011. Demand side management: demand response, intelligent energy systems, and smart loads. IEEE Trans. Ind. Inf. 7 (3), 381–388.

Qin, Q., Chen, Z., Wang, X., 2012. Overview of micro-grid energy management system research status. In: 2012 Power Engineering and Automation Conference.

Rajkumar, R., Lee, I., Sha, L., Stankovic, J., 2010. Cyber-physical systems. In: Proceedings of the 47th Design Automation Conference on—DAC'10.

Shi, Z., 2014. Non-worst-case response time analysis for real-time systems design. Georgia Institute of Technology, Atlanta, GA.

Shi, Z., Zhang, F., 2013. Predicting time-delays under real-time scheduling for linear model predictive control. In: 2013 International Conference on Computing, Networking and Communications (ICNC).

Subramanian, A., Garcia, M., Dominguez-Garcia, A., Callaway, D., Poolla, K., Varaiya, P., 2012. Real-time scheduling of deferrable electric loads. In: 2012 American Control Conference (ACC).

Tsai, W., Qi, G., 2012. DICB: dynamic intelligent customizable benign pricing strategy for cloud computing. In: 2012 IEEE Fifth International Conference on Cloud Computing.

Tsai, W., Sun, X., Shao, Q., Qi, G., 2010. Two-tier multi-tenancy scaling and load balancing. In: 2010 IEEE 7th International Conference on E-Business Engineering.

Wang, X., Shi, Z., Zhang, F., Wang, Y., 2015. Dynamic real-time scheduling for human-agent collaboration systems based on mutual trust. Cyber-Phys. Syst., 1 (2–4), 76–90.

Zhang, F., Shi, Z., 2009. Optimal and adaptive battery discharge strategies for cyber-physical systems. In: 48th IEEE Conference on Decision and Control, pp. 6232–6237.

Zhang, F., Szwaykowska, K., Wolf, W., Mooney, V., 2008. Task scheduling for control oriented requirements for cyber-physical systems. In: 2008 Real-Time Systems Symposium.

Zhang, F., Shi, Z., Wolf, W., 2009. A dynamic battery model for co-design in cyber-physical systems. In: 29th IEEE International Conference on Distributed Computing Systems Workshops.

Zhang, F., Shi, Z., Mukhopadhyay, S., 2013. Robustness analysis for battery-supported cyber-physical systems. ACM Trans. Embed. Comput. Syst. 12 (3), 1–27.

Author Index

Note: Page numbers followed by *f* indicate figures, and *t* indicate tables.

Subject Index

Note: Page numbers followed by *f* indicate figures, and *t* indicate tables.

Printed in the United States
By Bookmasters